INVESTIGATING Basic College Mathematics

Laura Bracken
Lewis-Clark State College

Hazel McKenna
Utah Valley State College

Australia • Canada • Mexico • Singapore • Spain
United Kingdom • United States

Publisher: *Bob Pirtle*
Sponsoring Editor: *Jennifer Huber*
Assistant Editor: *Rachael Sturgeon*
Editorial Assistant: *Jonathan Wegner*
Marketing Manager: *Leah Thomson*
Marketing Assistant: *Maria Salinas*
Project Manager, Editorial Production: *Janet Hill*
Print/Media Buyer: *Vena Dyer*
Permissions Editor: *Sue Ewing*
Production Service: *Martha Emry*
Illustrator: *Atherton Customs*
Cover Designer: *Irene Morris*
Cover photo: *Conrad Zobel/Corbis*
Compositor: *G & S Typesetters, Inc.*
Cover and Interior Printer:
Courier Corporation-Kendallville,Inc

Printed in the United States of America

3 4 5 6 7 06 05

For more information about our products,
contact us at:
Thomson Learning Academic Resource Center
1-800-423-0563

For permission to use material from this text,
contact us by:
Phone: 1-800-730-2214
Fax: 1-800-730-2215
Web: http://www.thomsonrights.com

The photos on the following pages were provided by the authors: 13, 28, 54, 93, 95, 120, 123, 324, 327, 341, 343, 377, 378, 393, 438, 471, 498, 682, 741.

Library of Congress Cataloging-in-Publication Data
Bracken, Laura.
Investigating basic college mathematics /
Laura Bracken, Hazel McKenna.
p. cm.
Includes index.
ISBN 0-03-034494-8 (pbk.)
1. Mathematics. I. McKenna, Hazel. II. Title.
QA39.3 .B72 2003
510–dc21 2002016417

Brooks/Cole–Thomson Learning
511 Forest Lodge Road
Pacific Grove, CA 93950
USA

Asia
Thomson Learning
5 Shenton Way #01-01
UIC Building
Singapore 068808

Australia
Nelson Thomson Learning
102 Dodds Street
South Melbourne, Victoria 3205
Australia

Canada
Nelson Thomson Learning
1120 Birchmount Road
Toronto, Ontario M1K 5G4
Canada

Europe/Middle East/Africa
Thomson Learning
High Holborn House
50/51 Bedford Row
London WC1R

Latin America
Thomson Learning
Seneca, 53
Colonia Polanco
11560 Mexico D.F.
Mexico

Spain
Paraninfo Thomson Learning
Calle/Magallanes, 25
28015 Madrid, Spain

In honor of my Dad, a kind and encouraging father, and a good man.

—Laura Bracken

For Kirsty and Emily—
the next generation.

—Hazel McKenna

Table of Contents

4 Decimals 417

5 Expressions and Equations 541

6 Rates, Ratios, and Percents 625

Appendix 1: Geometry

Appendix 2: Using Technology

Index of Applications

This index organizes the context or setting of application problems by section (for example, in the entry "Sec. 1.4: 25," we are referring to problem number 25 in Section 1.4). To make concepts more accessible to students, many skill practice and word problems are set in familiar contexts or in an academic discipline. Even though students may not be ready to do truly authentic problem solving, they can begin to see the importance of mathematics in their disciplines and in their daily lives.

To the Instructor

Investigating Basic College Mathematics is a flexible worktext that supports a variety of teaching methods and prepares students for their first course in algebra. The book engages students in active learning, allowing them to describe their understandings. It provides a comprehensive foundation in prealgebra concepts. This is accomplished through a blend of lecture-type *Discussions* and *Investigations* that help students take ownership of their learning. We wrote this text using activities and problems that we knew motivated our students and which were developed in our classrooms. We see our students changing from passive note-takers into energized and involved learners. In the spirit of the *Standards* of the American Mathematical Association of Two-Year Colleges (AMATYC), we use authentic applications and student-centered activities to teach the fundamental understandings needed for success in algebra.

CONTENT

The text is organized around the topics of whole numbers, integers, fractions, and decimals. Students learn to evaluate, simplify, graph, and solve. Problem solving and plane geometry are integrated throughout the text. Basic mathematics courses at different institutions range from three to five credits. We have included some topics, for example unit analysis, in such a way that they can be omitted without disturbing the rest of the course. Work with measurement systems is stressed throughout. If students are to do relevant work with applications, we believe they must have a basic understanding of units of measurement.

Chapter 1: Whole Numbers

Students work with basic concepts of number sense and organization of numbers and data including sets, number lines, place value, relative order, tables, and bar graphs. The axioms and algorithms of addition, subtraction, multiplication, and division of whole numbers are thoroughly reviewed, exponents are introduced, and the order of operations is used to evaluate expressions. Within the context of measurement, rounding and estimation is taught. The fundamental geometric concepts of angle, circle, degree, area, and volume are integrated. A five-step problem-solving strategy is used to help students develop their skills and confidence in solving word problems. Here the concepts of variable and equation have a just-in-time introduction. However, variables are limited to a single variable isolated on the left side of an equation.

Chapter 2: Integers

The work done with whole numbers is extended to the integers, including addition, subtraction, multiplication, division, exponents, and the order of operations. The concepts of opposite and absolute value are introduced. Students continue to become familiar with the concept of variable by solving problems using formulas. An optional section on line graphs (2.7) is also included.

Chapter 3: Fractions

The chapter begins with foundation work in multiples, divisors, prime and composite numbers, and factorization. Using concrete models, students investigate arithmetic operations with fractions and mixed numbers.

Chapter 4: Decimals

With a conceptual understanding of fractions, students are ready to study decimal notation. Using the concrete model of money, students extend the algorithms for addition, subtraction, multiplication, and division of integers to decimal numbers. They begin work with irrational numbers, including pi, allowing for extensive investigation of geometric formulas. Optional sections on unit analysis and scientific notation are also found in this chapter.

Chapter 5: Expressions and Equations

Students now use their skills with arithmetic to evaluate and simplify algebraic expressions. The properties of equality are used to solve equations in one variable. Equations are restricted to those with variables on one side of the equation. The final section integrates solving of equations and work with applications.

Chapter 6: Rates, Ratios, and Percents

With a conceptual understanding of fractions and decimals and the skills to solve simple equations, students are ready to study rates, ratios, and percents and their applications. An optional section on circle graphs is also included.

PEDAGOGICAL APPROACH

Developmental students and their instructors are a diverse group. Features in this text support the wide range of approaches used to teach basic mathematics and prealgebra to college students. Content is presented to students in *Investigations* and *Discussions.*

Investigations

The *Investigations* can be completed in cooperative groups of students, as teacher-led whole-class activities, or by a student working individually at home or in a self-paced lab setting. The purpose of the *Investigations* is to engage students in active learning, moving away from the passive approach that might have contributed to past failure. Unlike pure discovery learning, these are highly structured activities that often require the student to look for patterns and develop algorithms. The process of analyzing and describing helps students understand and retain important concepts and procedures. The *Investigations* are written in a worktext format to permit quick assessment of work. Each chapter of the text includes an *Investigation.* Additional *Investigations* are found in the *Instructor's Resource Manual* and may be reproduced for student use.

Discussions

The *Discussions* are a more traditional presentation of definitions and algorithms and include detailed examples. The *Discussions* can support lecture or video presentations, assist students in a self-paced setting, and help students complete exercises. *Investigations* and *Discussions* offer different perspectives for students of fundamental concepts.

Problem Solving

Integrated in both *Investigations* and *Discussions* is the five-step problem-solving strategy. Following the lead of AMATYC and the National Council of Teachers of Mathematics (NCTM), we ask students to apply concepts to the process of problem solving in almost every section. Extensive research and class testing was used to develop a large selection of relevant word problems. In our experience, students overcome their fear and loathing of word problems through extensive practice and success. Using the five-step strategy, students organize the information in the problem and express the relationships in both words and algebraic equations. Since this is difficult at first for most students, we begin with straightforward, even obvious, problems. Once students master the approach, they are empowered to solve more difficult problems.

How to Use This Text

If you choose to teach predominantly with lecture, this book can be used the same way as any textbook. Assigning reading to be completed before class allows the student to be better prepared to ask questions on more difficult concepts. Such an assignment might include an *Investigation* to be done at home. Lecturing can include discussion of concepts and, perhaps, leading the students as a class through an *Investigation.* If you choose to include cooperative group activities, a class period will be a blend of *Investigations,* coaching, and lecturing over certain topics. As the class works, you will walk about the room, answering questions and checking work. Convenient grading rubrics are included in the *Instructor's Resource Manual.* You may discover that the class shares a common struggle or misunderstanding and choose to interrupt them during the activity for a few minutes of explanation or to complete an additional example. When the activity concludes, you may move the students directly into another activity or use a summary of the findings to move into a short lecture on another topic.

FEATURES

Text

Investigating Basic College Mathematics contains several important and unique features. Included in the text narrative are *Sneak Previews, Examples,* and *Check Your Understanding* problems. The objectives from the *Sneak Preview* are included in the student text to support a brief "this is where we are going" and "this is where we've been" summary at the beginning and end of class. As a summary of what students are expected to do at the completion of the section, these also can serve as a study aid. The problems in the *Examples* provide detailed step-by-step explanations. We were careful to include every step to help students working on their own. The *Check Your Understanding* problems provide student practice with immediate feedback. These can also be used for additional in-class practice following an *Investigation* or lecture.

Exercises

The *Exercises* are a blend of skill practice and application problems set in relevant contexts. Except for the word problems, the exercises are arranged in a worktext format to facilitate efficient grading. In the skill practice problems, the same numbers are often repeated, accompanied by a slight change in operations or grouping. When the numbers are the same but the result is different, students have an opportunity to see the effect of such changes. The word problems are factual and authentic, drawn from the media, other disciplines, and reference resources. These real-life applications are often motivating for developmental students. However, we believe that a balance of problems is essential. Students should not be misled into thinking that the only valuable mathematics is applied mathematics.

At the end of each set of problems, we include a unique feature called *Error Analysis.* Students are asked to describe the error made in a problem and then redo the problem correctly. The errors are representative of common student mistakes at this level. Our students often comment on the insight they gain from completing these problems. Also included are a limited number of *Review Questions.* These questions reinforce concepts and skills previously learned, particularly those prerequisite to the next section.

End-of-Chapter Materials

Included in the end-of-chapter materials are *Read, Reflect, and Respond* selections, a *Glossary and Procedures* list, a *Chapter Review,* and a chapter *Test.* The *Read, Reflect, and Respond* selections ask students to thoughtfully consider ideas about learning and mathematics and to reflect on their own practices. Often developmental math students often have barriers to learning mathematics that have little to do with mathematical processes. These barriers include ineffective study skills and negative attitudes about mathematics. However, there is often little time to discuss these barriers in the classroom. The *Read, Reflect, and Respond* selections give students additional information about learning and studying, mathematics history, and the usefulness of mathematics. Students have the opportunity to think about how they might change their previous behaviors and attitudes. If you assign and assess these selections, you can privately connect with your students using brief written comments. If you have a writing-across-the-curriculum initiative, this allows you to integrate writing in a nonthreatening and interesting way.

Student Study Aids

New terms in both the *Investigations* and the *Discussions* are in bold type. Students are asked to keep a list of algorithms and concepts in their own words in the *Glossary and Procedures* section, which can be useful for learning and reviewing. You may wish to require students to prepare note cards or other study aids for these terms. The *Chapter Review* may be assigned as a study aid for the chapter test. Students may also benefit from using the chapter *Test* as a timed practice test before the class exam. All of the answers to the odd-numbered problems in the exercises, chapter reviews, and chapter tests are listed as an appendix.

Calculators

We do not mark exercises or examples as intended for completion with a calculator. Although AMATYC strongly supports the appropriate use of calculators at this level of mathematics, all departments and instructors do not share this position. Some departments also are constrained by the restriction of calculators on state competency exams. Instructors teaching under these constraints often prefer that calculator use not be specified by the text for particular problems. We leave the decision of appropriate technology use at this level to you and your department. On a personal level, we both require students to be able to perform the basic arithmetic algorithms without calculators on "reasonable problems." However, our views of "reasonable problems" differ considerably. We both agree that using calculator support in word problems allows students to concentrate on the process of problem solving instead of being mired in arithmetic errors.

SUPPLEMENTS

For the Instructor

Annotated Instructor's Edition (0-534-39307-1) This special version of the complete student text contains a Resource Integration Guide as well as answers printed next to the respective exercises.

Test Bank (0-534-40503-7) The test bank includes 8 tests per chapter as well as 3 final exams. The tests are made up of a combination of multiple-choice, free-response, true/false, and fill-in-the-blank questions.

Instructor's Resource Manual (0-534-40501-0) The Instructor's Resource Manual provides additional comments, suggestions, and materials for teaching with the text. It also contains the complete worked out solutions to all of the problems in the text.

BCA Testing (0-534-40505-3) With a balance of efficiency and high performance, simplicity, and versatility, *Brooks/Cole Assessment* gives you the power to transform the learning and teaching experience. This revolutionary, internet-ready testing suite is text-specific and allows instructors to customize exams and track student progress in an accessible, browser-based format. BCA offers full algorithmic generation of problems and free-response mathematics. No longer are you limited to multiple-choice or true/false test questions. The complete integration of the testing and course management components simplifies your routine tasks. Test results flow automatically to your gradebook and you can easily communicate to individuals, sections, or entire courses.

Text-Specific Videotapes (0-534-40506-1) This set of videotapes is available free upon adoption of the text. Each tape offers one chapter of the text and is broken down into 10 to 20 minute problem-solving lessons that cover topics and concepts in each section of the chapter.

For the Student

Student Solutions Manual (0-534-40500-2) The student solutions manual provides worked out solutions to the odd-numbered problems in the text.

Investigations Manual (0-534-40502-9) The Investigations Manual contains additional exercises related to the text that can be worked in a group setting or in a self-paced environment. This manual helps the student to further learn about and understand the concepts and skills presented in the text.

Website **www.brookscole.com/mathematics**
The Brooks/Cole Mathematics Resource Center offers book-specific student and instructor resources, discipline specific links, and a complete catalog of Brooks/Cole mathematics products.

BCA Tutorial Instructor and Student Versions This text-specific, interactive tutorial software is delivered via the web (at http://bca.brookscole.com) and is offered in both student and instructor versions. Like *BCA Testing,* it is browser based, making it an intuitive mathematical guide even for students with little technological proficiency. So sophisticated, it's simple, *BCA Tutorial* allows students to work with real math notation in real time, providing instant analysis and feedback. The tracking program built into the instructor version of the software enables instructors to carefully monitor student progress.

Interactive Video Skillbuilder CD (0-534-40504-5) Think of it as portable office hours! The Interactive Video Skillbuilder CD-ROM contains more than 8 hours of video instruction. The problems worked during each video lesson are shown next to the viewing screen so that student can try working them before watching the solution. To help students evaluate their progress, each section contains a 10-question Web quiz (the results of which can be emailed to the instructor) and each chapter contains a chapter test, with answers to each problem on each test. The CD also includes MathCue Tutorial software. This dual-platform software presents and scores problems and tutor students by displaying annotated, step-by-step solutions. Problem sets may be customized as desired.

ACKNOWLEDGMENTS

This text would not have been possible without the help of many people. At Harcourt, we thank Emily Barosse, Bill Hoffman, Angus MacDonald, and Kelley Tyner. Jan Kerman, our developmental editor, did a great job of shepherding this manuscript in the midst of the merger of our publisher with Brooks/Cole–Thomson Learning. At Brooks/Cole, we thank our sponsoring editor, Jennifer Huber, and her editorial assistant, Jonathan Wegner. Thanks also to Ellen Sklar and Martha Emry, who provided continuity and encouragement in design and production.

At Lewis-Clark State College, thanks are due to our colleagues Ed Miller and Shelley Hansen. A special thanks goes to Masoud Kazemi, who undertook the final accuracy check. Annie and Charles Petersen, Barbara Hayes, Harriet Husemann, and Diana Ames also helped a great deal in the process of writing this text.

At Utah Valley State College, thanks to John Jarvis, who helped make it possible to write the manuscript by allowing for an ideal teaching schedule. Thanks to Darren Wiberg for writing the *Student Solutions Manual* that accompanies this text, and to Carole Sullivan for accuracy checking his work.

On a personal note, Hazel thanks her partner, Melana Walker, her mother Jean McKenna, and great friends Cathy Morris, Deborah Marrot, and J. C. Cole for the love, support, and encouragement to write this book. Finally, thanks to Morgan and Sammy for their unconditional love and always wagging tails.

Finally, we are grateful to the following instructors who reviewed the text and made suggestions and comments: Marwan Abu-Sawwa, Florida Community College at Jacksonville; Raul Aparicio, Blinn College–Brenham Campus; Carol Barnett, St. Louis Community College–Meramec; Sharon Edgmon, Bakersfield College; Mitchel Fedak, Community College of Allegheny County–Boyce Campus; Greg Goodhart, Columbus State Community College; Celeste Hernandez, Richland College; Susann Kyraizopoulos, Devry Institute of Technology; Jeff Morford, Henry Ford Community College; Ted Panitz, Cape Cod Community College; Karen Stewart, College of the Mainland; Sharon Testone, Onondaga Community College.

Investigating Basic College Mathematics is a text that is flexible enough to meet the needs of instructors with a variety of teaching methods and student populations. We know this because of our own experiences: Our students and our teaching styles are very different and we both teach successfully with this text. We welcome your comments and suggestions.

Hazel McKenna
mckennha@uvsc.edu

Laura Bracken
bracken@lcsc.edu

To the Student

You are beginning an important task, learning the math you need for your future. It is not enough to simply want to succeed. You must put in the time and effort that is necessary for you to learn. Learning differs from person to person, and it is important for you to discover how you learn best. Attending class, doing homework, asking questions, preparing for tests, and asking for extra help when you do not understand are all part of this required effort. Learning math does not have to be done by yourself. In fact, the support and insights of other students can really help you learn math. If possible, complete the following activity with four of your fellow students.

INVESTIGATION Positive and Negative Experiences

In this first investigation we will explore how a group can work together to learn mathematics. When investigations are done in a group, we should take turns reading the information aloud. This helps students who learn best when they hear information. *In general, a new paragraph means a new reader.*

Some topics will be review for some group members but will be new topics for others. Those of us who are learning a concept for the first time should not compare our speed and progress to others. This is not a race. We do not get extra points for being done first. We can take the time to help each other.

One way to learn mathematics is to work in cooperative groups. Such groups are most effective when everyone participates. Desks or chairs should be arranged to allow for convenient interaction. We should speak in voices that are clear and easy to hear, but not so loud that other groups are distracted. We must work together on the same problem. Our instructor will monitor our progress to make sure we are going fast enough. We may ask the instructor a question only if the entire group has discussed the question and cannot agree on an answer. This ensures that we work together and that we learn to use each other as resources. We need to be ready to interrupt our group work to listen to our instructor provide extra examples or clarification. When we reach a "stop sign," we need to have our work checked by our instructor.

Each individual member brings different experiences in mathematics to this class. Our goal is to learn with each other, drawing on each other's strengths and helping each other. To begin this process, we will break our group into pairs. Each of us will describe to our partner *the most positive experience* we ever had in learning mathematics and *the most negative experience* we ever had in learning mathematics. Our partner will take notes about our experiences.

We will then rejoin our other group members. Each person will tell the group about the positive and negative experiences of his or her partner. Finally, we will decide on one positive and one negative experience to report to the whole class. One person should volunteer to be the "reporter."

1. Names of your group members (first and last):

2. Positive experiences of your partner:

3. Negative experiences of your partner:

4. Positive experiences of your group to report:

5. Negative experiences of your group to report:

Your personal attitudes about math can affect your success in learning math. Although you do not have to like math to learn it, it is important to think about your personal attitudes and their impact on your actions. Before you go any further in this course, we strongly recommend that you complete the following reading and answer the questions.

READ, REFLECT, AND RESPOND Assessing Your Attitudes

READ AND REFLECT

Thinking about your attitudes toward learning mathematics can help you discover barriers that may be holding back your progress. Begin by filling out the following questionnaire.

If you Strongly Agree, score 5.
If you Agree Most of the Time, score 4.
If you Agree Sometimes, score 3.
If you Disagree Most of the Time, score 2.
If you Strongly Disagree, score 1.

___ **1.** I feel sick when I go to math class.
___ **2.** When math teachers talk, I find myself unable to concentrate.
___ **3.** Although I attend my other classes regularly, I often miss math class.
___ **4.** I know I am probably going to have to take this class over.
___ **5.** I can do all the homework fine but when it comes to a test I just go blank.
___ **6.** I never get math the first time anyone explains it to me.
___ **7.** I often leave my math homework until the very last.
___ **8.** I've never been good at math.
___ **9.** I just dread going to class, I always feel so stupid.
___ **10.** I really worry that the teacher will call on me for an answer.
___ **TOTAL**

If your score is below 30, your attitudes toward mathematics are probably not affecting your success in math classes. The closer your score approaches 50, the more your feelings about math are preventing you from learning math.

Math anxiety is a term educators use for the collection of feelings described in the questionnaire statements. Most people experience some anxiety when learning something new and difficult, a normal feeling that actually increases concentration and motivation to study. However, too much worry can prevent learning. High levels of math anxiety may cause students to do things that stop learning such as skipping class or avoiding homework. Before or during class, anxious students may experience an uncomfortable increase in heart rate, get a pounding headache, feel sweaty or chilled, or become nauseous.

People with math anxiety can learn to control and diminish their fear enough to do well in math classes. However, good intentions are only the first step. Changing patterns of behavior and attitudes requires thoughtful practice. Some suggestions follow. Sometimes only one change in behavior can result in a huge reduction in math anxiety. If these suggestions do not help, seek the advice of a trained counselor at your school. These professionals are trained to help identify the causes of your fears and help you control them.

SUGGESTION 1 Discover the sources of your feelings about math.

As described in the following true stories, math anxiety may begin with a particular experience.

When Randi was in third grade, she moved to a school where her teacher seated the kids depending on their score on their last math test. Randi had to take a test the second day she was at her new school. She did very poorly because she hadn't learned double digit multiplication yet. She was moved to the end of the very last row. She stayed near that seat the rest of the year, embarrassed and frightened that she would never catch up. Now she is an intensive care nurse. She still worries that she will make a mistake on a dosage calculation and hurt a patient.

Mark's favorite subject was math until he had to learn his multiplication facts. He tried to remember them on the timed tests but the harder he tried, the worse he did. His parents made him work on them every night. The other kids got to do fun stuff but he had to keep doing flash cards to learn his tables. He never did get them. Math just got worse then because multiplication and division depended on knowing the facts. Mark now hates math and avoids it whenever possible.

Hiro has struggled with math all his life. People seem to expect him to be good at math just because he is Japanese. This really bothers him. He is worried that his teacher will expect him to learn new procedures with only one explanation.

When Nicole started algebra in ninth grade, she just did not get it. There were so many steps that she couldn't remember. When she asked questions, her teacher acted like she was wasting his time. Her dad tried to help her but he got impatient with her, too. At the end of the year, Mr. Anderson gave her a "D" instead of the "F" she should have received. He told her he did not want to put up with her for another year and she would probably just get married anyway. Nicole got a D in her next math class, too.

Laurie was always good at writing but never very good at math. Her brother was really good at math. Her mom always told her that it was the same way in her family. Her dad told her not to worry, just to try her best. Laurie avoided all classes that required math like chemistry. She figured there was no point in working hard at something that wasn't her thing anyway.

John quit school when he was seventeen. He wanted to move out and he needed a full time job to do that. After a while, his girlfriend talked him into getting his GED. He did the math for it with the help of a tutor. Now John wants to become a police officer which requires completion of a college math course. He is very scared of failing; he doesn't remember how to do much.

Many students in this class fear or even hate mathematics. These feelings often began with a negative experience like these. The first step in controlling fear of mathematics may be to identify where and when it started. Fortunately, attitudes can change. Past experiences can be left behind.

SUGGESTION 2 Encourage yourself; tell yourself that you can succeed.

Start to believe you can succeed by telling yourself that you can succeed. Talk to yourself before every class, before every homework assignment, and before every quiz or test. Try saying:

- I am a person who does many things well. Math may be difficult for me but I can succeed at it, too.
- I will judge myself by my own standards. I decide if I have succeeded or failed. I may learn slower than others do but the important thing is that I can learn.
- I am going to do the work I need to do to succeed, even if I do not enjoy it.
- I am willing to ask for help because I want to succeed. I want to succeed more than I am worried about looking stupid.
- I do not have to understand everything the first time it is explained to me.

SUGGESTION 3 Visualize yourself succeeding.

Identify those people in your math class that you think or know are successful. How do they act? How do they interact with other students? How do they interact with the instructor? Where do they sit? What did they bring with them to class? Now zero in on such a person who seems to be somewhat like you. Perhaps they are the same age or in the same situation. Perhaps they are from your culture. This is a successful person who shares some of your experiences and values. Visualize yourself as a successful person in math class. See yourself doing the things needed to succeed. Do this before class, before you do homework, when you feel fear settling in your stomach, whenever you have the urge to procrastinate preparing for class.

SUGGESTION 4 Evaluate the way you study and prepare for math class and tests.

Different people need to use different strategies for preparing for tests. When students find an effective way to study, class is not so confusing; homework is possible to complete; tests are less frightening. The more success you have, the less you fear; the anxiety goes down more. Do not depend on luck to do well; prepare carefully and thoughtfully. Ask your instructor for help and guidance.

Think about your own attitudes toward math. Are you afraid of math? If so, can you think of any event or events in your life where this attitude began? To build new patterns of success in mathematics that will help dissipate your fear, start encouraging yourself. Get into the habit of talking and thinking positively. Visualize yourself being successful in mathematics and then start acting like that person. Talking and thinking positively are important but not sufficient. You need to study effectively for math tests. This may require you to make some changes in the way you study.

RESPOND

1. What is your total score on the survey in this reading?

2. Describe your feelings/attitudes about math and learning math.

3. Describe the positive and/or negative experiences that have influenced your attitudes about learning mathematics.

4. Some students struggle to find enough time in the day to do their homework and study for their classes. Describe the priority college has in your life right now. Compare its importance to other activities in which you participate including parenting, work, athletics, and so forth.

5. Describe any past attitudes about learning math or habits you followed in learning math that you plan to change.

6. Predict how well you will do in this class. Explain why you think so.

We both teach students that, like you, need to learn math. We wrote this book so we could do a better job of teaching those students. We hope that it helps you learn as well. Remember that you need to want to learn and you need to work hard to be successful.

If you have suggestions to make this textbook better, let us know by e-mail.

Hazel McKenna
mckennha@uvsc.edu

Laura Bracken
bracken@lcsc.edu

INVESTIGATING
Basic College Mathematics

1

Whole Numbers

If basic mathematics is the concepts, skills, and experiences of arithmetic and geometry, then we have been learning basic mathematics since we were born. In this course, we will continue that process, using numbers to represent a wide variety of quantities, from ten toes to a million miles. Numbers will be added, subtracted, multiplied, and divided. We will look for patterns and relationships between quantities. We will analyze information that is presented in charts, tables, and graphs.

We will begin by thinking about numbers and what they represent. Numbers will be used to count and to measure. We will closely examine the operations of addition, subtraction, multiplication, and division. We will also learn how to represent unknown numbers with variables and solve problems.

CHAPTER OUTLINE

Section 1.1 Introduction to Whole Numbers

LEARNING TOOLS

CD-ROM SSM VIDEO

Sneak Preview

Numbers are part of our lives from an early age. We use numbers for counting and measuring without formal education or training. However, understanding mathematics requires a more in-depth knowledge of numbers. In this section, we will review some fundamental ideas about numbers.

After completing this section, you should be able to:

1) Distinguish between digits, whole numbers, and natural numbers.

2) Use set notation to describe a set of numbers.

3) Identify the place value of a digit in a number.

4) Write a whole number in expanded form.

5) Write whole numbers using word names or place value names.

6) Order two or more whole numbers using an inequality.

7) Graph a whole number on a number line.

8) Round a whole number to a given place value.

9) Name and classify an angle.

DISCUSSION

Our Number System

Numbers are an important part of our daily lives. We often use them without thinking. However, all the procedures we know for working with numbers depend on certain characteristics of our number system. To fully understand these procedures, we need to be familiar with these characteristics.

The numbers we use are written using symbols called **digits:** 0, 1, 2, 3, 4, 5, 6, 7, 8, and 9. Numbers can be classified by the number of digits: 65 is a two-digit number while 145 is a three-digit number. In the United States, commas are used to separate every group of three digits counting from the right: 15,619,215.

Mathematicians call a group of numbers or objects a **set.** Each member of the set is called an **element.** One way to represent a set is to list the elements inside a pair of braces. This is called **set notation.** For example, the set of digits can be represented as:

$$\text{Digits} = \{0, 1, 2, 3, 4, 5, 6, 7, 8, 9\}.$$

An important set of numbers is the **natural numbers.** (The natural numbers are also called the **counting numbers** because these are the numbers we use to count.) We can use N to represent the set of natural numbers in set notation: $N = \{1, 2, 3, 4, 5, \ldots\}$. The three dots at the end of the list of natural numbers are called **ellipses.** These dots show that the set of numbers goes on forever, following the same pattern. Since the natural numbers never end, this set is an example of an **infinite set.** If we include zero with the natural numbers, we have a new infinite set called the **whole numbers:** $W = \{0, 1, 2, 3, 4, \ldots\}$. A set that is not infinite is **finite.** A finite set of numbers does not go on forever. For example, the set $\{1, 2, 3\}$ is a finite set with exactly three elements.

The value of a digit in a number depends on its position to the left of a starting point called the **decimal point.** The position of each digit is called its **place value.** Each group of three digits is also given a group name.

	MILLIONS GROUP			THOUSANDS GROUP			ONES GROUP			
Billions	Hundred millions	Ten millions	Millions	Hundred thousands	Ten thousands	Thousands	Hundreds	Tens	Ones (units)	Decimal point
1,000,000,000	100,000,000	10,000,000	1,000,000	100,000	10,000	1,000	100	10	1	.

The value of each digit depends on its place value. In the number 527, the digit 7 represents 7 ones and has a value of 7. The digit 2 represents 2 tens and has a value of 20. The digit 5 represents 5 hundreds and has a value of 500. We can think of 527 as 500 + 20 + 7. 527 is called the **place value form** of this number; 500 + 20 + 7 is called the **expanded form.**

Numbers can also be expressed in words, without using digits. For example, 527 can be rewritten "five hundred twenty-seven." We work from left to right, writing the numbers in each group followed by the group name. Notice that the name of the ones group is not included. Also, unlike the way we might say a number in conversation, the word "and" is not used in writing the word name of a whole number. Similarly, a word name can be rewritten as a number in place value form.

Example 1 **Write nine hundred twenty-three thousand, seven hundred four in place value form.**

nine hundred twenty-three thousand, seven hundred four

923 in the thousands group, 704 in the ones group

923,704

Notice that 0 is used as a placeholder in the tens place since there are no tens in the number.

Example 2 **Write the word name of 8,926,041.**

Eight million, nine hundred twenty-six thousand, forty-one.

Example 3 **Write 6,453,008 in expanded form.**

6,000,000 + 400,000 + 50,000 + 3,000 + 0 + 0 + 8

CHECK YOUR UNDERSTANDING

1. Identify the place value of the digit 5 in the whole number 76,854.

2. Write the number 7,450,075 in expanded form.

3. Write the word name for the number 7,008,620.

4. Write "seventy-four million, two hundred thirty-eight thousand, nine hundred five" as a number in place value form.

Answers: 1. tens place; 2. 7,000,000 + 400,000 + 50,000 + 0 + 0 + 70 + 5; 3. seven million, eight thousand, six hundred twenty; 4. 74,238,905

DISCUSSION

Comparing the Value of Two Numbers

When comparing the value of two numbers, there are only two possible relationships: The numbers are **equal** or they are **not equal.** In words, we say, "six is equal to six." In symbols, we write $6 = 6$. In words, we say, "six is not equal to two." In symbols, we write, $6 \neq 2$. A statement that has an equals sign $(=)$ is called an **equation.** Whole number equations can be true or false: $6 = 6$ is a true equation; $5 = 9$ is a false equation.

When two numbers are not equal, one number is **greater than** the other number. The symbol $>$ is used to represent "greater than." We say, "eight is greater than two." Using symbols, we write $8 > 2$. Or, we could say "two is **less than** eight." The symbol $<$ is used to represent "less than." Using symbols, we write $2 < 8$. Since $8 > 2$ means exactly the same thing as $2 < 8$, these are **equivalent** statements. The smaller end of the $>$ symbol or the $<$ symbol always points to the smaller number. Statements that contain a $<$ or $>$ sign are called **inequalities.** Whole number inequality statements can be true or false. The inequality $7 > 3$ is true. However, $4 < 2$ is false. We can also write **compound inequalities** that compare the value of more than two numbers. For example, we say "1 is less than 4 which is less than 8." In symbols, $1 < 4 < 8$.

Inequalities are used to compare measurements. A measurement includes a number and a **unit of measurement.** For example, a football field is 100 yards long. The unit of measurement is *yards.* Measurements with the same unit can be directly compared using $>$ or $<$. For example, 10 feet $<$ 15 feet and 78 pounds $>$ 24 pounds. However, we cannot write 34 feet $<$ 50 pounds because we cannot directly compare feet and pounds. These units measure different things.

Example 4 **The distance from Salt Lake City, Utah, to Denver, Colorado, is 537 miles. The distance from Philadelphia, Pennsylvania, to Charlotte, North Carolina, is 545 miles. Write an inequality that compares the two distances.**

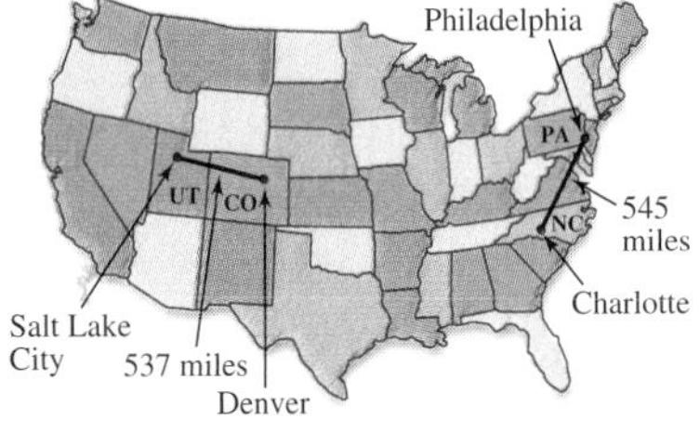

537 miles $<$ 545 miles or 545 miles $>$ 537 miles

CHECK YOUR UNDERSTANDING

Rewrite the inequality statements using $<$.

1. $9 > 8$ **2.** 9 feet $>$ 2 feet **3.** $6 > 0$

Identify the statements as true or false.

4. $\$4 < \$7$ **5.** $9 \neq 9$ **6.** $3 > 6$ **7.** $4 > 0$

Fill in the blank with $<$ or $>$ to make a true statement.

8. 6 meters __ 9 meters **9.** 15 __ 7 **10.** 91 __ 112

Write a compound inequality that orders the numbers from smallest to largest.

11. 53; 34; 73; 44 **12.** \$213; \$253; \$193; \$235

Answers: 1. $8 < 9$; *2.* 2 feet $<$ 9 feet; *3.* $0 < 6$; *4.* true; *5.* false; *6.* false; *7.* true; *8.* $<$; *9.* $>$; *10.* $<$; *11.* $34 < 44 < 53 < 73$; *12.* $\$193 < \$213 < \$235 < \$253$.

DISCUSSION

Lines and Number Lines

A point is drawn on paper with a dot. A dot on paper can be large or small. A **point** in mathematics has no size; it just represents a position. A **line** is straight and extends forever in both directions. It is made up of an infinitely large set of points. Since points have no size, a line has no width. (Of course, a line looks like it has width but this is just a picture of an idea.) Two lines that never cross each other are called **parallel lines.**

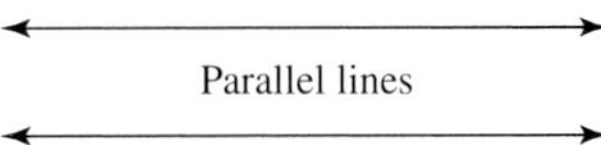

A **number line** looks like a ruler. Each mark on the number line pictured below represents a whole number. The numbers are written from left to right, smallest to largest. Since the whole numbers are an infinite set that continue on forever, a number line is written with an arrow on the right side showing that the number line also goes on forever. The number line for whole numbers begins at 0. In Chapter 2, we will consider the numbers to the left of 0.

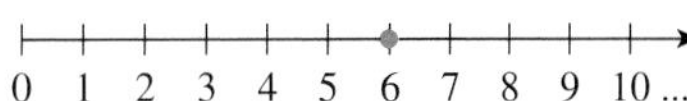

We can express a number in words (six), with a symbol (6), or as a graph on a number line. A graph of a number is a point drawn on the line directly above the specified number.

Example 5 **Graph the number 6 on a number line.**

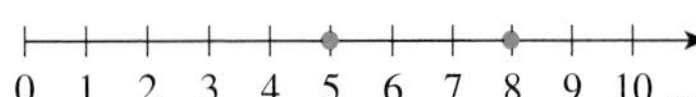

We can also graph two numbers on a number line. If a number is to the right of another number, it is *greater than* that number. If a number is to the left of another number, it is *less than* that number.

Example 6 **Graph 5 and 8 on a number line. Write two inequalities that compare the value of these numbers.**

0 1 2 3 4 5 6 7 8 9 10 ...

Since 8 is to the right of 5, $8 > 5$. Since 5 is to the left of 8, $5 < 8$.

CHECK YOUR UNDERSTANDING

1. Graph 2 and 7 on a number line. Write two inequalities that compare the value of these numbers.

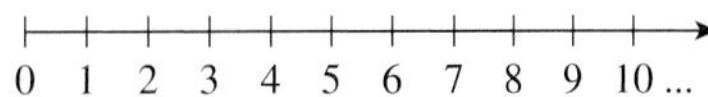

2. Graph 0 and 10 on a number line. Write two inequalities that compare the value of these numbers.

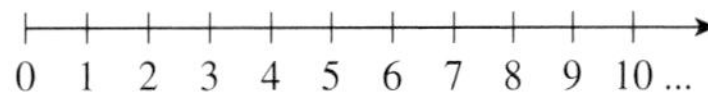

DISCUSSION

Angles, Circles, and Degrees

Imagine a straight line. Pick a point on that line. Now erase all the points on the line to one side of that point. What remains is called a **ray.** The point at the beginning of the ray is called its **endpoint.**

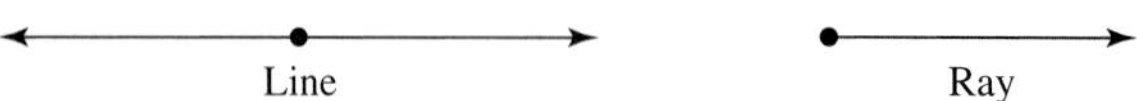

Rays are often named by naming the endpoint and another point on the line. For example, a ray might have an endpoint named A and another point named B. This ray can be named $\overrightarrow{AB}$.

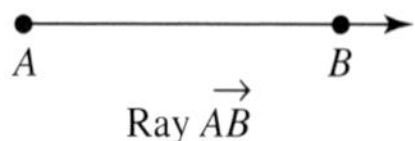

Ray $\overrightarrow{AB}$

The notation $\overrightarrow{AB}$ is used to show that the ray starts at the endpoint A and travels through the point B.

If we draw two different rays that have the same endpoint, we create an **angle.** Angles can be named using the letters of three points with the endpoint in the middle. Angles can also be named using just the endpoint. The symbol for angle is $\angle$.

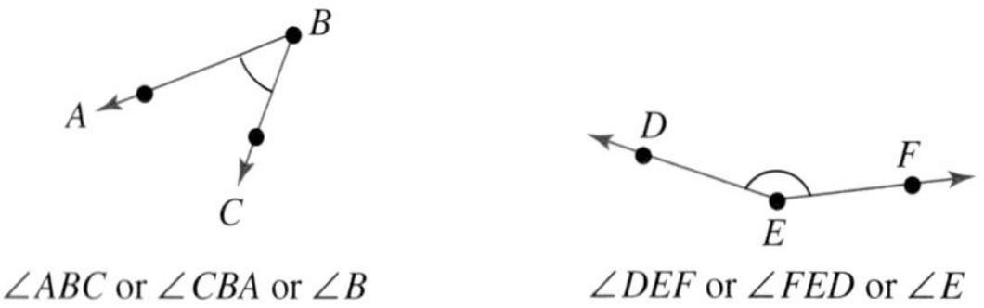

$\angle ABC$ or $\angle CBA$ or $\angle B$ $\quad$ $\angle DEF$ or $\angle FED$ or $\angle E$

A **plane** is a flat surface with no thickness that extends forever. The floor of a room can be thought of as part of a plane. The plane does not stop at the walls of the room but goes on forever.

A **circle** is the set of all points in a plane that are at the same distance from a point called the **center.** The distance around any circle can be divided into 360 equal parts. Each of these parts is called a **degree.** If we start at 0 on the right side of the circle, the point directly opposite 0 degrees will be 180 degrees away. When we arrive back at the starting point, we will have traveled 360 degrees. The symbol for degree is °. The degree symbol is written to the right of the measure of the angle. So, 360 degrees can be written 360°.

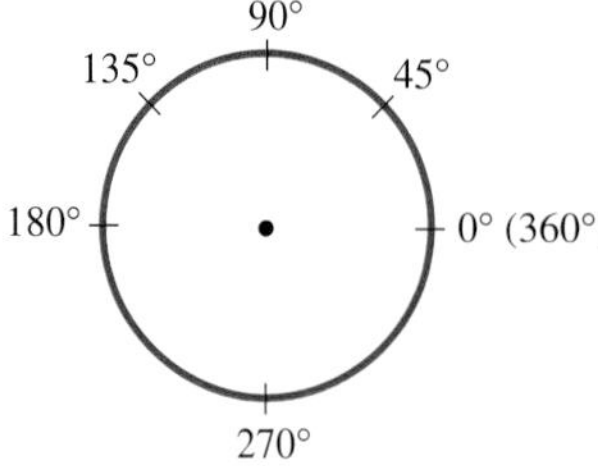

If the endpoint of an angle is the center of a circle, we can measure the angle in degrees.

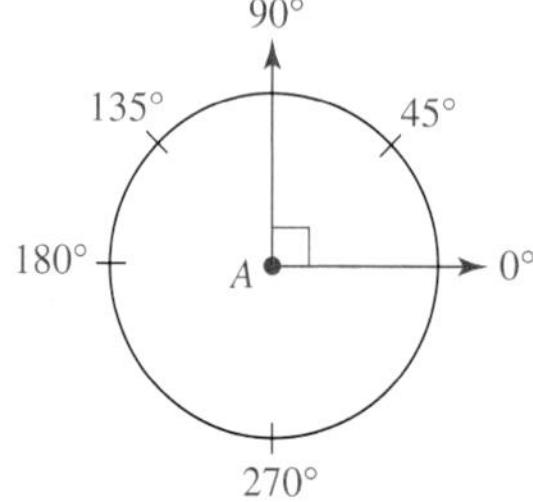

The measure of $\angle A$ is 90°. Any angle that measures 90° is called a **right angle.** Any angle that has a measure less than 90° is called an **acute angle.** Any angle that has a measure greater than 90° but less than 180° is called an **obtuse angle.** An angle that is exactly 180° is called a **straight angle.**

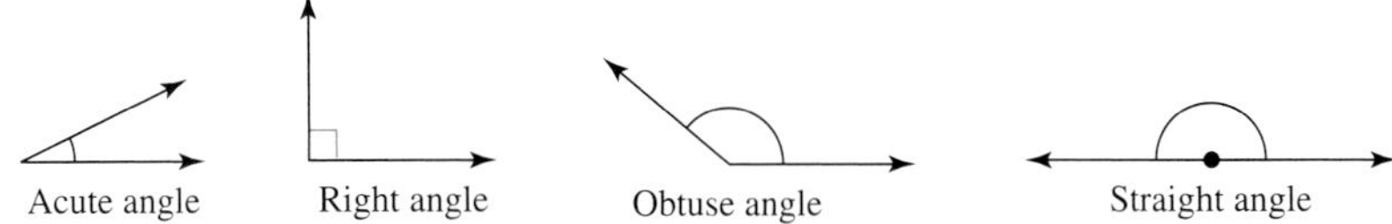

When drawing a right angle, we show that it measures exactly 90° by drawing a box in the corner of the angle.

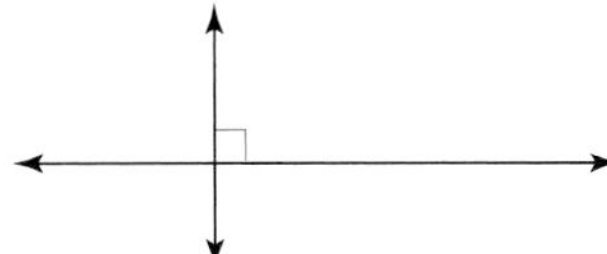

Example 7 **A 45° Y-branch pipe fitting allows two pipes to be attached to a single pipe. The pipes meet in a 45° angle. Is this angle acute, right, or obtuse?**

Since $45° < 90°$, this angle is acute.

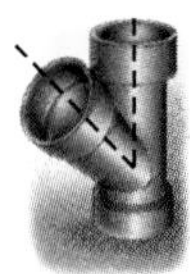

CHECK YOUR UNDERSTANDING

1. Name this angle:

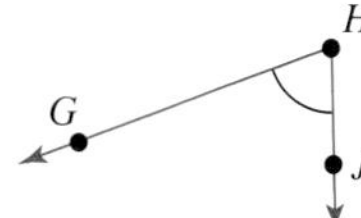

2. What is the difference between a ray and a line?

3. Classify an angle with measure 145° as acute, obtuse, right, or straight.

Answers: 1. Either $\angle GHJ$ or $\angle JHG$ or $\angle H$; 2. A line extends forever in both directions; a ray has an endpoint and only extends forever in one direction; 3. Obtuse.

DISCUSSION

Estimation and Rounding

Estimates are alternatives to exact measurements. For example, a hotel shuttle driver does not need to tell guests that the hotel is exactly 3 miles 200 yards 3 feet 2 inches from the airport pickup location. Instead, an estimate of 3 miles is sufficient. Generally, estimates are quicker to determine than exact answers.

Estimates can be used to check the reasonability of an exact answer to a problem. An answer is *reasonable* when its estimate is close to its exact value, and the answer makes sense. Estimates are also used when exact answers are not required. **Rounding** is often used in making an estimate. Rounding approximates a measurement to a particular place value. For example, the body length of the rodent in the picture below is being measured for a biology field study.

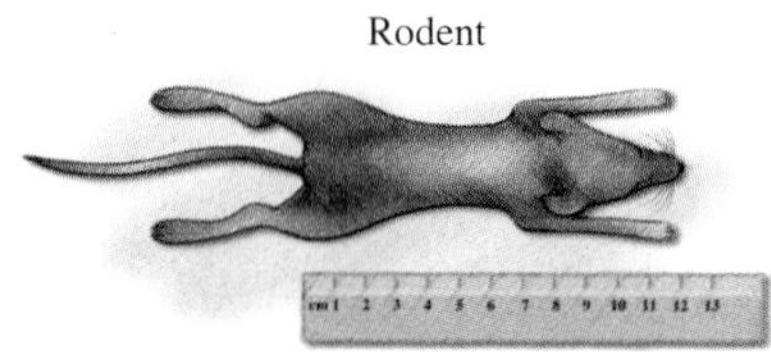

The end of the rodent is placed at the 0-centimeter mark of the ruler. Notice that the nose is closer to the 12-centimeter mark than to the 13-centimeter mark. An estimate of the body length of the rodent is 12 centimeters.

To round a measurement, choose the value closest to the actual measurement. If the measurement is exactly halfway between two values, the value is closest neither to the smaller measurement nor to the larger measurement. Which value should be used? There are different sets of rules that decide this question. However, in many everyday situations, it is best to select a larger value than a smaller value. So, when the measurement is exactly halfway between two values, use the higher value.

Example 8 **Round 639 feet to the nearest ten.**

To round to the nearest ten, choose either 630 feet or 640 feet. Since 639 feet is closer to 640 feet than to 630 feet, round to 640 feet.

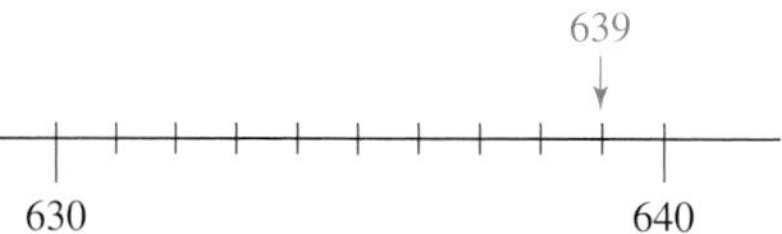

Example 9 **Round 639 quarts to the nearest hundred.**

To round to the nearest hundred, choose either 600 quarts or 700 quarts. Since 639 quarts is closer to 600 quarts than to 700 quarts, round to 600 quarts.

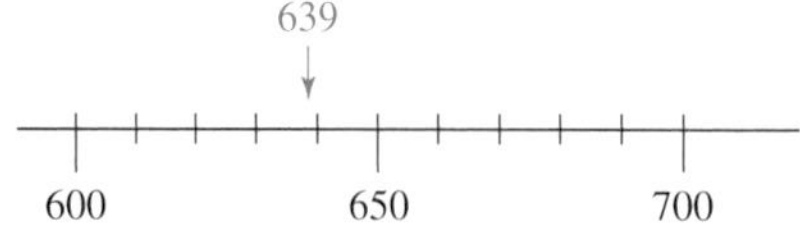

Example 10 **Round 2,500 miles to the nearest thousand.**

To round to the nearest thousand, choose either 2,000 miles or 3,000 miles. 2,500 miles is exactly halfway between 2,000 miles and 3,000 miles. When this happens, we will always round up to the higher number. So, 2,500 miles rounded to the nearest thousand is 3,000 miles.

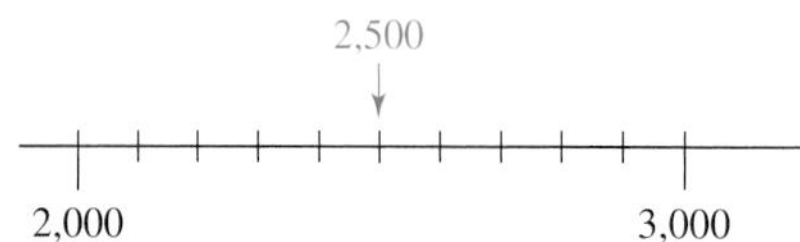

CHECK YOUR UNDERSTANDING

Round each measurement to the given place value.

MEASUREMENT	1,439 GRAMS	4,520 KILOMETERS	7,611 LIGHT-YEARS	\$9,873
Thousands				
Hundreds				
Tens				

Answers: Thousands: 1,000 grams; 5,000 kilometers; 8,000 light-years; \$10,000
Hundreds: 1,400 grams; 4,500 kilometers; 7,600 light-years; \$9,900
Tens: 1,440 grams; 4,520 kilometers; 7,610 light-years; \$9,870

We can also round a number by following a set of rules.

PROCEDURE

To round a whole number to a given place value:

1. Pick out the digit in the place value to which you are rounding; this is the **rounding digit.**
2. If the digit to the right of the rounding digit is less than 5, do not change the rounding digit. If the digit to the right of the rounding digit is 5 or greater, increase the value of the rounding digit by one.
3. Replace all the digits to the right of the rounding digit with zeros.

Example 11 **Round 7,365 to the nearest hundred.**

1. The rounding digit is 3.
2. The digit to the right of the rounding digit is 6. Since 6 is greater than 5, increase the rounding digit to 4.
3. Replace the digits to the right of the rounding digit with zeros; the rounded number is 7,400.

Example 12 **Round 73,926 to the nearest hundred.**

1. The rounding digit is 9.
2. The digit to the right of the rounding digit is 2. Since 2 is less than 5, the rounding digit is not changed.
3. Replace the digits to the right of the rounding digit with zeros; the rounded number is 73,900.

Example 13 **Round 12,971 to the nearest hundred.**

1. The rounding digit is 9.
2. The digit to the right of the rounding digit is 7. Since 7 is greater than 5, increase the rounding digit by one to 10. Since 10 is a two-digit number, it cannot be placed directly in the hundreds place. However, 10 hundreds is the same as 1 thousand. Increase the 2 in the thousands place by one to 3; replace the 9 in the hundreds place with 0. This is an example of **carrying.**

3. Replace the digits to the right of the rounding digit with zeros; the rounded number is 13,000.

Example 14 **Round 4,285,999 to the nearest ten thousand.**

1. The rounding digit is 8.
2. The digit to the right of the rounding digit is 5. Increase the rounding digit by one to 9.
3. Replace the digits to the right of the rounding digit with zeros; the rounded number is 4,290,000.

Example 15 **Round 3 to the nearest ten.**

1. There is no obvious rounding digit. Since there is no value in the tens place, we can think of this number as 03. So, the rounding digit is 0.
2. The digit to the right of the rounding digit is 3. Since 3 is less than 5, the rounding digit does not change.
3. Replace the digits to the right of the rounding digit with zeros; the rounded number is 0.

CHECK YOUR UNDERSTANDING

Round each number to the nearest hundred.

1. 561

2. 78,012

3. 9,886,254

4. 47

5. 391

Round each number to the nearest million.

6. 276,887

7. 8,265,964

8. 9,789,543

Answers: 1. 600; 2. 78,000; 3. 9,886,300; 4. 0; 5. 400; 6. 0; 7. 8,000,000; 8. 10,000,000.

EXERCISES SECTION 1.1

1. All digits are whole numbers. Give an example of a whole number with four digits.

2. There is one number that is an element of the set of whole numbers but is not an element of the set of natural numbers. What is this number?

3. Is the set {2, 4, 6, 8} a finite set or an infinite set? Explain why.

4. Write the set of whole numbers in set notation (use set braces and ellipses).

Identify the place value of the given digit in the number 6,985,741.

5. 7

6. 6

7. 1

8. 8

9. 9

10. 4

Identify the place value of the digit 4 in each number.

11. 34

12. 8,497

13. 4,837,002

14. 143,726,005

15. Write one thousand eighty-nine in place value form.

16. In a telephone conversation, a contractor gave the following bid for a job: "seven thousand, six hundred ninety-five dollars." Write this bid as a number. Include the dollar sign.

Write the word name of each number.

17. 5,981

18. 772,074

19. 502

20. Tabor is writing a check for a down payment on a house for $8,993. What word name should be written on the check?

21. Between 1891 and 1900, 3,687,564 people immigrated to the United States. Write the word name of this number.

Write each number in expanded form.

22. 187

23. 8,913,775

24. 79

25. 943,933

Graph each number on a number line.

26. 2

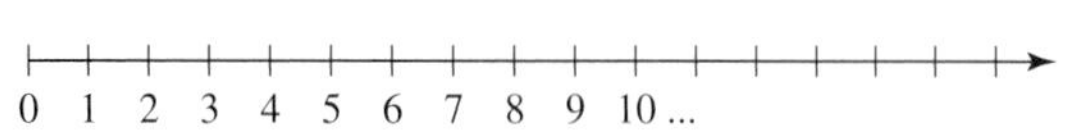

27. 8

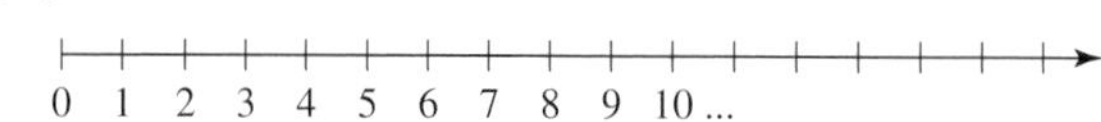

Identify each equality or inequality as true or false. A number line may be helpful.

28. $5° < 9°$

29. $87 > 98$

30. 786 inches < 985 inches

31. $56 = 74$

32. 85 liters < 85 liters

33. $94{,}997 \neq 94{,}990$

Rewrite each inequality with a < symbol.

34. $63 > 34$

35. $72 > 21$

36. $7 > 2$

Rewrite each inequality with a > symbol.

37. $34 < 45$

38. $82 < 101$

39. $8 < 14$

40. Mount Everest is 29,028 feet high. Mount McKinley is 20,320 feet high. Write an inequality statement that compares the height of Mount Everest to the height of Mount McKinley. Make sure you include the units of measurement in the inequality.

Mount Everest *(© Bill Ruthven/ Mount Everest Foundation)*

Mount McKinley *(© 2002 Kennan Ward)*

41. Write the numbers in order from smallest to largest using < signs: 7,284; 7,874; 7,964; 6,999; 7,003.

42. A tour company advertised the following tour prices:

Taste of Britain:	$750	*Best of Italy:*	$1,499
Best of Ireland:	$1,025	*Best of Spain:*	$1,425
European Highlights:	$1,090		

Write the prices in order from largest to smallest using > signs. Include the $ sign in each price.

43. Estimate the height of the British telephone booth below to the nearest foot.

44. Estimate the length of the scissors below to the nearest inch.

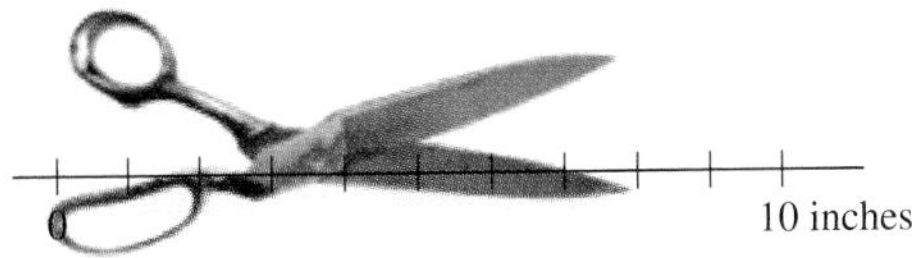

45. In the following newspaper article, circle the numbers that are estimates and put a box around the numbers that are exact.

"It takes more than 22 hours for a taxpayer who itemizes deductions and has some investment income to finish this year's required Internal Revenue Service forms, the tax agency estimates. That's three hours longer than last year. The U.S. tax code now stands at more than 1.5 million words and climbing thanks to 1,260 changes enacted by Congress and signed by President Clinton in the last two years alone. For this year's filing season, the IRS had to develop 11 new forms and revise 177 others." *(Source: Curt Anderson, Associated Press, 2/19/99.)*

Round each number to the given place value.

46. 9,008 to the nearest thousand

47. 8,992 to the nearest hundred

48. 19,534 to the nearest thousand

49. 88,001 to the nearest ten

50. 7,876 to the nearest ten thousand

51. 255 to the nearest thousand

52. 87 to the nearest ten

53. 4,001 to the nearest hundred

54. 23 to the nearest hundred

55. 79 to the nearest hundred

56. 456 to the nearest ten

57. 5,901 to the nearest ten

58. The population of the United States by official census was 248,709,873 people in 1990. Round this information to the nearest ten million.

59. In 1992, the Internal Revenue Service collected $1,120,799,558,875 in income tax. Round this figure to the nearest billion dollars.

60. In 1994, JFK International Airport in New York City had 28,799,275 passenger arrivals and departures. Round this to the nearest ten thousand.

61. The U.S. Department of the Treasury determined that it had the following number of bills in circulation. Round each number to the nearest million.

BILLS IN CIRCULATION

TYPE OF BILL	NUMBER OF BILLS	NUMBER OF BILLS ROUNDED TO NEAREST MILLION
$1	3,571,913,726	
$5	987,814,668	
$10	1,136,337,194	
$20	2,579,310,289	
$50	1,530,339,488	
$100	765,169,744	
$1,000	1,544,720	
$10,000	347	

62. Describe how to name an angle. Include an example. Name the angle by listing the letters of an endpoint, the vertex, and the other endpoint.

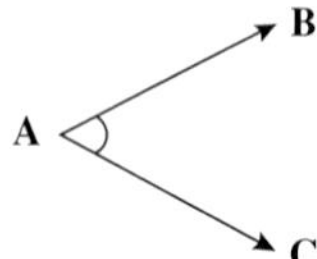

63. When building a house, the walls and floors are usually at right angles. What is the measure of the angle between a wall and the floor?

ERROR ANALYSIS

64. Problem: Round \$156,872 to the nearest hundred thousand.

Answer: The digit in the hundred thousands place is the 5. The rounding digit is 6, which is greater than 5. So, increase the rounding digit by 1 to 6. Replace all digits after the rounding digit with zeros: \$160,000.

Describe the mistake made in solving this problem.

Redo the problem correctly.

65. Problem: Write an inequality that compares 999; 99; 9,900; 990.

Answer: $99 > 990 > 999 > 9{,}900$

Describe the mistake made in solving this problem.

Redo the problem correctly.

Section 1.2 Numbers, Tables, and Bar Graphs

LEARNING TOOLS

CD-ROM SSM VIDEO

Sneak Preview

In order to make decisions or solve problems, we need information. Information is often organized in tables. Information can also be presented visually as a graph.

After completing this section, you should be able to:

1) Organize information in a table.

2) Present information as a bar graph.

3) Obtain information from a bar graph.

DISCUSSION

Tables and Bar Graphs

Bar graphs are pictures that show the relationship between two sets of information. They are commonly used in business, technical work, and the mass media. Bar graphs begin with a collection of data.

Example 1 **Science faculty members at a college are asking for additional positions based on student enrollment. As part of their presentation, they need to show the number of students with declared majors in different fields of study. They know that there are 326 majors in Arts, 745 majors in Science, 835 majors in Engineering, 438 majors in the Humanities, 745 majors in Social Sciences, and 486 majors in Education.**

This **data** can be organized into a table by matching up the number of students with their declared major. This is an example of **paired data.** The table is referred to as a **frequency table** since it shows the number of occurrences in each category.

DECLARED MAJORS BY FIELD OF STUDY	NUMBER OF STUDENTS
Arts	326
Science	745
Engineering	835
Humanities	438
Social Sciences	745
Education	486

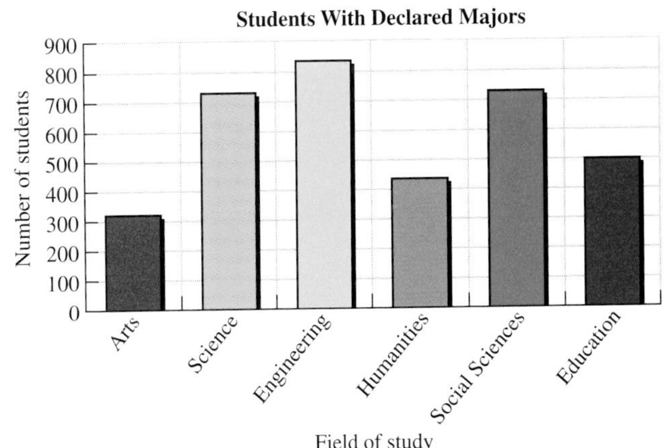

A bar graph can be a more powerful visual presentation of the data than the table. A bar graph is constructed by drawing a **vertical** line | and a **horizontal** line — that meet at one corner. Each of these lines is called an **axis** and is labeled with the name of the information, along with any measurement unit used. Notice that the rectangles on the graph do not touch, and all have the same width.

This bar graph visually shows the number of students who have chosen majors in different fields of study. For any set of data, the **mode** is the value or category that occurs most often. In this example, the field of study chosen by the most students is engineering; it is the mode of this data.

On a bar graph that shows the number of occurrences for each category or value, the mode is the category or value that has the largest rectangle. If a set of data has two modes, it is said to be **bimodal.**

PROCEDURE

To find the mode of a collection of data:

Determine the value or category that occurs most often.

Bar graphs can also be drawn with horizontal rectangles. To convert the previous bar graph into a horizontal graph, the information on each axis is reversed: *Field of study* is placed on the vertical axis and *number of students* is placed on the horizontal axis.

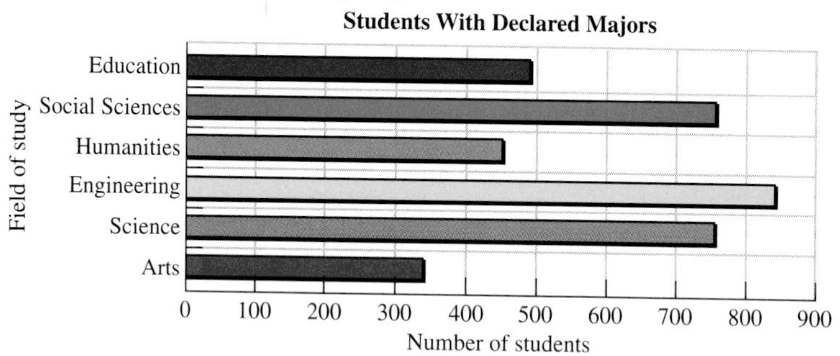

Example 2 **The noon temperatures on a spring day were recorded in 20 different cities across Texas: 68°F, 72°F, 82°F, 81°F, 74°F, 81°F, 77°F, 81°F, 82°F, 79°F, 68°F, 77°F, 68°F, 84°F, 81°F, 85°F, 78°F, 81°F, 80°F, 77°F. Organize this information in a table, construct a bar graph, and identify the mode.**

Temperature in °F	68	72	74	77	78	79	80	81	82	84	85
Number of recordings	3	1	1	3	1	1	1	5	2	1	1

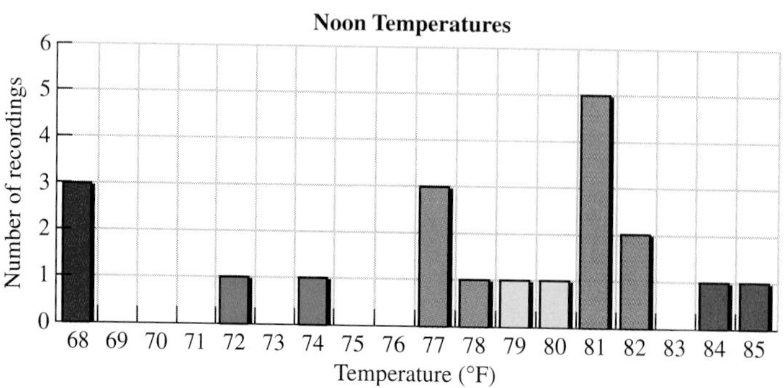

Since the temperature with the most recordings is 81°F, the mode is 81°F.

Notice that temperatures with no recordings are shown on the graph. Every value between the lowest and the highest temperature is included on the temperature axis.

CHECK YOUR UNDERSTANDING

The bar graph shows the number of doctors who own land that is used to grow tobacco.

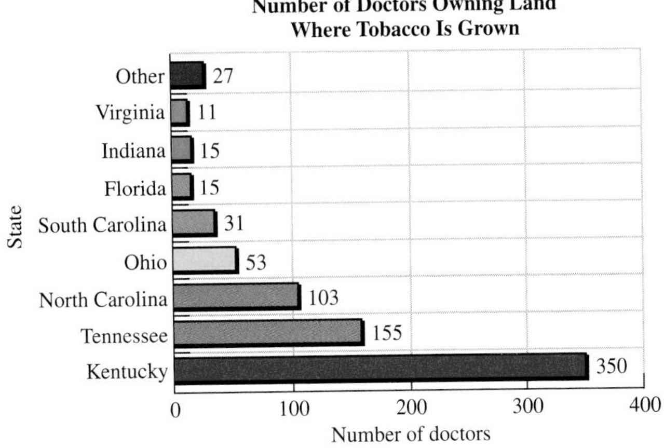

(Source: Department of Agriculture; State Medical Rosters)

1. Which state has the largest number of doctors who own land used for growing tobacco?

2. How many doctors own land used for growing tobacco in South Carolina?

3. How many doctors own tobacco-producing land in Kentucky?

4. Which states have the same number of doctors owning tobacco-producing land?

Answers: 1. Kentucky; 2. 31; 3. 350; 4. Florida and Indiana.

EXERCISES SECTION 1.2

1. A kennel worker weighed each dog that came in to board over Labor Day weekend.

a. Round the weight of each dog to the nearest 10 pounds.

NAME	WEIGHT (POUNDS)	ROUNDED WEIGHT
Aerni	45	50
Barnes	71	70
Beastie	86	90
Champ	97	100
Escamilla	71	70
Flo	67	70
Harley	45	50
Pirate	61	60
Sugar	71	70
Chuckie	92	90
Duke	48	50
Paddy	12	10
Sarge	52	50
Mitzi	49	50
Sam	38	40
Ruby	47	50
Rudy	70	70
Luka	74	70

b. Complete the frequency table.

ROUNDED WEIGHT	NUMBER OF OCCURRENCES
10	1
20	0
30	0
40	1
50	6
60	1
70	6
80	0
90	2
100	1

c. Present this information as a bar graph with vertical rectangles.

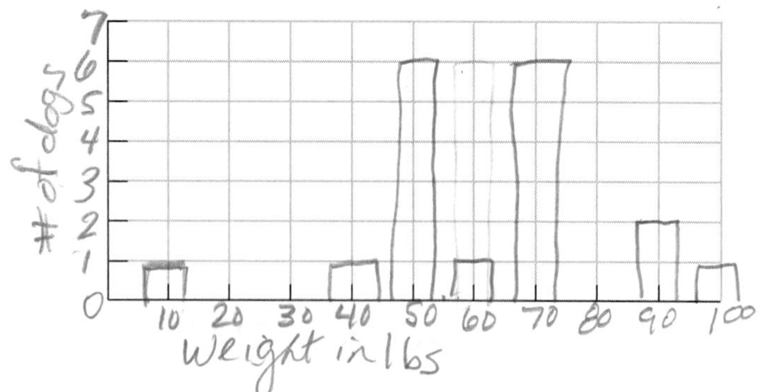

d. What is the mode of the rounded weights? 50 lbs & 70 lbs

2. Adrian makes sales calls as assigned by his regional manager, records his weekly mileage, and e-mails the mileage to his manager once a week: August 2, 187 miles; August 9, 278 miles; August 16, 918 miles; August 23, 956 miles; August 30, 267 miles; September 6, 278 miles; September 13, 988 miles; September 20, 278 miles; September 27, 187 miles; October 4, 187 miles.

a. Organize this information in a table.

b. Present this information as a bar graph with vertical rectangles. One axis should show the date; the other axis should show the number of miles.

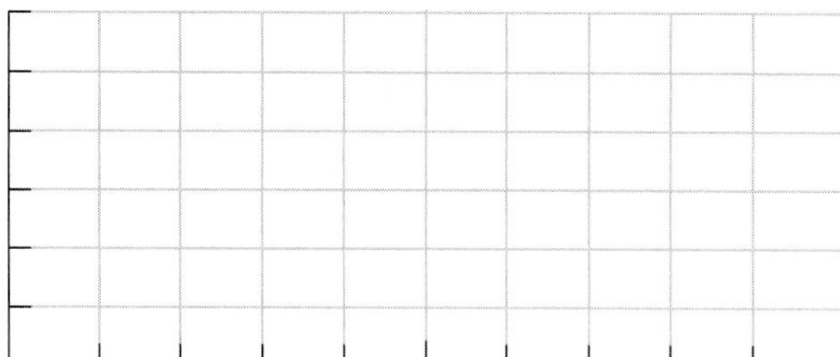

c. Construct a frequency table for this information.

d. Find the mode of Adrian's weekly mileage.

e. Present the information in the frequency table as a bar graph with vertical rectangles. One axis should show the mileage; the other axis should show the number of occurrences.

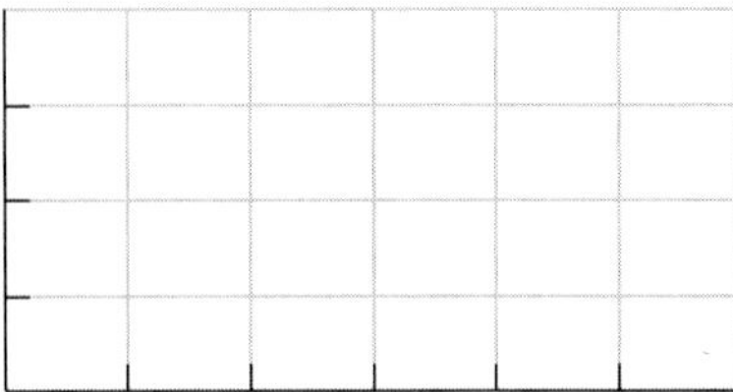

3. A cashier at a college bookstore entered the value of returned books into a spreadsheet and used the spreadsheet to round the value of each book to the nearest ten dollars.

a. Round the value of each book in the table to the nearest ten dollars.

VALUE OF BOOK ($)	92	45	76	17	27	45	31	98	89	65	17	27	45	98	62	17	70	46
ROUNDED VALUE ($)																		

b. Present this rounded information as a frequency table.

ROUNDED VALUE OF BOOK ($)	10	20	30	40	50	60	70	80	90	100
NUMBER OF OCCURRENCES										

c. Present the rounded information as a bar graph.

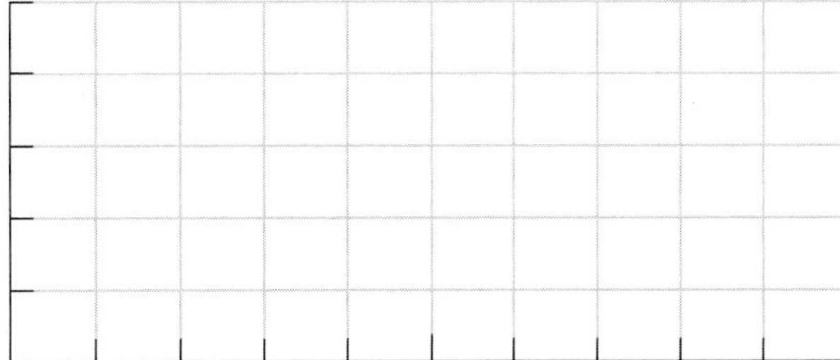

d. Find the mode of the exact value of returned books.

e. Find the mode of the rounded value of returned books.

4. The bar graph shows the population of the United States from 1910 to 2000.

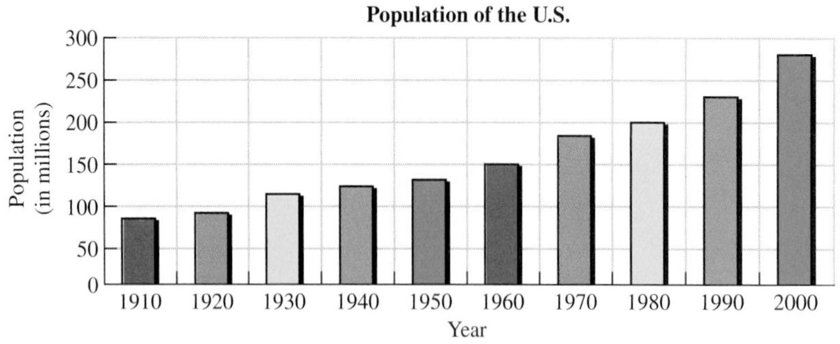

(Source: U.S. Census)

a. In what time period did the population go over 100 million for the first time?

b. In what time period did the population exceed 200 million for the first time?

c. What was the population in 1960?

5. The bar graph shows the number of human rights violations recorded in 55 nations by Amnesty International in 1997.

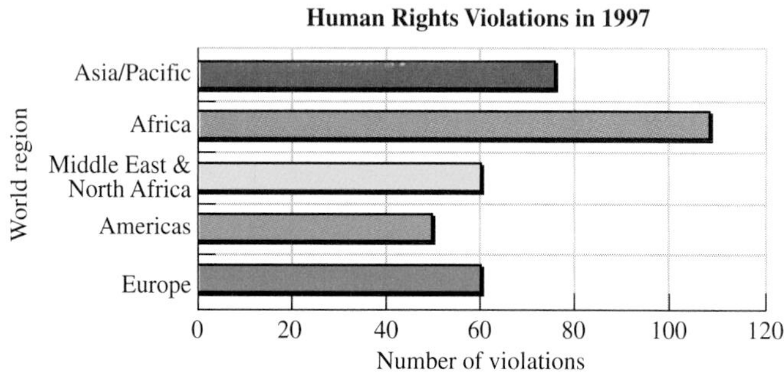

(Source: Amnesty International Annual Report 1998)

a. Which region accounted for the most violations?

b. To the nearest ten, how many violations were recorded in the Americas?

c. To the nearest ten, how many violations were recorded in Asia or the Pacific?

6. 3M Company has a diverse product base. In 1997, sales of consumer and office products produced \$2,600,000,000 of revenues. Sales of health care products produced \$2,480,000,000 of revenues. Sales of connecting and insulating products produced \$1,740,000,000 of revenues. Sales of tape products produced \$2,080,000,000 of revenues. *(Source: 3M Company)*

a. Round each number to the nearest hundred million dollars. Organize the original data and the rounded data in a table.

TYPE OF PRODUCT	REVENUES	ROUNDED REVENUES

b. Present the rounded data as a bar graph.

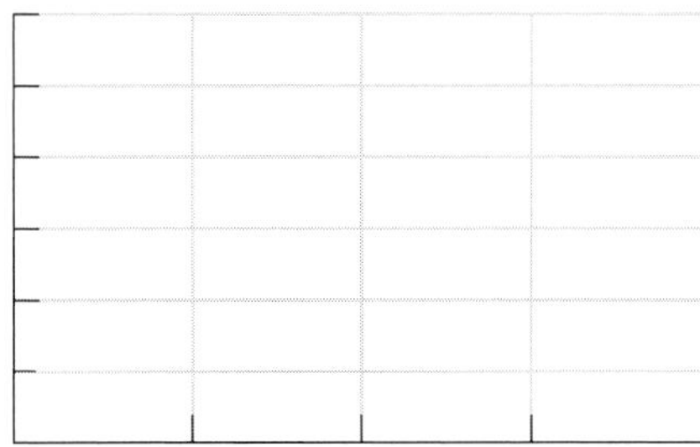

7. According to the American Medical Association, 79% of all U.S. physicians are White, 3% are African-American, 5% are Hispanic, 10% are Asian, and 3% are "Other."

a. Present this information in a table.

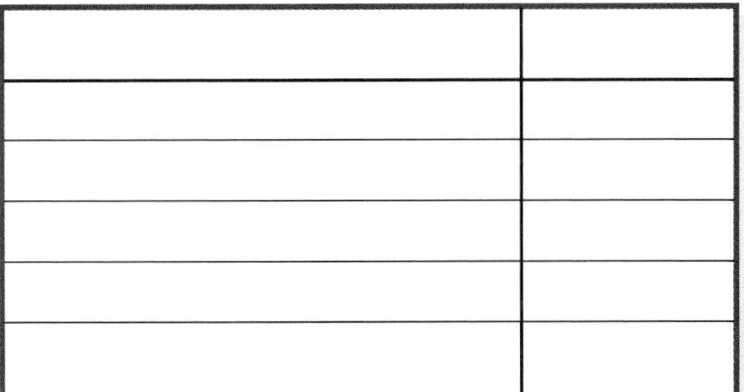

b. Present this information as a bar graph.

8. John Elway of the Denver Broncos retired from football in 1999. He went out on top as the all-time NFL winningest quarterback. The table shows some of his career statistics for regular seasons.

YEAR	TOUCHDOWNS	PASSING YARDS	RUSHING YARDS
'83	7	1,663	146
'84	18	2,598	237
'85	22	3,891	253
'86	19	3,485	257
'87	19	3,198	304
'88	17	3,309	234
'89	18	3,051	244
'90	15	3,526	258
'91	13	3,253	255
'92	10	2,242	94
'93	25	4,030	153
'94	16	3,490	235
'95	26	3,970	176
'96	26	3,328	249
'97	27	3,635	218
'98	22	2,806	94

a. Present the number of touchdowns by year in a bar graph.

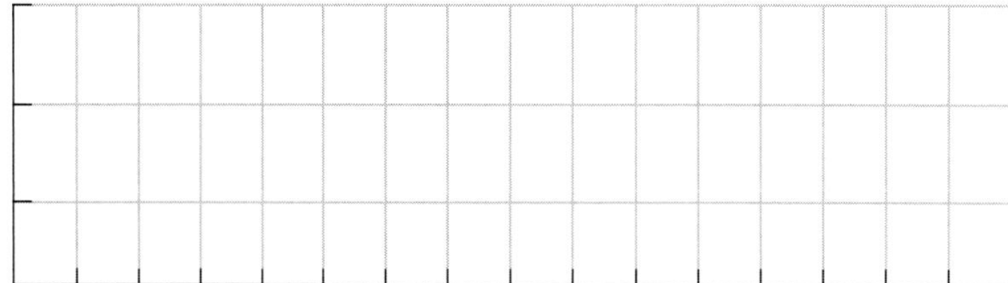

b. Present the passing yards by year in a bar graph.

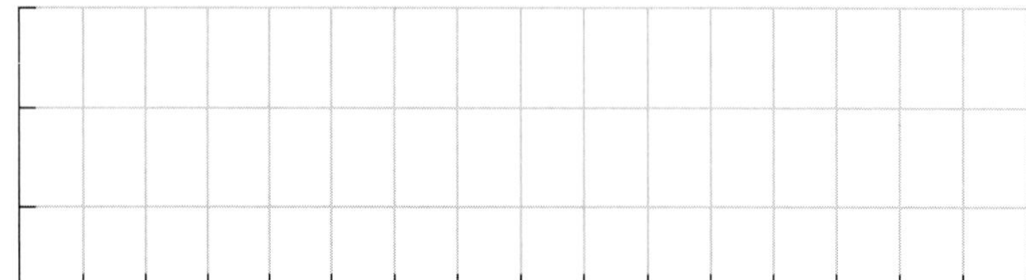

c. Present the rushing yards by year in a bar graph.

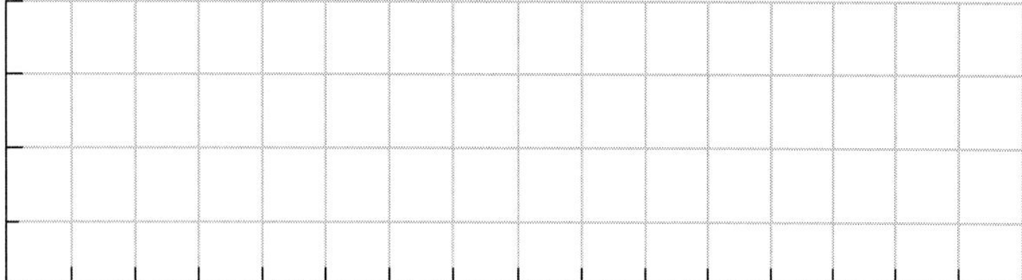

d. Find the mode of touchdowns.

e. What year(s) did he have the greatest number of passing yards?

f. What year(s) did he have the lowest number of rushing yards?

9. Summit Adventure Travel offers guided ascents of mountains to adventure travelers. In Bolivia, an ascent of Huayna Potoshi (19,870 feet) costs $2,950 while an ascent of Illimani (21,201 feet) costs $2,850. A trip that includes climbing two Ecuador volcanoes, Cotopaxi (19,348 feet) and Chimborazo (18,701 feet), costs $2,850. In Mexico, an ascent of Pico de Orizaba (18,701 feet) costs $1,950. In France, climbing Mt. Blanc (15,767 feet) costs $2,495.
a. Construct a bar graph that compares the different heights of these mountains.

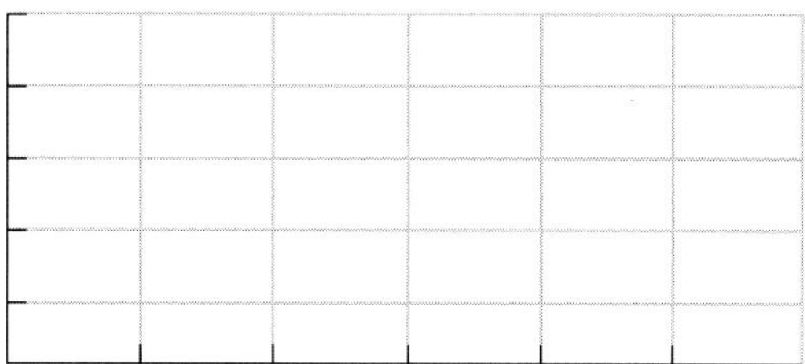

b. Construct a bar graph that compares the different costs of guided ascents of these mountains.

10. In 2000, the cost of attending Lewis-Clark State College (tuition, fees, housing, and books) was about $10,252. The average loan debt on completing a bachelor's degree was $4,000. At Washington State University, the average cost of attending was $12,820. The average loan debt on completing a bachelor's degree was $18,567. At the University of Idaho, the cost of attending was $10,150. The average loan debt on completing a bachelor's degree was $18,600.

Lewis-Clark State College, Idaho

a. Organize this data in a table.

b. Draw a bar graph that compares the cost of attending each school.

c. Draw a bar graph that compares the average loan debt at completion of a bachelor's degree at each school.

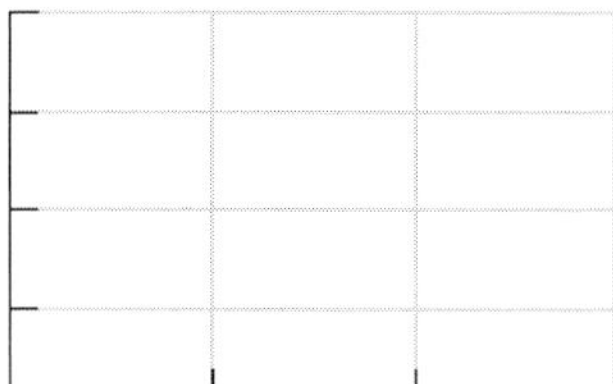

11. Donor kidneys for kidney transplants are most often obtained from people who have died. However, it is no longer unusual for a living spouse, relative, or even a stranger to donate a kidney to a person in need. The United Network for Organ Sharing reported the following information on the numbers of living kidney donors by year: 1994—3,008 donors; 1995—3,360 donors; 1996—3,606 donors; 1997—3,856 donors; 1998—4,154 donors.

a. Present this information in a table.

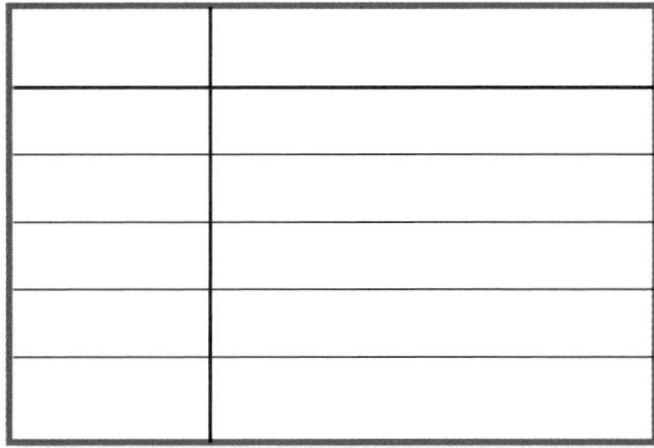

b. Present this information using a bar graph.

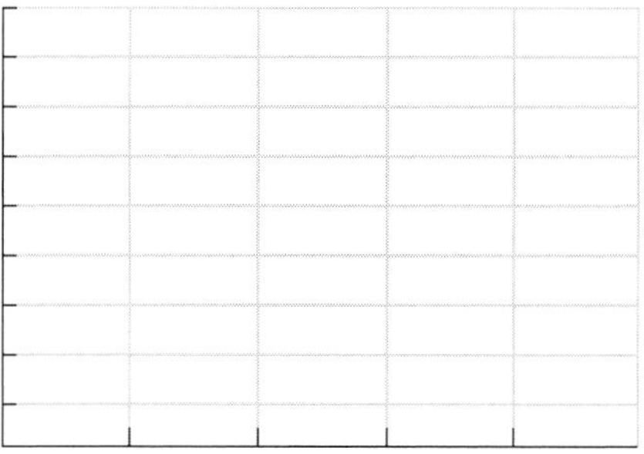

12. The top ten fastest growing companies in Europe, their country, their sector of business, and their revenues in 1999 were Realtech AG (Germany, IT and Internet services, \$17,400,000), B.U.W. Unternehmensgruppe (Germany, IT and Internet services, \$4,600,000), Lernout and Hauspie Speech Products Ltd. (Belgium, voice recognition, \$203,200,000), Key Tech Products Ltd. (Ireland, electronics, \$14,100,000), Brokat Infosystems AG (Germany, IT and Internet services, \$31,600,000), Progressive Computer Recruitment Ltd. (Britain, IT and Internet services, \$91,100,000), Genion Fahrzeugtechnik GmbH (Germany, manufacturing, \$6,200,000), Iona Technologies PLC (Ireland, IT and Internet services, \$110,200,000), Adam Associates (Britain, IT and Internet services, \$9,700,000), and Dyson Appliances (Britain, manufacturing, \$271,100,000).

a. Construct a bar graph that shows the number of these companies located in each country.

b. Construct a bar graph that shows the number of these companies in each sector of business.

c. Construct a bar graph that shows the 1999 revenues of each company.

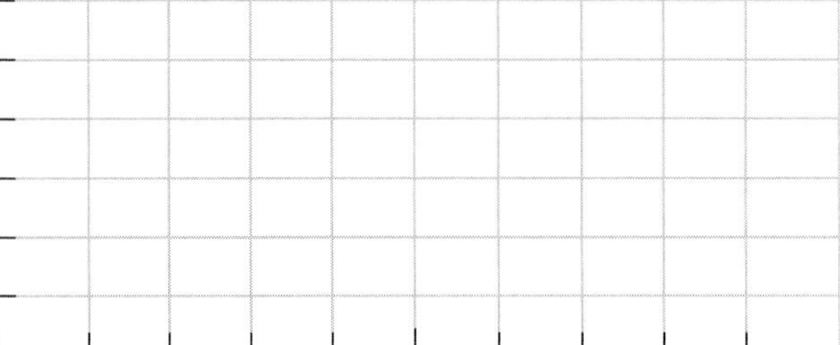

13. On Super Tuesday, there are 11 Republican primaries with 554 delegates at stake. California (162), Connecticut (25), Georgia (54), Maine (14), Maryland (31), Massachusetts (37), Missouri (35), New York (101), Ohio (69), Rhode Island (14), and Vermont (12) all choose their delegates to the Republican convention on this day. Construct a bar graph that shows the number of delegates at stake in each state.

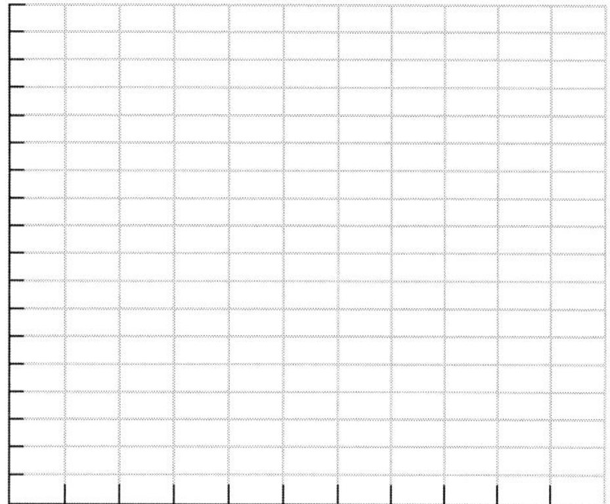

14. Complaints of sexual harassment in the workplace reported by the U.S. Equal Opportunity Commission rose between 1993 and 1997. In 1993, 11,908 complaints were received. In 1994, 14,420 complaints were received. In 1995, 15,549 complaints were received. In 1996, 15,342 complaints were received. In 1997, 15,880 complaints were received. Construct a bar graph that shows the number of complaints received each year.

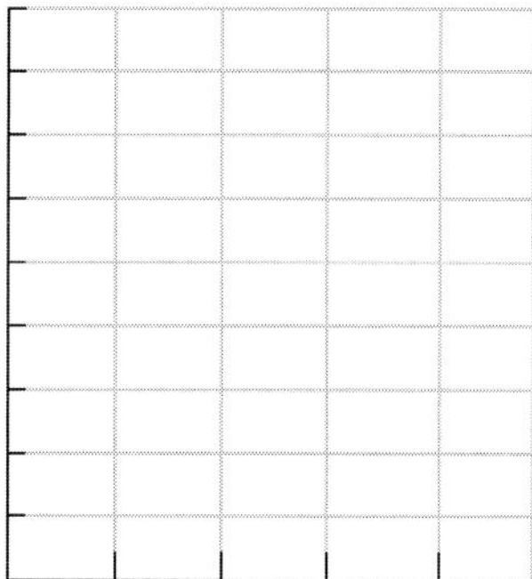

15. A table in a magazine advertisement showed these guaranteed annual premiums for a guaranteed level term period for term life insurance for women.

AGE	10-YEAR	15-YEAR	20-YEAR	25-YEAR	30-YEAR
35	$103	$138	$163	$200	$225
40	$133	$175	$203	$253	$290
45	$190	$238	$283	$348	$405
50	$255	$308	$388	$495	$643
55	$360	$418	$585	$1,280	$2,618
60	$503	$608	$880	$5,923	$5,923
65	$818	$983	$1,320	$7,858	$7,858
70	$1,363	$1,833	$3,820	$10,003	$10,003
75	$2,613	$5,158	$10,440	$14,613	$14,613

a. Construct a bar graph that shows the change in price for a 10-year policy as the age of a woman increases.

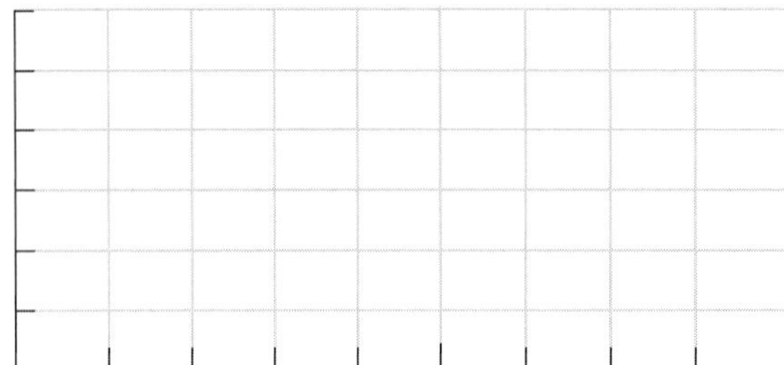

b. Construct a bar graph that shows the difference in price by the length of the policy for a 35-year-old woman.

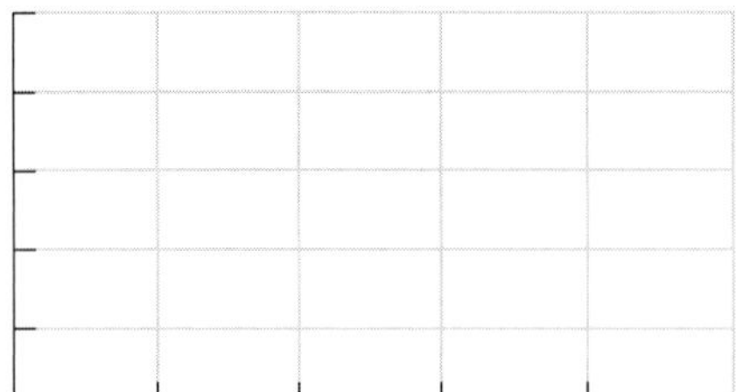

ERROR ANALYSIS

16. Problem: Average daily river flow is important information for flood control. A manager is holding a press conference and wishes to show the average daily river flow of the Salmon River at White Bird, Idaho, in a bar graph. The data is in the table below. The manager rounded the river flows before creating the bar graph.

DATE	4/23	4/24	4/25	4/26	4/27
RIVER FLOW IN CUBIC FEET OF WATER PER SECOND	18,250	17,339	18,061	21,037	24,294

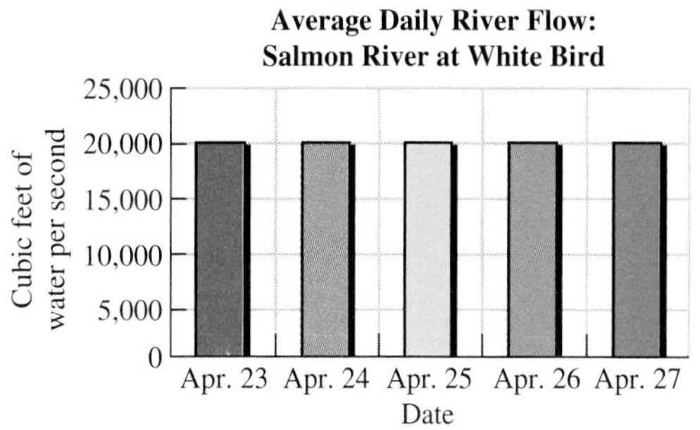

a. If the manager did not round the river flows before graphing, how would this change the bar graph?

b. The manager decided not to round the river flows. An assistant suggested that the manager change the scale on the graph to divisions of 1,000 instead of 5,000. How would this change the appearance of the graph?

17. Problem: Bar graphs are frequently used in the media and for advertising. The following bar graph was used to promote Security Bank's certificate of deposit.

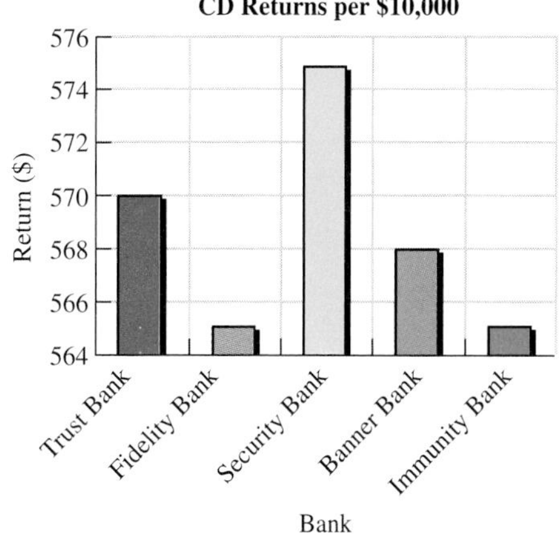

On a $10,000-certificate of deposit, the difference in yields among the five banks is about $10. However, an advertisement for Security Bank's certificate of deposit makes it look like a much better investment. How could Trust Bank change the scale of this graph so that it is clear to investors that their certificate of deposit is almost as good a value as Security Bank's certificate?

REVIEW

18. Write 5,637 in expanded form.

19. Write 40,002 in expanded form.

20. Round 7,921 to the nearest hundred.

21. Round 7,921 to the nearest ten.

22. Round 7,921 to the nearest thousand.

Section 1.3 Addition and Subtraction

LEARNING TOOLS

CD-ROM SSM VIDEO

Sneak Preview

Addition and subtraction are familiar arithmetic operations. We need to review the basic principles of addition and subtraction of whole numbers so that we can extend these principles to arithmetic with other numbers in later chapters.

After completing this section, you should be able to:

1) Add and subtract whole numbers using a number line.

2) Recognize and use the commutative and associative properties of addition.

3) Add and subtract like measurements.

4) Estimate sums and differences.

5) Add and subtract multidigit whole numbers.

6) Check exact answers using an estimate.

7) Calculate supplementary and complementary angles.

DISCUSSION

Addition, Subtraction, and Number Lines

When we add, we combine the value of two or more numbers to get a **sum.** The symbol for addition is $+$. On a number line, the sum of two natural numbers will be found to the right of the largest number. The sum is greater than the largest number being added. When zero is added to another whole number, the sum is equal to the original whole number.

For example, to add 3 and 4, begin by graphing 3 on a number line. From 3, move to the right 4 units. (This is usually shown using a line with an arrowhead that points to the right.) The line ends at 7, the sum of 3 and 4.

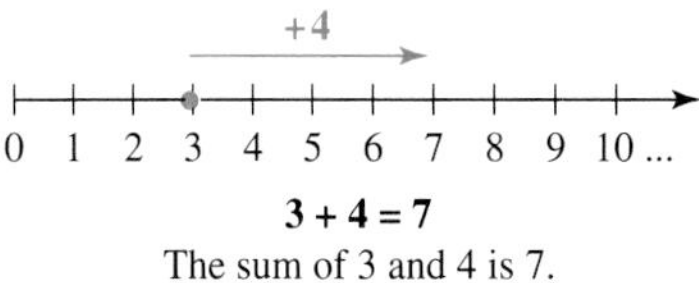

3 + 4 = 7

The sum of 3 and 4 is 7.

Subtraction is the opposite of addition. Subtraction results in a **difference.** The symbol for subtraction is $-$. The difference of two natural numbers will be found to the left of the larger number on the number line. Subtracting zero from a whole number will result in the original whole number.

For example, to find the difference of 8 and 3, begin by graphing the larger number, 8, on the number line. From 8, move to the left 3 units, drawing an arrowed line. The line ends at 5, the difference of 8 and 3.

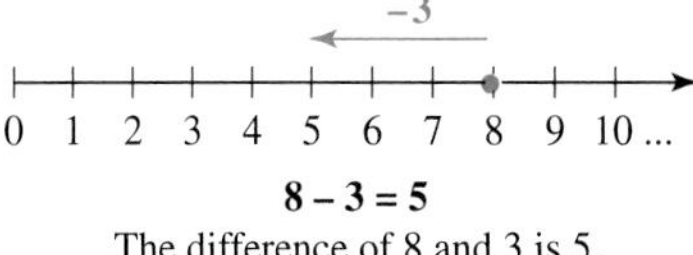

8 − 3 = 5

The difference of 8 and 3 is 5.

If the difference is to be a whole number, we cannot subtract a larger number from a smaller number. For example, if we try to subtract $4 - 7$, we would end up to the left of zero on the number line where there are no whole numbers.

CHECK YOUR UNDERSTANDING

Add or subtract using a number line.

1. $6 - 1$

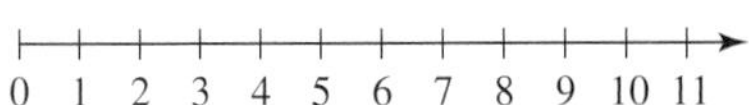

2. $4 + 4$

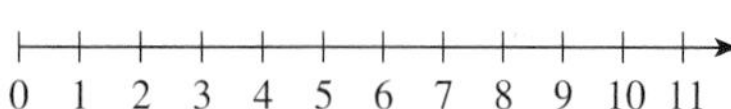

3. $10 - 9$

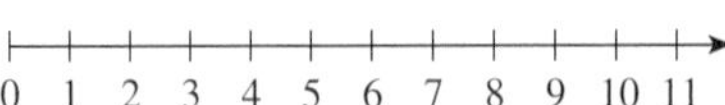

4. $7 - 3$

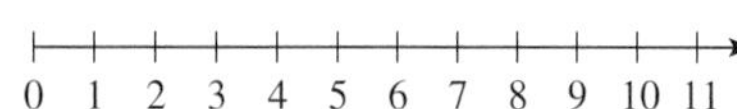

Answers: 1. 5; 2. 8; 3. 1; 4. 4.

DISCUSSION

The Commutative and Associative Properties of Addition

Axioms are the fundamental principles of our number system, a sort of constitution upon which all the other rules of the number system must be based. Two important axioms of our number system are the **commutative property of addition** and the **associative property of addition.**

The commutative property of addition says that the *order* in which numbers are added can be changed and the sum will remain the same. For example, $2 + 3 = 5$ and $3 + 2 = 5$. The order of addition of 2 and 3 does not change the sum, 5. The commutative property works for any numbers. We can rewrite the commutative property using letters of the alphabet to represent any numbers. When letters are used in such a way, they are called **variables.** Variables are just stand-ins for numbers.

PROPERTY

The commutative property of addition

In words: "The order in which numbers are added does not affect the sum."

In symbols: $a + b = b + a$

Parentheses ( ) are examples of grouping symbols and can be used to determine which numbers are added first. Arithmetic inside the parentheses must be completed first.

Example 1 **Evaluate (2 + 6) + 3.**

$(2 + 6) + 3$

$= 8 + 3$ *Add inside the parentheses first.*

$= 11$

The word **evaluate** means "do the arithmetic." Writing down each addition instead of doing the addition mentally is an example of "showing intermediate steps."

Example 2 **Evaluate 2 + (6 + 3).**

$2 + (6 + 3)$
$= 2 + 9$ *Add inside the parentheses first.*
$= 11$

Using parentheses to show the sequence of addition is called **grouping.** Parentheses are an example of **grouping symbols.** As this example shows, changing the grouping of numbers being added together does not change their sum. This is called the **associative property of addition,** another axiom of our number system.

PROPERTY

The associative property of addition

In words: "When adding more than 2 numbers, the grouping does not affect the sum."

In symbols: $(a + b) + c = a + (b + c)$

We can use grouping to add and subtract large numbers. To do this we rewrite the numbers using place value word names.

Example 3 **Evaluate 3,000,000 + 5,000,000.**

3,000,000 + 5,000,000
= 3 million + 5 million *Rewrite numbers using the word name "million."*
= (3 + 5) million *Group the digits together.*
= 8 million *Add the digits.*
= 8,000,000 *Rewrite in place value form.*

Example 4 **Evaluate 500 − 200.**

500 − 200
= 5 hundred − 2 hundred *Rewrite numbers using the word name "hundred."*
= (5 − 2) hundred *Group the digits together.*
= 3 hundred *Add the digits.*
= 300 *Rewrite in place value form.*

Example 5 **Subtract 200 from 2,400.**

To rewrite the numbers, choose the largest possible place value. We cannot choose thousands because 200 is less than a thousand. The largest possible place value that can be used to rewrite both numbers is hundreds.

2,400 − 200
= 24 hundred − 2 hundred
= (24 − 2) hundred
= 22 hundred
= 2,200

This alternative way of adding by grouping is not necessarily faster than other ways to add. However, it will help us add or subtract measurements and later to add and subtract variables.

CHECK YOUR UNDERSTANDING

Rewrite each sum using digits and place value words. Add. Change the answer back to place value form.

1. 300 + 100 **2.** 6,000 + 2,000 **3.** 70,000 + 20,000

4. 1,500 + 200 **5.** 500,000 + 300,000 **6.** 2,000 + 500

Answers: 1. 3 hundred + 1 hundred = 4 hundred = 400; 2. 6 thousand + 2 thousand = 8 thousand = 8,000; 3. 7 ten thousands + 2 ten thousands = 9 ten thousands = 90,000; 4. 15 hundred + 2 hundred = 17 hundred = 1,700; 5. 5 hundred thousand + 3 hundred thousand = 8 hundred thousand = 800,000; 6. 20 hundred + 5 hundred = 25 hundred = 2,500.

DISCUSSION

Addition and Subtraction of Measurements

We can use grouping to add or subtract measurements with the same units. We group the numbers together, add or subtract, and keep the original unit of measurement. This is called **combining like units.**

Example 6 **Two lines are drawn below. One line is 2 inches long. The other line is 4 inches long. What is the total length of these lines?**

2 inches 4 inches

Total length = 2 inches + 4 inches
= (2 + 4) inches *Group the numbers.*
= 6 inches *Keep the original unit of measurement.*

Example 7 **Kate is a 10-year-old girl who has been lethargic and lacking appetite and energy. After blood tests revealed that Kate has an iron deficiency, her family doctor recommended that Kate take one ferrous sulfate tablet daily. Each tablet contains 65 milligrams of iron. Kate also takes one regular vitamin supplement every day. This vitamin contains 9 milligrams of iron. What is the total amount of iron Kate is taking per day?**

Total iron = 65 milligrams + 9 milligrams
= (65 + 9) milligrams *Group the numbers.*
= 74 milligrams of iron *Keep the original unit of measurement.*

We can only directly add or subtract measurements that have the same units. For example, we cannot add 96 fluid ounces of Pepsi directly to 2 liters of Pepsi. To add these measurements, one of them has to first be converted into the other unit.

The distance around a shape or figure is called its **perimeter.** A **rectangle** is a geometric figure with four sides. The opposite sides are parallel and have the same length. The longer two sides are called the length of the rectangle and the shorter two sides are called the width.

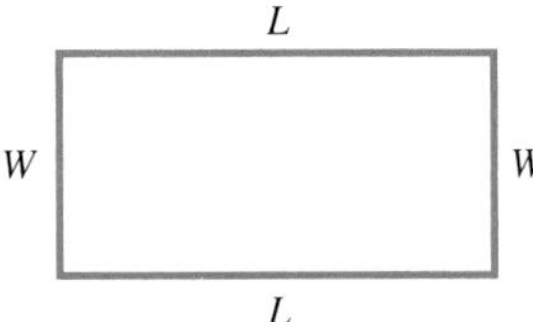

Perimeter = Length + Width + Length + Width

Example 8 **Find the perimeter of a rectangle that has a length of 8 meters and a width of 5 meters.**

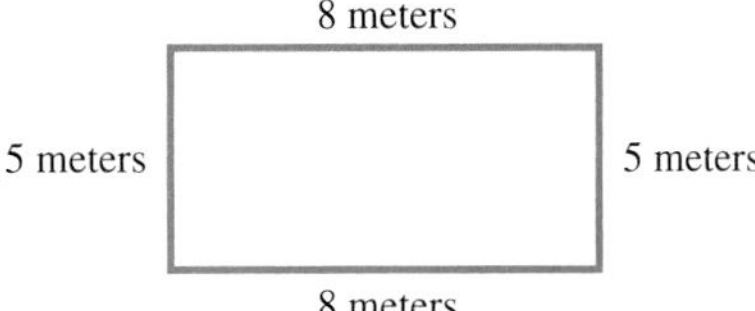

Perimeter = Length + Width + Length + Width
= 8 meters + 5 meters + 8 meters + 5 meters
= (8 + 5 + 8 + 5) meters
= 26 meters

A **triangle** is a figure with three sides. An **isosceles triangle** is a triangle in which at least two of the three sides are equal in length. The perimeter of a triangle is equal to the sum of the length of each side of the triangle.

Example 9 **Find the perimeter of the isosceles triangle shown.**

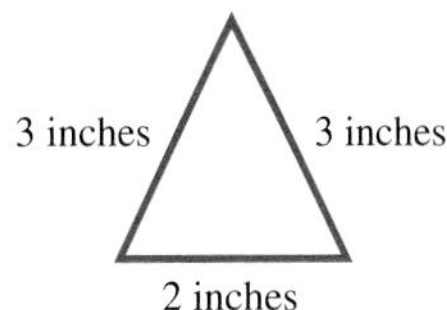

Perimeter = Sum of the length of the 3 sides
= 3 inches + 3 inches + 2 inches
= (3 + 3 + 2) inches
= 8 inches

The perimeter of any figure can be found by determining the distance around the figure. When the figure has straight sides, the perimeter is the sum of the lengths of all the sides.

Example 10 **To estimate the amount of concrete to be used in footings for a house, a contractor needs to determine the perimeter of the foundation. What is the perimeter of the foundation?**

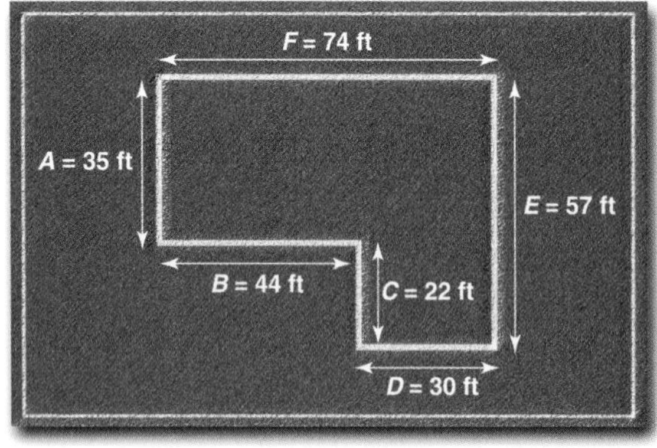

$$
\begin{aligned}
\text{Perimeter} &= \text{Sum of the lengths of the sides} \\
&= 74 \text{ ft} + 57 \text{ ft} + 30 \text{ ft} + 22 \text{ ft} + 44 \text{ ft} + 35 \text{ ft} \\
&= (74 + 57 + 30 + 22 + 44 + 35) \text{ ft} \\
&= 262 \text{ ft}
\end{aligned}
$$

The perimeter of the foundation is 262 feet.

CHECK YOUR UNDERSTANDING

Add or subtract the measurements. A number line may be helpful.

1. 6 feet + 10 feet

2. 18 quarts − 9 quarts

3. 2,000 acres + 5,000 acres

4. 1,500 horsepower − 1,200 horsepower

5. Find the perimeter of this figure:

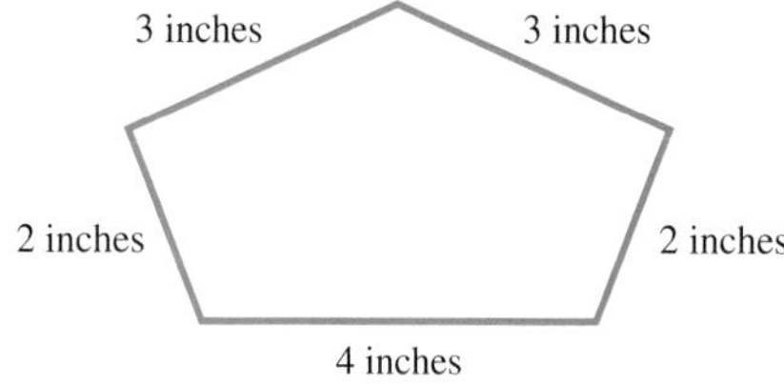

Answers: 1. 16 feet; 2. 9 quarts; 3. 7,000 acres; 4. 300 horsepower; 5. 14 inches.

DISCUSSION

Estimation

In many situations, we do not need an exact answer. An estimate is a close, but not necessarily exact answer. Estimates are also used to determine if an exact answer is *reasonable.* An exact answer is considered reasonable when the estimate is close and the answer makes sense.

PROCEDURE

To estimate a sum or a difference:

1. Determine the highest place value of the largest number.
2. Round each number to this highest place value.
3. Add or subtract the rounded numbers.

Example 11 **Estimate the difference of 674 and 89.**

The larger number is 674 with a highest place value of *hundreds*.

Round each number to the nearest hundred.

674 rounds to 700. 89 rounds to 100.

The estimated difference is 700 − 100 = 600.

Answer: 674 − 89 ≈ 600 *The ≈ sign means "approximately equal to."*

Example 12 **Estimate the sum of 34,905 and 259.**

The larger number is 34,905 with a highest place value of *ten thousands*.

Round each number to the nearest ten thousand.

34,905 rounds to 30,000. 259 rounds to 0.

The estimated sum is 30,000 + 0 = 30,000.

Answer: 34,905 + 259 ≈ 30,000

It may seem strange to "throw away" the 259 in this last example. However, 259 really is very small compared to 30,000 and isn't big enough to matter in the estimated sum. That is why we round 259 to 0 in the estimate.

Example 13 **Estimate the sum of 532,987 + 3,876,990 + 1,003,980.**

The largest number is 3,876,990 with a highest place value of *millions*.

Round each number to the nearest million.

532,987 rounds to 1,000,000. 3,876,990 rounds to 4,000,000.

1,003,980 rounds to 1,000,000.

The estimated sum is 1,000,000 + 4,000,000 + 1,000,000 = 6,000,000.

Answer: 532,987 + 3,876,990 + 1,003,980 ≈ 6,000,000.

Example 14 **The manufacturer's list price for a new truck is $27,999. A 2-year-old truck of the same model is advertised for $18,789 in the classified ads. Estimate the difference in price between the two trucks.**

The larger number is $27,999 with a highest place value of *ten thousands*.

Round $27,999 to $30,000. Round $18,789 to $20,000.

The estimated difference in prices is $30,000 − $20,000 = $10,000.

Answer: $27,999 − $18,789 ≈ $10,000.

CHECK YOUR UNDERSTANDING

Estimate the sum or difference.

1. 347 + 945

2. 2,980 + 3,876

3. 156,900 + 456,976 + 569,023

4. 578 − 56

5. 74,934 − 4,976

6. The high temperature in Bakersfield, California, was 94°F. The low temperature was 43°F. Estimate the difference between the high and low temperatures.

Answers: 1. 300 + 900 = 1,200; 2. 3,000 + 4,000 = 7,000; 3. 200,000 + 500,000 + 600,000 = 1,300,000; 4. 600 − 100 = 500; 5. 70,000 − 0 = 70,000; 6. 90°F − 40°F = 50°F.

DISCUSSION

Carrying in Addition

The methods we use to add or subtract numbers greater than ten are based on place value. The value of each digit in a number depends on its position. The number 234 represents 2 hundreds + 3 tens + 4 ones. This is the **expanded form** of the number. Similarly, 451 represents 4 hundreds + 5 tens + 1 one. To add the numbers, we position the numbers vertically, lining up the same place values. We then add the digits in each place value:

$$\begin{array}{r} 2 \text{ hundreds} + 3 \text{ tens} + 4 \text{ ones} \\ + \; 4 \text{ hundreds} + 5 \text{ tens} + 1 \text{ one} \\ \hline 6 \text{ hundreds} + 8 \text{ tens} + 5 \text{ ones} \end{array}$$

We usually do not add numbers in expanded form. Instead, we position the numbers vertically and add each column of numbers. We may not think at all about the fact that we are adding hundreds, tens, and ones.

$$\begin{array}{r} 234 \\ +\,451 \\ \hline 685 \end{array}$$

When the sum of the digits in a column is greater than 9, we need to **carry.** Each position in a number can contain only one digit and numbers greater than 9 have at least two digits. When we carry, we transfer value from one position to another. This transfer can be seen by adding the numbers in expanded form.

Example 15 **Add in expanded form: 28 + 63.**

In expanded form, we write

$$\begin{array}{r} 2 \text{ tens} + 8 \text{ ones} \\ + \; 6 \text{ tens} + 3 \text{ ones} \\ \hline 8 \text{ tens} + 11 \text{ ones} \end{array}$$

This is *not* 811.

Since only one digit is allowed in each place value position, we rewrite the 11 ones as 1 ten + 1 one.

$$\begin{aligned} & 8 \text{ tens} + 11 \text{ ones} \\ &= 8 \text{ tens} + 1 \text{ ten} + 1 \text{ one} \\ &= 9 \text{ tens} + 1 \text{ one} \\ &= 91 \end{aligned}$$

When we carry, we are transferring value from one place value to another. When we add vertically, we can show this transfer by writing the extra 1 ten above the tens column as a small number.

$$\begin{array}{r} {\scriptstyle 1} \\ 28 \\ +\,63 \\ \hline 91 \end{array}$$

People educated somewhere other than the United States may write the "carry" number somewhere else, perhaps at the bottom of the column. Adding is not done exactly the same way everywhere in the world.

PROCEDURE

To add whole numbers:

1. Write the numbers in columns so that the place values line up.
2. Add the digits in each place value starting in the ones column.
3. Write the sum of the digits in each place value directly below the place value. If the sum of the digits is greater than 9, write the ones digit directly below the place value and carry the tens digit into the next place value.

Example 16 **Add 276 + 354.**

$$\begin{array}{r} {}^{1\,1} \\ 276 \\ +\ 354 \\ \hline 630 \end{array}$$

Add the numbers in the ones place: 6 + 4 = 10.
Write the 0 under the ones column, carry the 1 into the tens.
Add the numbers in the tens place: 7 + 5 + 1 = 13.
Write the 3 under the tens column and carry the 1 into the hundreds.
Add the digits in the hundreds place: 2 + 3 + 1 = 6.
Write the 6 under the hundreds.

So, 276 + 354 = 630.

Example 17 **Find the sum of 287 pounds, 965 pounds, and 267 pounds. Check the answer by finding an estimated sum.**

The exact sum is:

$$\begin{array}{r} {}^{2\,1} \\ 287 \\ 965 \\ +\ \ 267 \\ \hline 1519 \end{array}$$

In the ones column: 7 + 5 + 7 = 19.
Write the 9 in the ones column and carry the 1 into the tens column.
In the tens column: 8 + 6 + 6 + 1 = 21.
Write the 1 in the tens column and carry the 2 into the hundreds column.
In the hundreds column: 2 + 9 + 2 + 2 = 15.
Write the 5 in the hundreds column and the 1 in the thousands column.

An estimate of the sum is:

300 pounds + 1,000 pounds + 300 pounds = 1,600 pounds *Round each number to the nearest hundred.*

Since the estimated sum (1,600 pounds) is quite close to the exact sum (1,519 pounds), we can be reasonably confident that the exact sum is correct.

So, the sum of 287 pounds, 965 pounds, and 267 pounds is 1,519 pounds.

Estimates allow us to catch errors we make in doing the exact addition. Estimating is one strategy for checking an answer for reasonability.

CHECK YOUR UNDERSTANDING

Find the exact sum. Check each sum by finding an estimate.

1. 36 + 84

2. 376 kilograms + 634 kilograms

3. 7,934 inches + 34 inches + 6,886 inches + 658 inches

4. 8,776 + 356 + 457

Answers: 1. 120 (≈ 120); 2. 1,010 kilograms (≈ 1,000 kilograms); 3. 15,512 inches (≈ 16,000 inches); 4. 9,589 (≈ 9,000).

Borrowing in Subtraction

We can subtract numbers by first writing each number in expanded form.

Example 18 **Evaluate 87 − 62.**

In expanded form, we write

$$\begin{array}{r} 8 \text{ tens} + 7 \text{ ones} \\ \underline{-\,(6 \text{ tens} + 2 \text{ ones})} \\ 2 \text{ tens} + 5 \text{ ones} \end{array}$$

Subtract the digits in each column.

2 tens + 5 ones = 25

So, 87 − 62 = 25.

In this example, the digits in the first number are larger than the digits in the second number. If this is not the case, we must transfer value from one column to another. This process is called **borrowing.** Borrowing is the opposite process of carrying.

Example 19 **Evaluate 72 − 47.**

In expanded form we write

$$\begin{array}{r} 7 \text{ tens} + 2 \text{ ones} \\ \underline{-\,(4 \text{ tens} + 7 \text{ ones})} \end{array}$$

Since 2 ones is smaller than 7 ones, we need to transfer value from the tens place to the ones place.

$$\begin{array}{l} \quad 7 \text{ tens} + 2 \text{ ones} \\ = 6 \text{ tens} + 10 \text{ ones} + 2 \text{ ones} \\ = \mathbf{6 \text{ tens} + 12 \text{ ones}} \end{array}$$

Borrow *a ten from the 7 tens, rewriting it as 10 ones. Transfer this value to the ones column: 10 ones + 2 ones = 12 ones. Reduce the digit in the tens column by one to 6.*

After borrowing, the subtraction is

$$\begin{array}{r} \mathbf{6 \text{ tens} + 12 \text{ ones}} \\ \underline{-\,(4 \text{ tens} + 7 \text{ ones})} \\ 2 \text{ tens} + 5 \text{ ones} \end{array} = 25$$

So, 72 − 47 = 25.

As with addition, we can subtract numbers that are not in expanded form.

$$\begin{array}{r} {\scriptstyle 6\ 12} \\ \not{7}\not{2} \\ \underline{-\,47} \\ 25 \end{array}$$

Borrow 1 from the 7 to leave 6.

Take the borrowed 1 ten into the ones column as 10 ones for a total of 12 ones.

Now subtract: 12 − 7 = 5.

PROCEDURE

To subtract whole numbers:

1. Write the numbers in columns so that the place values match up.
2. Subtract the digits in each place value starting from the ones place and moving left.
3. If a digit cannot be subtracted, borrow 1 from the next column up, add 10 to the upper digit in the current column, and then subtract.

Example 20 **Find the difference of 264 days and 176 days. Check the difference by finding an estimate.**

The exact difference is:

$$\begin{array}{r} {}^{1\ 15\ 14} \\ \not{2}\not{6}\not{4} \\ -\ 176 \\ \hline 88 \end{array}$$

6 cannot be subtracted from 4. Borrow 1 from the 6 in the tens column which leaves 5. Adding 10 ones to 4 gives 14. Subtract: 14 − 6 = 8.
In the tens column, 7 cannot be subtracted from 5. Borrow 1 from the 2 in the hundreds column, leaving 1. In the tens column, we add the borrowed 10 tens to the 5 resulting in 15. Subtract: 15 − 7 = 8. Finally, in the hundreds column, 1 − 1 = 0.

We can check the exact answer of 88 days by finding an estimate.

Round each measurement: 264 days rounds to 300 days.
176 days rounds to 200 days.

Subtract the rounding estimates: 300 days − 200 days = 100 days.

This is close to the exact answer of 88 days. The exact answer is reasonable.

Checking exact answers by estimation does not guarantee that the exact answer is correct. However, an estimate that is close to the exact answer tells us that it is likely that we did not make a mistake in calculating the exact answer.

CHECK YOUR UNDERSTANDING

Calculate the exact difference. Check this answer by finding an estimate.

1. 76 − 35 **2.** 93 meters − 69 meters **3.** 761 − 495 **4.** 5,925 feet − 2,859 feet

Answers: 1. 41 (≈ 40); 2. 24 meters (≈ 20 meters); 3. 266 (≈ 300); 4. 3,066 feet (≈ 3,000 feet).

DISCUSSION

Adding and Subtracting Angle Measurements

Two angles are said to be **supplementary angles** if the sum of the measure of the two angles is 180°.

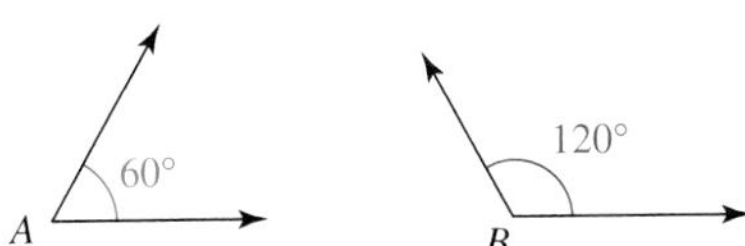

$\angle A$ and $\angle B$ are supplementary angles since 60° + 120° = 180°. This means that the combined angles form a straight angle.

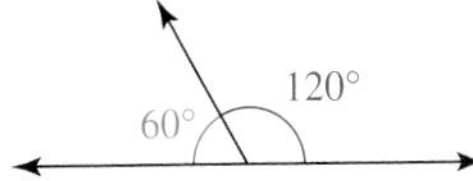

Example 21 **Angles *C* and *D* are supplementary angles. The measure of angle *C* is 88°. Calculate the measure of angle *D*.**

$$\begin{aligned} \angle D &= 180° - \angle C \\ &= 180° - 88° \\ &= 92° \end{aligned}$$

If the sum of the measure of two angles is 90°, the angles are **complementary angles.**

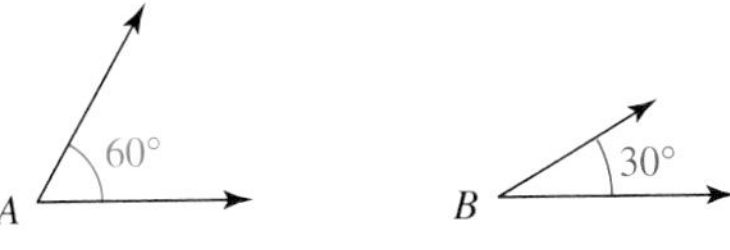

$\angle A$ and $\angle B$ are complementary angles since $60° + 30° = 90°$. This means that the combined angles form a right angle.

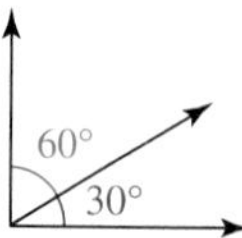

Example 22 **Angles X and Y are complementary angles. The measure of angle X is 17°. Calculate the measure of angle Y.**

$$\begin{aligned}\angle Y &= 90° - \angle X\\ &= 90° - 17°\\ &= 73°\end{aligned}$$

CHECK YOUR UNDERSTANDING

The measure of $\angle B$ is 34°. The measure of $\angle C$ is 56°.

1. Are $\angle B$ and $\angle C$ complementary angles? Explain.

2. Find the measure of an angle that is supplementary to $\angle B$.

3. Find the measure of an angle that is supplementary to $\angle C$.

Answers: 1. Yes. The sum of the measures of these two angles is 90°; 2. 146°; 3. 124°.

DISCUSSION

Adding and Subtracting English System Measurements

The SI (metric) system is the measurement system used by most of the world. However, since the United States still uses the English system, it is necessary to learn this system as well. The table that follows shows some of the units of the English system.

UNITS OF THE ENGLISH SYSTEM OF MEASUREMENT	
LENGTH	**TIME**
12 inches (in.) = 1 foot (ft)	60 seconds (sec) = 1 minute (min)
3 feet = 1 yard (yd)	60 minutes = 1 hour (hr)
5,280 feet = 1 mile (mi)	24 hours = 1 day
1,760 yards = 1 mile	7 days = 1 week
LIQUID VOLUME	**WEIGHT**
3 teaspoons (t) = 1 tablespoon (T)	16 ounces (oz) = 1 pound (lb)
16 tablespoons (T) = 1 cup	2,000 pounds = 1 ton
2 cups (c) = 1 pint (pt)	
2 pints = 1 quart (qt)	
4 quarts = 1 gallon (gal)	

Only measurements with the same units can be directly added or subtracted. To add or subtract these measurements, combine the numbers and keep the original unit of measurement. This is called **combining like units.**

Example 23 **Add 3 feet + 5 feet.**

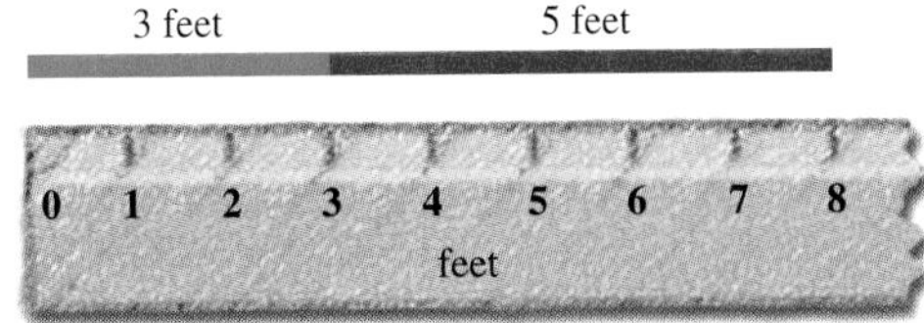

3 feet + 5 feet
= (3 + 5) feet
= 8 feet

Example 24 **Subtract 80°F − 3°F.**

80°F − 3°F
= (80 − 3)°F
= 77°F

Many measurements in the English system consist of more than one measurement unit. For example, height is usually reported in feet and inches. These measurements have **mixed units.** Measurements with mixed units are added or subtracted by combining like units.

Example 25 **Add 4 days 20 hours 5 minutes to 1 day 2 hours 43 minutes.**

4 days	20 hr	5 min
+ 1 day	2 hr	43 min
5 days	22 hr	48 min

When we add whole numbers, we carry when the sum is greater than 9. When we add mixed English measurements, we carry when the sum is equal to or greater than the unit to the immediate left.

Example 26 **Add 4 feet 7 inches to 3 feet 9 inches.**

1
4 ft 7 in.
+ 3 ft 9 in.
8 ft 4 in.

When we add the inches, the sum is 16 inches. Since there are 12 inches in one foot, 16 inches is larger than the unit to the immediate left. To carry, rewrite 16 inches as 1 foot 4 inches. Write down 4 inches in the sum and carry the 1 foot to the next column to the left.
Add: 1 foot + 4 feet + 3 feet = 8 feet.

Example 27 **Find the perimeter of the arrow:**

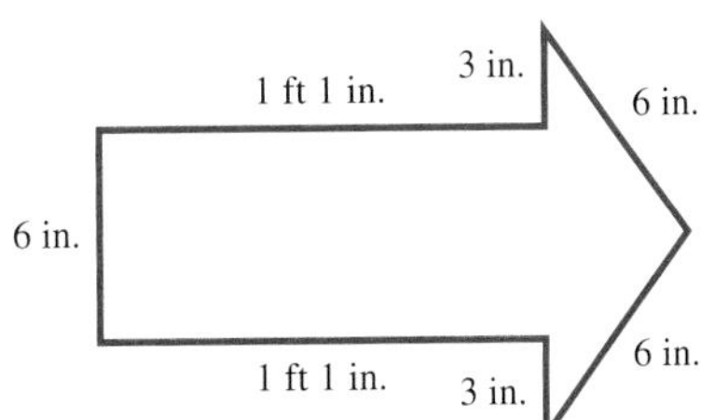

Perimeter = Sum of the lengths of the sides
= 1 ft 1 in. + 3 in. + 6 in. + 6 in. + 3 in. + 1 ft 1 in. + 6 in.
= 2 ft 26 in.
= 4 ft 2 in.

When we subtract mixed measurements, borrowing is necessary if the measurements in the second number are larger than the corresponding measurements in the first number.

Example 28 **Subtract 3 gallons 1 quart 1 pint − 1 gallon 2 quarts 1 cup.**

	3 gallons	1 quart	1 pint	0 cups
−	1 gallon	2 quarts	0 pints	1 cup

Include 0 cups in the top measurement since there are cups in the bottom measurement. Include 0 pints in the bottom measurement since there are pints in the top measurement.

Some of the numbers in the second measurement are larger than the corresponding numbers in the first measurement. We need to transfer value from one part of the measurement to another by borrowing. For example, value needs to be transferred into the cups column from the pints column. Since 1 pint = 2 cups, we transfer 2 cups into the cups column and decrease the pints column by 1.

	3 gallons	1 quart	~~1~~ 0 pint	**2 cups**
−	1 gallon	2 quarts	0 pints	1 cup

We also need to transfer 1 gallon into the quarts column. Since 1 gallon = 4 quarts, we will transfer 4 quarts. Adding 4 quarts to the 1 quart that is already there, we write 5 quarts in the quarts column and replace the 3 gallons with 2 gallons.

	~~3~~ 2 gallons	**5 quarts**	0 pints	**2 cups**
−	1 gallon	2 quarts	0 pints	1 cup

Now all the numbers in the top row are larger than the corresponding numbers in the bottom row. We are ready to subtract the measurements in each column.

	2 gallons	5 quarts	0 pints	2 cups
−	1 gallon	2 quarts	0 pints	1 cup
	1 gallon	3 quarts	0 pint	1 cup

The difference between the two liquid measurements is 1 gallon 3 quarts 1 cup.

CHECK YOUR UNDERSTANDING

Add or subtract the measurements.

1. 3 ft 4 in. + 5 ft 10 in.

2. 5 hr 40 min 30 sec + 5 hr 35 min 55 sec

3. 3 yd 5 ft 8 in. − 2 ft 5 in.

4. 5 lb 8 oz − 2 lb 12 oz

Answers: 1. 9 ft 2 in.; 2. 11 hr 16 min 25 sec; 3. 3 yd 3 ft 3 in.; 4. 2 lb 12 oz.

EXERCISES SECTION 1.3

Add or subtract using a number line.

1. $4 + 7$

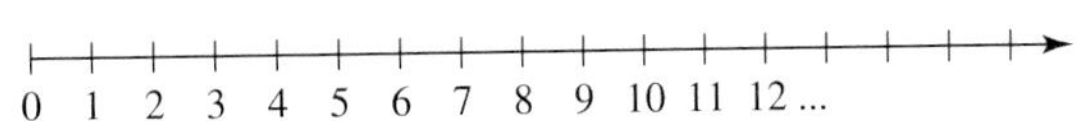

2. $11 - 7$

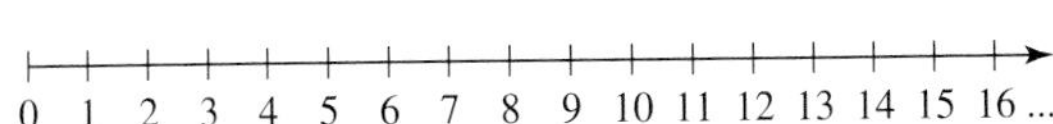

3. $15 - 6$

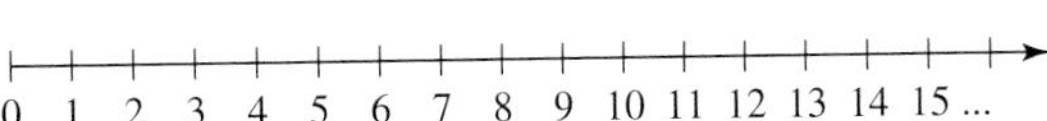

4. $4 + 8$

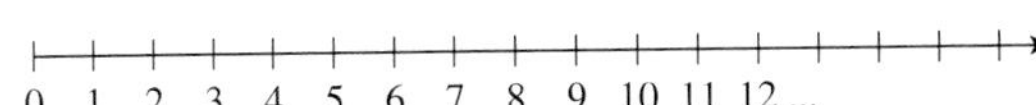

Add or subtract.

5. $2{,}300 + 4{,}000$

6. $25{,}000 - 1{,}000$

7. $(6 + 3) + 2$

8. $6 + (3 + 2)$

9. $16 + 3 + 2$

10. $2 + 3 + 16$

11. State the commutative law of addition in words. Give an example using numbers.

12. State the associative law of addition in words. Give an example using numbers.

13. An axiom is an accepted principle of our number system. The associative property of addition is an axiom: $a + (b + c) = (a + b) + c$. An axiom is not true if we can find even one **counterexample** that contradicts the axiom. There is no associative property of subtraction: $a - (b - c) \neq (a - b) - c$. We can show this by finding a counterexample, an example that shows that the associative property of subtraction is not true.

a. Subtract $(10 - 7) - 3$.

b. Subtract $10 - (7 - 3)$.

c. Is $(10 - 7) - 3 = 10 - (7 - 3)$ a counterexample to an associative property of subtraction? Explain.

14. What is the difference between 9 and 2?

15. What is the sum of 4 and 5?

16. Subtract 5 from 8.

17. Increase 3 by 5.

18. Add 4,000 to the difference of 6,000 and 3,000

19. Subtract 300 from the sum of 400 and 700.

Add or subtract the measurements. If the measurements are not like measurements, write "cannot be combined" as the answer. Include any measurement units in the sum or difference.

20. 700 meters − 300 meters

21. 12 light-years + 5 light-years

22. 23 pounds + 4 pounds

23. \$8,000,000 − \$2,000,000

24. 17 feet + 4 meters

25. 17 feet + 4 feet

26. 2 volts + 5 volts + 3 volts

27. 12 miles + 6 miles − 3 miles

Calculate the perimeter.

28. A rectangle of length 5 feet and width 3 feet.

29. An isosceles triangle with two sides of length 5 meters and the third side of length 6 meters.

30.

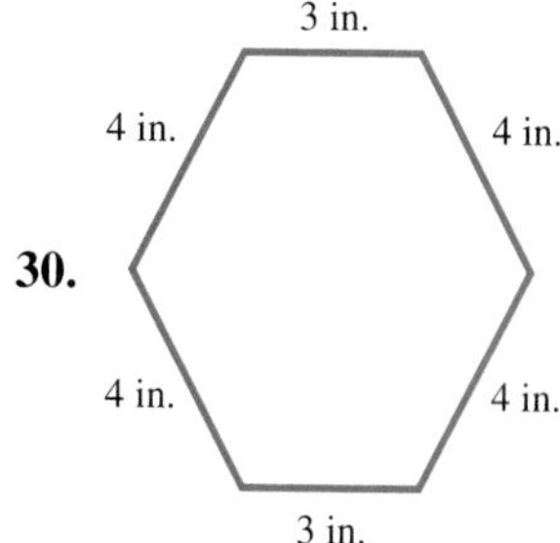

31.

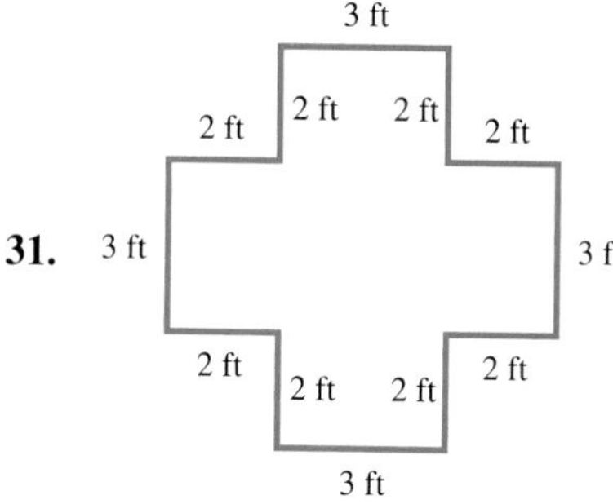

32. Use a ruler to measure this line in millimeters. Round to the nearest millimeter.

____

33. Use a ruler to measure this line in inches. Round to the nearest inch.

____

34. Round 7,988,020 to the nearest ten thousand.

35. Round 87 to the nearest ten.

36. Round 9,008 to the nearest thousand.

37. Round 9,999,234 to the nearest hundred thousand.

38. Round 19,534 to the nearest thousand.

39. Round 88,001 to the nearest ten.

40. Round 8,992 to the nearest hundred.

41. In Ypsilanti Township, Michigan, an anonymous donor gave $2,500,000 to send 2,500 fifth-graders to space camp. Round the donation to the nearest million.

42. Sales of bottled water in the United States in 1998 totaled $4,300,000. Round this amount to the nearest million.

43. Worldwide sales for the antihistamine *Claritin* were $2,300,000,000 in 1998. Round this amount to the nearest billion.

Determine the highest place value of the largest number. Round each number to this highest place value. Add or subtract the rounded numbers to find the estimated sum. Find the exact sum. Show your work including any carrying or borrowing.

44. $3,451 + $2,385
  a. Higher place value: **b.** Round $3,451 **c.** Round $2,385
  d. Estimated sum: **e.** Exact sum:

45. $598 + $3,210
  a. Higher place value: **b.** Round $598 **c.** Round $3,210
  d. Estimated sum: **e.** Exact sum:

46. 57,281 + 4,253
  a. Higher place value: **b.** Round 57,281 **c.** Round 4,253
  d. Estimated sum: **e.** Exact sum:

47. $31,415 + $7,153
  a. Higher place value: **b.** Round $31,415 **c.** Round $7,153
  d. Estimated sum: **e.** Exact sum:

48. 4,710 − 2,536
  a. Higher place value: **b.** Round 4,710 **c.** Round 2,536
  d. Estimated difference: **e.** Exact difference:

49. 387 − 291

a. Higher place value: **b.** Round 387 **c.** Round 291

d. Estimated difference: **e.** Exact difference:

50. 1,000 − 837

a. Higher place value: **b.** Round 1,000 **c.** Round 837

d. Estimated difference: **e.** Exact difference:

51. 78,432 − 26,791

a. Higher place value: **b.** Round 78,432 **c.** Round 26,791

d. Estimated difference: **e.** Exact difference:

52. Sallee is considering buying a new truck. A local dealership is advertising a new Ford F-250 for $26,756. However, another dealership has a 1-year-old truck of the same model for $22,093. Estimate the difference in price of the two trucks.

53. Karen earned $42,715 last year. Sue earned $34,361. Estimate their combined income.

Add or subtract. Show all carrying or borrowing.

54. 5 feet 6 inches + 3 feet 9 inches

55. 5 feet 6 inches − 3 feet 9 inches

56. 3 days 2 hours 54 minutes + 2 days 5 hours 21 minutes

57. 3 days 2 hours 54 minutes − 2 days 5 hours 21 minutes

58. 9 gallons 2 quarts 1 pint 1 cup + 6 gallons 2 cups

59. 9 gallons 2 quarts 1 pint 1 cup − 6 gallons 2 cups

60. 4 pounds 13 ounces + 2 pounds 5 ounces

61. 4 pounds 7 ounces − 2 pounds 8 ounces

62. 6 hours 5 minutes − 4 hours 46 minutes

63. 4 miles 300 feet − 2 miles 700 feet

64. 5 weeks 5 hours − 6 days 8 hours

65. 3 gallons 1 pint + 2 gallons 3 quarts 1 pint

66. Explain why two right angles are supplementary angles.

67. Explain why two right angles are not complementary angles.

68. Can two acute angles be supplementary angles? Explain why.

69. Can two acute angles be complementary angles? Explain why.

70. Explain why two obtuse angles cannot be supplementary angles.

71. Explain why two obtuse angles cannot be complementary angles.

72. Angles A and B are complementary. The measure of angle A is 34°. Calculate the measure of angle B.

73. Angles X and Y are supplementary. The measure of angle X is 76°. Calculate the measure of angle Y.

74. Find the perimeter of a rectangle with a length of 5 ft 8 in. and a width of 2 ft 7 in.

75. Find the perimeter of a triangle with side lengths of 3 ft 8 in., 4 ft 2 in., and 2 ft 7 in.

76. As part of the landscaping, a city park has a garden that is shaped like an arrow to show the way to the picnic area. Usually, the garden is planted with petunias. Find the perimeter of the garden.

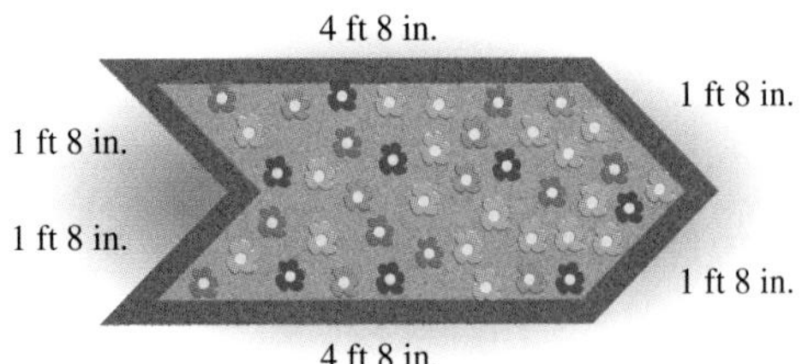

77. A square always has four equal sides. Each side of a square dance floor measures 35 ft 4 in. Find the perimeter of this dance floor.

78. A rectangular poster for a Jethro Tull concert measures 24 inches by 36 inches. Find the perimeter of the poster.

79. A rectangular poster of a Monet painting is 2 feet by 3 feet and has a frame that is 1 inch wide around it.

a. Draw a diagram of the frame and poster.

b. What is the outer perimeter of the frame?

80. The plans for a house usually indicate the sizes of doors and windows equal to the size of the *finished* opening. When a carpenter is framing a door opening, 5 inches must be added to the width of the door and 3 inches must be added to the height of the door shown on the plans to allow for finishing trim. A plan shows door openings that are 2 feet 6 inches wide and 6 feet 6 inches high.

a. Calculate the height of the opening that the carpenter will frame for this door.

b. Calculate the width of the opening that the carpenter will frame for this door.

c. Find the perimeter of the door opening.

81. A rectangular swimming pool is 50 meters long and 25 meters wide. A concrete deck that is 10 meters wide surrounds the pool.

a. Draw a diagram that illustrates the pool and deck.

b. Calculate the perimeter of the swimming pool.

c. Calculate the outer perimeter of the concrete deck.

82. The front of a tent has an opening shaped like an isosceles triangle. The bottom of the triangle has a length of 6 ft 4 in. One of the other sides has a length of 8 ft 2 in. What is the perimeter of this opening?

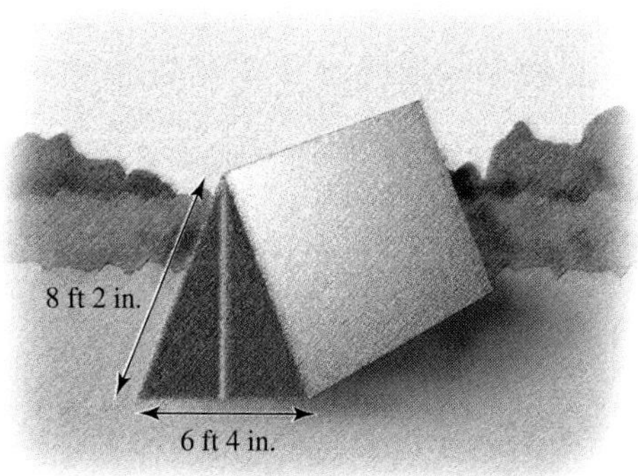

ERROR ANALYSIS

83. **Problem:** Estimate the sum of 45 and 32.

Answer: 45 + 32 = 77
77 rounds to 100.
The estimate of the sum is 100.

Describe the mistake made in solving this problem.

Redo the problem.

84. **Problem:** Estimate the sum of $1,299 and $1,533.

Answer: The estimated sum of $1,299 and $1,533 is $4,000 since $1,299 rounds to $2,000 and $1,533 rounds to $2,000.

Describe the mistake made in solving this problem.

Redo the problem.

85. **Problem:** Estimate the difference of \$1,299 and \$463.

Answer: The estimated difference of \$1,299 and \$463 is \$800 since \$1,299 rounds to \$1,300 and \$463 rounds to \$500.

Describe the mistake made in solving this problem.

Redo the problem.

REVIEW

86. Using set notation, write the set of all whole numbers that are less than 7.

87. Write the numbers in order from smallest to largest using $<$ signs:
65,001; 56,100; 65,010; 65,100.

Section 1.4 Problem Solving

LEARNING TOOLS

CD-ROM SSM VIDEO

Sneak Preview

Solving word problems is a part of applied mathematics. The tools and concepts of mathematics are used to find the solution to problem situations. A five-step strategy can be used to organize the process of reading, analyzing, and solving a problem.

After completing this section, you should be able to:

1) Assign a variable to represent an unknown.

2) Use the problem-solving strategy to solve word problems.

DISCUSSION

Equations

When we solve a problem, we are finding the answer to a question. Something in the problem is "unknown" and we want to know what that "unknown" is. Often we can solve a problem by writing and solving an equation that includes the "unknown." An **equation** is simply two statements that are equal.

Example 1 **Charles has \$400 saved. He wants to buy a computer that costs \$1,200. Write a word equation and an equivalent math equation that can be used to find out how much more money he needs to buy the computer.**

Word equation: Money needed = Price of computer − Money saved

The unknown in this problem is the amount of money Charles still needs to buy the computer. The words "money needed" can be replaced with a letter, m. This letter is called a **variable.** The words "price of computer" can be replaced with \$1,200 and "money saved" can be replaced with \$400. This changes the word equation into a math equation.

Math equation: $m = \$1{,}200 - \$400$

Example 2 **During the week of July 8, 1997, the top ten movies had a total box office gross of \$173 million. *Men in Black* grossed \$84 million. Write a word equation and an equivalent math equation that can be used to calculate the total box office gross of the other nine movies.**

Word equation: Gross of others = Total gross − Gross of *Men in Black*

Math equation: $g = \$173 \text{ million} - \$84 \text{ million}$

When we find the value of an unknown in an equation, we have **solved** the equation. For some problems, we may be able to find the solution to the problem without actually writing an equation. However, practicing writing equations for these easier problems will prepare us to write and solve equations for more complex problems later.

CHECK YOUR UNDERSTANDING

Write a word equation and a math equation for each problem situation.

1. The top-scale salary for flight attendants at America West is $23,800. Top-scale flight attendants at rival Southwest Airlines earn $41,200. What is the difference in the two salaries?

2. Sticks & Bowls is a noodle shop on Connecticut Avenue in Washington, D.C. A noodle bowl with spicy black sauce and seafood topping costs $5. A side dish of *nori* (roast seaweed) costs $1. An order of green tea costs $1. What is the cost of lunch if all three items are ordered?

Answers: 1. Difference in salaries = Salary at Southwest − Salary at America West, d = $41,200 − $23,800;
2. Cost of lunch = Cost of noodles + Cost of nori + Cost of tea, c = $5 + $1 + $1.

DISCUSSION

The Five-Step Problem-Solving Strategy

Many real-life problems are solved using mathematics. Reading, organizing and prioritizing information, and identifying relationships are also required. To help you organize your problem solving, a five-step strategy is suggested. This strategy will be used in all example word problems. Each step has a purpose.

Five-Step Strategy for Solving Problems

STEP 1: ***I know . . .***

Read the problem.

Identify the main question; what am I trying to find?

Assign a variable to the unknown.

Organize the other information given in the problem using words, numbers, and equals signs. Drawings or tables may be helpful.

STEP 2: ***Write equations.***

Write a *word equation* using the words from the "I know" step.

Write a *math equation* that is equivalent to the word equation.

STEP 3: ***Solve the math equation.***

Find the value of the unknown in the math equation.

STEP 4: ***Check the solution.***

Find an estimate to check the exact solution.

Think about whether this solution makes sense as an answer to the problem.

STEP 5: ***Answer the question.***

Answer the question presented in the original problem in an English sentence. Do not use mathematical symbols in this answer.

The following example problems are solved using the five-step strategy. Many of these problems are not difficult; the answers may be obvious. They are chosen so that the emphasis can be on learning and practicing the five-step strategy rather than struggling with the problem itself. Later, when problems to be solved are more difficult, we will be ready to use the strategy.

Example 3 **In early 1999, the Federal Aviation Administration ordered inspections of older Boeing 727 aircraft for fatigue cracking on skin joints. The required visual inspection cost was \$960 per airplane. The low frequency eddy current inspection test for cracks not visible to the naked eye was \$1,920 per airplane. What was the total cost of inspections per airplane?**

STEP 1: ***I know . . .***

Identify the main question; what am I trying to find? Assign a variable to the unknown. Organize the other information given in the problem using words and numbers.

c = Total cost per airplane
\$960 = Cost of visual inspection
\$1,920 = Cost of eddy test

STEP 2: ***Write equations.***

Write a word equation using the words from the "I know" step.

Total cost per airplane = Cost of visual inspection + Cost of eddy test

$$c = \$960 + \$1{,}920$$

Write a math equation that is equivalent to the word equation.

STEP 3: ***Solve the math equation.***

Find the value of the unknown in the math equation.

$$\begin{array}{r} \overset{1}{} \\ \$1{,}920 \\ +\quad 960 \\ \hline \$2{,}880 \end{array}$$

$c = \$2{,}880$

STEP 4: ***Check the solution.***

Find an estimate to check the exact solution. Think about whether this solution makes sense as an answer to the problem.

Estimate: $c \approx \$1{,}000 + \$2{,}000 = \$3{,}000$, which is close to \$2,880. So, our exact answer is reasonable.

STEP 5: ***Answer the question.***

Answer the question presented in the original problem in an English sentence.

The total cost per airplane for inspections was \$2,880.

Example 4 **Below is a copy of Fred Smith's paycheck. The left column shows the money earned, his gross income. The right column shows deductions removed from his gross income. What is Fred's gross income for this paycheck?**

SMITH, FRED B 043098		528-11-0001	South Side Hospital
Regular salary	1774	FICA-OASDI	623
Merit pay	256	FICA-MQGE	21
Sick leave	38	FED W/H-E	288
Personal leave	0	NY W/H	96
Overtime	281	HEALTH HMO	17
		401k1	50
		403b	100
Gross Income ________		Total deductions ________	
		Net pay ________	

STEP 1: ***I know . . .***

g = Gross income
\$256 = Merit pay
\$0 = Personal leave
\$1,774 = Regular salary
\$38 = Sick leave
\$281 = Overtime

STEP 2: ***Write equations.***

Gross income = Regular salary + Merit pay + Sick leave + Personal leave + Overtime

$g = \$1{,}774 + \$256 + \$38 + \$0 + \$281$

STEP 3: ***Solve the math equation.***

```
  1 2 1
 $1774
 $ 256
 $  38
+$ 281
 $2349
```

$g = \$2{,}349$

STEP 4: ***Check the solution.***

Estimate $g \approx \$2{,}000 + \$0 + \$0 + \$0 = \$2{,}000$, which is close to \$2,349. Our exact answer is reasonable.

STEP 5: ***Answer the question.***

Fred's gross income is \$2,349.

Example 5 **A triangle has three sides and three angles. The sum of the measures of the angles always equals 180°. Elwood is designing a swing set. The side supports of the swing set make a triangle with the ground. The angles made by the supports and the ground equal 70°. What is the measure of the angle at the top?**

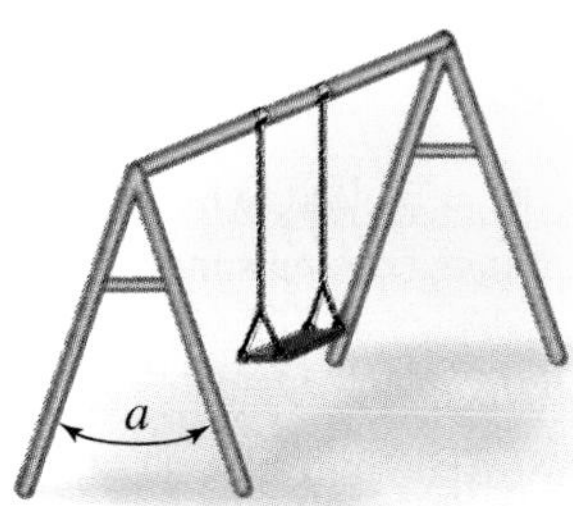

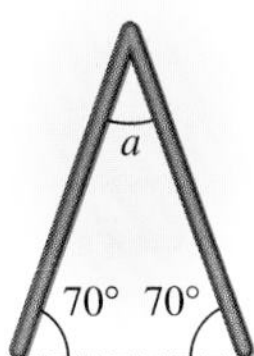

I know . . .

a = Top angle   180° = Total degrees   70° = Angle with ground

Equations

Top angle = Total degrees − Ground angle − Ground angle

$a = 180° - 70° - 70°$

Solve

$a = 180° - 70° - 70°$

$a = 110° - 70°$

$a = 40°$

Check

An estimate does not tell us if this answer is reasonable; 200° − 100° − 100° = 0°. However, we can think about whether an answer of 40° makes sense. We know that 70° + 70° = 140°. So we need 40° to get a total of 180°, exactly what our answer says. Sometimes, checking can involve doing the problem another way and getting a similar answer.

Final answer

The top angle formed by the side supports of the swing set is 40°.

CHECK YOUR UNDERSTANDING

Use the five-step strategy to solve the following problems.

1. Karen lives in Lexington, Kentucky. She is saving airline frequent flyer miles to qualify for four free round-trip tickets to Europe. She currently has 148,900 miles in her account. She is going on a business trip next week to Las Vegas, which will net her another 1,686 miles each way. She needs 200,000 miles to receive her free tickets to Europe. After her trip, how many miles does she still need for her free tickets?

2. Abdou has saved $10,000 for college. His tuition for the year is $5,792. His books will cost him $528. How much money will remain for his living expenses?

3. The sum of the angles of any triangle is 180°. If two of the angles in a triangle each measure 20°, what is the measure of the third angle?

Answers: 1. Karen still needs 47,728 miles; 2. Abdou has $3,680 left for living expenses; 3. The third angle measures 140°.

INVESTIGATION The Five-Step Problem-Solving Strategy

In this investigation, we will not necessarily solve a given problem. For example, in this first problem, we will only write the word equation and the math equation and will not find the answer to the problem.

The demand for electrical power in California is lower in the winter than in the summer. The peak demand in winter is 32,000 megawatts. The peak demand in summer is 45,000 megawatts. How much higher is the peak demand in summer than the peak demand in winter? *(Source: Orange County Register, 2/10/2001).*

I know . . .

d = Difference in demand
45,000 megawatts = Demand in summer
32,000 megawatts = Demand in winter

1. Write a word equation using only the words found in the *I know* . . . step.

2. Rewrite this word equation as a math equation.

In some problems, more information is given than is needed to solve the problem. Extra information is sometimes called **extraneous information.** Extraneous information does not have to be included in the ***I know . . .*** step.

In 1999, marijuana plants that were equivalent to about 995,100 pounds of marijuana were removed from the United States national forest land. In comparison, the U.S. Customs Service seized 989,368 pounds of marijuana along the Southwest border in 1999 and the Border Patrol confiscated about 1,200,000 pounds. How much less marijuana was seized by the U.S. Customs Service than was found in the national forests?

3. Complete the *I know* . . . step for this problem.

m = ______________________________
995,100 pounds = ______________________________
989,368 pounds = ______________________________

4. What extraneous information is included in this problem?

5. Write a word equation using only the words found in the *I know* . . . step.

6. Rewrite this word equation as a math equation.

To solve problems, we often write equations. If the reasoning we use to build the equations is wrong, the solutions will also be wrong. When this happens, we need to be ready to throw the equations out and try again. In each of the following problems, an error has been made in reasoning.

Biologists introduced Canada geese to the Puget Sound, Washington, area about 30 years ago. There are now an estimated 25,000 Canada geese living in urban areas around Puget Sound, and that number is expected to increase to more than 80,000 in the next decade. The birds are blamed for polluting lakes, scattering bacteria-laden droppings, and flying into airplanes. Wildlife managers believe that current population levels need to be reduced by 3,500. How many birds will there be after this reduction?

I know . . .

b = Remaining birds
25,000 = Current population
80,000 = Expected population
3,500 = Reduction needed

Equations

Remaining birds = Expected population − Reduction needed
$b = 80{,}000 - 3{,}500$

7. Describe the error in reasoning made in writing these equations.

Hundreds of Amtrak passengers were forced onto buses for travel when a fire damaged a 657-foot long, 100-foot high bridge over the Nisqually River. Flames reached as high as 20 feet above the bridge, according to witnesses. Firefighters were forced to fight the flames by air, dumping bucket after bucket of water on it. How high were the flames above the river?

I know . . .

h = Height of flames above river
100 feet = Height of bridge
657 feet = Length of bridge
20 feet = Height of flames above bridge

Equations

Height of flames above river = Height of bridge − Height of flames above bridge
$h = 100 \text{ feet} - 20 \text{ feet}$

8. Describe the error in reasoning made in writing these equations.

When we solve a problem, we need to check the answer. There are two parts to checking an answer. First, we can use an estimate to make sure of our exact arithmetic. Second, we can think about the solution and make sure it makes sense.

In February 2000, the professors at the University of California at Santa Cruz voted for a new requirement: letter grades. In the past, students received narrative essays evaluating their work. However, only 231 professors voted on the decision with 77 opposed. There are more than 588 professors eligible to vote. How many of the professors voted to support the change?

I know . . .

p = Professors voting yes
588 = Eligible professors
77 = Professors voting no
231 = Professors who voted

Equations

Professors voting yes = Eligible professors − Professors voting no
$p = 588 - 77$

Solve $p = 588 - 77$
$p = 511$

9. Think about the solution. Is it a reasonable answer to the question in the problem? Explain why it is or is not reasonable.

Now, use all five steps of the problem-solving strategy to solve the following word problems. You can probably solve these problems without using the strategy. However, the main purpose of this work is to learn how to use the problem-solving strategy. Later, the problems will be more complicated. This practice with easier problems will prepare you to solve the more complicated problems to come.

10. Julie is a sales representative for a large pharmaceutical company. Because she drives a company car, she keeps a daily log of her mileage. The car odometer read 62,753 at the end of work yesterday. At the end of work today, the odometer read 63,018. How many miles did Julie drive today?

I know . . .

Word equation

Math equation

Solve

Check

Final answer

11. Almost 900,000,000 people speak Mandarin Chinese, the most spoken language on Earth. The next most spoken language is English with more than 322,000,000 speakers. If the world's population is about 5,100,000,000 people, how many people speak languages other than English and Mandarin Chinese?

I know . . .

Word equation

Math equation

Solve

Check

Final answer

Answers: 10. Julie drove 265 miles today. 11. Other languages are spoken by about 3,878,000,000 people.

EXERCISES SECTION 1.4

Solve each word problem using the five-step problem-solving strategy. The main purpose of these exercises is to learn how to use the problem-solving strategy. The process seems involved for these relatively straightforward problems, but it is a good way to learn the method. Concentrate on doing each step correctly.

1. Juanita needs 120 course credits to obtain her degree. She currently has 74 credits. How many more credits does she need to graduate?

$74 + A = 120$

$120 - 74 = A$

$120 - 74 = 45$

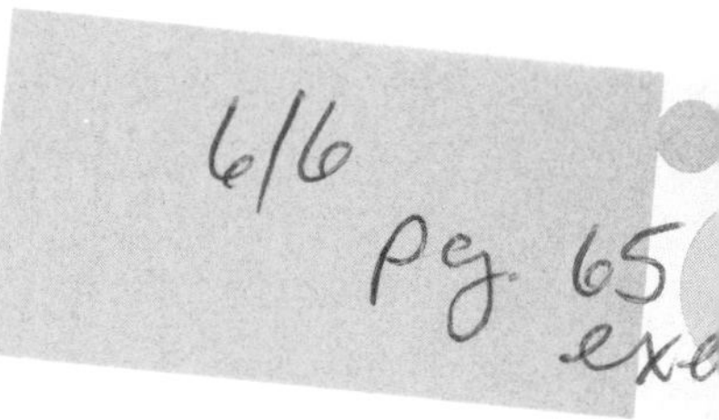

2. Heather is buying a new house and must choose either a 20- or a 30-year mortgage. Both mortgages have the same interest rate but have different monthly payment amounts. For the 30-year loan, her payment would be \$1,023 a month, while with the 20-year loan her monthly payment would be \$1,204. What is the difference in the monthly payment amounts for the two mortgages?

$1,204 - 1,023 = \$181$

3. Chris is considering buying a high-definition television (HDTV) set. The price of a 56-inch projection television set is \$5,499. The set-top box that can tune in the digital signals is an additional \$1,499. What is the total cost of the set and the box?

$5,499 + 1,499 = 6,998$

4. Nichole needs to buy a calculator, graph paper, and a ruler for her engineering class. The calculator costs $158, the paper is $8, and the ruler is $3. If she buys all three items, how much will she spend?

$158.00
$8.00
$3.00
$169.00

5. Mitch is driving from Norfolk, Virginia, to Philadelphia, Pennsylvania, for a conference. He is then going to drive to Baltimore, Maryland, to visit his mom before returning to Norfolk. If it is 261 miles from Norfolk to Philadelphia, 103 miles from Philadelphia to Baltimore, and 257 from Baltimore to Norfolk, how many miles will he drive on his trip?

261
103
261
625

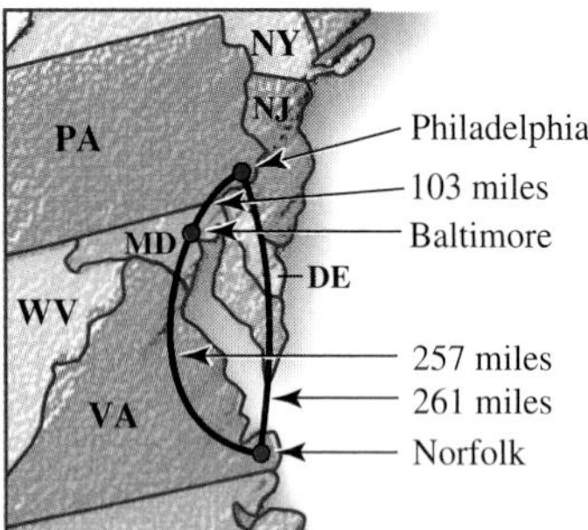

6. Manuel must decide whether to finance his house with a thirty-year mortgage or a twenty-year mortgage. Over thirty years, the total payments with principal and interest will be $368,280. Over twenty years, less interest must be paid. The total payments will be $288,960. How much money will he save by choosing the twenty-year mortgage instead of the thirty-year mortgage?

368,280
− 288,960
79,320

7. A newspaper ad reads:

SPA WAREHOUSE
Spa Covers to fit Hot Springs Spa!
New—$175, Scratch & Scuff—$99.

How much money can be saved by buying a "scratch & scuff" cover?

$175.00
− 99.00
$76.00

8. The most popular pet in America is the cat. In 1997, Americans owned 59,000,000 cats. Americans also owned about 53,000,000 dogs. How many cats and dogs did Americans own?

9. In 1998, the charitable organizations that received the most donations were the Salvation Army ($1,200,000,000), the YMCA ($493,000,000), the American Red Cross ($490,000,000), the American Cancer Society ($488,000,000), the Fidelity Investments Charitable Gift Fund ($490,000,000), Harvard University ($428,000,000), and Catholic Charities USA ($425,000,000). What is the total amount of these donations?

10. Using Fred's paycheck from Example 4 in this section, calculate Fred's net pay.

11. The population of Salt Lake City, Utah, was 20,768 in 1880. By 1980, the population had grown to 163,033. What was the increase in population over the 100 years from 1880 to 1980?

12. Antarctica is the highest continent in the world with an average altitude of 7,500 feet. Even though the Himalaya Mountains are located in Asia, this continent has an average altitude of only 2,900 feet. What is the difference in average altitudes of the two continents?

13. In 1997, the annual hours worked per person in the United States was 1,966 hours. In 1980, the annual hours worked per person in the United States was 1,883 hours. Calculate the difference in annual hours worked per year in 1997 and 1980.

14. New York City, the largest city in the United States, had a population of 7,420,166 in 1998. The second largest city, Los Angeles, had 3,822,610 fewer residents than New York in 1998. What was the population of Los Angeles in 1998?

15. A study by the Harvard School of Public Health showed that 20 percent of all college students can be classified as "frequent" binge drinkers, consuming about 18 drinks a week. How many drinks would a "frequent" binge drinker consume in 2 weeks?

16. In 1995, 2,700,000,000 direct-mail credit-card solicitations were received by Americans compared to the 3,450,000,000 such solicitations received in 1998. What was the increase in direct-mail credit-card solicitations?

17. Of the four military services, the Navy and Air Force have the greatest shortages of pilots. In 1999, the Navy had 6,572 pilots, 1,077 short of their requirement. How many pilots should the Navy have had in 1999?

18. Water boils at 212°F and freezes at 32°F. What is the difference between these temperatures?

19. Sleeping bags sold for backpacking are rated for warmth, fill weight, and total weight. The fill weight refers to how much insulation is in the sleeping bag and is important for backpackers who may be sleeping in cold conditions. The total weight is important to backpackers because they want to minimize how much they must carry as they hike. The *Chief* sleeping bag is filled with fortrel polyester and weighs a total of 2 pounds 12 ounces. The *Polar* is filled with Hollofil II and weighs 3 pounds 8 ounces. What is the difference in weight between the two bags?

20. Both the *Chief* and *Polar* sleeping bags are intended for use by children. The *Chief* sleeping bag is filled with fortrel polyester and weighs a total of 2 pounds 12 ounces. The *Polar* is filled with Hollofil II and weighs 3 pounds 8 ounces. If a parent has to carry them both in his backpack, what is the total weight of the two sleeping bags?

21. In 1783, in Versailles, France, a sheep, a rooster, and a duck were lifted up into the air in the world's first hot air balloon. Later that same year, Jean Francois Pilatre de Rozier and the Marquis d'Arlandes achieved the first manned free flight over France. In 1978, a team from the United States first crossed the Atlantic Ocean in a hot air balloon. In 1999, Bertrand Piccard from Switzerland and Brian Jones from Great Britain completed the first around-the-world hot air balloon crossing in 20 days. How many years passed between the first hot air balloon flight and the around-the-world hot air balloon flight?

22. A McDonnell Douglas MD-11 passenger aircraft carries a maximum of 297 passengers. With a wingspan of 52 meters and length of 61 meters, its maximum allowable take-off weight is 280,300 kilograms. Its cruising speed is 880 kilometers per hour. Its maximum range for long-distance flights is 10,000 kilometers. In contrast, a Boeing 767-300 ER carries a maximum of 226 passengers. With a wingspan of 48 meters and length of 55 meters, its maximum allowable take-off weight is 172,400 kilograms. Its cruising speed is 850 kilometers per hour and its maximum range is 8,100 kilometers. How many more passengers can a McDonnell Douglas MD-11 carry than a Boeing 767-300 ER?

23. The florist industry has been dramatically changed by vendors on the Internet who sell flowers directly from the grower to customers. Proflowers.com has growers fill its online orders so that customers receive the flowers within a day or two of being picked. As reported in *Time* magazine, "During the February rush, a dozen long-stemmed roses are selling for $48 (including shipping)—$37 cheaper than the same order at FTD.com and $46 cheaper than the same order at 1-800-Flowers.com, the two leaders in online sales" (*Time,* February 21, 2000). Calculate the price of a dozen long-stemmed roses at FTD.com.

24. A study released by the Harvard School of Public Health in 2000 reported that about 3,178 of 14,000 surveyed students were frequent binge drinkers. How many surveyed students were not frequent binge drinkers?

25. It costs Washington State farmers about $1,660 in total operating expenses to grow an acre of potatoes. Of that, only $22 is for fuel expenses. How much is used for nonfuel expenses?

26. Judy is a volunteer on a work crew that is raking the lawns of elderly and disabled people. Since the sun is very hot, she takes her handkerchief out of her pocket and folds one corner over to the opposite corner to make a triangle. The sum of the measures of the three angles in a triangle is always 180°. One angle of this triangle measures 45°. Another angle is a right angle. What is the measure of the third angle?

27. A county spent \$803,785 to stabilize and begin repairs on a mudslide that closed a county road. The Federal Highway Administration was asked to reimburse the county for its expenses. The first payment to the county was \$66,037. A total of \$346,265 more in reimbursement was promised to the county. How much of the county's costs will not be reimbursed?

28. Wolves in the Gros Ventre River drainage in Wyoming corralled about 2,500 elk into an area that can sustain only about 600 elk. How many elk needed to be removed from the area to prevent starvation?

29. The suggested retail price on a Sealy Posturepedic King mattress set is \$1,199. The sale price is \$899. The final clearance price is \$749. What is the difference between the suggested retail price and the final clearance price? This difference is known as the **discount.**

30. The Fox Network had record numbers of viewers tune in for the special *Who Wants to Marry a Multimillionaire?* in February, 2000. In the first 30 minutes of the 2-hour program, 10 million people were watching. This grew to 12,300,000 people in the second 30-minute period. By the third 30-minute period, which included the swimsuit competition, 18,900,000 people had tuned in. In the last 30-minute period, in which the marriage took place, 22,800,000 people were watching. What was the increase in viewers between the third viewing period and the last viewing period?

31. In 1997, General Motors had a net income of $6,698,000,000. The net income of Ford Motor Company was $6,920,000,000. What was the total net income of these two companies?

32. As the cost of an overnight stay in a hospital has risen, the number of outpatient visits has also risen. In 1990, there were 368,174 outpatient visits to hospitals. By 1996, there were 505,455 visits. Calculate the increase in outpatient visits between 1990 and 1996.

Jefferson University Hospital, Philadelphia
(© 2001 Beth Strauss, Tyler Photograph)

33. The largest ticket revenues from concerts in 1997 were earned by the Rolling Stones ($89,300,000), U2 ($79,900,000), Fleetwood Mac ($36,300,000), Metallica ($34,100,000), and Brooks & Dunn/Reba McEntire ($33,000,000). Calculate the total ticket revenues generated by these five concert tours.

34. A square is a shape with four sides of equal lengths. It has four right angles. What is the total measure of the angles in a square?

35. An isosceles triangle has three angles. Two of the angles have the same measure of 55°. The measure of the third angle is 70°. What is the total measure of the angles in this triangle?

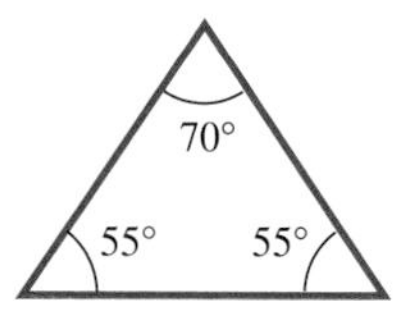

36. The distance around a square is called its perimeter. A square patio has four sides. Each side is 21 feet long. What is the perimeter of the square?

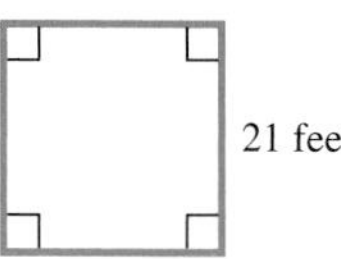

ERROR ANALYSIS

37. Problem: The Portland *Oregonian* (February 27, 2000) reported, "In 1999, single-family residential construction permits rose to 1,231,811 in the United States. . . . Here's a look at permits in the West for some metro areas."

METRO AREA	SINGLE-FAMILY PERMITS	METRO AREA	SINGLE-FAMILY PERMITS
Las Vegas	21,623	Boise	5,342
Portland–Vancouver	10,684	San Francisco	1,681
Sacramento, Calif.	10,275	Salem	1,659
Seattle–Bellevue–Everett	9,617	Spokane	1,529
Los Angeles–Long Beach	7,789	Medford–Ashland	1,298
Salt Lake City–Ogden	7,493	Eugene–Springfield	1,161

Source: The Meyers Group.

Determine how many building permits were issued outside the West.

Answer: ***Equations***

Permits outside the West = Total permits − Permits in the West

$P = 1{,}231{,}811 - (21{,}623 + 10{,}684 + 10{,}275 + 9{,}617 + 7{,}789 + 7{,}493 + 5{,}342 + 1{,}681 + 1{,}659 + 1{,}529 + 1{,}298 + 1{,}161)$

Describe the mistake made in solving this problem.

38. Problem: The U.S. Senate approved a measure to give hog farmers an extra $250 million in direct payments to help them survive a sharp downturn in pork prices. The money was added to a $1,900 million emergency-spending bill that included aid for the country of Jordan and hurricane-Mitch-battered Central America. What was the total of the spending bill after the addition of aid to farmers?

Answer: ***I know . . .***

t = Total of spending bill
$1,900,000 = Original amount
$250,000,000 = Hog payments

Word equation

Total of spending bill = Original amount + Hog payments

Math equation

$t = \$1{,}900{,}000 + \$250{,}000{,}000$

Solve

$t = \$251{,}900{,}000$

Describe the mistake made in solving this problem.

Redo the problem correctly.

REVIEW

39. Explain how to estimate the sum of two whole numbers.

40. Explain why we cannot add 44 inches and 30 quarts.

41. State the commutative property of addition using variables.

42. Explain the meaning of this symbol: $\approx$.

Section 1.5 Multiplication of Whole Numbers

LEARNING TOOLS

CD-ROM SSM VIDEO

Sneak Preview

Multiplication is a familiar arithmetic operation. We will review the basic principles of multiplication of whole numbers so that these principles can be extended to arithmetic with other numbers and variables.

After completing this section, you should be able to:

1) Rewrite multiplication as multiple additions.
2) Multiply single-digit whole numbers.
3) Use the distributive property.
4) Find a product of a whole number and a multiple of ten.
5) Estimate products.
6) Multiply multidigit whole numbers.
7) Multiply a measurement by a whole number.
8) Convert a measurement to an equivalent measurement in different units.

DISCUSSION

Multiplication Is Multiple Additions

Multiplication of whole numbers is another way to represent repeated addition. For example, the addition $3 + 3 + 3 + 3 + 3 = 15$ can also be represented as $3 \times 5 = 15$. Similarly, $2 + 2 + 2 + 2 = 8$ can be represented as $2 \times 4 = 8$. When two numbers are multiplied together, the result is called a **product**. The numbers being multiplied are called **factors.** The **notations** for multiplication include $\times$, $\cdot$, or parentheses.

Example 1 **Write the product of 5 and 3.**

Using the symbol $\times$	$5 \times 3 = 15$		
Using the symbol $\cdot$	$5 \cdot 3 = 15$		
Using one set of parentheses	$5(3) = 15$	or	$(5)3 = 15$
Using two sets of parentheses	$(5)(3) = 15$		

Parentheses are an example of **implied multiplication.** The multiplication is implied since no actual multiplication symbol is written.

An important axiom of multiplication is that any whole number multiplied by zero results in a product of zero. For example, $0 \cdot 8 = 0$ and $4(0) = 0$.

PROPERTY

Zero product property

In words: The product of zero and any number is zero.

In symbols: $a(0) = 0(a) = 0$

The multiplication tables list the products of the whole numbers between 0 and 12. In elementary school, students are required to memorize these products. Memorizing the multiplication tables allows us to quickly multiply larger numbers. Unfortunately, students who have difficulty memorizing "their tables" sometimes think that this means they are not good at math. Actually, it only means that they have difficulty memorizing. Knowing the "times tables" does increase the speed at which estimated and exact products can be calculated.

CHECK YOUR UNDERSTANDING

Use repeated addition to complete the multiplication table. Some entries are completed.

×	0	1	2	3	4	5	6	7	8	9	10	11	12
0	0							0					
1		1											
2	0					10						22	
3								21					
4													48
5						25							
6									48				
7													
8				24									
9											90		
10										90			
11						55							
12													144

From memory or the multiplication tables, find each product:

1. 2×7 **2.** (6)(8) **3.** 9(12) **4.** $11 \cdot 7$

5. $9 \cdot 9$ **6.** 12(4) **7.** 2×0 **8.** $0 \cdot 11$

9. (5)(7) **10.** 3(6) **11.** (0)(7) **12.** (8)(1)

Answers: 1. 14; 2. 48; 3. 108; 4. 77; 5. 81; 6. 48; 7. 0; 8. 0; 9. 35; 10. 18; 11. 0; 12. 8.

The **commutative property of multiplication,** an axiom of our number system, states that changing the order of multiplication does not change the product. For example, the products $5 \cdot 10$ and $10 \cdot 5$ are both 50.

PROPERTY

The commutative property of multiplication

In words: Changing the *order* of multiplication does not change the product.

In symbols: $a \cdot b = b \cdot a$

When we multiply three numbers together, there are two possible groupings for the numbers. However, the product will be the same no matter how we group the factors.

PROPERTY

The associative property of multiplication

In words: Changing the *grouping* of factors does not change the product.

In symbols: $(a \cdot b) \cdot c = a \cdot (b \cdot c)$

Example 2 **Demonstrate the associative property of multiplication using parentheses and the numbers 2, 3, and 4.**

One grouping: $(2 \cdot 3) \cdot 4$
$= 6 \cdot 4$
$= 24$

A different grouping: $2 \cdot (3 \cdot 4)$
$= 2 \cdot 12$
$= 24$

Even if the grouping is changed, the product of the three numbers is the same, 24.

The multiplication tables also show that the product of any number and 1 is the original number. For example, $7 \cdot 1 = 7$ and $12 \cdot 1 = 12$. The number 1 is given the name **multiplicative identity** because it does not change the identity of the other factor.

CHECK YOUR UNDERSTANDING

Each example demonstrates an axiom. Identify the axiom.

1. $6 + 4 = 4 + 6$

2. $3(1 \cdot 2) = (3 \cdot 1)\ 2$

3. $3 \times 2 = 2 \times 3$

4. $8 \cdot 0 = 0$

5. $2 + (6 + 3) = (2 + 6) + 3$

6. $3(1 \times 2) = 3(2 \times 1)$

Answers: 1. Commutative property of addition; 2. Associative property of multiplication; 3. Commutative property of multiplication; 4. Zero product property; 5. Associative property of addition; 6. Commutative property of multiplication.

DISCUSSION

Patterns in Products

The products of 10 in the multiplication tables follow an interesting pattern:

×	0	1	2	3	4	5	6	7	8	9	10	11	12
10	0	10	20	30	40	50	60	70	80	90	100	110	120

Each product of a factor and 10 is just the digits of the factor followed by a single zero.

A similar pattern applies to multiplication by any number with only one digit that is not zero.

×	0	1	2	3	10	20	300	400	5,000
10	0	10	20	30	100	200	3,000	4,000	50,000
100	0	100	200	300	1,000	2,000	30,000	40,000	500,000
300	0	300	600	900	3,000	6,000	90,000	120,000	1,500,000

This pattern suggests a way to multiply such numbers using only the multiplication facts from the times tables.

PROCEDURE

To multiply numbers that consist of single digits followed by zeros:

1. Multiply the single digits together.
2. Write zeros following this product equal to the total number of zeros in all factors.

Example 3 **Multiply (60)(70).**

$6 \cdot 7 = 42$ *Multiply the single digits together.*

60 has one zero; 70 has one zero, for a total of two zeros.

$(60)(70) =$ **4,200** *Write zeros following the product equal to the total number of zeros in both factors.*

Example 4 **Multiply 200 · 80.**

$2 \cdot 8 = 16$ *Multiply the single digits together.*

200 has two zeros; 80 has one zero, for a total of three zeros.

$200 \cdot 80 =$ **16,000** *Write zeros following the product equal to the total number of zeros in both factors.*

Example 5 **Find the product of 5,000 and 80,000.**

$5 \cdot 8 = 40$ *Multiply the single digits together.*

5,000 has three zeros; 80,000 has four zeros, for a total of 7 zeros.

$(5{,}000)(80{,}000) =$ **400,000,000** *Write zeros following the product equal to the total number of zeros in both factors.*

CHECK YOUR UNDERSTANDING

Find the product.

1. $30 \cdot 20$ **2.** $40(80)$ **3.** $(600)(70{,}000)$

4. $100 \cdot 50$ **5.** $(3)(90)$ **6.** $4{,}000(50{,}000)$

Answers: 1. 600; 2. 3,200; 3. 42,000,000; 4. 5,000; 5. 270; 6. 200,000,000.

DISCUSSION

Estimating Products

When estimating a sum or difference, each number is first rounded to the highest place value of the largest number. This makes estimation easier because we end up adding only the largest place values. The digits in the smaller place values are simply ignored in the estimate.

Example 6 **Estimate the sum of 1,563 + 24 + 3,299.**

The largest place value in any of these numbers is the thousands place.

Round each number to the thousands place.

1,563 rounds to 2,000.

24 rounds to 0.

3,299 rounds to 3,000.

Add the rounded numbers: 2,000 + 0 + 3,000 = **5,000.**

Because 24 is too small to significantly affect the sum, we round it to 0 so that it has no effect on the estimate. This also occurs with the 63 in the number 1,563 and the 99 in the number 3,299.

In multiplication, we must use different rules for rounding than we use in addition. If we rounded small numbers to 0, the zero product property tells us that the resulting estimate would have to be zero. For example, the estimated product of 500 and 10 would be 0 since 10 rounded to the hundreds place is 0. To avoid having estimates of 0, we instead round each factor to its own highest place value. Each rounded factor will only have one digit that is not zero.

PROCEDURE

To estimate a product:

1. Round each factor to its highest place value.
2. Multiply the rounded factors.

Example 7 **Estimate the product of 43 and 3.**

Round 43 to its highest place value: 40

Round 3 to its highest place value: 3

Multiply the rounded factors: $40 \cdot 3 = 120$

$(43)(3) \approx 120$

Example 8 **Estimate the product of 64 and 59.**

Round 64 to its highest place value: 60

Round 59 to its highest place value: 60

Multiply the rounded factors: $(60)(60) = 3{,}600$

$64(59) \approx 3{,}600$

Example 9 **Estimate the product of 3,476 and 49,987.**

Round 3,476 to its highest place value: 3,000

Round 49,987 to its highest place value: 50,000

Multiply the rounded factors: $3{,}000(50{,}000) = 150{,}000{,}000$

$(3{,}476)(49{,}987) \approx 150{,}000{,}000$

Example 10 **Estimate the product of 3 and 124.**

Round 3 to its highest place value: 3

Round 124 to its highest place value: 100

Multiply the rounded factors: $3 \cdot 100 = 300$

$3(124) \approx 300$

Example 11 **Estimate the product of 8 and 11.**

Round 8 to its highest place value: 8

Round 11 to its highest place value: 10

Multiply the rounded factors: $8 \cdot 10 = 80$

$8(11) \approx 80$

CHECK YOUR UNDERSTANDING

Estimate each product.

1. $3{,}467 \cdot 7{,}193$

2. $(72{,}091)\ (3{,}891)$

3. $(282{,}001)\ (671)$

4. $15 \cdot 2$

5. $(800{,}001)(2{,}997)$

6. $(0)(143{,}872)$

Answers: 1. 21,000,000; 2. 280,000,000; 3. 210,000,000; 4. 40; 5. 2,400,000,000; 6. 0.

DISCUSSION

The Distributive Property

The distributive property is another axiom of our number system.

PROPERTY

The distributive property of multiplication over addition

In words: When multiplying a number by a sum, you can add first and then multiply or you can multiply each number first and then add. The final answer will be the same.

In symbols: $a \cdot (b + c) = a \cdot b + a \cdot c$

The distributive property is illustrated in the following example.

Example 12 **Mike and Judy are buying large pepperoni pizzas. The price per pizza is \$9. Mike is buying four pizzas for their son's birthday party. Judy is buying six pizzas for the biology club meeting. Calculate the total cost of the pizzas if they pay separately or if they pay together.**

If they pay separately:

Total cost = Mike's cost + Judy's cost
= Cost per pizza · Mike's pizzas + Cost per pizza · Judy's pizzas
= \$9 · 4 + \$9 · 6
= \$36 + \$54
= \$90

If they pay together:

Total cost = Cost per pizza · (Mike's + Judy's pizzas)
= \$9 · (4 + 6)
= \$9 · 10
= \$90

It does not matter whether Mike and Judy pay separately or together. In each case, the total cost is \$90. This is an example of the distributive property.

Example 13 **Evaluate 5(1 + 3).**

$5(1 + 3)$	*1 and 3 are like terms; they can be combined.*
$= 5 \cdot 4$	*Add the numbers inside the parentheses.*
$= 20$	*Multiply.*

Example 14 **Evaluate 5(1 + 3) using the distributive property.**

$5(1 + 3)$	
$= 5 \cdot 1 + 5 \cdot 3$	*Distribute the 5 to the 1 and the 3.*
$= 5 + 15$	*Multiply.*
$= 20$	*Add.*

CHECK YOUR UNDERSTANDING

Evaluate using the distributive property.

1. 3(7 + 1) **2.** 2(4 + 5) **3.** 7(2 + 4) **4.** 6(3 + 5)

Answers: 1. 24; 2. 18; 3. 42; 4. 48.

DISCUSSION

Multiplication Using the Distributive Property

The distributive property is often used in mental multiplication of larger numbers.

Example 15 **Azar runs a business out of her home. She pays \$20 a month for her Internet connection. She mentally calculates her Internet expense for a year like this: "*There are 12 months in a year which is the same as 10 months + 2 months. My Internet expenses are \$20 a month. I multiply \$20 by 10 months (\$200) and I multiply \$20 by the leftover 2 months (\$40). The total expenses are \$200 plus \$40 which equals \$240.*"**

Azar first mentally rewrote 12 months as (10 + 2) months. She then multiplied by 20 using the distributive property.

\$20 · 12
= \$20(10 + 2) *Rewrite 12 in expanded form as 10 + 2.*
= (\$20 · 10) + (\$20 · 2) *Use the distributive property $a(b + c) = ab + ac$.*
= \$200 + \$40
= \$240

We can find products of larger numbers using the distributive property. This is the method that an abacus uses to find products. However, most children are taught a different way to multiply that requires less room. Although this method also depends on the distributive property, it may not be obvious that the property is being used.

Example 16 **Multiply 5 and 724.**

Strategy 1
Write each factor in expanded form. Use the distributive property.

5(724)
= 5(700 + 20 + 4) *Write 724 in expanded form.*
= 5 · 700 + 5 · 20 + 5 · 4 *Apply the distributive property.*
= 3,500 + 100 + 20 *Multiply.*
= 3,620 *Add.*

Strategy 2
Multiply without expanding the numbers.

$$\begin{array}{r} {\scriptstyle 1\,2} \\ 724 \\ \times \quad 5 \\ \hline 3620 \end{array}$$

Multiply 5 · 4 = 20. Write down the 0 in the ones place; carry the 2 to above the tens column. Multiply 5 · 2 = 10. Add the carried 2 to get a sum of 12. Write the 2 down below in the tens column and carry the 1 to the hundreds column. Multiply 5 · 7 = 35. Add the carried 1 for a sum of 36. Write the 36 down below. The product is 3,620.

Example 17 **Calculate (234)(52).**

Strategy 1
Write each factor in expanded form. Use the distributive property.

(234)(52) *Since both 234 and 52 can be expanded, the distributive property must be used twice.*
= 234(50 + 2) *Write 52 in expanded form: 52 = 50 + 2.*
= 234(50) + 234(2) *Use the distributive property.*
= (200 + 30 + 4)(50) + (200 + 30 + 4)(2) *Write 234 in expanded form.*
= 200(50) + 30(50) + 4(50) + 200(2) + 30(2) + 4(2) *Use the distributive property.*
= 10,000 + 1,500 + 200 + 400 + 60 + 8 *Multiply.*
= 12,168 *Add.*

Strategy 2

Multiply without expanding the numbers.

	Start from the right:
$\overset{1\,2}{234}$	*2 · 4 = 8. Write the 8.*
× 52	*2 · 3 = 6. Write the 6.*
468	*2 · 2 = 4. Write the 4.*
+ 11700	*Place a 0 in the ones place and multiply by 5:*
12168	*5 · 4 = 20. Write the 0 carry the 2.*
	5 · 3 = 15. Add the carried 2 to get 17. Write the 7; carry the 1.
	5 · 2 = 10. Add the carried 1 to get 11. Write 11.
	Add to get the final answer: (234)(52) = 12,168.

This second strategy was developed so that multiplication could be done quickly and accurately. Although we now have calculators and computers to do routine calculations, there are still times when we need to be able to calculate the exact product of large numbers by hand. Exact products can be checked quickly with an estimate. This is especially true when a calculator is used. Estimating catches errors made when entering numbers into a calculator.

Example 18 **The exact product of 234 and 52 is 12,168. Check this product with an estimate.**

234 rounds to 200. 52 rounds to 50.

200 · 50 = 10,000.

Is 10,000 close enough to our exact answer of 12,168? This is a matter of judgment; there are no set rules. However, if the exact and estimated numbers have the same highest place value (the ten thousands place), the exact answer is usually acceptable.

Example 19 **A math department is setting up a new computer lab with 32 machines. Each computer sells for $1,285. What is the total cost of the computers?**

I know . . .

c = Total cost of computers

$1,285 = Cost of each computer

32 = Number of computers

Equations

Total cost of computers = Cost of each computer · Number of computers

c = ($1,285)(32)

Solve

```
   2 1
   1 1
  $1285
×    32
   2570
+ 38550
 $41120    (1,285)(32) = $41,120
```

Check

Estimate:

c = $1,285 · 32

≈ $1,000 · 30

= $30,000

This is close enough to the exact answer of $41,120. The answer is reasonable.

Answer

The total cost of the computers is $41,120.

CHECK YOUR UNDERSTANDING

Write each number in expanded form and multiply using the distributive property.

1. $3 \cdot 14$ **2.** $70 \cdot 52$ **3.** $20 \cdot 170$ **4.** $9 \cdot 742$

Multiply without expanding the numbers. Check each product with an estimate.

5. $3 \cdot 14$ **6.** $70 \cdot 52$ **7.** $20 \cdot 170$ **8.** $9 \cdot 742$

Answers: 1. $3(10 + 4) = 3 \cdot 10 + 3 \cdot 4 = 42$; 2. $70(50 + 2) = 70 \cdot 50 + 70 \cdot 2 = 3{,}640$; 3. $(20)(100 + 70) = 20 \cdot 100 + 20 \cdot 70 = 3{,}400$; 4. $9(700 + 40 + 2) = 6{,}678$; 5. 42 (estimate is 30); 6. 3,640 (estimate is 3,500); 7. 3,400 (estimate is 4,000); 8. 6,678 (estimate is 6,300).

DISCUSSION

Working With Measurements

A single inch is an example of a **unit measurement.** When we measure the length of a steel bar with a ruler that is measured in inches, we are counting the number of one-inch unit measurements contained in its length.

A steel bar that is 6 inches long contains 6 unit measurements of length 1 inch each.

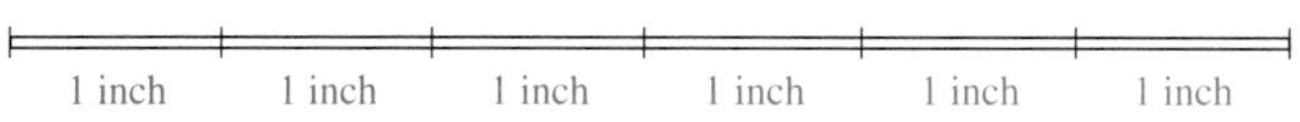

The total length of the steel bar can be found by *adding* the unit measurements together.

6 inches

Total length = 1 inch + 1 inch + 1 inch + 1 inch + 1 inch + 1 inch = 6 inches

The total length can also be found by *multiplying* the unit length by 6: 6(1 inch) = 6 inches. The unit of measurement is not changed.

PROCEDURE

To multiply a measurement by a whole number:

1. Multiply the whole numbers.
2. Write the same unit of measurement.

Example 20 **Write the length of an airport runway that is 2,520 feet long as the product of a whole number and a unit length.**

2,520 feet = (2,520)(1 foot)

Example 21 **A video takes 95 minutes to view. Write this viewing time as a product of a whole number and a unit time.**

95 minutes = (95)(1 minute)

Example 22 **A bottle contains 750 milliliters of wine. How much wine is contained in four bottles?**

Total wine = Number of bottles · Amount of wine in each bottle
= 4(750 milliliters)
= 3,000 milliliters

CHECK YOUR UNDERSTANDING

1. Alejandro takes 3 milliliters of liquid medicine each day. How much will he take in 12 days?

2. Lana's dogs eat 7 cups of dry dog food each day. How much will they eat in a week?

Answers: 1. Alejandro will take 36 milliliters of medicine; 2. Lana's dogs eat 49 cups of dog food in a week.

In the English system, a measurement can be a combination of units: "6 feet 3 inches" is called a **mixed measurement.** This measurement has an **implied plus sign:** 6 feet 3 inches = 6 feet + 3 inches. We use the distributive property to multiply a mixed measurement by a whole number.

Example 23 **Multiply 5 feet 4 inches by 2.**

2(5 feet 4 inches)	
= 2 (5 feet + 4 inches)	*Put in the implied plus sign.*
= 2 · 5 feet + 2 · 4 inches	*Use the distributive property.*
= 10 feet + 8 inches	*Multiply.*
= 10 feet 8 inches	*Eliminate the plus sign; addition is again implied.*

We do not have to rewrite the expression with the plus sign. We can simply distribute, mentally recognizing that there is an implied plus sign between measures. Carrying may also be required.

Example 24 **Find the product of 3 and 2 hours 42 minutes 21 seconds.**

3(2 hr 42 min 21 sec)	
= 6 hr 126 min 63 sec	*Use the distributive property.*
= 6 hr 126 min (60 + 3) sec	*Rewrite 63 sec as (60 + 3) sec.*
= 6 hr (126 + 1) min 3 sec	*Rewrite 60 sec as 1 min; carry.*
= 6 hr 127 min 3 sec	
= 6 hr (60 + 60 + 7) min 3 sec	*Rewrite 127 min as (60 + 60 + 7) min.*
= (6 + 1 + 1) hr 7 min 3 sec	*Rewrite 60 min as 1 hr; carry.*
= 8 hr 7 min 3 sec	

Example 25 **A solar cell module has a length of 1 foot 2 inches. If four modules are to be placed end-to-end on the roof of a snack shop at a ski hill to generate electricity, what is the total length of the modules?**

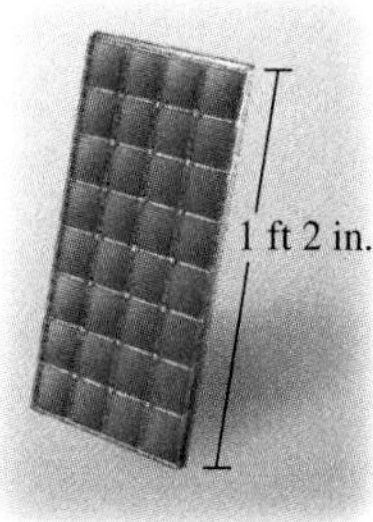

I know . . .
L = Length of modules
1 foot 2 inches = Length of one module
4 = Number of modules

Equations

Total Length = Number of modules · Length of one module

$L = 4(1 \text{ foot } 2 \text{ inches})$

Solve

$L = 4(1 \text{ foot } 2 \text{ inches})$

$= (4 \cdot 1 \text{ foot}) + (4 \cdot 2 \text{ inches})$

$= 4 \text{ feet } 8 \text{ inches}$

Check

We can estimate the length by rounding the length of each module to 1 foot. The estimated total length is then 4(1 foot) = 4 feet, which is close to the exact answer. The answer seems reasonable.

Answer

The total length of the four modules is 4 feet 8 inches.

CHECK YOUR UNDERSTANDING

1. What is the combined length of five pipes that are each 6 feet 2 inches long?

2. A birthday cake weighs 2 pounds 3 ounces. What is the weight of 5 such cakes?

3. A bucket holds 2 gallons 1 quart of water. Three buckets of water are poured on a newly planted tree. How much total water was poured on the tree?

Answers: 1. The combined length of the pipes is 30 feet 10 inches; 2. The weight of 5 birthday cakes is 10 pounds 15 ounces; 3. A total of 6 gallons 3 quarts of water was poured on the tree.

DISCUSSION

Conversion of Measurements

When working with measurements, we may need to convert a measurement to an equivalent measurement in a different unit. For example, a measurement in quarts may need to be converted into an equivalent measurement in pints. We multiply the measurement in quarts by the number of pints in a quart. The number of pints in a quart is an example of a **conversion factor.** Conversion factors can be looked up in references such as dictionaries, encyclopedias, and textbooks. A table of conversion factors for the English measurement system is found in Section 1.3.

PROCEDURE

To convert a measurement in a larger unit to an equivalent measurement in smaller units:

1. Write the measurement as a multiple of a unit measurement.
2. Replace the unit measurement with the conversion factor and multiply.

Example 26 **How many pints are equivalent to 8 quarts?**

Conversion factor: *1 quart = 2 pints*

8 quarts
= 8(1 quart) *Write 8 quarts as a multiple of the unit measurement 1 quart.*
= 8(2 pints) *Replace 1 quart with 2 pints.*
= 16 pints *Multiply.*

Example 27 **How many inches are there in 5 feet?**

Conversion factor: *1 foot = 12 inches*

5 feet
= 5(1 foot) *Write 5 feet as a multiple of the unit measurement 1 foot.*
= 5(12 inches) *Replace 1 foot with 12 inches.*
= 60 inches *Multiply.*

Example 28 **U.S. Forest Service hiking trails are usually measured in miles. The trail to the top of Trapper Peak in the Bitterroot Mountains of Montana is about 6 miles long. Cody needs to know the length of the trail in feet so that she can compare the length to the gain in altitude. How long is this hiking trail in feet? (5,280 feet = 1 mile)**

I know . . .
L = Length of trail in feet
Conversion factor: 1 mile = 5,280 feet
Number of miles on trail = 6

Courtesy of USDA Forest Service

Equations
Length = Number of miles on trail · Conversion factor
$L = 6 \cdot 5{,}280$ feet

Solve
$L = 6 \cdot 5{,}280$ feet
$= 31{,}680$ feet

Check
Estimate by rounding:
$L = (6)(5{,}280) \approx (6)(5{,}000) = 30{,}000$, which is close to our exact answer; this is reasonable.

Answer
The length of the trail up Trapper Peak is 31,680 feet.

CHECK YOUR UNDERSTANDING

1. In the United Kingdom, weight is measured in stones. One stone is equivalent to 14 pounds. If Ben weighs 14 stones, what is his weight in pounds?

2. Bill came up from Texas to fly-fish on the Deschutes River in Oregon. He caught a 3-pound trout. How many ounces did the trout weigh? (1 pound = 16 ounces)

3. A rainbow flag displayed in a parade was 4 yards long. What was its length in inches? (1 yard = 36 inches)

Answers: 1. Ben weighs 196 pounds; 2. The trout weighed 48 ounces; 3. The length of the flag was 144 inches.

EXERCISES SECTION 1.5

Rewrite the addition as multiplication. Find the product.

1. $3 + 3 + 3 + 3$

2. $5 + 5$

3. State the commutative property of multiplication in words.

4. State the associative property of multiplication in words.

5. Define *axiom.*

6. Identify the axiom demonstrated in this example: $5 + 4 = 4 + 5$

7. Identify the axiom demonstrated in this example: $6 \cdot (3 \cdot 4) = (6 \cdot 3) \cdot 4$

Evaluate.

8. $(3)(8)$

9. 8×3

10. $3(1 + 9)$

11. $1(5 + 5)$

12. $5(1 + 1)$

13. $8(0 + 2)$

Evaluate using the distributive property.

14. $4(8 + 3)$

15. $2(3 + 7)$

16. $8(8 + 3)$

17. $12(1 + 4)$

18. $6(3 + 3)$

19. $3(2 + 4)$

Evaluate.

20. $30 \cdot 600$

21. $500(10)$

22. $8{,}000 \cdot 20$

23. $70 \cdot 40$

24. $(100)(80)$

25. $(60)900$

26. $40(200{,}000)$

27. $100 \cdot 1{,}000$

28. $(100)(100)$

Solve. Use the five-step problem-solving strategy.

29. The monthly special at a rapid oil change business is $30 for an oil change and car wash. Calculate the revenue from 4,000 specials.

30. A local charity raises funds by capturing businesspeople and bringing them to "jail" at a shopping mall. The prisoners are not allowed to leave until someone bails them out with a donation to charity of $50. If 80 businesspeople agree to be captured, how much money could the charity raise from this fundraiser?

Estimate the product.

31. $55 \cdot 410$

32. $64 \cdot 53$

33. $48 \cdot 930$

34. $789 \cdot 4{,}918$

35. $2 \cdot 55$

36. $1{,}001 \cdot 4{,}482$

37. $53 \cdot 2{,}971$

38. $33{,}801 \cdot 35{,}801$

39. $5{,}500 \cdot 25$

40. $383(2{,}199{,}888)$

41. How are the rules for rounding when estimating products and the rules for rounding when estimating sums different?

Solve. Use the five-step problem-solving strategy.

42. Carol purchased 20 railroad ties for landscaping her yard. Each railroad tie weighs about 275 pounds. Estimate the weight of the load if she puts the ties in the bed of her pickup truck.

43. A jug of two percent milk costs about $2. Patti buys 48 jugs of milk each week for her daycare business. Estimate the cost of buying milk for a year if there are 52 weeks in a year.

44. A zip disk can store 100 megabytes of data. A package contains six disks. If Joao buys four packages of zip disks, how many megabytes of data can be stored on these disks?

45. In 1999, directors of the San Diego County Water Authority increased the fee for hooking a new home up to the water system to $1,871. Estimate the fees collected from a new housing development with 380 homes.

Calculate the exact product.

46. (21)(52)

47. 4(83)

48. (33)(39)

49. (99)(13)

50. (378)(44)

51. (690)(79)

52. $232 \cdot 88$

53. 99(99)

54. $37 \cdot 4$

Solve. Use the problem-solving strategy.

55. CarCheck is a "blackbox" for cars that connects to the outlet for the engine's onboard diagnostic system. Values for speed, engine revolutions, engine loads, and throttle position are set in the box; the number of times the preset limits are exceeded is recorded. Each CarCheck costs $295. A landscaping company is planning to install them on all 25 of its pickup trucks that are driven by its employees. Calculate the cost of this purchase.

56. The health insurance costs for county employees are paid in full by the county government. Employees must pay extra for family coverage. If the cost for one month of single employee coverage is $167, what will be the annual cost of coverage?

57. A landfill will take a city's trash at a cost of $53 per ton. If the annual amount of trash to be disposed at the landfill is about 14,987 tons, what is the annual cost of disposal?

58. A yurt is a semipermanent circular tent, often constructed with a wooden floor and canvas walls. At Winchester State Park, a yurt can be rented for $30 a night for the first four occupants and a $6 reservation fee. Additional occupants are charged $4 each per night. Mitch, Bill, Ray, Wally, and Warren decide to rent a yurt for 7 days to do some trout fishing. What will be the total cost to rent the yurt?

59. The NFL football stadium in Baltimore, Maryland, has 1,074 toilets. Before the stadium opened, engineers conducted a "Super Flush" practice to simulate the toilets being flushed during halftime. If each toilet requires about 3 gallons to flush, what would be the total amount of water flushed if all the toilets were flushed simultaneously?

60. Referring to Problem 59, if each of the 1,074 toilets cost an average of $125 for labor and materials to install, calculate the total cost of the toilets.

61. Jerry is writing an essay on morality in American politics. His word processing software counts 696 words on the one page of text he has already written. If he writes a total of 13 similar pages, will he meet the required 10,000-word minimum for the essay?

Multiply.

62. 3(4 feet 2 inches)

63. 7(2 hours 5 minutes 6 seconds)

64. 5(3 pounds 1 ounce)

65. 3(4 yards 1 foot 3 inches)

Convert the measurements.

66. 35 feet = ? inches

67. 220 pounds = ? ounces

68. 14 gallons = ? quarts

69. 8 miles = ? feet

70. 2 hours = ? minutes

71. 4 yards = ? feet

72. The brain of an average man weighs 3 pounds. Calculate the total weight of the brains of an 18-man football team in ounces.

73. A cubic foot of wet, compact snow weighs about 20 pounds. How much would 10 cubic feet of snow weigh in ounces?

74. Surveyors often measure in *chains* and then convert that measurement to feet (1 chain = 66 feet). A boundary of a piece of land measures 48 chains. What is the length of the boundary in feet?

Courtesy of Penn State Surveying Program

ERROR ANALYSIS

75. **Problem:** Multiply 30(600 feet).

Answer: 1,800 feet

Describe the mistake made in solving this problem.

Redo the problem correctly.

76. **Problem:** Convert 10 days to seconds.

Answer: Conversion factors: 1 day = 12 hours
1 hour = 60 seconds

10 days
= 10(12 hours)
= 120 hours
= 120(60 seconds)
= 7,200 seconds

Describe the mistake made in solving this problem.

Redo the problem correctly.

REVIEW

77. Multiply 10(1,000).

78. Multiply (100)(1,000,000).

79. State the associative property of addition in words.

80. Use set notation to write the set of whole numbers.

Section 1.6 Exponents

LEARNING TOOLS

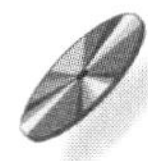  

CD-ROM SSM VIDEO

Sneak Preview

Repeated addition can be represented with multiplication. Repeated multiplication can be represented using exponents.

After completing this section, you should be able to:

1) Rewrite an expression with repeated multiplication using exponents.

2) Evaluate exponential expressions.

3) Multiply exponential numbers using the product rule of exponents.

4) Multiply whole numbers by powers of ten.

DISCUSSION

Introduction to Exponents and Exponential Notation

Multiplication represents repeated addition. Instead of writing $4 + 4 + 4$, we can write $4 \cdot 3$. These expressions are equivalent since $4 + 4 + 4 = 12$ and $4 \cdot 3 = 12$.

Repeated multiplication can be represented using exponential notation. Rather than writing $4 \cdot 4 \cdot 4$, we write 4^3. This is read "4 raised to the power 3," "4 to the 3rd," or "4 cubed" and is an example of an **exponential expression.** The small number written above and to the right is called the **exponent.** The other number is called the **base.** In this example, 4^3, the exponent is 3 and the base is 4. These expressions are equivalent: $4 \cdot 4 \cdot 4 = 64$ and $4^3 = 64$.

base $\longrightarrow 4^3 \nwarrow$ **exponent**

Writing exponential expressions as repeated multiplications of the base and the number 1 reveals a pattern:

$4^5 = 4 \cdot 4 \cdot 4 \cdot 4 \cdot 4 \cdot 1$ *Remember that multiplying by 1 does not change the product.*

$4^4 = 4 \cdot 4 \cdot 4 \cdot 4 \cdot 1$ *As the exponent decreases, the number of times the base is multiplied also decreases.*

$4^3 = 4 \cdot 4 \cdot 4 \cdot 1$

$4^2 = 4 \cdot 4 \cdot 1$

$4^1 = 4 \cdot 1$

$4^0 = ?$

If the pattern is to continue, then 4^0 must equal 1. By definition, any base (except for zero) raised to the power 0 is equal to 1.

Axioms are used to organize a number system. To make the system complete and logical, definitions are used. Sometimes a concept or situation is left undefined to avoid contradictions. For example, 0^0 is left undefined and not said to be equal to any particular number.

PROCEDURE

To evaluate an exponential expression:

1. If the exponent $= 1$, the value of the expression is equal to the base.
2. If the exponent $= 0$ and the base $\neq 0$, the value of the expression is 1.
3. If the exponent $= 0$ and the base $= 0$, the value of the expression is undefined.

   0^0 is undefined.
4. If the exponent is any natural number, the value of the expression is equal to the base multiplied by itself the exponent number of times.

Example 1 **Evaluate 2^3.**

$2^3 = 2 \cdot 2 \cdot 2$ *The base 2 is multiplied by itself three times.*
$= 8$

Example 2 **Evaluate 10^4.**

$10^4 = 10 \cdot 10 \cdot 10 \cdot 10$ *The base 10 is multiplied by itself four times.*
$= 10{,}000$

Example 3 **Evaluate 361^0.**

$361^0 = 1$ *By definition, any nonzero number raised to the zero power is 1.*

Example 4 **Evaluate 8^1.**

$8^1 = 8$ *A number with an exponent of 1 is equivalent to that number.*

Example 5 **Evaluate 0^0.**

0^0 is undefined.

Example 6 **Evaluate 5^2.**

$5^2 = 5 \cdot 5$ *The base is multiplied by itself two times.*
$= 25$

CHECK YOUR UNDERSTANDING

Evaluate each expression.

1. 5^2 **2.** 4^3 **3.** 34^1 **4.** 6^0 **5.** 3^4 **6.** 2^5

Answers: 1. 25; 2. 64; 3. 34; 4. 1; 5. 81; 6. 32.

DISCUSSION

Powers of Ten

An exponential number with a base of 10 is called a **power of ten.**

$10^0 = 1$	*Except for zero, any base raised to the 0 power = 1.*
$10^1 = 10$	*Any base raised to the power 1 equals the base.*
$10^2 = 100$	*$10 \cdot 10 = 100$*
$10^3 = 1{,}000$	*$10 \cdot 10 \cdot 10 = 1{,}000$*
$10^4 = 10{,}000$	*$10 \cdot 10 \cdot 10 \cdot 10 = 10{,}000$*
. . .	*Ellipsis shows that this pattern continues.*
$10^n = 1$ followed by n zeros	*n is a natural number.*

This pattern shows that any number that is a 1 followed by zeros can be rewritten as an exponential expression with a base of 10 and an exponent equal to the number of zeros in the number. So, 1,000,000,000 can be rewritten as 10^9 since there are nine zeros in 1,000,000,000. We can multiply a whole number by a power of ten in the same way.

Example 7 **Rewrite 700,000 as the product of a whole number and a power of ten.**

$700{,}000$
$= 7 \cdot 100{,}000$
$= 7 \cdot 10^5$

Example 8 **Write 3,900 as the product of a whole number and a power of ten.**

$3{,}900$
$= 39 \cdot 100$
$= 39 \cdot 10^2$

The symbol for multiplication used in powers of ten is usually not a dot, ·, but a ×. For example, in September 1995, the U.S. Department of the Treasury reported that the national debt was nearly \$5,000,000,000,000. This amount could be rewritten using powers of ten as \$$5 \times 10^{12}$.

Example 9 **Write 21×10^7 in place value form.**

21×10^7
$= 210{,}000{,}000$

CHECK YOUR UNDERSTANDING

Rewrite the measurement as the product of a whole number and a power of ten.

1. 40,000 miles **2.** 60,000,000 feet **3.** 850,000 watts

Rewrite the measurement in place value form, without using exponents.

4. 4×10^8 calories **5.** 3×10^5 degrees Celsius **6.** 55×10^3 feet

7. An empty Boeing 747 jetliner weighs about 16×10^4 kilograms.

Answers: 1. 4×10^4 miles; 2. 6×10^7 feet; 3. 85×10^4 watts; 4. 400,000,000 calories; 5. 300,000 degrees Celsius; 6. 55,000 feet; 7. 160,000 kilograms

DISCUSSION

Simplifying Exponential Expressions

When exponential expressions have the same base, their product can be simplified by keeping the common base and adding the exponents.

Example 10 **Simplify $4^3 \cdot 4^2$. Write in exponential notation.**

We can find the product by multiplying:

$4^3 \cdot 4^2$
$= (4 \cdot 4 \cdot 4)(4 \cdot 4)$
$= 4^5$

Or, we can find the product by adding the exponents:

$4^3 \cdot 4^2$
$= 4^{(3+2)}$
$= 4^5$

Example 11 **Simplify $3^4 \cdot 3^2 \cdot 3^1$. Write in exponential notation.**

We can find the product by multiplying:

$3^4 \cdot 3^2 \cdot 3^1$
$= (3 \cdot 3 \cdot 3 \cdot 3)(3 \cdot 3)(3)$
$= 3^7$

Or, by adding the exponents:

$3^4 \cdot 3^2 \cdot 3^1$
$= 3^{(4+2+1)}$
$= 3^7$

Product Rule of Exponents

In words: To multiply exponential expressions with the same base, add the exponents and keep the common base.

In symbols: $a^m \cdot a^n = a^{m+n}$

The product rule of exponents applies only to exponential expressions that have the same base.

Example 12 **Evaluate $2^4 \cdot 3^7$.**

$2^4 = 2 \cdot 2 \cdot 2 \cdot 2 = 16$
$3^7 = 3 \cdot 3 \cdot 3 \cdot 3 \cdot 3 \cdot 3 \cdot 3 = 2{,}187$
$(16)(2{,}187) = 34{,}992$

We see that $2^4 \cdot 3^7 = 34{,}992$. We cannot say that $2^4 \cdot 3^7 = 6^{11}$ since $6^{11} = 362{,}797{,}056$. The only way to evaluate the product of two numbers with different bases is to actually do the multiplications.

CHECK YOUR UNDERSTANDING

Use the product rule of exponents to simplify. Write the answer in exponential form.

1. $7^5 \cdot 7^9$ **2.** $17^8 \cdot 17$ **3.** $5^3 \cdot 5^4 \cdot 5^5$

4. $2^3 \cdot 2^3 \cdot 2^3$ **5.** $2^0 \cdot 2^7$ **6.** $3^7 \cdot 3^4 \cdot 3^0$

Answers: 1. 7^{14}; 2. 17^9; 3. 5^{12}; 4. 2^9; 5. 2^7; 6. 3^{11}.

EXERCISES SECTION 1.6

Rewrite the multiplication using exponential notation.

1. $3 \cdot 3 \cdot 3 \cdot 3$

2. $0 \times 0 \times 0 \times 0 \times 0$

3. $(5)(5)$

4. $10 \cdot 10 \cdot 10 \cdot 10 \cdot 10 \cdot 10 \cdot 10$

5. $1 \cdot 1 \cdot 1 \cdot 1 \cdot 1 \cdot 1 \cdot 1 \cdot 1 \cdot 1 \cdot 1 \cdot 1$

Evaluate. Write in place value form.

6. 2^3

7. 1^{12}

8. 4^2

9. 10^3

10. 5^1

11. 6^0

12. 0^6

13. 5^2

14. 342^0

15. 8^3

16. 3^4

17. 0^0

18. The product rule of exponents can be used to find the product $2^3 \cdot 2^5$. The product rule of exponents cannot be used to find the product $2^3 \cdot 4^5$. Explain why this is true.

Simplify. Write in exponential form.

19. $2^3 \cdot 2^1$

20. $3^2 \cdot 3^2$

21. $8^1 \cdot 8^2$

22. $(10^{23})(10^2)$

23. $(10^3)(10^8)$

24. $(10)(10^7)$

25. $5^2 \cdot 5^3 \cdot 5^5$

26. $7 \cdot 7^0 \cdot 7^6$

27. $0^7 \cdot 0^8 \cdot 0^6$

Evaluate. Write in place value form.

28. $(3^0)(5^0)(2^0)$

29. $(5^1)(3^1)(2^2)$

30. $4^2 \cdot 0^5$

31. $(2^4)(1^3)$

32. $(4^2)(2^3)$

33. 3×10^4

34. $(67)(10^3)$

35. 210×10^2

36. $(40)(10^5)$

Write the measurement in place value form. Include the units in the measurement.

37. The thrust of the *Saturn V* rocket engines is about 74×10^5 pounds.

38. A fiber optic wire carries about 17×10^8 bits per second.

39. The moon has a mass of about 7×10^{22} kilograms.

40. An elephant has a mass of about 2×10^4 pounds.

41. There are about 3×10^9 possible ways the first four moves can be played in chess.

42. There are 6×10^6 smell cells in our noses.

43. In four months, a pair of flies could theoretically produce 191×10^{19} descendants.

44. There are 602×10^{21} atoms in a mole of carbon, which weighs about 13 grams.

45. The average number of quills on a porcupine is about 25×10^3.

46. An average-sized woman wearing stiletto heels applies a pressure to the floor of 11×10^2 pounds per square inch.

47. The speed of light is 3×10^8 meters per second.

48. The speed of light is 3×10^{10} centimeters per second.

49. The speed of light is 186×10^3 miles per second.

ERROR ANALYSIS

50. **Problem:** Evaluate $3^2 \cdot 5^4$. Write in exponential form.

Answer:

$3^2 \cdot 5^4$
$= (3 \cdot 5)^{2+4}$
$= 15^6$

Describe the mistake made in solving this problem.

Redo the problem correctly.

51. **Problem:** Evaluate $(4^2)(4^3)$. Write in exponential form.

Answer:

$$(4^2)(4^3)$$
$$= 4^{2 \cdot 3}$$
$$= 4^6$$

Describe the mistake made in solving this problem.

Redo the problem correctly.

REVIEW

52. What is a *conversion factor?* Give an example.

53. Change 4 feet into inches.

54. Change 50 pounds into ounces.

55. Change 3 gallons into quarts.

Section 1.7 Using the SI (Metric) System of Measurement

LEARNING TOOLS

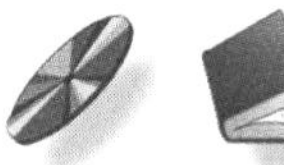

CD-ROM SSM VIDEO

Sneak Preview

The SI (metric) system of measurement is the primary system of measurement used in every industrial country in the world except for the United States.

After completing this section, you should be able to:

1) Convert a metric measurement into an equivalent measurement in different units.

2) Calculate the area of a rectangle.

3) Calculate the volume of a cuboid.

DISCUSSION

Introduction to the SI (Metric) System of Measurement

When the United States organized its government, it developed its own monetary system based on powers of ten. However, the English system of measurement that was based on tradition was still used. The rest of the industrial world now uses the SI (Système International d'Unités) system of measurement. This is the modern (1960) version of the metric system and is based on powers of ten. The original purpose of the SI system was to standardize measurements between countries and make conversions easier.

The SI system has seven basic units: **meter** (length), **kilogram** (mass), **Kelvin** (temperature), **ampere** (electric current), **mole** (quantity), **candela** (luminosity/brightness), and **second** (time). (Except for scientific work, most temperatures worldwide are still measured in the Celsius temperature scale. Celsius temperatures are exactly 273 degrees less than Kelvin temperatures.)

All other measurement units can be built from the seven basic units. For example, liquids are often measured in liters. One thousand liters is the amount of liquid that is contained in a box that is 1 meter high, 1 meter wide, and 1 meter long. We say that the unit of liter is derived from the unit for length, meter.

Units that are smaller or larger than the basic units can be built by adding a **prefix** in front of the unit. Each prefix multiplies the size of the basic unit by a power of ten. For example, the prefix *kilo* means 1,000 or 10^3. So, 1 *kilo*meter is equal to 1,000 meters. Prefixes that make the basic units smaller include *deci, centi, milli, micro,* and *nano. Nano*meters are excellent units for measuring very small distances such as the wavelengths of visible light. There are 10^9 nanometers in a meter.

METRIC PREFIXES FOR UNITS LARGER THAN THE BASIC UNIT

1,000,000,000	10^9	giga
1,000,000	10^6	mega
1,000	10^3	kilo
100	10^2	hecto
10	10^1	deka

Writing a prefix in front of a unit multiplies the unit by the given number. For example, a kilometer is equal to 1,000 meters.

Changing a measurement into an equivalent measurement with different units can be done by multiplying by a conversion factor. Selected conversion factors are listed in the following table. An abbreviation for each unit is also included.

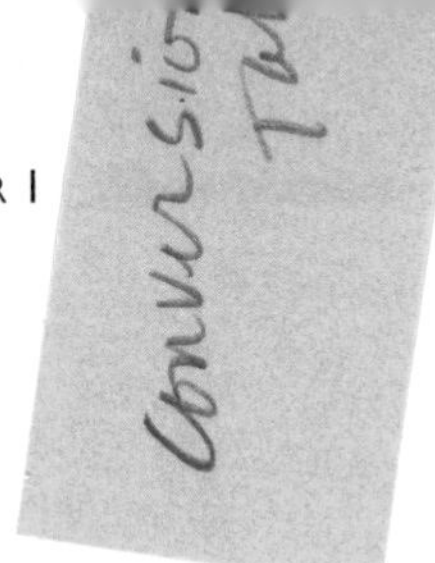

CONVERSION FACTORS

LENGTH

1 centimeter (cm) = 10 millimeters (mm)
1 meter (m) = 1,000 millimeters (mm)
1 meter (m) = 100 centimeters (cm)
1 meter (m) = 10 decimeters (dm)
1 meter (m) = 1,000,000 micrometers (μm)
1 meter (m) = 1,000,000,000 nanometers (nm)
1 kilometer (km) = 1,000 meters (m)

MASS

1 gram (g) = 1,000 milligrams (mg)
1 gram (g) = 100 centigrams (cg)
1 kilogram (kg) = 1,000 grams (g)
1 metric tonne (t) = 1,000 kilograms (kg)

PROCEDURE

To convert from a measurement in larger units to an equivalent measurement in smaller units:

Replace the measurement unit with the equivalent conversion factor and multiply.

Example 1 **Change a mass of 5 kilograms into grams.**

5 kg	
= 5(1 kg)	*Rewrite as a product of a number and a unit.*
= 5(1,000 g)	*Replace 1 kg with 1,000 g.*
= 5,000 g	*Multiply.*

Example 2 **Change a length of 70 meters into centimeters.**

70 m	
= 70(1 m)	*Rewrite as a product of a number and a unit.*
= 70(100 cm)	*Replace 1 m with 100 cm.*
= 7,000 cm	*Multiply.*

If we do not have a needed conversion factor, we can do the conversion in a series of steps with conversion factors that we know.

Example 3 **Change a mass of 4 kilograms into milligrams.**

We do not know a conversion factor for kilograms and milligrams. However, we do know these conversion factors: 1 kilogram = 1,000 grams and 1 gram = 1,000 milligrams.

4 kg	
= 4(1,000 g)	*Replace kg with 1,000 g.*
= 4,000 g	*Multiply.*
= 4,000(1,000 mg)	*Replace g with 1,000 mg.*
= 4,000,000 mg	*Multiply.*

CHECK YOUR UNDERSTANDING

Change each measurement into the given unit of measurement.

1. 6 kilograms; grams

2. 6 kilometers; millimeters

3. 2 kilograms; centigrams

4. 10,000 grams; milligrams

5. 6 meters; centimeters

6. 2 meters; nanometers

Answers: 1. 6,000 grams; 2. 6,000,000 millimeters; 3. 200,000 centigrams; 4. 10,000,000 milligrams; 5. 600 centimeters; 6. 2,000,000,000 nanometers.

Example 4 **A computer processing chip runs at a speed of 400 megahertz (MHz). A faster chip operates at a speed of 1 gigahertz (GHz). If there are 1,000 megahertz in each gigahertz, what is the difference in speeds of the two chips?**

I know . . .

d = Difference in speed

400 MHz = Speed of slow chip

1 GHz = Speed of fast chip

1 GHz = 1,000 MHz (conversion factor)

Equations

Difference in speed = Speed of fast chip − Speed of slow chip

$$d = 1\text{ GHz} - 400\text{ MHz}$$

Solve

$$\begin{aligned} d &= 1\text{ GHz} - 400\text{ MHz} \\ &= 1(1{,}000\text{ MHz}) - 400\text{ MHz} \\ &= 1{,}000\text{ MHz} - 400\text{ MHz} \\ &= 600\text{ MHz} \end{aligned}$$

We cannot directly subtract unlike measures so convert GHz to MHz: Replace GHz with the equivalent conversion factor 1,000 MHz and multiply.

Check

The difference should be less than 1 GHz because that is the speed of the fastest chip. A GHz is 1,000 MHz and our difference is 600 MHz, so the answer seems reasonable.

Answer

The difference in speed of the two chips is 600 megahertz.

CHECK YOUR UNDERSTANDING

Add or subtract. Conversion of measurements may be necessary.

1. 6 grams − 4 grams

2. 6 g − 4 mg

3. 15 km − 9,452 m

4. 2 m − 15 cm

5. 3 tonnes − 800 kg

Answers: 1. 2 grams; 2. 5,996 mg; 3. 5,548 m; 4. 185 cm; 5. 2,200 kg.

INVESTIGATION Area and Volume

Area is a measure of surface. In other words, area tells the amount of surface contained in a two-dimensional (flat) shape. **Figure** is another word used by mathematicians to describe a shape.

A **square** is a two-dimensional figure that has four equal sides and four equal angles. Each side of the square below is 1 centimeter long. The measure of each angle in a square is 90°. The **area** covered by this square is 1 **square centimeter** (abbreviated 1 cm^2).

1 cm

A **rectangle** is a four-sided figure whose angles all measure 90°. The opposite sides of the rectangle are equal in length. (A square is just a rectangle where all the sides are equal.)

Each of the squares in Figure 1.1 has a length of 1 cm and a width of 1 cm. The area of each individual square is 1 square centimeter.

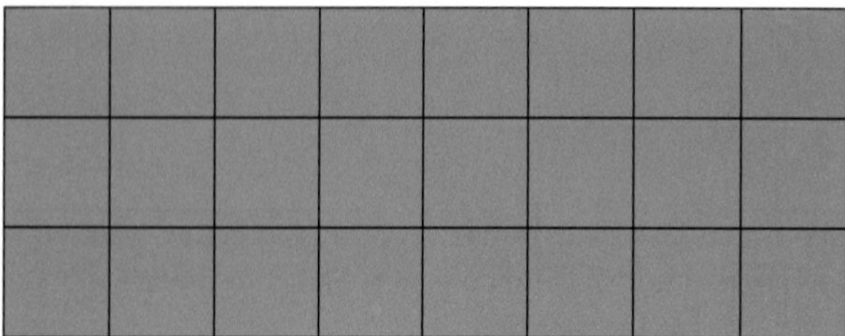

Figure 1.1

1. How many squares with an area of 1 square centimeter are found in the rectangle in Figure 1.1?

2. The area of a figure is equal to the number of square units required to cover its surface. What is the area of the rectangle in Figure 1.1?

Counting individual squares is not a practical way to determine the area of a rectangle. Instead, we multiply the number of squares that make up the width by the number of squares that make up the length. This is an alternate way to find the total number of square centimeters contained in a rectangle.

3. Calculate the area of the rectangle in Figure 1.1 by multiplying its length in centimeters by its width in centimeters.

Calculate the area of each rectangle with the given length and width.

4. Length = 3 centimeters Width = 12 centimeters

5. Length = 10 yards Width = 15 yards

6. Length = 10,000 miles Width = 500 miles

7. Length = 3 feet Width = 4 inches

Answers: 4. 36 square cm or 36 cm^2; 5. 150 square yards or 150 yd^2; 6. 5,000,000 square miles or 5,000,000 miles^2; 7. 144 square inches or 144 in.^{2}.

Area is a measure of the amount of space on a flat, two-dimensional surface. **Volume,** in contrast, is a measure of the amount of space in a three-dimensional object.

A **cube** is a three-dimensional figure that has equal height, length, and width. The space occupied by the cube is its **volume.** The volume of a cube that has a height, length, and width of 1 centimeter is 1 cubic centimeter or 1 cm^3. This measurement can also be abbreviated as cc. A unit equal to a cubic centimeter is the milliliter (ml). For example, a standard can of Pepsi contains 355 milliliters of liquid. The manufacturers could label the can as 355 cm^3 of liquid.

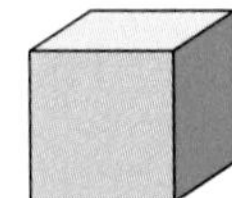

Three-dimensional rectangles are called **cuboids** or **rectangular solids.** The number of cubes in a cuboid can be found without directly counting each individual cube. Instead, the length is multiplied by the width and this product is multiplied by the height.

8. The height of this cuboid is 2 cm. The length is 4 cm. The width is 1 cm. How many 1-cm cubes are contained in this cuboid?

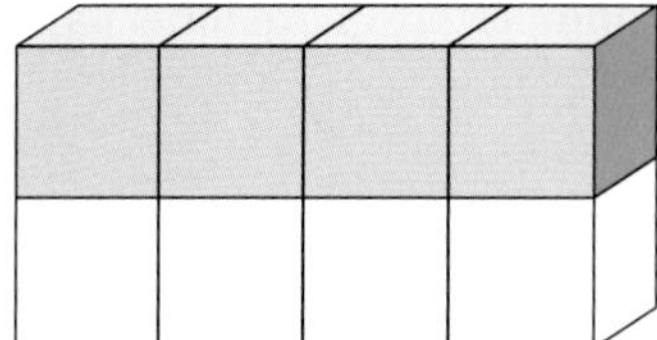

Since the product of (length)(width)(height) tells us how many cubes are in a cuboid, it can be used to calculate the volume of a cuboid. Usually, it is written as a formula: $V = lwh$. Each distance must be in the same unit of measurement.

Calculate the volume of the cuboids with the given measurements.

LENGTH	WIDTH	HEIGHT	VOLUME
9. 3 cm	12 cm	4 cm	
10. 10 yards	15 yards	8 yards	
11. 2 feet	3 inches	5 inches	
12. 10,000 miles	500 miles	75,000 miles	

Answers: 9. 144 cm^3; 10. 1,200 yards^3; 11. 360 inches^3; 12. 375,000,000,000 miles^3.

DISCUSSION

Area and Volume

Area is a measure of the amount of space on a two-dimensional, flat surface and is measured in square units, written *unit*^{2}.

This square has an area of 1 cm^2 (read "one square centimeter"). It is an example of a **unit square,** a square whose sides each have a measure of 1 unit. (Read "one square centimeter.")

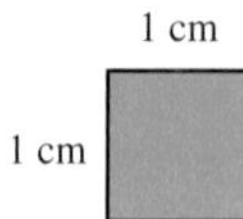

This rectangle is 3 cm long and 2 cm wide. It is made up of 6 squares that are each 1 cm long and 1 cm wide and have an area of 1 cm^2.

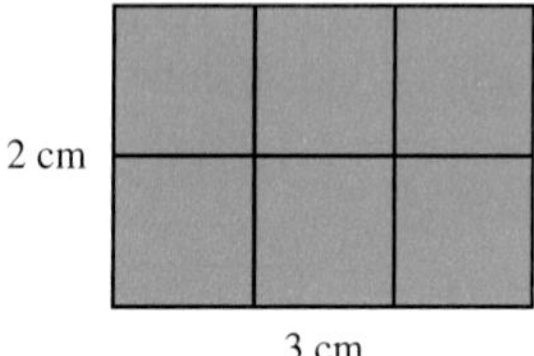

To find the area of this rectangle we can add the area of all the unit squares.

Area of rectangle = Sum of area of unit squares
$= 1\text{ cm}^2 + 1\text{ cm}^2 + 1\text{ cm}^2 + 1\text{ cm}^2 + 1\text{ cm}^2 + 1\text{ cm}^2$
$= 6\text{ cm}^2$ *Combine like terms.*

The area of the rectangle is also equal to the product of the length and the width of the rectangle.

Area of rectangle = Length of rectangle · Width of rectangle
$= 3\text{ cm} \cdot 2\text{ cm}$
$= (3 \cdot 2)(\text{cm} \cdot \text{cm})$ *The Associative Property allows regrouping of factors.*
$= 6\text{ cm}^2$ *This area is equal to the sum of the areas of the individual squares.*

Multiplying the length by the width is a more efficient method than counting unit squares.

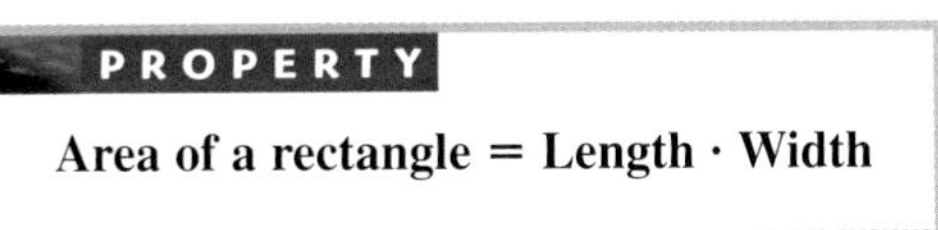

Example 5 **Calculate the area of a rectangle with length 8 meters and width 3 meters.**

Area = Length · Width
= 8 meters · 3 meters
= 24 meters^2 or 24 m^2

In order to calculate a useful area of a rectangle, the length and width must be in the same units. It may be necessary to convert one of the measurements before calculating the area.

Example 6 **A groundskeeper is ordering fertilizer for the turf on a football field that is 120 yards by 160 feet. Because the turf is mowed frequently and receives heavy use, it has special fertilizer requirements. He needs to know the area of the field to order the correct amount of fertilizer.**

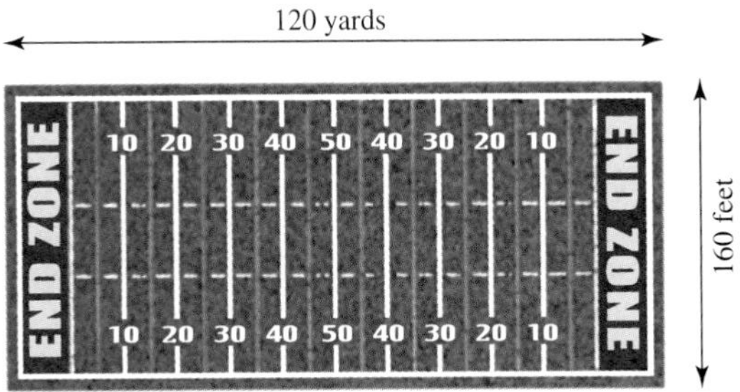

I know . . .
A = Area of field
120 yards = Length of field
160 feet = Width of field
1 yard = 3 feet (conversion factor)

Equations
Area of field = Length of field · Width of field
A = 120 yards · 160 feet

Solve
A = 120 yard · 160 feet *The length and width must be in the same unit.*
= 120(3 feet) · 160 feet *Convert yards to feet by replacing "yards" with the conversion factor "3 feet."*
= 360 feet · 160 feet
= 57,600 ft^2

Check

Estimate the product of 360 feet and 160 feet. 360 ft rounds to 400 ft. 160 ft rounds to 200 ft. (400 feet)(200 feet) = 80,000 ft^2. This is close to the exact answer of 57,600 ft^2. The answer is reasonable.

Answer

The area of the football field is 57,600 square feet.

CHECK YOUR UNDERSTANDING

1. Find the area of a rectangle with length 4 m and width 2 m.

2. Find the area of a square with length 21 cm.

3. Calculate the area of a rectangle with a length of 16 cm and a width of 2 m.

Answers: 1. 8 m^2; 2. 441 cm^2; 3. 3,200 cm^2.

Volume is a measure of the amount of space inside a three-dimensional object and is measured in cubic units, written *unit*^{3}.

This cube has a volume of 1 cm^3 (read "one cubic centimeter"). It is an example of a **unit cube:** a cube whose sides each have a measure of 1 unit.

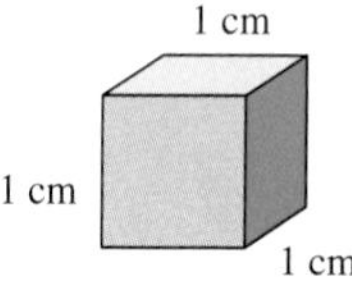

If we place six cubes with length 1 centimeter next to each other in two rows of three, we form a box. The box is 3 centimeters long, 2 centimeters wide, and 1 centimeter high. Such a box is an example of a **cuboid** or **rectangular solid.** (A cube is a cuboid with equal length, width, and height.)

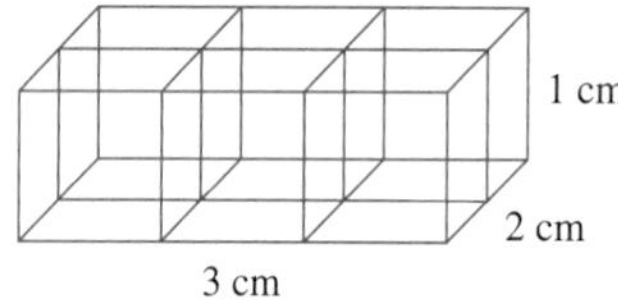

To find the volume of this cuboid, we can add the volume of all the unit cubes.

$$
\begin{aligned}
\text{Volume of cuboid} &= \text{Sum of volume of unit cubes} \\
&= 1\ \text{cm}^3 + 1\ \text{cm}^3 + 1\ \text{cm}^3 + 1\ \text{cm}^3 + 1\ \text{cm}^3 + 1\ \text{cm}^3 \\
&= 6\ \text{cm}^3 \quad \textit{Combine like terms.}
\end{aligned}
$$

The volume of the cuboid is also equal to the product of the length, width, and height of the cuboid.

$$
\begin{aligned}
\text{Volume of cuboid} &= \text{Length of cuboid} \cdot \text{Width of cuboid} \cdot \text{Height of cuboid} \\
&= 3\ \text{cm} \cdot 2\ \text{cm} \cdot 1\ \text{cm} \\
&= (3 \cdot 2 \cdot 1)(\text{cm} \cdot \text{cm} \cdot \text{cm}) \\
&= 6\ \text{cm}^3 \quad \textit{This volume is equal to the sum of the volume of the individual cubes.}
\end{aligned}
$$

PROPERTY

Volume of a cuboid = Length · Width · Height

In the SI system, volume is measured in cubic units such as cm^3 and m^3. It also is measured in units based on the liter: 1,000 liters is equivalent to 1 cubic meter; 1 milliliter is equivalent to 1 cubic centimeter (cm^3). Another abbreviation used for cubic centimeter is cc. The auto industry measures engine volume in either cc (a 350-cc motorcycle) or in liters (a 4-liter engine). Health care professionals may refer to liquid medicine volume in cubic centimeters using cc as the abbreviation. For example, an order for liquid Tylenol for a child might read "*5 cc every 4 hours.*" This is equivalent to "*5 ml every 4 hours.*" In the English system, liquid volume is measured in gallons, quarts, pints, cups, tablespoons, and teaspoons. Solid volume is measured in cubic units like cubic feet or cubic inches.

PROPERTY

Equivalent measures

$1 \text{ cm}^3 = 1 \text{ ml}$

$1 \text{ cm}^3 = 1 \text{ cc}$

$5 \text{ cm}^3 = 1 \text{ teaspoon}$

Example 7 **Find the volume of a cube with a side of 4 inches.**

$$\begin{aligned} \text{Volume} &= \text{Length} \cdot \text{Width} \cdot \text{Height} \\ &= 4 \text{ in.} \cdot 4 \text{ in.} \cdot 4 \text{ in.} \\ &= (4 \cdot 4 \cdot 4)(\text{in.} \cdot \text{in.} \cdot \text{in.}) \quad \textit{The associative property allows us to regroup.} \\ &= 64 \text{ in.}^3 \end{aligned}$$

Example 8 **Find the volume of a cuboid that is 2 feet by 18 inches by 14 inches.**

To calculate volume, all measurements must be in the same units. Convert 2 feet into inches using the conversion factor 1 foot = 12 inches.

$$\begin{aligned} & 2 \text{ feet} \\ &= 2(12 \text{ inches}) \quad \textit{Replace "feet" with "12 inches."} \\ &= 24 \text{ inches} \end{aligned}$$

$$\begin{aligned} \text{Volume} &= \text{Length} \cdot \text{Width} \cdot \text{Height} \\ &= 24 \text{ in.} \cdot 18 \text{ in.} \cdot 14 \text{ in.} \\ &= 6{,}048 \text{ in.}^3 \end{aligned}$$

CHECK YOUR UNDERSTANDING

1. Find the volume of a cuboid with length 6 inches, width 3 inches, and height 2 inches.

2. Find the volume of a cube with side length of 5 millimeters.

3. Find the volume of a rectangular briefcase that measures 24 inches by 12 inches by 5 inches.

Answers: 1. 36 inches^3; 2. 125 millimeters^3; 3. 1,440 inches^3.

EXERCISES SECTION 1.7

Convert the measurement.

1. 40 kilometers = ? meters

2. 40 kilograms = ? grams

3. 40 liters = ? milliliters

4. 325 meters = ? centimeters

5. 2 liters = ? milliliters

6. 160 grams = ? milligrams

7. 25 liters = ? milliliters

8. 325 meters = ? millimeters

9. The Mammoth-Flint Ridge cave system in Kentucky has more than 531 kilometers of passageways. Change this distance into meters.

10. The average distance of Uranus from the Sun is 2,870,990,000 kilometers. Change this distance into meters.

11. The length of Montrell's finger is 8 centimeters. Change this length into millimeters.

12. A mineral sample has a mass of 14 grams. Change this mass into milligrams.

13. Human nerve cells can transport mitochondria down a long narrow tube called the axon. The axon may be as long as 1 meter. The mitochondria can travel 36 micrometers in 1 minute. There are 1,000 nanometers in 1 micrometer. How many nanometers can the mitochondria travel in 1 minute?

14. The unmanned probe *Pioneer 10* was launched into space on March 3, 1972. From the liftoff date to April 25, 1983, it traveled about 4,580,000,000 kilometers into space. How many meters did it travel?

Calculate the area of the rectangle.

15. length = 4 cm; width = 32 cm

16. length = 10 m; width = 65 m

17. length = 2 inches; width = 6 inches

18. length = 4 yards; width = 2 ft

19. length = 2 meters; width = 85 cm

20. length = 1 mile; width = 455 yd

21. length = 62 inches; width = 37 inches

22. length = 55 yd; width = 135 ft

23. Some Native Americans in south central Ohio lived in enclosures that had embankments for defense. One such enclosure was located on the left bank of Paint Creek, a tributary of the Scioto River. The enclosure was square with each side being 1,008 feet in length. What was the area of this enclosure? *[Source: Brine, L. (1996).* The ancient earthworks and temples of the American Indians. *Hertforshire, England: Oracle Publishing Ltd. Originally published in 1894.]*

24. The rectangular screen on a Texas Instruments TI-92 calculator measures about 9 centimeters long by 5 centimeters wide. What is the area of this screen?

25. A rectangular skating rink measures 200 feet by 85 feet. What is the area of the ice on this rink?

26. The Pullman Aquatics Center includes a rectangular main pool that is 45 feet wide and 75 feet long and a rectangular warm water/activity pool that is 15 feet wide by 58 feet long. What is the total area of the pool in the aquatics center?

27. What is the area of glass in a rectangular window that is 18 inches wide and 2 feet high?

Calculate the volume of the cuboid (rectangular solid).

28. length = 3 cm; height = 2 cm; width = 20 cm

29. length = 8 inches; height = 2 feet; width = 14 inches

30. length = 4 feet; height = 2 feet; width = 18 inches

31. width = 22 cm; height = 79 cm; length = 85 cm

32. Carole is insulating her attic with blow-in insulation to a depth of 6 inches. If the attic floor is rectangular and measures 40 feet wide and 60 feet long, what volume of insulation in cubic inches is needed?

33. A phone book is about 21 cm wide, 27 cm long, and 3 cm thick. What is the volume of this phone book?

34. Wong is making Jell-O in a 9-inch by 13-inch rectangular cake pan that is 2 inches deep. What volume of Jell-O can he make if he fills the entire pan?

35. A storage unit is 38 feet long, 12 feet wide, and 10 feet high. What is the volume of this unit?

36. A septic tank that is adequate for a house with seven or fewer people has a cuboid shape. It has a width of 3 feet, a depth of 4 feet, and a length of 6 feet. What is the capacity of this septic tank in cubic feet?

37. What is the volume of a bedroom that is 12 feet long and 10 feet wide with 9-feet-high ceilings?

38. A glass block is 8 inches wide, 8 inches long, and 8 inches high. What is the volume of the glass block?

Solve. Use the five-step problem-solving strategy.

39. The plastic bins used by a community recycling program measure 8 inches high by 2 feet long by 12 inches wide. Per week, 1,350 families each fill a bin with newspapers.

a. What volume of newspaper is collected by each family?

b. What is the total volume of newspaper collected?

40. A free advertising newspaper called *The Moneysaver* is distributed to 20,000 homes in a city. It is printed on both sides of sheets of newsprint that are 23 inches by 16 inches. On March 17, *The Moneysaver* was printed on 7 sheets of this kind of paper.

a. What was the total area of newsprint used for each newspaper? Calculate the total area of both sides of the paper.

b. What was the total area of newsprint used for 20,000 copies of the newspaper?

41. The *substance weight* of book paper is the weight in pounds of 500 sheets that measure 25 inches by 38 inches. For example, if a printing job specifies that 60-pound paper be used, this means that the paper must be of such weight that 500 sheets of 25-inch by 38-inch paper weighs 60 pounds.

a. What is the area of one sheet of paper that is 25 inches by 38 inches?

b. What is the total area of 500 sheets of paper that is 25 inches by 38 inches? Calculate the area of both sides of the paper.

c. A ream of paper is 500 sheets. What is the total area of 4 reams of paper that is 25 inches by 38 inches?

d. What is the weight of 4 reams of 60-pound paper?

42. A developer originally divided a rectangular piece of land into rectangular lots that were 100 feet wide and 120 feet long. She decided to redraw the boundaries so that each lot measured 90 feet wide and 100 feet long. Calculate the amount of area that was removed from each of the lots.

43. Standard size plywood panels are 4 feet wide and 8 feet long. Roberto is using plywood left over from another project for a new subfloor in his family room. The floor of the family room is 16 feet long and 12 feet wide.

a. Calculate the area of the family room floor.

b. Calculate the area of one side of a plywood panel.

c. Roberto has 7 plywood panels. Calculate the total area of one side of this plywood.

d. Does Roberto have enough plywood to do the job? Explain.

44. A rectangular kitchen table measures 90 inches long and 40 inches wide. Clair is making a tablecloth that will hang down 4 inches on each side of the table. Calculate the area of the tablecloth.

ERROR ANALYSIS

45. **Problem:** Calculate the volume in a box that is 40 centimeters long, 10 centimeters wide, and 150 millimeters high.

Answer:

$$\begin{aligned} \text{Volume} &= (\text{Length})(\text{Width})(\text{Height}) \\ &= (40\text{ cm})(10\text{ cm})(150\text{ mm}) \\ &= 60{,}000\text{ cm}^3 \end{aligned}$$

Describe the mistake made in solving this problem.

Redo the problem correctly.

46. **Problem:** How many milliliters are in a cubic meter?

Answer:

$$\begin{aligned} 1\text{ cubic meter} &= 100\text{ cubic centimeters} \qquad \text{since } 1\text{ meter} = 100\text{ cm} \\ &= 100\text{ milliliters} \qquad \text{since } 1\text{ cm}^3 = 1\text{ ml} \end{aligned}$$

Describe the mistake made in solving this problem.

Redo the problem correctly.

REVIEW

47. Complete the subtraction: $12 - 2 - 2 - 2 - 2 - 2 - 2$

48. Complete the subtraction: $12 - 5 - 5$

49. State the zero product property. Give an example.

50. What is a "power of ten"?

Section 1.8 Division of Whole Numbers

LEARNING TOOLS

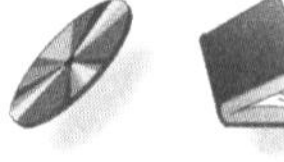

CD-ROM SSM VIDEO

Sneak Preview

Division is the opposite of multiplication. We will review the basic principles of division of whole numbers so that we can apply them to arithmetic with other numbers, variables, and measurements.

After completing this section, you should be able to:

1) Express division as multiple subtractions.

2) Rewrite a division problem as a multiplication problem.

3) Check the reasonability of the quotient of a division problem using multiplication.

4) Divide by powers of ten.

5) Divide a measurement by a whole number.

6) Calculate quotients of multidigit dividends and divisors.

DISCUSSION

Division

Just as multiplication is repeated addition, division is repeated subtraction. To divide 12 by 4, written in symbols as $12 \div 4$, we repeatedly subtract 4 from 12. We want to know how many times 4 is contained in 12.

$$\begin{array}{rl} 12 & \\ \underline{-\ \mathbf{4}} & \textit{One subtraction of 4.} \\ 8 & \\ \underline{-\ \mathbf{4}} & \textit{Two subtractions of 4.} \\ 4 & \\ \underline{-\ \mathbf{4}} & \textit{Three subtractions of 4.} \\ 0 & \textit{No more 4's remain.} \end{array}$$

By repeated subtraction, we find that there are 3 four's contained in 12. In symbols, we write $12 \div 4 = 3$. The answer to a division is called a **quotient.** The number that is being divided is called the **dividend,** and the number used to divide is called the **divisor.** In this example, 12 is the dividend, 4 is the divisor, and 3 is the quotient.

When the quotient is not a whole number, there is a **remainder.**

Example 1 **Find the quotient of $11 \div 5$ by repeated subtraction.**

$$\begin{array}{rl} 11 & \\ \underline{-\ \mathbf{5}} & \textit{One subtraction of 5.} \\ 6 & \\ \underline{-\ \mathbf{5}} & \textit{Two subtractions of 5.} \\ 1 & \textit{Remainder. This number is not big enough for another subtraction of the divisor.} \end{array}$$

Five divides into 11 two times with a remainder of 1. We write this quotient 2 R(1).

CHECK YOUR UNDERSTANDING

Use multiple subtractions to find each quotient.

1. $24 \div 8$ **2.** $24 \div 7$ **3.** $52 \div 6$ **4.** $31 \div 8$

Answers: 1. 3; 2. 3 R(3); 3. 8 R(4); 4. 3 R(7).

Subtraction "undoes" addition. For example, we can begin with the number 5. If we add 4, the sum of $5 + 4$ is 9. We can "undo" the addition of 4 by subtracting 4: $9 - 4 = 5$. We are back to our original number, 5. Because they undo each other, mathematicians say that subtraction and addition are **opposite operations.**

Similarly, multiplication and division are opposite operations. Division "undoes" multiplication. For example, $5 \cdot 4 = 20$. We can "undo" the multiplication by dividing: $20 \div 4 = 5$. We are back to our original number, 5.

Because multiplication and division are opposite operations, we can rewrite any division problem as a multiplication problem. $15 \div 5 = 3$ can be rewritten as $5 \cdot 3 = 15$. We can do this using any of the memorized products from the multiplication tables. For example, since we know that $2 \cdot 8 = 16$, we also know that $16 \div 8 = 2$ and $16 \div 2 = 8$.

Because multiplication and division are opposite operations, *zero can never be a divisor.* When we write $4 \div 0 =$ "quotient," then "quotient" $\cdot\ 0 = 4$. However, the zero product property states that any number multiplied by 0 must equal 0, not 4. So, there is no possible quotient that allows both multiplication and division to be opposite operations and the zero product property to be true. Because of this impossible situation, mathematicians call division by zero **undefined.**

CHECK YOUR UNDERSTANDING

Determine each quotient. Then, rewrite each problem as a multiplication problem.

1. $49 \div 7$ **2.** $18 \div 6$ **3.** $72 \div 9$

4. $0 \div 1$ **5.** $24 \div 0$ **6.** $100 \div 10$

Answers: 1. 7; $7 \cdot 7 = 49$; 2. 3; $6 \cdot 3 = 18$; 3. 8; $8 \cdot 9 = 72$; 4. 0; $1 \cdot 0 = 0$; 5. undefined; there is no multiplication problem that can be written; 6. 10; $10 \cdot 10 = 100$.

Because multiplication and division are opposite operations, we can check the reasonability of a quotient by writing an equivalent multiplication problem. If there is a remainder in the quotient, the divisor and the nonremainder part of the quotient are multiplied and the remainder is then added to this product.

Example 2 **Divide 14 ÷ 5 using multiple subtractions. Check the quotient by multiplication.**

$$\begin{array}{r l} 14 & \\ \underline{-\ 5} & \textit{One subtraction of 5} \\ 9 & \\ \underline{-\ 5} & \textit{Two subtractions of 5} \\ 4 & \textit{Remainder} \end{array}$$

The quotient is 2 R(4).

Check

Multiply the divisor, 5, by the nonremainder part of the quotient: $5 \cdot 2 = 10$.

Add the remainder to this product: $10 + 4 = 14$.

Since 14 was the original dividend, the quotient is correct.

PROCEDURE

To check a quotient with a remainder:

1. Multiply the divisor and the nonremainder part of the quotient.
2. Add the remainder to this product.
3. If this sum equals the dividend, the quotient is correct.

CHECK YOUR UNDERSTANDING

Use multiple subtractions to find each quotient. Check each quotient.

1. 24 ÷ 8 **2.** 24 ÷ 5 **3.** 52 ÷ 7 **4.** 31 ÷ 8

Answers: 1. 3; Check: $3 \cdot 8 = 24$; 2. 4 R(4); Check: $4 \cdot 5 + 4 = 20 + 4 = 24$; 3. 7 R(3); Check: $7 \cdot 7 + 3 = 49 + 3 = 52$; 4. 3 R(7); Check: $8 \cdot 3 + 7 = 24 + 7 = 31$.

DISCUSSION

Rounding Quotients

When a quotient has a remainder it is often good enough to round the answer to the nearest whole number. To do this we must look at the remainder and the divisor.

PROCEDURE

To round a quotient with a remainder to the nearest whole number:

1. If the remainder is closer to 0 than it is to the divisor, we round to the whole number part of the quotient.
2. If the remainder is closer to the divisor than it is to 0, we increase the whole number part of the quotient by 1.
3. If the remainder is exactly halfway between 0 and the divisor, we increase the whole number part of the quotient by 1.

Example 3 **Find the quotient of 21 ÷ 5, and round to the nearest whole number.**

21 ÷ 5 = 4 R(1)

The remainder is 1. The divisor is 5.

1 is closer to 0 than it is to 5, so we round 4 R(1) to 4.

Example 4 **Find the quotient of 24 ÷ 5, and round to the nearest whole number.**

39 ÷ 5 = 7 R(4)

The remainder is 4. The divisor is 5.

4 is closer to 5 than it is to 0, so we round 7 R(4) to 8.

Example 5 **Find the quotient of 33 ÷ 6, and round to the nearest whole number.**

33 ÷ 6 = 5 R(3)

The remainder is 3. The divisor is 6.

Three is exactly halfway between 0 and 6, so we round 5 R(3) to 6.

CHECK YOUR UNDERSTANDING

Find the quotient, then round to the nearest whole number.

1. 14 ÷ 3 **2.** 43 ÷ 9 **3.** 10 ÷ 4 **4.** 15 ÷ 7

Answers: 1. 4 R(2) ≈ 5; 2. 4 R(7) ≈ 5; 3. 2 R(2) ≈ 3; 4. 2 R(1) ≈ 2.

DISCUSSION

Division of a Measurement

When we multiply a measurement by a whole number, we multiply the numbers and keep the same unit of measurement. For example, 4(6 meters) = 24 meters. Similarly, to divide a measurement by a whole number, we divide the whole numbers and keep the same unit of measurement.

Example 6 **Divide 6 feet into three parts. Check by multiplication.**

6 feet ÷ 3
= (6 ÷ 3) feet
= 2 feet

Check

2 feet · 3 = 6 feet

Example 7 **A carpenter is building a 9-step staircase whose total height will be 72 inches. How high is each step?**

I know . . .

h = Height of each step

72 inches = Total height of staircase

9 = Number of steps

Equations

Height of each step = Total height of staircase ÷ Number of steps

h = 72 inches ÷ 9

Solve

h = 72 inches ÷ 9

= 8 inches

Check

Check by multiplication: 9(8 inches) = 72 inches

The answer is reasonable.

Answer

Each step will be 8 inches high.

A measurement can also be divided by a measurement with the same units. The quotient of the units is always 1. For example, feet ÷ feet = 1 just as 5 ÷ 5 = 1.

Example 8 **Each book of a set of reference books is 2 inches thick. How many of the books will fit on a bookshelf that is 18 inches long?**

I know . . .

b = Number of books

18 in. = Length of shelf

2 in. = Width of books

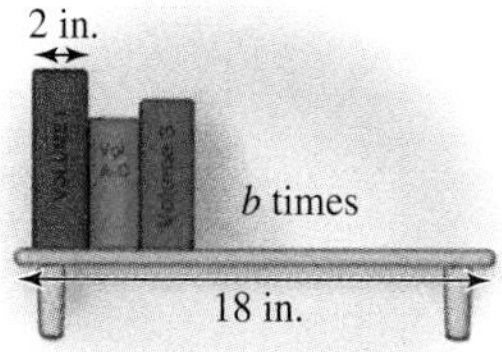

Equations

Number of books = Length of shelf ÷ Width of books

b = 18 in. ÷ 2 in.

Solve

b = 18 in. ÷ 2 in.

= (18 ÷ 2)(in. ÷ in.)

= 9(1) *2(9) = 18 and in. ÷ in. = 1*

= 9

Check

We can check by multiplying.

9 books that are each 2 inches wide have a total width of 9(2 in.) = 18 in.

Answer

Nine books will fit on the shelf.

CHECK YOUR UNDERSTANDING

Use the five-step problem-solving strategy to solve these problems.

1. An 18-inch board is to be cut into three equal pieces. Ignoring the loss of wood in cutting, calculate the length of each piece of board.

2. Karl Malone takes 12 steps to run from one end of the basketball court to the other where he dunks the ball. If he runs a distance of 72 feet, what is the average length of each step?

3. How many 10-pound bags of sand are needed for a child's sandbox that requires 180 pounds of sand?

Answers: 1. Each board will be 6 inches long; 2. The average length of Karl's step is 6 feet; 3. The sandbox will require 18 bags of sand.

Since division is the opposite operation of multiplication, the distributive property, $a(b + c) = ab + ac$, can also be applied to division. We can rewrite the distributive property using division instead of multiplication as $(a + b) \div c = (a \div c) + (b \div c)$. This application of the distributive property allows us to divide a mixed measurement that has an implied plus sign.

Example 9 **Divide 4 feet 6 inches by 2.**

(4 feet 6 inches) ÷ 2
= (4 feet + 6 inches) ÷ 2 — *Rewrite measurement with a plus sign.*
= (4 feet ÷ 2) + (6 inches ÷ 2) — *Apply the distributive property.*
= 2 feet + 3 inches
= 2 feet 3 inches — *The plus sign is implied.*

This quotient can be checked using multiplication and the distributive property.

Check
2(2 feet 3 inches) = 4 feet 6 inches — *Since the product of the quotient and divisor is the original dividend, the quotient is correct.*

Example 10 **Divide 5 pounds 10 ounces by 5.**

(5 pounds 10 ounces) ÷ 5
= (5 pounds + 10 ounces) ÷ 5 — *Rewrite the measurement with a plus sign.*
= (5 pounds ÷ 5) + (10 ounces ÷ 5) — *Apply the distributive property.*
= 1 pound + 2 ounces
= 1 pound 2 ounces — *The plus sign is implied.*

CHECK YOUR UNDERSTANDING

Use the distributive property to divide. Check each quotient by multiplication.

1. (8 years 12 days) ÷ 2

2. (18 pounds 6 ounces) ÷ 3

3. (9 feet 6 inches) ÷ 3

4. (24 hr 12 min) ÷ 6

Answers: 1. 4 years 6 days; Check: 2(4 years + 6 days) = 8 years + 12 days; 2. 6 pounds 2 ounces; Check: 3(6 pounds 2 ounces) = 18 pounds 6 ounces; 3. 3 feet 2 inches; Check: 3(3 feet 2 inches) = 9 feet 6 inches; 4. 4 hr 2 min; Check: 6(4 hr 2 min) = 24 h 12 min.

DISCUSSION

Division by a Power of Ten

To quickly multiply a whole number by a power of ten, additional zeros can be attached to the right of the last digit of the number. The power of ten can be written in exponential form or in place value form.

Example 11 **Write 376×10^4 in place value form.**

376×10^4
= 3,760,000 — *Attach four zeros after 376.*

Since division and multiplication are opposite operations, we can divide a whole number by a power of ten simply by removing zeros from the end of the number.

Example 12 **Evaluate 4,500 ÷ 10.**

4,500 ÷ 10
= 450 *Remove one zero from the end of the number.*

Example 13 **Evaluate 4,000,000 ÷ 10^3.**

4,000,000 ÷ 10^3
= 4,000 *Remove three zeros from the end of the number.*

Example 14 **One hundred co-workers at a post office purchased together the winning ticket of a lottery. The prize of $6,000,000 was divided equally among them. How much did each person win?**

I know . . .

a = Amount per person
$6,000,000 = Total prize
100 = Number of people

Equations

Amount per person = Total prize ÷ Number of people
a = $6,000,000 ÷ 100

Solve

a = $6,000,000 ÷ 100
a = $60,000 *Remove two zeros.*

Check

Check by multiplying:
$60,000 · 100 = $6,000,000 *Attach two zeros.*
The answer is reasonable.

Answer

Each person won $60,000.

CHECK YOUR UNDERSTANDING

Divide. Check by multiplication.

1. 300 ÷ 10
2. 5,000 ÷ 10^3
3. 45,000 ÷ 100
4. 5,900,000 ÷ 10^4
5. 402,000 ÷ 10^3

6. One hundred dogs at the local animal shelter eat an average of 500 pounds of dog food each week. On average, how much dog food does each dog eat each week?

Answers: 1. 30; Check: 10 · 30 = 300; 2. 5; Check: 5 · 10^3 = 5,000; 3. 450; Check: 450 · 100 = 45,000; 4. 590; Check: 590 · 10^4 = 5,900,000; 5. 402; Check: 402 · 10^3 = 402,000; 6. Each dog eats an average of 5 pounds of dog food.

DISCUSSION

Quotients of Numbers Ending in Zero

To multiply together two whole numbers that end with zeros, we multiply the digits that are not zero together. We then write the total number of zeros at the end of this product. For example, (200)(8,000) = 1,600,000 because $2 \cdot 8 = 16$ and there are a total of five zeros in the factors. Similarly, certain quotients can be found by doing a simple division and removing the correct number of zeros. However, the quotient cannot include a remainder.

PROCEDURE

To find the quotient of numbers that end in zero:

1. Divide the nonzero digits of the numbers. If the divisor is larger than the dividend, include enough zeros with the dividend to make the division possible.
2. Find the difference in the number of zeros between the divisor and the dividend. Write the number of remaining zeros after the quotient found in Step 1.

Example 15 **Find the quotient of 250,000 ÷ 500. Check by multiplication.**

250,000 ÷ 500
= (25 ÷ 5) followed by (4 − 2) zeros *Four zeros in 250,000; two zeros in 500.*
= 500

Check
(500)(500) = 250,000 *Quotient · Divisor = Dividend*

Example 16 **Divide 600,000 ÷ 120. Check by multiplication.**

600,000 ÷ 120
= (60 ÷ 12) followed by (4 − 1) zeros *Use 60 instead of 6 since 6 is not divisible by 12.*
= 5,000

Check
(5,000 · 120) = 600,000

Example 17 **Divide 9,900,000 ÷ 11. Check by multiplication.**

9,900,000 ÷ 11
= (99 ÷ 11) followed by (5 − 0) zeros
= 900,000

Check
(900,000)(11) = 9,900,000

CHECK YOUR UNDERSTANDING

Find each quotient. Check by multiplication.

1. 35,000 ÷ 700

2. 720,000 ÷ 9,000

3. 540 ÷ 90

4. 6,300,000 ÷ 70,000

5. 1,440,000 ÷ 1,200

6. 1,000 ÷ 500

Answers: 1. 50; Check: 50 · 700 = 35,000; 2. 80; Check: 80(9,000) = 720,000; 3. 6; Check: (90)(6) = 540; 4. 90; Check: 90 · 70,000 = 6,300,000; 5. 1,200; Check: 1,200 · 1,200 = 1,440,000; 6. 2; Check: 2 · 500 = 1,000.

DISCUSSION

Long Division

A quotient tells how many times a divisor can be subtracted from a dividend. Long division is an efficient way to do such division. Long division is set up using this symbol: $\overline{)\quad}$. The dividend is written inside the symbol, and the divisor is written to the left. As the digits of the quotient are determined, they are written above the line:

$$\begin{array}{r} \textit{quotient} \\ \textit{divisor}\overline{)\textit{dividend}} \end{array}$$

Unlike addition, subtraction, and multiplication, division requires estimation. We must estimate how many times a divisor goes into a digit or digits of the dividend. We work from left to right with the digits of the dividend, asking, "How many times can the divisor be subtracted from this digit?" If the digit is not large enough, we include the next digit to the right in the dividend.

Example 18 **Use long division to find the quotient of 74 ÷ 2. Check the quotient.**

$$\begin{array}{r} 37 \\ 2\overline{)74} \\ \underline{6\downarrow} \\ 14 \\ \underline{14} \\ 0 \end{array}$$

2 can be subtracted from 7 three times (3 · 2 = 6). Place a 3 in the quotient.

Subtract 6 from 7 for a difference of 1; bring down the 4 from the dividend.

2 can be subtracted from 14 seven times (2 · 7 = 14). Place a 7 in the quotient.

Subtract 14 from 14 for a difference of 0.

74 ÷ 2 = 37.

There is no remainder. In other words, 2 can be subtracted from 74 a total of 37 times.

To check the quotient, multiply the divisor and the nonremainder part of the quotient. Add the remainder to this product. The final answer should equal the original dividend.

Check

37 · 2 = 74 *The quotient, 37, is correct: 74 ÷ 2 = 37.*

Example 19 **Use long division to find the quotient of 328 ÷ 34.**

Begin by writing the problem like this:

34)328

We need to know how many times 34 can be subtracted from 328. We look at each digit in the dividend, working from left to right. Since 34 cannot be subtracted from 3, include the next digit in the dividend to the right, 2. However, 34 can still not be subtracted from 32 so include the next digit from the dividend, 8. Make a guess of how many times 34 can be subtracted from 328, knowing that 10 · 34 is 340, which is too large. Try 9 times.

```
     9
34)328
  -306
    22
```

Place the guess of 9 above the last digit of the dividend.

To test whether this guess is correct, multiply it by the divisor: 9 · 34 = 306. Write 306 under the dividend. Subtract 306 from 328 to get a remainder.

If the remainder is greater than 34, the guess of 9 is too small. If subtraction is impossible because the product of the guess and the divisor is too large, then the guess is too large.

The remainder is 22, which is less than 34, so the guess of 9 is correct.

So, 328 ÷ 34 = 9 R(22). *This means that 34 can be subtracted from 328 a total of 9 times with a remainder of 22.*

Check

34 · 9 + 22

= 306 + 22

= 328 *This equals the original dividend, 328. The quotient is correct.*

Example 20 **Use long division to find the quotient of 7,823 ÷ 27.**

```
     3
27)7,823
   81
```

27 cannot be subtracted from 7, so include the next digit in the dividend. How many times can 27 be subtracted from 78? Guess 3 times. Write the 3 above the 8 (the last digit of 78). Multiply: 3 · 27 = 81.

However, 81 is greater than 78, so the guess is too large.

```
     2
27)7,823
   54
   24
```

Try a guess of 2 instead of 3.

Place the 2 above the 8 of 78. Multiply: 2 · 27 = 54. Write the 54 below the 78. Subtract 54 from 78 to get a remainder of 24, which is less than 27. This means that the guess of 2 is correct; 27 can be subtracted from 78 only 2 times.

```
     289
27)7,823
   54↓
   242
   216↓
    263
    243
     20
```

Bring down the 2 from the dividend. How many times can 27 be subtracted from 242? Guess 8 times.

Multiply: 8 · 27 = 216. Write the 216 below the 242.

Subtract 216 from 242 to get a remainder of 26. This is less than 27 so the guess of 8 is correct.

Bring down the 3 from the dividend. How many times can 27 be subtracted from 263? Guess 9 times. Write the 9 on the top line above the 3. Multiply 9 · 27 = 243.

Subtract 243 from 263 to get a remainder of 20. This is less than 27 so the guess of 9 is correct. There are no more digits to bring down from the dividend.

The quotient of 7,823 and 27 is 289 R(20).

Check

27 · 289 + 20

= 7,803 + 20

= 7,823 *This is equal to the dividend. The quotient is correct.*

This method for long division is the one most commonly taught in the United States. Students who come from other countries may complete long division differently but will get the same quotient.

Example 21 **A 2-acre parking lot has seven equal rows of parking spaces and a total of 679 spaces. How many spaces are there in each row?**

I know . . .

s = Spaces per row

679 = Total spaces

7 = Number of rows

Equations

Spaces per row = Total spaces ÷ Number of rows

$s = 679 \div 7$

Solve

$s = 679 \div 7$

$= 97$ *Calculate by long division.*

$$\begin{array}{r} 97 \\ 7\overline{)679} \\ \underline{63}\downarrow \\ 49 \\ \underline{49} \\ 0 \end{array}$$

7 cannot be subtracted from 6 so include the next digit to the right in the dividend: 67.

Guess: 7 can be subtracted from 67 nine times.

Write the 9 above the 7 of 67. Multiply: 7 · 9 = 63.

Write the 63 below the 67 and subtract to get a remainder of 4.

Bring down the 9 from the dividend. 7 can be subtracted from 49 seven times.

Write the 7 above the 9. Multiply 7 · 7 = 49.

Write the 49 below the 49 and subtract to get a remainder of 0.

There are no more digits to bring down from the denominator. The quotient is 97. There is no remainder.

Check

Estimate the quotient: $s = 700 \div 7 = 100$, which is close to 97. Our answer is reasonable. Or, do an exact check by multiplication:

$97 \cdot 7 = 679$.

Answer

There are 97 spaces in each row in the parking lot.

CHECK YOUR UNDERSTANDING

Use long division to find the quotient. Check each quotient.

1. $265 \div 4$ **2.** $482 \div 2$ **3.** $5{,}910 \div 23$ **4.** $7{,}201 \div 37$

Answers: 1. 66 R(1); Check: 66 · 4 = 264 + 1 = 265; 2. 241; Check: 241 · 2 = 482; 3. 256 R(22); Check: 256 · 23 = 5,888 + 22 = 5,910; 4. 194 R(23); Check: 194 · 37 = 7,178 + 23 = 7,201.

EXERCISES SECTION 1.8

Write the division problem that is equivalent to the subtraction.

1. $40 - 10 - 10 - 10 - 10 = 0$

2. $60 - 8 - 8 - 8 - 8 - 8 - 8 - 8 = 4$

Calculate the quotient using multiple subtractions. There may be a remainder.

3. $15 \div 6$

4. $12 \div 4$

Find the quotient, then round to the nearest whole number.

5. $18 \div 3$

6. $18 \div 4$

7. $18 \div 10$

8. $10 \div 3$

9. $11 \div 3$

10. $12 \div 3$

11. $13 \div 3$

12. $13 \div 0$

13. $0 \div 13$

14. Why is division by zero undefined?

Evaluate.

15. $4{,}200 \div 10$

16. $4{,}200 \div 100$

17. $350{,}000 \div 10^3$

18. $2{,}800{,}000 \div 10{,}000$

19. $4{,}000{,}500 \div 0$

20. $75{,}500{,}000 \div 10^4$

Solve. Use the five-step problem-solving strategy.

21. When Grandma Irene died, she left $120,000 to be divided equally among her 100 cats and dogs for their perpetual care. How much money will be allocated to each animal?

22. A group of ten anthropology students is driving to Arizona in a van for a field trip. The total mileage of the trip will be about 4,200 miles. If the students split the driving responsibilities equally among them, how many miles will each student have to drive?

23. In June 2001, the national debt of the United States was about $5,645,002,000,000. If this was divided among the 100,000 wealthiest people in the United States, how much would each have to pay to retire the debt?

24. A bag of potato chips contains 63 grams of fat. If the bag is divided into nine servings, how many grams of fat are contained in each serving?

Evaluate.

25. 22 feet ÷ 11

26. 45 ounces ÷ 9

27. 36 hours ÷ 4

Simplify using the distributive property.

28. (12 + 84) ÷ 12

29. (72 + 56) ÷ 8

30. (10 + 15) ÷ 5

Evaluate.

31. 32 feet 4 inches ÷ 2

32. 8 gallons 2 quarts ÷ 2

33. 50,000 ÷ 5,000

34. 220,000,000 ÷ 110,000

35. 360 ÷ 60

36. 2,700,000 ÷ 30,000

37. 22,000 ÷ 1,100

38. 7,200 ÷ 90

Solve. Use the five-step problem-solving strategy.

39. A sausage recipe requires 6 pounds 4 ounces of ground pork. Glenn wants to divide the recipe into two parts. How much pork will be needed for each part of the recipe?

40. Eight identical solar photovoltaic modules measure 16 feet 8 inches when laid end-to-end. What is the length of each module?

41. To increase business, a Fort Myers, Florida, restaurant decided to distribute 4,500 tokens for free pie to students during spring break. Nine people were given an equal amount of tokens to distribute on the beach. How many tokens did each person have to distribute?

42. In her will, Granny left $24,000 in 3M Company stock to be divided evenly among her eight grandchildren. What was the value of stock inherited by each grandchild?

43. The average speed of a vehicle can be calculated by dividing the miles driven by the time spent driving the distance. Calculate the average speed of a truck that traveled 872 miles in 12 hours. Round to the nearest mile per hour.

44. The average donation to a new mathematics building fund can be calculated by dividing the total amount of donations by the number of donors. Calculate the average donation to a fund that has collected $832,718 from 923 donors. Round to the nearest dollar.

Divide using long division. Check your answer by multiplication.

45. 1,243 ÷ 9

46. 1,240 ÷ 9

47. 73,000 ÷ 90

48. 73,000 ÷ 82

49. 184 ÷ 34

50. 184 ÷ 38

51. 456 ÷ 3

52. 536 ÷ 4

53. 827 ÷ 7

54. 206 ÷ 12

55. 789 ÷ 21

56. 792 ÷ 15

57. 1,945 ÷ 26

58. 7,458 ÷ 14

59. 11,902 ÷ 42

Solve.

60. Alaska Airlines Frequent Flyer Mileage Plan awards one free round-trip coach ticket within the United States or between Canada and the United States for 20,000 earned air miles. Demian has 138,000 miles in her frequent flyer account.

a. How many free tickets has she earned?

b. How many miles are left over?

61. When Abraham Lincoln began to block the southern coastline, he had 26 steam powered ships to cover 4,000 miles, a distance greater than the distance between New York City and Liverpool, England.

a. If the distance was divided equally into whole miles, how many miles would each steamer have to patrol?

b. What distance would be left over?

62. This year 438 freshmen need to take English composition spring semester. If the classrooms have 35 chairs, how many sections of composition will need to be offered?

ERROR ANALYSIS

63. **Problem:** Divide 577 ÷ 7.

Answer:

$$\begin{array}{r} 81 \\ 7\overline{)577} \\ \underline{56} \\ 7 \\ \underline{7} \\ 0 \end{array}$$

The quotient is 81.

Describe the mistake made in solving this problem.

Redo the problem correctly.

64. **Problem:** 60,000,000 miles ÷ 10,000,000

Answer: 60 miles

Describe the mistake in the answer to this problem.

Redo the problem correctly.

REVIEW

65. What is a *variable*?

66. The Associative Property of Addition states that $a + (b + c) = (a + b) + c$. What is the purpose of the parentheses in this property?

67. State the Associative Property of Multiplication using variables.

Section 1.9 The Order of Operations

LEARNING TOOLS

CD-ROM SSM VIDEO

Sneak Preview

When an expression requires us to do more than one arithmetic operation, there is a standard order established by mathematicians in which to do the operations. This is called the order of operations.

After completing this section, you should be able to:

1) Describe the order of operations.

2) Evaluate or simplify expressions following the order of operations.

INVESTIGATION Order of Operations

The operations of arithmetic are addition, subtraction, multiplication, and division. When an expression has more than one operation, we need rules that tell the order in which to complete the operations and that tell us the purpose of parentheses. These rules are called the **order of operations.**

In each of the following problems, an expression and the evaluated expression are given. Describe the order in which operations are done in each problem.

	Expression	Evaluated Expression	Order
ex.	$4 + 12 \div 4$	$4 + 12 \div 4$	
		$= 4 + 3$	1. *division*
		$= 7$	2. *addition*
1.	$4 + 2 \cdot 3$	$4 + 2 \cdot 3$	
		$= 4 + 6$	1.
		$= 10$	2.
2.	$10 - 8 \div 2$	$10 - 8 \div 2$	
		$= 10 - 4$	1.
		$= 6$	2.
3.	$18 - 6 \cdot 2$	$18 - 6 \cdot 2$	
		$= 18 - 12$	1.
		$= 6$	2.

In Problems 1–3,

4. Which operation is done first: addition or multiplication?

5. Which operation is done first: addition or division?

6. Which operation is done first: subtraction or multiplication?

7. Which operation is done first: subtraction or division?

Expression	Evaluated Expression	Order
8. $7 \cdot 3^2$	$7 \cdot 3^2$	
	$= 7 \cdot 9$	1.
	$= 63$	2.
9. $4^2 \div 2$	$4^2 \div 2$	
	$= 16 \div 2$	1.
	$= 8$	2.

In Problems 8 and 9,

10. Which operation is done first: multiplication or evaluation of exponents?

11. Which operation is done first: division or evaluation of exponents?

Expression	Evaluated Expression	Order
12. $12 \div 2 \cdot 4$	$12 \div 2 \cdot 4$	
	$= 6 \cdot 4$	1.
	$= 24$	2.
13. $12 \cdot 2 \div 4$	$12 \cdot 2 \div 4$	
	$= 24 \div 4$	1.
	$= 6$	2.
14. $6 \div 2 \cdot 3 \cdot 4 \div 12$	$6 \div 2 \cdot 3 \cdot 4 \div 12$	
	$= 3 \cdot 3 \cdot 4 \div 12$	1.
	$= 9 \cdot 4 \div 12$	2.
	$= 36 \div 12$	3.
	$= 3$	4.

15. In Problems 12–14, which operation is done first: multiplication or division?

Expression	Evaluated Expression	Order
16. $12 + 2 - 4$	$12 + 2 - 4$	
	$= 14 - 4$	1.
	$= 10$	2.
17. $12 - 2 + 4$	$12 - 2 + 4$	
	$= 10 + 4$	1.
	$= 14$	2.
18. $6 - 2 + 3 - 4 + 12$	$6 - 2 + 3 - 4 + 12$	
	$= 4 + 3 - 4 + 12$	1.
	$= 7 - 4 + 12$	2.
	$= 3 + 12$	3.
	$= 15$	4.

19. In Problems 16–18, which operation is done first: addition or subtraction?

20. The order of operations are rules that tell us the order in which to evaluate exponents and do division, multiplication, addition, and subtraction. Based on the previous examples, list the order of operations.

Evaluate each expression. Show all intermediate steps.

21. $6 + 12 \div 2^2$

22. $4 \cdot 3 + 8 \div 2^2$

23. $2^3 + 4 \cdot 3 - 8 \div 2$

Answers: 21. 9; 22. 14; 23. 16.

Sometimes an expression includes parentheses ( ) or brackets [ ]. Always evaluate inside parentheses or brackets first, following the order of operations. Once everything inside the parentheses is evaluated, the parentheses are eliminated. If there is more than one set of parentheses, do the evaluating in the innermost set first (work from the inside out).

Evaluate. Follow the order of operations. Show all intermediate steps.

24. $(12 + 4) \div 2 + 9$

25. $25 - (6 + 2) \cdot 3$

26. $25 - 6 + 2 \cdot 3$

27. $[(25 - 6) + 2] \cdot 3$

28. $(12 + 35 + 7) \div 3$

Answers: 24. 17; 25. 1; 26. 25; 27. 63; 28. 18.

Evaluate. Follow the order of operations. Show all intermediate steps.

29. $15 + 20 \div 2$

30. $(10 + 20) \div 2$

31. $15 + (20 \div 2)$

32. $30 \div 6 + 4 \cdot 3$

33. $30 \div (6 + 4) \cdot 3$

34. $[(30 \div 6) + 4] \cdot 3$

35. $3^2 - 2 + 3 \cdot 8$

36. $2 \cdot (6 + 17 + 8)$

37. $48 \div 2^4 \cdot 6 \cdot 1$

38. $24 - 12 \div 3 + 1$

39. $12 \div 0$

40. $0 \div 12$

Answers: 29. 25; 30. 15; 31. 25; 32. 17; 33. 9; 34. 27; 35. 31; 36. 62; 37. 18; 38. 21; 39. undefined; 40. 0.

DISCUSSION

Order of Operations

When more than one arithmetic operation is found in an expression, there is more than one possible way to evaluate the expression. Depending on which order we choose to do the operations of multiplication and addition, we get a different value for the expression. Mathematicians have established a set of rules that specifies the order to do arithmetic operations. These rules are called the **order of operations.** These rules ensure that everyone will evaluate an expression in the same order and get the same answer.

PROCEDURE

To evaluate an expression following the order of operations:

1. **Grouping.** Operations inside of grouping symbols are done first, following rules 2–4. Grouping symbols include parentheses ( ), brackets [ ], and braces { }.
2. **Exponents.** Do the operations indicated by exponents from left to right as they appear in the expression.
3. **Multiplication or division.** Do all multiplication and divisions from left to right as they appear in the expression.
4. **Addition or subtraction.** Do all additions and subtractions from left to right as they appear in the expression.

Example 1 **Evaluate $3 \cdot 5 + 2 \cdot 9$.**

$3 \cdot 5 + 2 \cdot 9$

$= \mathbf{15} + 2 \cdot 9$ *Multiply or divide from left to right.*

$= 15 + \mathbf{18}$ *Multiply or divide from left to right.*

$= \mathbf{33}$ *Add or subtract from left to right.*

With practice, steps can be combined and less writing is required. However, at the beginning, each individual step should be written out. All of these examples are detailed and completed in individual steps. The result of each step is shown in bold type.

Example 2 **Evaluate $20 - 6 \div 2 + 3 \cdot 2$.**

$20 - 6 \div 2 + 3 \cdot 2$

$= 20 - \mathbf{3} + 3 \cdot 2$ *Multiply or divide from left to right.*

$= 20 - 3 + \mathbf{6}$ *Multiply or divide from left to right.*

$= \mathbf{17} + 6$ *Add or subtract from left to right.*

$= \mathbf{23}$ *Add or subtract from left to right.*

Example 3 **Evaluate $(3 - 2 + 4)^2$.**

$(3 - 2 + 4)^2$

$= (\mathbf{1} + 4)^2$ *Work inside parentheses. Add or subtract from left to right.*

$= \mathbf{5}^2$ *Work inside parentheses. Add or subtract from left to right.*

$= \mathbf{25}$ *Do the operations indicated by exponents from left to right.*

Example 4 **Evaluate $2^2 + 3^2 + 13 \div (5 \cdot 3 - 2)$.**

$2^2 + 3^2 + 13 \div (5 \cdot 3 - 2)$

$= 2^2 + 3^2 + 13 \div (\mathbf{15} - 2)$ *Work inside parentheses. Multiply or divide from left to right.*

$= 2^2 + 3^2 + 13 \div \mathbf{13}$ *Work inside parentheses. Add or subtract from left to right.*

$= \mathbf{4} + 3^2 + 13 \div 13$ *Do the operations indicated by exponents from left to right.*

$= 4 + \mathbf{9} + 13 \div 13$ *Do the operations indicated by exponents from left to right.*

$= 4 + 9 + \mathbf{1}$ *Multiply or divide from left to right.*

$= \mathbf{13} + 1$ *Add or subtract from left to right.*

$= \mathbf{14}$ *Add or subtract from left to right.*

Some expressions are complex, involving multiple sets of parentheses, some inside brackets, and many exponents. Although the expression takes many steps to evaluate, each individual step is just addition, subtraction, multiplication, or division. Careful and clear organization is essential.

Example 5 **Evaluate $[(3^3 \div 9)^2 + (4^2 + 6 \div 3)] \div (5^2 + 6 - 2^2)$.**

$[(3^3 \div 9)^2 + (4^2 + 6 \div 3)] \div (5^2 + 6 - 2^2)$

$= [(\mathbf{27} \div 9)^2 + (4^2 + 6 \div 3)] \div (5^2 + 6 - 2^2)$

$= [(\mathbf{3})^2 + (4^2 + 6 \div 3)] \div (5^2 + 6 - 2^2)$

$= [\mathbf{9} + (4^2 + 6 \div 3)] \div (5^2 + 6 - 2^2)$

$= [9 + (\mathbf{16} + 6 \div 3)] \div (5^2 + 6 - 2^2)$

$= [9 + (16 + \mathbf{2})] \div (5^2 + 6 - 2^2)$

$= [9 + \mathbf{18}] \div (5^2 + 6 - 2^2)$

$= \mathbf{27} \div (5^2 + 6 - 2^2)$

$= 27 \div (\mathbf{25} + 6 - 2^2)$

$= 27 \div (25 + 6 - \mathbf{4})$

$= 27 \div (\mathbf{31} - 4)$

$= 27 \div \mathbf{27}$

$= \mathbf{1}$

CHECK YOUR UNDERSTANDING

Evaluate.

1. $6 + 4 \div 2 - 3 \cdot 2$

2. $(5 - 3)^2$

3. $(4 \cdot 2 + 12 \div 4)^2 - 7 \cdot 3$

4. $2(5 - 1)^2 \div (24 \div 3)$

5. $24 \div 8 \cdot 2 + 8 \cdot 7 \div 4$

Answers: 1. 2; 2. 4; 3. 100; 4. 4; 5. 20.

DISCUSSION

The Arithmetic Mean

The **arithmetic mean,** also called the **average,** is a statistical term that measures central tendency. A statistician defines the arithmetic mean as the number that can replace all the numbers in a sum without changing the sum.

Example 6 **Use the definition of arithmetic mean to show that the arithmetic mean of the numbers 1, 3, 5, and 7 is 4.**

The sum of the numbers is $1 + 3 + 5 + 7 = 16$.

If 4 is the arithmetic mean, then the definition says we should be able to replace each of the numbers in the sum with 4 and still get the same sum.

$1 + 3 + 5 + 7 = 16.$ *Original sum is 16.*

$4 + 4 + 4 + 4 = 16.$ *Replacing each number with the arithmetic mean, 4, still results in a sum of 16.*

Since the sum is the same, **4** is the arithmetic mean of 1, 3, 5, and 7.

In practice, we usually find the arithmetic mean by dividing the sum of a set of numbers by the number of members in the set.

Arithmetic Mean

Arithmetic mean = (Sum of a set of numbers) ÷ (Number of members in the set)

The arithmetic mean is typically called the **average** in the media.

Example 7 **On January 27, 1998, the following high temperatures were reported: Albuquerque, New Mexico (54°F); Anchorage, Alaska (26°F); Atlanta, Georgia (54°F); Billings, Montana (46°F); New York City (42°F); Boston, Massachusetts (34°F); Sacramento, California (51°F); Minneapolis, Minnesota (27°F); Bismarck, North Dakota (26°F); and Philadelphia, Pennsylvania (40°F). What was the average high temperature in these cities?**

I know. . .

A = Average temperature in degrees

Temperatures in degrees = 54, 26, 54, 46, 42, 34, 51, 27, 26, 40

Number of temperatures = 10

Equations

Average temperature = (Sum of temperatures) ÷ (Number of temperatures)

$$A = (54 + 26 + 54 + 46 + 42 + 34 + 51 + 27 + 26 + 40) \div 10$$

Solve

$A = (54 + 26 + 54 + 46 + 42 + 34 + 51 + 27 + 26 + 40) \div 10$

$A = 400 \div 10$ *Add inside parentheses.*

$A = 40$ *Divide.*

Check

Estimate the average by rounding.

$A = (50 + 30 + 50 + 50 + 40 + 30 + 50 + 30 + 30 + 40) \div 10$

$A = 400 \div 10$

$A = 40$, equal to the exact answer of 40.

Also, a quick look at the temperatures shows a range of temperatures between 26°F and 54°F. We expect the average to be between these temperatures and it is.

Answer

The average high temperature in these cities was 40°F.

Averages can be calculated using data obtained from graphs or tables. In the following example, we can do considerably less writing by using multiplication to represent repeated additions of the same number.

Example 8 **The scores from a math test are displayed on the bar graph. Calculate the average score on the test.**

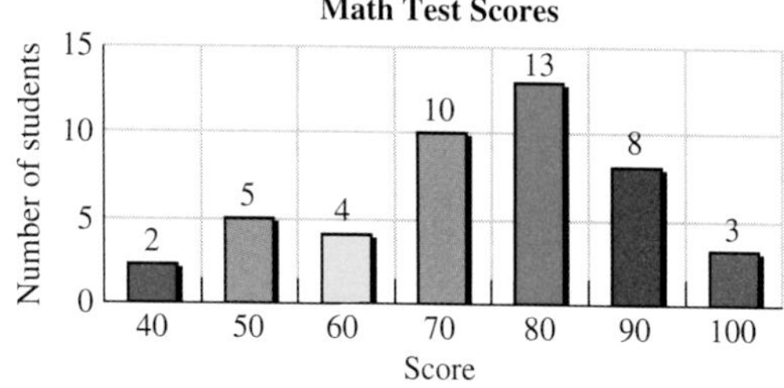

I know . . .

A = Average test score

2 = Students who scored 40

5 = Students who scored 50

4 = Students who scored 60

10 = Students who scored 70

13 = Students who scored 80

8 = Students who scored 90

3 = Students who scored 100

Total students = 2 + 5 + 4 + 10 + 13 + 8 + 3 = 45

Equations

Average = Sum of scores ÷ Number of students

$$A = (2 \cdot 40 + 5 \cdot 50 + 4 \cdot 60 + 10 \cdot 70 + 13 \cdot 80 + 8 \cdot 90 + 3 \cdot 100) \div 45$$

Solve

$$\begin{aligned} A &= (2 \cdot 40 + 5 \cdot 50 + 4 \cdot 60 + 10 \cdot 70 + 13 \cdot 80 + 8 \cdot 90 + 3 \cdot 100) \div 45 \\ &= (80 + 250 + 240 + 700 + 1{,}040 + 720 + 300) \div 45 && \textit{Multiply.} \\ &= 3{,}330 \div 45 && \textit{Add.} \\ &= 74 && \textit{Divide.} \end{aligned}$$

Check

We can do an estimate or we can look at the scores and see that they range from 40 to 100 with most of the scores being at 70, 80, and 90. An average score of 74 is reasonable because it is between 40 and 100, and it is in the area where most of the scores were earned. This kind of thinking about reasonability may not work if there are extreme values (for example, scores of 0 or 100) included in the information.

Answer

The average score on the test was 74.

CHECK YOUR UNDERSTANDING

Calculate the arithmetic mean.

1. 9, 13, 7, 21, 6, and 16.

2. 4, 7, 1, 1, 3, 12, 8, and 4.

3. Calculate the average earnings of the following drivers in the 1998 FedEx Championship Series of car racing.

DRIVER	AGE	EARNINGS
Alex Zanardi	31	$2,096,250
Al Unser, Jr.	35	$ 606,250
Andre Ribeiro	32	$ 567,250
Gil De Ferran	30	$ 567,250

Source: *Sports Illustrated*, March 9, 1998.

Answers: 1. 12; 2. 5; 3. $959,250.

EXERCISES SECTION 1.9

1. List the rules of the order of operations.

Evaluate. Show all intermediate steps.

2. $16 + 32 \div 4 + 3$

3. $(16 + 32) \div 4 + 3$

4. $(16 + 32) \div (4 \cdot 3)$

5. $5(13 - 9) + 6$

6. $5 \cdot 13 - 9 + 6$

7. $5 \cdot 13 - (9 + 6)$

8. $21 - 18 \div 3^2 + 9$

9. $21 + 18 \div 3^2 + 9$

10. $(21 + 18) \div 3 + 9^2$

11. $(21 + 18 \div 3) \div 3^2$

12. $4^2 \div 2 \cdot 4$

13. $4^2 \div (2 \cdot 4)$

14. $20 - 16 - 4 \div 4$

15. $3^3 \div 3^2$

16. $15 - 12 \div 2 + 9$

17. $[2(6 + 1) - 3^2] \cdot 4$

18. $0 \div 6 + 15$

19. $6 \div 0 + 15$

20. $6 \cdot 7 - 2 \cdot 3 + 8$

21. $2 \cdot 8$ feet $+ 2 \cdot 9$ feet

22. $6^2 - 6 \div 3$

23. $(5 - 2)^2 + (3 + 1)^2$

24. $(23 - 15 \div 5) \div 2^2$

25. $(15 \cdot 2 \div 3 - 2) \div (3^2 - 5)$

26. Christine Boskoff of Seattle hopes to become the first woman to summit the 14 mountains in the world higher than 26,400 feet. By 1998, she had climbed Lhotse (27,916 feet), Cho Oyu (26,906 feet), and Broad Peak (26,400 feet). What is the average height of these three mountains?

27. Everest (29,028 feet), K2 (28,250 feet), Kanchenjunga (28,208 feet), Lhotse (27,916 feet), Cho Oyu (26,906 feet), Broad Peak (26,400 feet), Makalu (27,766 feet), Dhaulagiri (26,795 feet), Manaslu (26,781 feet), Nanga Parbat (26,660 feet), Annapurna (26,545 feet), Gasherbrum I (26,470 feet), and Shishapangma (26,397 feet) are mountains that have altitudes greater than 26,000 feet. What is the average height of these mountains? If there is a remainder, round the answer to the nearest foot.

28. In 1994 Saturn sold 286,003 cars. In 1995, 285,674 cars were sold. In 1996, 278,574 cars were sold. What was the average number of cars sold in this 3-year period?

29. Wildfires raged over Florida during the Fourth of July in 1998. The Associated Press reported that more than 450,000 acres had burned in nearly 2,000 fires from Memorial Day to July 5. What was the average acreage burned per fire?

30. The number of live births in the United States for a 6-year period are recorded in the table. Calculate the average number of live births per year. If there is a remainder, round to the nearest whole number.

YEAR	BIRTHS
1991	4,110,907
1992	4,065,014
1993	4,000,240
1994	3,952,767
1995	3,899,589
1996	3,891,494

31. Bodie, California, is now a ghost town and state park. In 1879, discovery of gold boosted the population of Bodie to about 10,000 miners, gamblers, and other entrepreneurs. While the gold boom lasted, some 30 companies produced $400,000 in gold bullion per month. What was the average amount of bullion produced per company per month? If there is a remainder, round to the nearest dollar.

© Tom Benoit/Superstock

32. Write your own word problem that requires the calculation of an average. Solve the problem.

33. During the 1998 major league baseball season, Mark McGwire broke and surpassed Roger Maris' season home-run record of 61 home runs. Along the way, he also hit some of the longest home runs in baseball history. He hit a home run that was 545 feet long at Busch Stadium in St. Louis, a home run that was 538 feet long at the Kingdome in Seattle, and a home run that was 458 feet long in San Diego. What was the average length of these three home runs? If there is a remainder, round to the nearest foot.

34. In 1998, 5,000,000 jars of Skippy peanut butter bearing the likeness of Yankees' shortstop Derek Jeter were distributed. Over a 365-day year, what is the average number of jars that will be distributed each day? If there is a remainder, round to the nearest jar.

35. A nurse evaluator for a major health insurance provider is comparing family physician charges for a brief patient visit. Dr. Petersen charges $45 a visit; Dr. Clark charges $32 a visit; Dr. Hayes charges $39 a visit; Dr. Carrero charges $43 a visit; Dr. Pant charges $36 a visit. What is the average physician charge for such a visit?

36. The total energy in a molecule is the sum of the energy of the chemical bonds in the molecule. One form of glucose molecule has five carbon-carbon single bonds, seven carbon-hydrogen single bonds, four carbon-oxygen single bonds, one carbon-oxygen double bond, and four oxygen-hydrogen single bonds. Use the table of bond energies per mole to calculate the energy of a mole of glucose molecules.

TYPE OF BOND	BOND ENERGY IN KILOCALORIES PER MOLE
Carbon-carbon single bond	83
Carbon-hydrogen single bond	99
Carbon-oxygen single bond	84
Carbon-oxygen double bond	192
Oxygen-hydrogen single bond	111

ERROR ANALYSIS

37. Problem: Evaluate $1 + 4^2 - 18 \div 3$.

Answer:

$$
\begin{aligned}
&1 + 4^2 - 18 \div 3 \\
&= 1 + 8 - 18 \div 3 \\
&= 1 + 8 - 6 \\
&= 9 - 6 \\
&= 3
\end{aligned}
$$

Describe the mistake made in solving this problem.

Redo the problem correctly.

38. Problem: Calculate the average of the numbers: 5, 7, 9, 11, 13.

Answer:

$$
\begin{aligned}
\text{Average} &= 5 + 7 + 9 + 11 + 13 \\
&= 45
\end{aligned}
$$

Describe the mistake in the answer to this problem.

Redo the problem correctly.

39. **Problem:** Makbuhl recorded the time he spent in the tutoring center each day. Calculate the average amount of time he spent on each visit to the tutoring center.

Monday:	4 hours
Tuesday:	20 minutes
Wednesday:	4 hours
Thursday:	20 minutes

Answer: Average = (4 + 20 + 4 + 20) ÷ 4
= 48 ÷ 4
= 12 minutes

Describe the mistake made in solving this problem.

Redo the problem correctly.

REVIEW

40. Estimate the product of (15 feet)(5 feet).

41. Estimate the sum of 54 + 26 + 54 + 46 + 42 + 34 + 51 + 27 + 26 + 40.

42. Estimate the product of (352 feet)(159 feet).

READ AND REFLECT

Effective studying for a mathematics test is done in small amounts and it is done frequently. Effective studying does not mean "looking over" notes or problems. Simply doing the homework once is not enough either. Effective studying requires practice and then more practice in weaker areas. Although "cramming" for an exam may work in some disciplines, it is disastrous in math. The concepts you cram and forget will be immediately needed to do the work in the next chapter. Study time should include completion of homework, completing the glossary and procedure list, and learning concepts and procedures.

Math tests generally measure your understanding of important concepts and your skill at doing procedures such as multiplication or simplifying an algebraic expression. You also are asked to use these concepts and procedures to solve application problems. Each area requires a different kind of studying. How can you study concepts, terms, and important ideas? Try these suggestions.

- **Keep a glossary of concepts and terms.**

As each new concept or term in the text is introduced, it is printed in **bold.** They are also included in the glossary at the end of each chapter. Explain each of these concepts or terms in your own words. Add any other terms to the glossary that are new to you. Look through the Check Your Understanding boxes and exercises for questions that begin with "explain," "state," or "describe." Make sure these concepts are included in your personal glossary.

- **Study the concepts and terms in the glossary every time you do homework.**

People learn in different ways. So, everyone should not study the same way. If you learn best by listening, read the concepts and definitions into a tape recorder so that you can listen to them repeatedly. If you are a visual learner, make flash cards with the concept name on one side and its meaning on the other. Kinesthetic and visual learners may benefit from writing definitions or explanations of concepts over and over again. Some students find they learn easier if they walk around the room as they say definitions. If you learn best with others, study with a classmate or a friend. Study the glossary every time you do homework. Do not wait until just before the test.

- **Keep a list of procedures.**

Each new procedure in the text is included in the list at the end of the chapter. Write an explanation in your own words of how to do the procedure. You will remember the procedure much better if you do not just copy the words from your text or class notes. Keep procedures from previous chapters that were difficult for you on the current chapter list.

- **Study the list of procedures every time you do homework.**

You can learn procedures by memorizing the steps. Memorizing steps can be done the same way concepts and terms are learned. Or, you can practice the procedure over and over until the steps are automatic. This approach is similar to practicing a tennis serve or parallel parking until you can do them without much thought. Also, highlight two or three challenging problems in each homework section. When you do the next assignment, go back and redo these problems. Do the Check Your Understanding problems from the previous section as well. This will help to reinforce material you've recently learned.

- **Use a system to organize your work with word problems.**

Word problems often have extra information that is not needed to solve the problem. Sometimes more than one calculation is needed to get the answer. Moving from the problem to the answer can seem overwhelming. Organize your problem-solving with a strategy such as the five-step strategy in this text. This helps you break an overwhelming problem into small tasks that you can do. Begin using this strategy on fairly easy problems. By the time the word problems get more difficult, your use of the strategy will be automatic.

- **Do lots of word problems.**

Solving word problems requires some "insight"—that part of our intelligence that sees the relationship between two quantities. Insight improves with practice; practice helps us recognize common relationships and ways to approach solving a problem. If you are struggling, the best way to get better at doing word problems is to do even more problems. Do extra problems in the exercises that are not assigned by your instructor. Redo problems from previous exercises until the relationships seem automatic. Hoping to do well or wishing for luck are not effective strategies. Practice and more practice is what makes the difference. Do not pay attention to how much other people study. You have to do enough so that you can be successful.

- **Final preparation for a test.**

In math, most studying for a test is completed on a daily basis. However, you can also check your knowledge right before a test by completing the Chapter Review. Analyze your mistakes. Do extra problems from the section exercises in weak areas. Do the Chapter Test with a time limit. When the time is up, correct your test. Redo problems you missed. For extra practice, redo Check Your Understanding questions throughout the chapter. Make sure that you can explain each concept in the glossary from memory. If you have been preparing well throughout the chapter, this work should confirm that you are ready for the exam. You may also find a weak area that needs more practice before the exam.

All of this takes time and effort, but it's worth it. Make sure you have it scheduled into your regular routine. Effective studying leads to good test scores and passing grades. Imagine the last day of class, taking the final exam, and feeling confident. This can happen if you study regularly and effectively throughout this course.

RESPOND

1. Describe how you have studied for tests in previous math classes. Has this been effective?

2. Realistically, how many hours can you spend per week studying for this class?

3. Describe the best way for you to memorize terms.

Set aside several blank pages in a notebook for a glossary and procedures list. As you complete your homework assignment, write a definition or explanation for new terms or concepts and include an example. Explain how to do each new procedure in your own words. Include the page and number of example problems from homework assignments that illustrate this algorithm for later review and practice. Review every time you do homework. You may find it helpful to use note cards for reviewing.

Section	Terms and Concepts/Procedures
1.1	digit; set; element; natural numbers; counting numbers; ellipses; infinite set; finite set; whole numbers; place value; decimal point; place value form of a number; expanded form of a number; equal numbers; not equal numbers; greater than; less than; equivalent statements; equality; inequality; compound inequality; unit of measurement; point; line; number line; parallel lines; ray; endpoint; angle; plane; circle; degree; right angle; acute angle; obtuse angle; straight angle; estimate; rounded measurement; rounding digit 1. Graph a number on a number line. 2. Write an inequality using $>$ or $<$. 3. Write a compound inequality. 4. Round a whole number to a given place value. 5. Name an angle. 6. Classify an angle as acute, right, obtuse, or straight.
1.2	data; table; bar graph; axis; horizontal axis; vertical axis; mode; paired data 1. Create a bar graph from paired data.
1.3	sum; difference; axiom; commutative property of addition; associative property of addition; variable; grouping symbol; parentheses; like units; combining like units; carrying; borrowing; measurement with mixed units; complementary angles; supplementary angles 1. Add or subtract whole numbers using a number line. 2. Add or subtract like measurements. 3. Estimate a sum or difference. 4. Find the exact sum of two numbers. Carrying may be necessary. 5. Find the exact difference of two numbers. Borrowing may be necessary. 6. Add or subtract mixed English measurements.
1.4	equation; solution of an equation; reasonable solution of a problem 1. Solve a word problem using the five-step problem-solving strategy.
1.5	product; factor; implied multiplication; zero product property; commutative property of multiplication; associative property of multiplication; multiplicative identity; distributive property; conversion factor 1. Rewrite multiplication as multiple additions. 2. Evaluate an expression using the distributive property. 3. Multiply a whole number by a multiple of ten. 4. Estimate the product of numbers.

Section	Terms and Concepts/Procedures—cont'd
	5. Multiply multidigit whole numbers. **6.** Multiply a measurement by a whole number. **7.** Multiply a mixed measurement by a whole number. **8.** Convert a measurement into an equivalent measurement in different units.
1.6	exponential expression; exponent; base; power of ten; product rule of exponents **1.** Rewrite an exponential expression as repeated multiplication. **2.** Evaluate an exponential expression. **3.** Multiply exponential expressions using the product rule of exponents. **4.** Multiply a whole number by a power of ten.
1.7	SI system; prefix; area; figure; square; rectangle; cube; volume; cuboid; rectangular solid; unit square; unit cube **1.** Calculate the area of a rectangle. **2.** Calculate the volume of a cuboid.
1.8	quotient; dividend; divisor; remainder; opposite operations **1.** Rewrite division as multiple subtractions. **2.** Check the reasonability of the quotient of a division problem. **3.** Divide by a power of ten. **4.** Divide a measurement by a whole number. **5.** Divide using long division.
1.9	order of operations; evaluate; grouping symbol; arithmetic mean; average **1.** Evaluate an expression following the order of operations. **2.** Calculate the arithmetic mean of a given set of information.

REVIEW EXERCISES PREPARING FOR THE CHAPTER 1 TEST

1. **Complete the Review Exercises section by section.** If you do not know how to complete a problem, refer back to that section in the text and study your completed homework exercises and the text examples. After you complete a section, check your answers with the provided answers. Redo the problems you did incorrectly immediately. Analyze your mistakes. Are you going too fast and making computation mistakes? Do you understand the notation in the formulas? If you cannot figure out what you are doing wrong, consult your instructor.

2. **Complete extra practice in difficult concepts.** Do the Check Your Understanding exercises again and/or selected homework problems. Retake any quizzes given in class. Ask a tutor or instructor to provide extra practice problems. Use the included software to generate more practice problems.

3. **Complete the Chapter Test.** The Chapter Test in the text may be quite different in format and style from the test given to you by your instructor. However, the concepts will be the same. Complete the test in an environment similar to your classroom. Time yourself. Do not check the provided answers until you are totally finished. Redo any problems that you did not do correctly. It may be necessary to review the concept again.

REVIEW EXERCISES CHAPTER 1

SECTION 1.1 INTRODUCTION TO WHOLE NUMBERS

1. Identify the place value of 4 in the number 3,541.

2. Write the place value form of three thousand, two hundred seventy-five.

3. Write the word name of 951.

4. Write 20,713 in expanded form.

5. Rewrite the inequality statement $15 > 4$ with a $<$ symbol.

6. Write an inequality that compares 84 and 92.

7. Write a compound inequality that compares 35, 57, and 41.

8. Write the set of whole numbers using set notation.

9. Round 2,059 to the nearest ten.

10. Round 2,059 to the nearest hundred.

11. Round 2,059 to the nearest thousand.

12. Round 2,059 to the nearest million.

13. The total fall 1996 enrollment of Native American and Alaskan Native students in institutions of higher education was 134,000 students. Round this number to the nearest hundred thousand.

14. The average cost of tuition, room, and board for private 4-year universities was $23,520 for the 1996–1997 academic year. Round this amount to the nearest thousand.

SECTION 1.2 NUMBERS, TABLES, AND BAR GRAPHS

15. Present the following information in a table and in a bar graph. Place the names of the cities on the horizontal axis and the population on the vertical axis. Estimated population of the world's ten largest cities in 2000: Tokyo (28,000,000); Mexico City (18,100,000); Bombay (18,000,000); Sao Paulo, Brazil (17,700,000); New York City (16,600,000); Shanghai (14,200,000); Lagos, Nigeria (13,500,000); Los Angeles (13,100,000); Seoul (12,900,000); and Beijing (12,200,000).

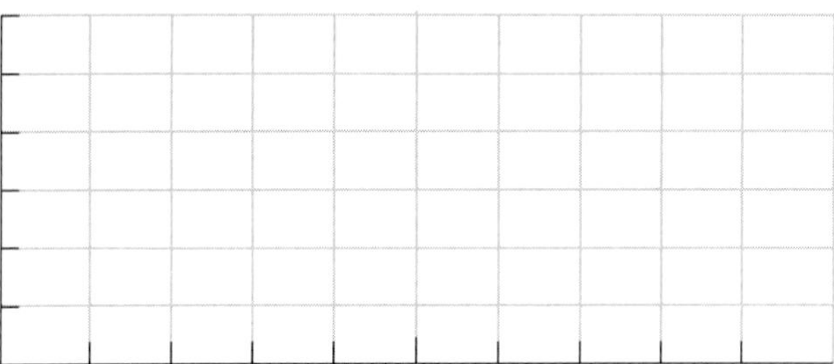

16. Use the following bar graph to estimate the total people killed in avalanches in the Alps during the last week of February 1999.

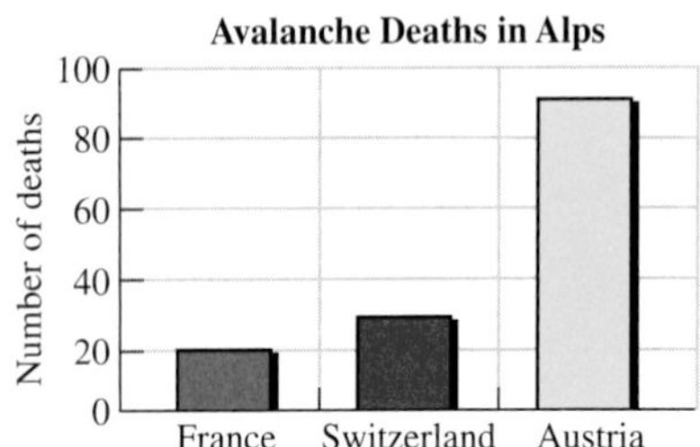

17. The following table shows the number of students scoring each of the grades in a class. Use the information to draw a bar graph and find the mode of the data.

GRADE	NUMBER OF STUDENTS
A	5
B	7
C	12
D	8
F	3

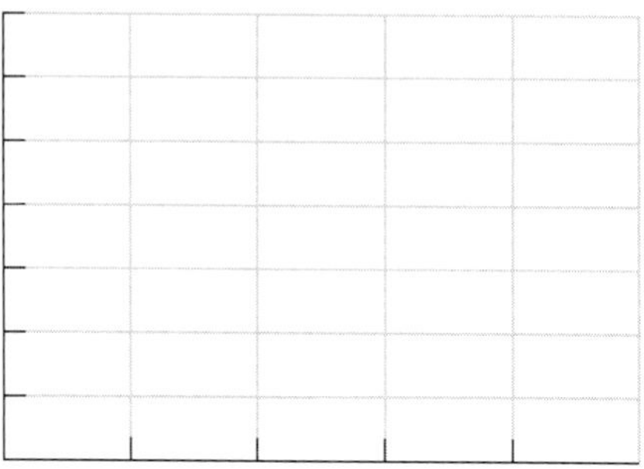

SECTION 1.3 ADDITION AND SUBTRACTION

18. State the commutative law of addition in your own words.

19. State the associative law of addition in your own words.

Evaluate.

20. 5 feet + 3 feet + 2 feet

21. 20,000 pounds − 800 pounds

22. Round 299 to the nearest hundred.

23. Round 387 to the nearest thousand.

Estimate the sum or difference.

24. 80,999 − 57,466

25. 10,652 + 903

26. 5,881 − 2

27. In 1999, the rap music label Def Jam had a value of $250,000,000. The rock music label Interscope had a value of $320,000,000. Estimate the combined value of the two record labels.

28. Plano, Texas, is a suburb of Dallas. In 1970, the population was 18,000. By 1990, the population had exploded to 130,000 people. Estimate the growth in population between 1970 and 1990.

Evaluate. Show all carrying or borrowing.

29. 1,319 − 568

30. 40,727 + 60,333

31. 1,004 − 799

32. 8,355,200 + 70,816,900

33. 87 + 55 + 30 + 117

34. 10 yd 2 ft + 3 yd 2 ft

35. 10 feet 7 inches + 3 feet 8 inches

36. $54° + 23°$

37. $180° - 93°$

38. $90° - 71°$

39. 2 hours 13 minutes − 1 hour 26 minutes

40. $119° + 24°$

41. Angles *Q* and *R* are supplementary. The measure of angle *Q* is 59°. What is the measure of angle *R*?

42. Angles *S* and *T* are complementary. The measure of angle *S* is 59°. What is the measure of angle *T*?

43. Find the perimeter of a rectangle that has a width of 3 cm and a length of 8 cm.

44. Find the perimeter of the garden shown in the diagram.

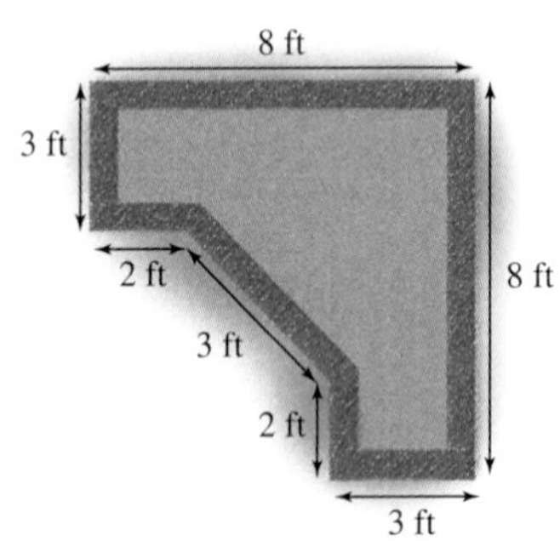

SECTION 1.4 PROBLEM SOLVING

Use the five-step problem-solving strategy.

45. In 1998, Dell Computer had revenues of \$18,200,000,000 and profits of \$1,460,000,000. What was the difference between the total revenues and the profits of this company?

46. In 1997, nearly 9,440 people died in vehicle rollovers. Of these deaths, 1,482 occurred in sport utility vehicles (SUVs). How many deaths did not occur in SUVs?

47. Eli Lilly is a pharmaceutical company whose drugs include Evista (treatment of osteoporosis) with annual sales of \$138,000,000 and Prozac (antidepressant) with annual sales of \$2,800,000,000. What is the combined annual sales of these two drugs?

SECTION 1.5 MULTIPLICATION OF WHOLE NUMBERS

48. Rewrite $4 + 4 + 4$ as a multiplication problem. Find the product.

49. State the commutative property of multiplication in words.

50. What is the product of zero and any number?

51. Evaluate: (5,000)(200)

52. Evaluate: $80{,}000 \cdot 10$

Solve using the five-step problem-solving strategy.

53. A flea can lay up to 500 eggs in its 5-week life span. How many eggs could be laid by 800 fleas in their life spans?

54. Certified wild ginseng can sell for up to $200 a pound. How much would 80 pounds of certified wild ginseng cost at a price of $200 a pound?

Estimate the product.

55. 307(54)

56. (4,928)(8)

Solve using the five-step problem-solving strategy.

57. A new diaper that contains strips treated with petrolatum and stearate was introduced by Procter & Gamble in 1999. A jumbo package of 36 such diapers costs about $13. If a baby needs 2 packages of diapers a week and there are 52 weeks in a year, estimate the cost of this kind of diapers for the baby in a year.

Calculate the exact product.

58. (399)(72)

59. $5 \cdot 209$

Solve using the five-step problem-solving strategy.

60. In February 1999, a jury awarded $730 to each of 75,000 plaintiffs from Pennsylvania to buy safer air bags for Chrysler vehicles made between 1988 and 1990. What was the total amount of this award?

61. The annual Uniontown all-you-can-eat sausage dinner costs $8 for adults. In 1999, 1,582 adult tickets were sold. What was the value of these tickets?

62. A bottle of milk contains 64 fluid ounces. How many fluid ounces are contained in six such bottles of milk?

63. A roll of wrapping paper is 6 feet 2 inches. What is the total length of eight such rolls?

Evaluate.

64. $5 \cdot 18$ inches

65. $60 \cdot 60$ seconds

66. 5(6 hours 10 minutes 4 seconds)

67. 2(3 pounds 4 ounces)

SECTION 1.6 EXPONENTS

68. Rewrite (8)(8)(8)(8) as an exponential expression.

Evaluate.

69. 4^2

70. 3^3

71. 19^0

Evaluate. Write in exponential form.

72. $3^6 \cdot 3^1$

73. $5^7 \cdot 5^4$

74. $(2^3)(2^4)(2^0)$

Write the measurement in place value form.

75. A round-the-world balloon flight was a total distance of 18×10^3 miles.

76. The state and federal government spent $\$48 \times 10^7$ to purchase 12 square miles of the Headwaters Forest and two other redwood groves from the Pacific Lumber Company and set the properties aside as public preserves.

SECTION 1.7 USING THE SI (METRIC) SYSTEM OF MEASUREMENT

Convert the measurements.

77. 16 liters = ? milliliters

78. 8,521 meters = ? centimeters

79. Calculate the area of a rectangle that has a length of 35 inches and a width of 24 inches.

80. Calculate the volume of a cuboid with length 8 centimeters, height 17 centimeters, and width 2 centimeters.

Solve using the five-step problem-solving strategy.

81. A slide of mud, fallen trees, and heavy brush in a logging area was 150 feet long and 50 feet wide. What was the area of this slide?

82. As a rising river threatened to flood a city park, volunteers shoveled sand into empty bags. Filled with sand, the bags measure about 6 inches by 2 feet by 3 feet and were used to build a temporary levee to hold back the water. What is the volume of sand needed to fill 1,000 such sandbags?

SECTION 1.8 DIVISION OF WHOLE NUMBERS

83. Calculate the quotient using multiple subtraction: 28 ÷ 7.

Find the quotient. Check.

84. 25 ÷ 8

85. 50,000 ÷ 10

86. 180,000 ÷ 600

Find the quotient using long division, then round to the nearest whole number.

87. 2,970 ÷ 3

88. 2,970 ÷ 30

89. 842 ÷ 11

Evaluate.

90. 36 feet ÷ 9

91. 18 feet 9 inches ÷ 3

Solve using the five-step problem-solving strategy.

92. The average heartbeat of an animal per minute can be calculated by dividing the heartbeats counted in a certain number of minutes by the number of minutes. A chicken's heart beat 18,641 times in 21 minutes. Find the average heartbeat of the chicken per minute. Round to the nearest heartbeat per minute.

93. Florida's citrus industry is trying to prevent Mediterranean fruit flies from entering the state. In 1998, two adult insects and five larvae were discovered in a suburb north of Miami International Airport. The state released 13,000,000 sterile male Mediterranean fruit flies each week for 12 weeks. The sterile males were expected to mate with any females that survived ground spraying in March. The target was a 25-square-mile area. Calculate the total number of sterile male flies released per square mile.

94. Grandpa is moving into a small apartment from his home of many years and is dividing his campaign button collection among his six grandchildren. If he has 1,409 buttons, how many will each grandchild receive? Will there be any left over?

SECTION 1.9 THE ORDER OF OPERATIONS

Evaluate. Show intermediate steps.

95. $100 + 200 \div 5 - 15$

96. $84 \div 3 \cdot 2 + 1 \cdot 2$

97. $(2 + 3)^2 - 4 + 6$

98. $24 \div (8 - 2) \cdot 2$

Calculate the average.

99. 12; 14; 7; 9; 8

100. 62; 36; 71; 31

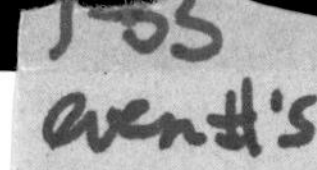

TEST CHAPTER 1

1. Explain the difference of the meaning of the wo[illegible] word "number."

2. Use set notation to write the set of natural numbers.

3. Identify the place value of 3 in the number 5,872,301.

4. Write 2,301 in expanded form.

5. Write the word name of 5,695.

6. Write two inequalities that show the relationship of 1 and 8.

7. Explain why an inequality cannot be written to compare 4 feet and 7 quarts.

8. Write the place value name of two thousand, five hundred sixty-five.

9. Round 9,516 to the tens place.

10. Round 527 to the hundreds place.

11. Round 527 to the thousands place.

12. Name this angle:

13. Graph the number 7 on a number line.

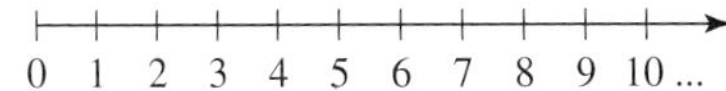

14. Name the axiom shown in this example: $3 + 7 = 7 + 3$.

15. The weight of seasoned building timbers depends on the tree used. Ash weighs 45 pounds per cubic foot; birch weighs 32 pounds per cubic foot; cherry weighs 44 pounds per cubic foot; hickory weighs 48 pounds per cubic foot; white pine weighs 28 pounds per cubic foot; hard maple weighs 42 pounds per cubic foot; white maple weighs 33 pounds per cubic foot.

a. Build a table from this information.

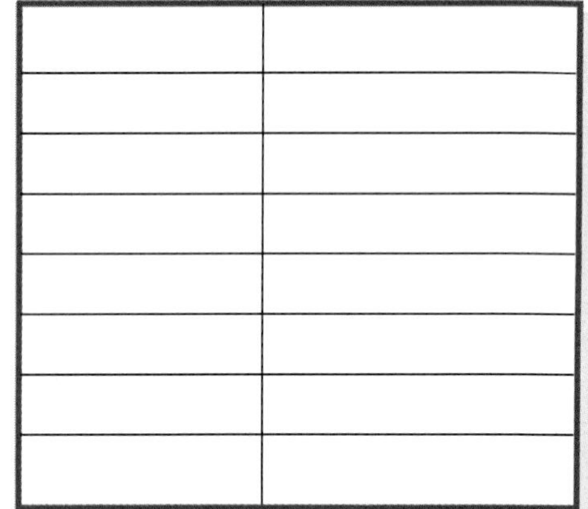

b. Present this information in a bar graph.

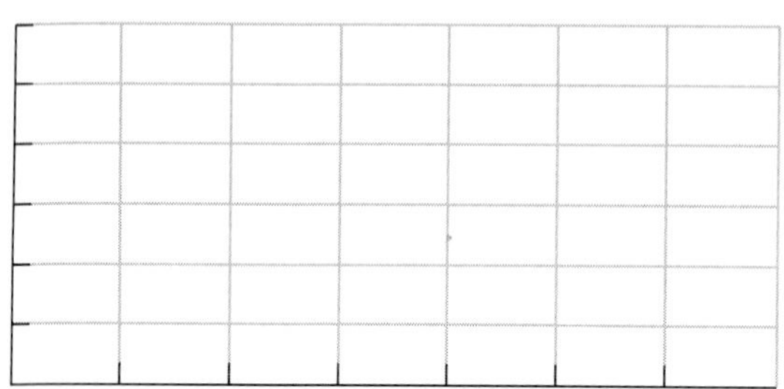

16. Use the bar graph to answer the questions.

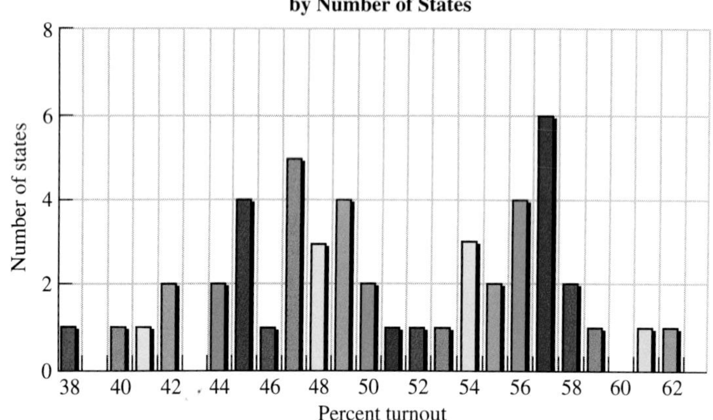

a. How many states had voter turnout of 56%?

b. How many states had voter turnouts less than 50%?

c. What is the mode?

17. Subtract using a number line: $9 - 4$

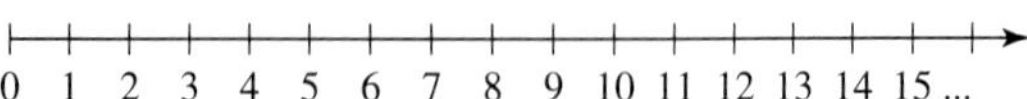

18. Subtract 15 gallons − 9 gallons.

For Problems 19 and 20, follow the five-step problem-solving strategy.

19. In his job as a real estate agent, Luis uses a minimum of 300 minutes of long distance time, 700 minutes of cellular phone time, and 1,200 minutes of local phone time a month. What is the minimum total time he spends on the phone per month?

20. A Toyota manufacturing plant in Putnam County, West Virginia, has the capacity to build 500,000 engines and 360,000 transmissions. How many more engines can be built than transmissions?

21. Explain what is necessary for two measurements to be *like measurements.*

22. Divide 12 pounds 8 ounces by 4.

23. Add 3 feet 6 inches + 2 feet 7 inches.

24. Estimate the sum 4,523 + 267 + 8,509.

25. Evaluate $(5 - 3)^2 \cdot 5 + 8 \div 2$.

26. Add 362 + 849.

27. Subtract 1,030 − 677.

28. Rewrite the multiplication 3×9 using addition.

29. Evaluate 6(9).

30. Simplify 7(8 + 9).

31. Find the product of 40,000 and 700,000.

32. Estimate the product of 8,420 and 25,901.

33. Evaluate (32)(971).

34. Evaluate 2(3 feet 5 inches).

35. Change 7 feet into inches.

36. Rewrite $9 \times 9 \times 9 \times 9$ as an exponential expression.

37. Evaluate 5^3.

38. Evaluate 7^0.

39. Evaluate and write in exponential form $(3^5)(3^0)(3^7)$.

40. Write in place value form 15×10^6 tourists.

41. Convert 185 kilometers into meters.

42. Convert 76 centimeters into millimeters.

Solve using the five-step problem-solving strategy.

43. The first drive-in movie was opened by Richard Hollingshead on June 6, 1933. The screen measured 40 feet by 30 feet. What was the area of the screen?

44. Simone has built a brick rectangular planting box in the front of her home. The inside of the box measures 2 feet high, 6 feet long, and 2 feet wide. What is the volume of dirt that will be required to fill the planting box?

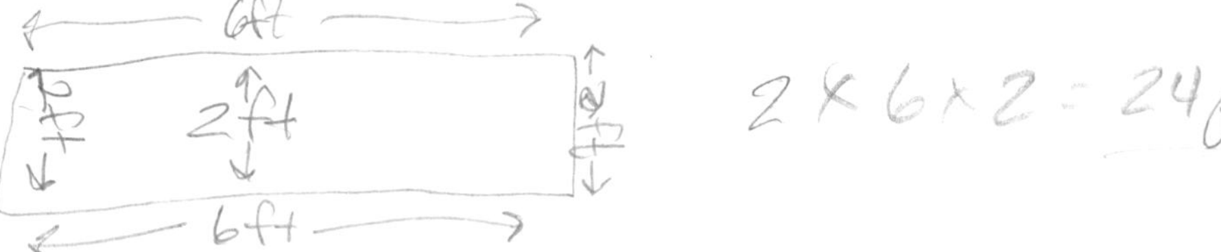

45. Write 25 ÷ 8 = 3 R(1) using multiple subtractions.

46. Write 24 ÷ 8 = 3 as a multiplication problem.

47. Evaluate 39 ÷ 9. Check the quotient.

48. Evaluate (16 feet 8 inches) ÷ 2.

49. Evaluate 450,000,000 ÷ 100.

50. Evaluate, then round to the nearest whole number 37 ÷ 7.

51. Evaluate using long division 86,789 ÷ 114.

Solve using the five-step problem-solving strategy.

52. About 2,190,000 Zippo lighters of World War II fame are still made every year. If there are 365 days in a year, calculate the average number of Zippos made per day.

53. A reflective pond is included in the design for a conservatory at an arboretum. Each side of the pond is 4 feet in length. Calculate the perimeter of the pond.

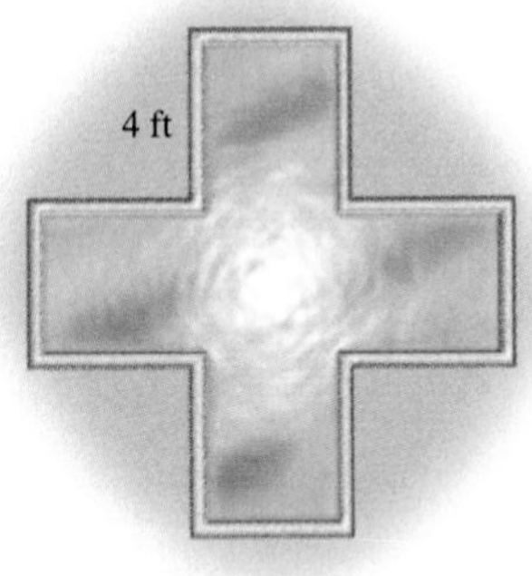

54. Angles A and B are complementary. The measure of angle A is 23°. What is the measure of angle B?

55. Angles K and L are supplementary. The measure of angle K is 19°. What is the measure of angle L?

2

Integers

Above-zero temperatures, certificates of deposit, gains, and checking accounts with sufficient funds are represented by positive numbers. Subzero temperatures, debts, losses, and overdrawn checking accounts are represented by negative numbers. The integers are the set of numbers that include the whole numbers and their opposites.

CHAPTER OUTLINE

Section 2.1 Introduction to Integers

LEARNING TOOLS

Sneak Preview

The integers are a set of numbers that include all the whole numbers and their opposites. Like whole numbers, the integers can be graphed, ordered, and combined. In this chapter we will examine the arithmetic operations of addition, subtraction, multiplication, and division of integers. We will also look at line graphs and introduce some basic formulas.

After completing this section, you should be able to:

1) Describe the set of integers.
2) Graph an integer on a number line.
3) Order two or more integers using inequality signs.
4) Identify the opposite of an integer.
5) Identify the absolute value of an integer.
6) Solve an equation that includes absolute value.
7) Round an integer to a specified place value.

DISCUSSION

Positive Numbers, Negative Numbers, and the Integers

Numbers found to the right of zero on the number line are called **positive numbers.** The natural numbers $N = \{1, 2, 3, \ldots\}$ are all positive numbers. Usually the + sign in front of a positive number is **implied** rather than written: $+6$ is usually written 6 and $+12$ is usually written 12. All positive numbers are greater than zero. The further from zero a positive number is on the number line, the greater its value. For example, 10 is further from zero than 5. So, $10 > 5$.

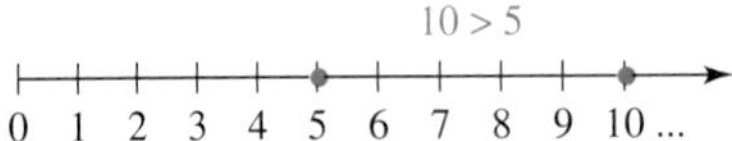

Negative numbers are less than zero and are located to the left of zero on the number line. Negative numbers are always written with a − sign. For example, negative nine is written -9. The further a negative number is from zero on the number line, the smaller its value. For example, -5 is further from zero than -1. So, $-5 < -1$. This means that numbers that are to the left of a given number are less than the given number. Numbers that are to the right of a given number are greater than the given number. -5 is to the left of -1 and is less than -1.

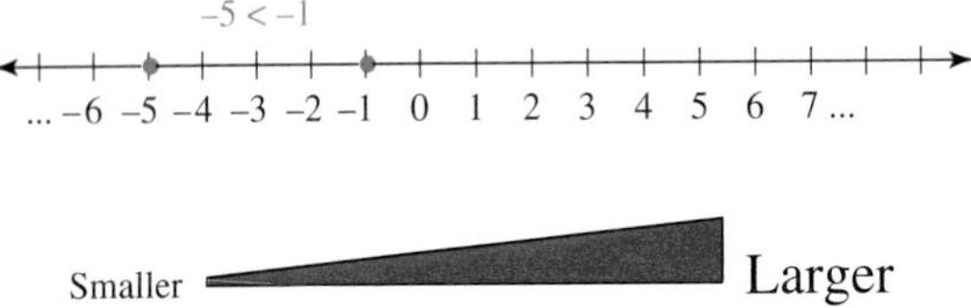

Negative numbers are used to represent situations such as debts, overdrawn bank accounts, a depth under the surface of the ocean, a distance behind the line of scrimmage in a football game, and subzero temperatures. Ten degrees above zero Fahrenheit is written 10°F. Ten degrees below zero Fahrenheit is written -10°F. A loss of \$15 is represented by $-$\$15. A depth of 130 feet below sea level is written -130 feet.

We can compare the size of such measurements by considering their place on the number line.

Example 1 **Write an inequality that compares the measurements $-2°C$ and $-4°C$.**

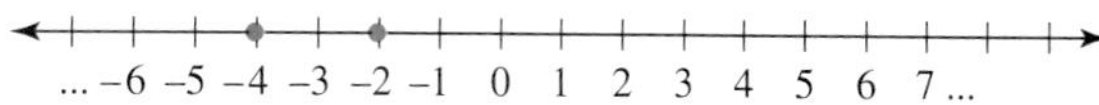

Since -2 lies to the right of -4, $-2°C > -4°C$.

Alternatively, since -4 lies to the left of -2, $-4°C < -2°C$.

Example 2 **Write an inequality that compares the measurements -353 meters and -381 meters.**

Since -353 is to the right of -381 on the number line (-353 is closer to zero), -353 meters $>$ -381 meters. Alternatively, -381 meters $<$ -353 meters.

The **integers** are the set of numbers $\{\ldots, -5, -4, -3, -2, -1, 0, 1, 2, 3, 4, \ldots\}$. The integers include negative numbers, positive numbers, and zero. Zero is neither positive nor negative.

CHECK YOUR UNDERSTANDING

Write each measurement using an integer.

1. A decrease in pressure of 4 atmospheres

2. A weight loss of 12 pounds

3. A deceleration of 20 miles per hour

4. A depth of 4 feet

5. An increase in appraised value of a house of $12,000

Place a < or > sign between the numbers to make a true statement.

6. -2 7

7. -4 -2

8. -9 -11

9. -321 -371

10. 84°F -23°F

11. $-1{,}563$ $-1{,}532$

Answers: 1. −4 atmospheres; 2. −12 lb; 3. −20 mph; 4. −4 ft; 5. $12,000; 6. $<$; 7. $<$; 8. $>$; 9. $>$; 10. $>$; 11. $<$.

DISCUSSION

Opposites

The − sign in front of a negative number represents either the word "negative" or the word "opposite." For example, -6 can be read "negative six" or "the opposite of six."

As shown in the diagram below, the **opposite** of a number is the same distance from zero, but in the other direction on the number line.

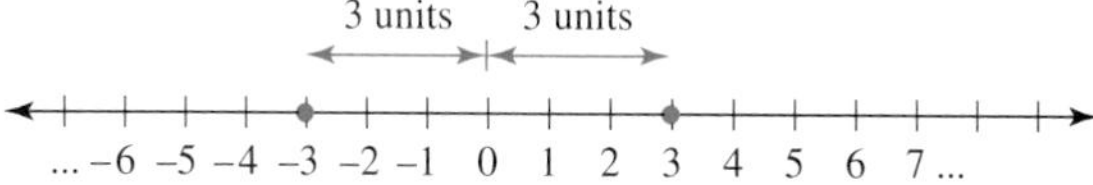

The **opposite** of 3 is −3; 3 is three units from zero and −3 is three units from zero in the other direction on the number line.

The opposite of a positive number is always a negative number.

$$-(+8) = -8$$

The opposite of a negative number is always a positive number.

$$-(-41) = 41$$

When the $-$ sign is directly in front of the number, it is usually read as "negative."

For example, -7 is usually read "negative 7." When the $-$ sign is in front of a set of parentheses, it is usually read as "opposite." For example, $-(-3)$ is usually read "the opposite of negative 3." On a calculator, the opposite of a number is often found by pushing the +/− key. More instruction on using a scientific calculator is found in Appendix 2.

The opposite of a measurement is on the other side of a **reference point.** For example, a drop of 18 points in the Dow Jones Industrial Average is the opposite of an increase of 18 points. The reference point is the starting position of the Dow Jones Industrial Average at the beginning of the trading day.

CHECK YOUR UNDERSTANDING

State the opposite of each number or measurement.

1. a gain of 10 yards

2. a decrease of 10°C

3. a decrease in SAT score of 200 points

4. 10 miles north

5. 72°F

6. -45

Answers: 1. a loss of 10 yards; 2. an increase of 10°C; 3. an increase in SAT score of 200 points; 4. 10 miles south; 5. −72°F; 6. +45 or 45.

To evaluate an expression with multiple $-$ signs or parentheses, rewrite it as an equivalent expression with the fewest possible symbols.

Example 3 **Read and then evaluate $-(-4)$.**

We read $-(-4)$ as "the opposite of negative four."

$-(-4) = 4$

Example 4 **Read and then evaluate $-(+12)$.**

We read $-(+12)$ as "the opposite of positive twelve."

$-(+12) = -12$

Example 5 **Read and then evaluate -81.**

We read -81 as "negative eighty-one." We cannot rewrite -81 with fewer symbols.

Example 6 **Read and then evaluate $-(2)$.**

We read $-(2)$ as "the opposite of two." It can be rewritten as -2.

Example 7 **Read and then evaluate $+1$.**

We read $+1$ as "positive one" and can rewrite it as 1. The $+$ sign is now implied.

CHECK YOUR UNDERSTANDING

Read and then evaluate:

1. -3 **2.** $-(+17)$ **3.** $-(9)$ **4.** $-(-16)$

5. $+12$ **6.** $-(0)$ **7.** $-(-(-10))$

Answers: 1. "negative 3," −3; 2. "the opposite of positive 17," −17; 3. "the opposite of 9," −9; 4. "the opposite of negative 16," 16; 5. "positive 12," 12; 6. "the opposite of 0," 0; 7. "the opposite of the opposite of negative 10," −10.

DISCUSSION

Absolute Value

The **absolute value** of a number is the distance the number is from zero on the number line. Absolute value ignores the direction the number is from zero on the number line, so the absolute value of a number is always positive. The symbol for absolute value is a pair of vertical lines around the number; the absolute value of 3 is written $|3|$.

Example 8 **What is the absolute value of -3?**

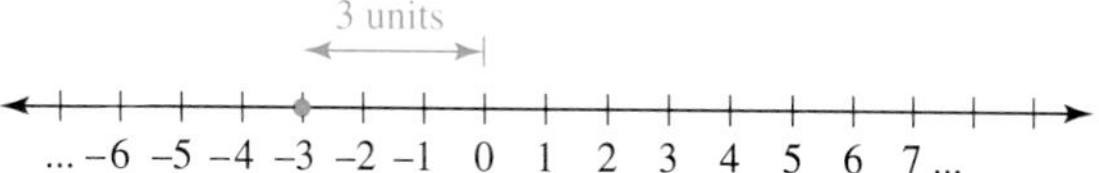

$|-3| = 3$ since -3 is a distance of 3 units from 0 on the number line.

Example 9 **Evaluate $|4|$.**

$|4| = 4$ since 4 is a distance of 4 units from 0 on the number line.

Example 10 **Evaluate $|0|$.**

$|0| = 0$ since 0 is a distance of 0 units from 0 on the number line.

Example 11 **Evaluate $-|-2|$.**

Work from the inside out. Simplify the absolute value first. Then, find the opposite.

$-|-2|$

$= -(2)$ *The absolute value of −2 is 2.*

$= -2$ *The opposite of 2 is −2.*

Example 12 **Evaluate $-|9|$.**

$-|9|$

$= -(9)$ *The absolute value of 9 is 9.*

$= -9$ *The opposite of 9 is −9.*

CHECK YOUR UNDERSTANDING

Evaluate.

1. $|8|$ **2.** $|-15|$ **3.** $|-45|$ **4.** $|0|$

5. $-|92|$ **6.** $-|-56|$ **7.** $-|-318|$

Answers: 1. 8; 2. 15; 3. 45; 4. 0; 5. −92; 6. −56; 7. −318.

DISCUSSION

Equations With Absolute Value

The definition of absolute value can be used to solve absolute value equations. Remember that the solution of an equation is the number or numbers that can replace the variable and make the equation true.

Example 13 **Solve the equation $|x| = 4$.**

The unknown, x, has an absolute value equal to 4. So, x must be exactly 4 units from zero on the number line. There are two such numbers, 4 and −4.

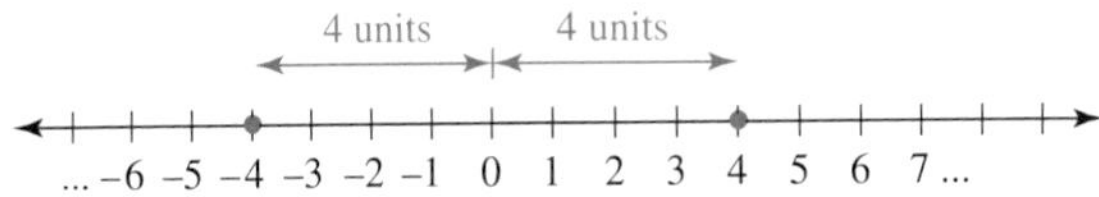

We check each solution separately:

Check	***Check***
$\lvert x\rvert = 4$	$\lvert x\rvert = 4$
$\lvert \mathbf{4}\rvert = 4$	$\lvert \mathbf{-4}\rvert = 4$
$4 = 4$ *True.*	$4 = 4$ *True.*

We say that the solutions of the equation $|x| = 4$ are $x = 4$ and $x = -4$. Or, we can write the solutions as $x = \pm 4$. The symbol $\pm$ is read "positive or negative."

Example 14 **Solve the equation $|y| = 127$.**

There are two solutions to this equation: $y = -127$ and $y = 127$.

Check	***Check***
$\lvert y\rvert = 127$	$\lvert y\rvert = 127$
$\lvert \mathbf{127}\rvert = 127$	$\lvert \mathbf{-127}\rvert = 127$
$127 = 127$ *True.*	$127 = 127$ *True.*

Example 15 **Solve the equation $|h| = -2$.**

This equation has no solution; absolute values can never be less than zero.

CHECK YOUR UNDERSTANDING

Solve each equation. Check all solutions.

1. $|x| = 3$ **2.** $|y| = 9$ **3.** $|t| = -6$ **4.** $|x| = 0$

Answers: 1. $x = 3$ or $x = -3$; 2. $y = 9$ or $y = -9$; 3. no solution; 4. $x = 0$.

DISCUSSION

Rounding Integers

Rounding approximates a number to a particular place value. The same approach used for rounding whole numbers can be used to round integers. However, for negative numbers, rounding is done using the absolute value of the number. After rounding, the negative sign is replaced.

PROCEDURE

To round an integer:

1. Find the absolute value of the integer.
2. Round this absolute value to the given place value using the rounding rules for whole numbers.
3. Attach the sign of the original number to this rounded value.

Example 16 **Round −56, 924 to the nearest thousand.**

Find the absolute value: $|-56{,}924| = 56{,}924$.

Round: 56,924 rounds to 57,000. *6 is the rounding digit.*

Attach the negative sign: $-56{,}924$ rounds to $-57{,}000$.

Rounding $-56{,}924$ to the nearest thousand determines if $-56{,}924$ is closer to $-56{,}000$ or $-57{,}000$. As the number line shows, $-56{,}924$ is closer to $-57{,}000$.

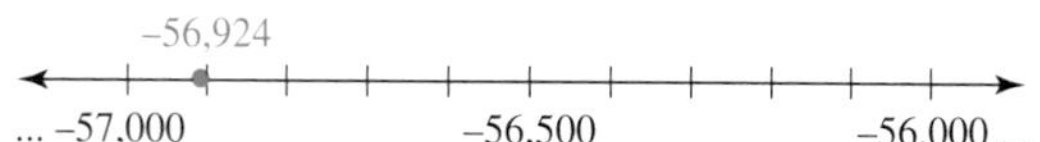

Example 17 **Round −595,942 to the nearest ten thousand.**

Find the absolute value: $|-595{,}942| = 595{,}942$.

Round: 595,942 rounds to 600,000. *The first 9 is the rounding digit.*

Attach the negative sign: $-600{,}000$.

Example 18 **Round −465 to the nearest thousand.**

Find the absolute value: $|-465| = 465$.

465 rounds to 0. *465 is closer to 0 than it is to 1,000.*

Since 0 is neither positive nor negative, -465 rounds to 0.

CHECK YOUR UNDERSTANDING

Round each number to the indicated place value.

1. 67; tens place

2. −89; tens place

3. −976; hundreds place

4. −562,917; ten thousands place

5. −379,500; thousands place

6. −132; thousands place

Answers: 1. 70; 2. −90; 3. −1,000; 4. −560,000; 5. −380,000; 6. 0.

EXERCISES SECTION 2.1

Represent each quantity as an integer.

1. A gain of \$2 per share in the price of Minnesota Mining and Manufacturing stock.
2. A decrease in the temperature of a swimming pool by 5°F.
3. A drop in value of a classic Mustang convertible by \$500.
4. An increase in weight of a 4-H lamb of 4 pounds.
5. The elevation of Mt. Whitney, California, is 14,494 feet above sea level.
6. A check written on a checking account reduces the balance by \$125.
7. A gain of 40 yards on a passing play in football.
8. A deposit made to a checking account increases the balance by \$400.
9. A flight from Seattle to New York crosses three time zones and results in a time gain of 3 hours.
10. Changing from standard time to daylight savings time results in a loss of 1 hour.

Graph each pair of numbers on its own number line.

11. 0 and 4

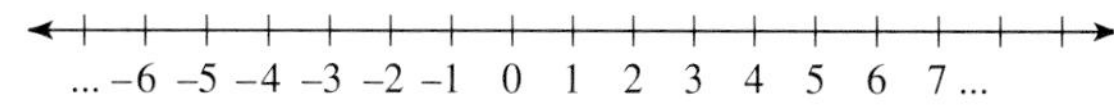

12. 0 and −4

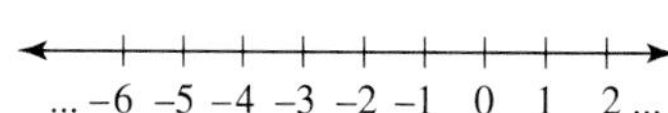

13. −4 and 4

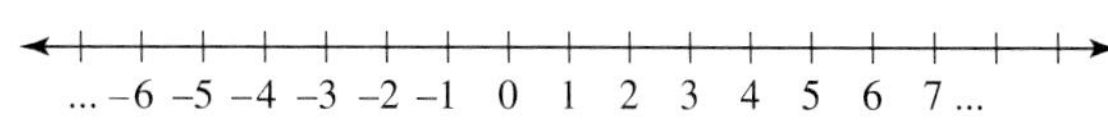

14. 2 and −10

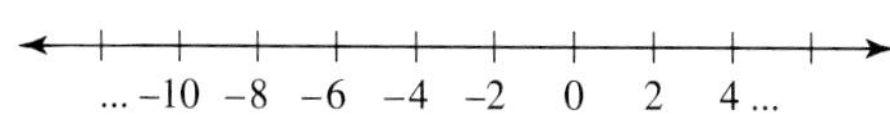

15. −2 and 10

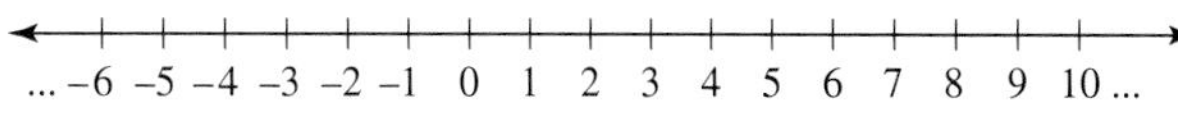

16. −2 and −10

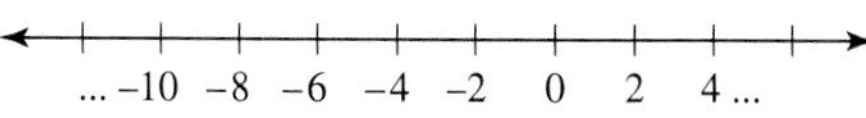

Put $<$, $>$, or $=$ between the two numbers or measurements to make a true statement.

17. 24   −24
18. 24   24
19. −24   24
20. 0   24
21. 0   −24
22. −13°F   −18°F
23. \$13   −\$18
24. −\$13   \$18
25. 0°C   −13°C

26. 0 ounces   13 ounces

27. -359 -362

28. 5,938 -19

29. \$589 $-\$590$

30. -40 10

31. $-1{,}018$ $-1{,}000$

State the opposite.

32. An increase of 10

33. 30 miles north

34. Five degrees below zero

35. A depth of 20 meters

Translate the expression into words, then evaluate.

36. $-(-3)$

37. $-1{,}005$

38. $-(+53)$

39. $-(-45)$

40. $-(-(-(-6)))$

41. $-(-8{,}927)$

42. $-(-(-891))$

43. $-(-(-3))$

Evaluate.

44. $|-3|$

45. $-|-13|$

46. $|19|$

47. $-|791|$

48. $|7|-|-3|$

49. $|-5|+|-6|$

Solve the equation. Check the solution(s).

50. $|x| = 8$

51. $12 = |t|$

52. $|t| = -5$

53. $|s| = 0$

Round each integer to the indicated place values.

	INTEGER	TENS	HUNDREDS	THOUSANDS
54.	438			
55.	−438			
56.	−1,965			
57.	8,053			
58.	−8,053			

ERROR ANALYSIS

59. **Problem:** Round −567 to the nearest thousand.
Answer: −567 rounds to −600.

Describe the mistake made in solving this problem.

Redo the problem correctly.

60. **Problem:** Solve $|x| = 9$.
Answer: The solution is $x = -9$.

Describe the mistake made in solving this problem.

Redo the problem correctly.

REVIEW

61. Using a number line, add 6 + 3.

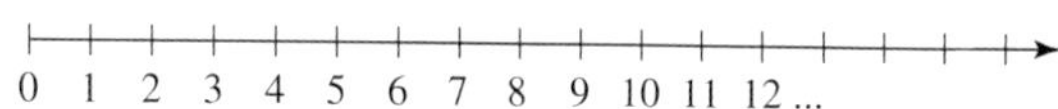

62. State the commutative property of addition.

63. State the associative property of addition.

64. Add $2 + 8 + 5 + 13$.

Section 2.2 Addition of Integers

LEARNING TOOLS

CD-ROM SSM VIDEO

Sneak Preview

When we add two integers, we can either add two positive numbers, two negative numbers, or a positive and a negative number. From our work with whole numbers, we already know how to add two positive numbers.

After completing this section, you should be able to:

1) Add integers.

2) Evaluate expressions by adding integers.

3) Solve equations that include addition of integers.

4) Solve word problems that include addition of integers.

INVESTIGATION 1 Addition of Integers

Addition is combining the value of two or more numbers to get a total or a sum. When we add a *positive* number to another number, we increase the number by a certain amount. On a number line, the sum will be found to the right of the original number.

EXAMPLE A: Show the sum of −1 and 5 using a number line.

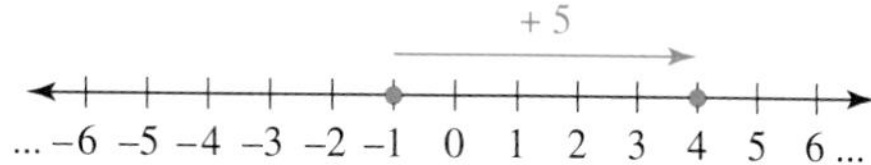

Step 1) Graph −1 on the number line.

Step 2) Move to the *right* 5 units.

Step 3) The sum is 4, the result of adding 5 to −1.

Draw a number line to determine each sum.

1. 3 + 4

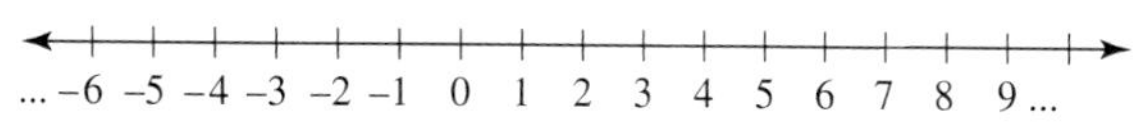

2. −2 + 6

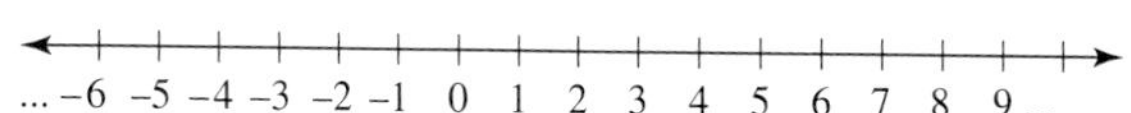

When we add a *negative* number to another number, this decreases the number by a certain amount. The sum will be found to the left of the original number on the number line. If the negative number immediately follows a plus sign, the negative number is written inside parentheses. For example, the sum of 6 and negative 8 is written 6 + (−8). The only purpose of the parentheses is to separate the addition sign from the negative number.

EXAMPLE B: Show the sum of 6 and −8 using a number line.

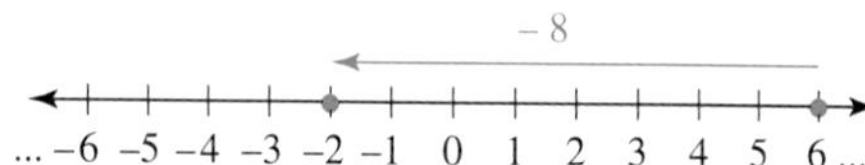

Step 1) Graph 6 on the number line.

Step 2) Move to the *left* 8 units.

Step 3) The sum is −2, the result of adding −8 to 6.

Draw a number line to determine each sum.

3. $3 + (-4)$

... −6 −5 −4 −3 −2 −1 0 1 2 3 4 5 6 7 8 9 ...

4. $-1 + (-4)$

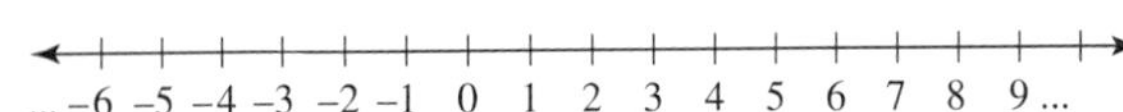

If we do not use a number line to add integers, we need a way to determine the sign of the sum. Fill in the blank in each sentence with one of these words: *positive, negative, add,* or *subtract.*

5. $6 + 2 = 8$ The sum of a positive number and a positive number is always a __________ number.

6. $6 + (-2) = 4$ The sum of a larger positive number and a smaller negative number is a __________ number.

7. $-6 + 2 = -4$ The sum of a larger negative number and a smaller positive number is a __________ number.

8. $-6 + (-2) = -8$ The sum of a negative number and a negative number is always a __________ number.

9. Explain how to determine the sign of the sum of two integers.

Evaluate.

10. $5 + (-3)$

11. $-8 + 15$

12. $-15 + (-6)$

13. $-15 + 6$

14. $-20 + (-30)$

15. $20 + 30$

16. $-20 + 30$

17. $20 + (-30)$

Answers: **10.** *2;* **11.** *7;* **12.** *−21;* **13.** *−9;* **14.** *−50;* **15.** *50;* **16.** *10;* **17.** *−10.*

DISCUSSION

Addition of Integers

When a positive number is added to a positive number, the sum is found to the right of the first number on the number line. However, when a negative number is added to a positive number, the sum is found to the left of the positive number.

Example 1 **Add $6 + (-5)$.**

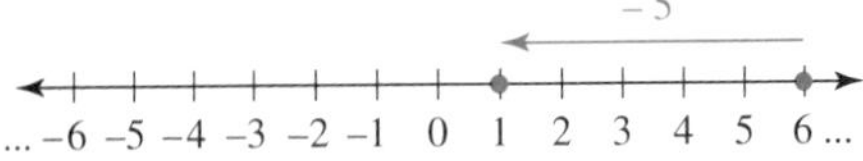

$6 + (-5) = 1$ *The parentheses around (−5) keep the + and − signs separated.*

Using a number line is not a practical way to add large integers. Instead, we can use the concept of absolute value to find the correct sum.

PROCEDURE

To add integers:

1. If the numbers have the same sign, add the absolute values of the numbers and attach the original sign to this sum.
2. If the numbers have different signs, subtract the smaller absolute value from the larger absolute value. Attach the original sign of the number that has the larger absolute value to this difference.

Example 2 **Add $-5°C + (-6°C)$.**

These numbers have the same sign.

Absolute values: $|-5| = 5$; $|-6| = 6$.
Add the absolute values: $5 + 6 = 11$.

The original sign was $-$ so the sum will be negative: $-5 + (-6) = -11$.

Example 3 **Add $-43 + 48$.**

These numbers have different signs.

Absolute values: $|-43| = 43$; $|48| = 48$.
Subtract the smaller absolute value from the larger absolute value: $48 - 43 = 5$.

Since $|48| > |-43|$, attach the positive sign from 48: $-43 + 48 = 5$.

Example 4 **Add $-1{,}943 + 212$.**

These numbers have different signs.

Absolute values: $|-1{,}943| = 1{,}943$; $|212| = 212$.
Subtract the smaller absolute value from the larger absolute value: $1{,}943 - 212 = 1{,}731$.

Since $|-1{,}943| > |212|$, attach the negative sign of $-1{,}943$: $-1{,}943 + 212 = -1{,}731$.

CHECK YOUR UNDERSTANDING

Evaluate.

1. $5 + (-8)$

2. $-7 + 14$

3. $4 + (-5) + 7$

4. $-2 + (-7)$

5. $-3°C + (-9°C)$

6. $3°C + (-9°C)$

Answers: 1. −3; 2. 7; 3. 6; 4. −9; 5. −12°C; 6. −6°C.

When adding three or more integers, we can choose either of two methods. In the first method, we just follow the order of operations and add from left to right. In the second method, we use the commutative and associative properties of addition to regroup and reorder the numbers before adding.

PROCEDURE

To add three or more integers:

Method 1 Add the numbers two at a time from left to right.

Method 2 Add all the positive numbers to get a positive sum. Add all the negative numbers to get a negative sum. Add the two sums.

Example 5 **Add $7 + (-5) + (-8) + 9 + 2 + (-7)$.**

Method 1

Add the numbers two at a time from left to right.

$$\begin{aligned} & 7 + (-5) + (-8) + 9 + 2 + (-7) & \\ &= 2 + (-8) + 9 + 2 + (-7) & 7 + (-5) = 2 \\ &= -6 + 9 + 2 + (-7) & 2 + (-8) = -6 \\ &= 3 + 2 + (-7) & -6 + 9 = 3 \\ &= 5 + (-7) & 3 + 2 = 5 \\ &= -2 & \end{aligned}$$

Method 2

Add the positive numbers; add the negative numbers. Add these sums.

$$\begin{aligned} & 7 + (-5) + (-8) + 9 + 2 + (-7) \\ &= (7 + 9 + 2) + (-5 + (-8) + (-7)) \\ &= 18 + (-20) \\ &= -2 \end{aligned}$$

Example 6 **Solve the equation $T = 8 + 9 + (-5) + (-4)$.**

$$\begin{aligned} T &= 8 + 9 + (-5) + (-4) \\ &= (8 + 9) + (-5 + (-4)) \\ &= 17 + (-9) \\ &= 8 \end{aligned}$$

Example 7 **Solve the equation $P = \$14$ million + $8 million + (−$7 million).**

$$\begin{aligned} P &= \$14 \text{ million} + \$8 \text{ million} + (-\$7 \text{ million}) \\ &= \$(14 + 8 + (-7)) \text{ million} \\ &= \$(22 + (-7)) \text{ million} \\ &= \$15 \text{ million} \end{aligned}$$

Example 8 **The profits and losses in millions of dollars per quarter of an export business are shown in the bar graph. Calculate the total profit or loss over the 2-year period.**

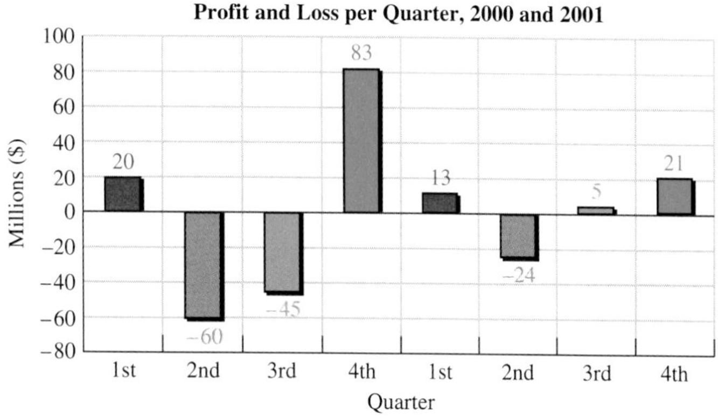

I know . . .

T = Total profit or loss in millions of dollars

\$20 = 1st quarter 2000 profit	−\$60 = 2nd quarter 2000 loss
−\$45 = 3rd quarter 2000 loss	\$83 = 4th quarter 2000 profit
\$13 = 1st quarter 2001 profit	−\$24 = 2nd quarter 2001 loss
\$5 = 3rd quarter 2001 profit	\$21 = 4th quarter 2001 profit

Equations

Total profit = Sum of profits + Sum of losses

$$T = (\$20 + \$83 + \$13 + \$5 + \$21) + (-\$60 + (-\$45) + (-\$24))$$

Solve

$T = (\$20 + \$83 + \$13 + \$5 + \$21) + (-\$60 + (-\$45) + (-\$24))$

$= (\$142) + (-\$129)$

$= \$13$ *Since \$13 is positive, it is a profit.*

Check

Estimate the total profit by rounding each profit or loss to the tens place.

$T = (\$20 + \$80 + \$10 + \$10 + \$20) + (-\$60 + (-\$50) + (-\$20))$

$= \$140 + (-\$130)$

$= \$10$ *This is close to our exact answer of \$13; the answer is reasonable.*

Answer

The total profit over the two-year period was \$13 million.

The solution to the equation was $T = \$13$. However, in the *I know* step, we said that T was the total profit or loss in millions of dollars. The final answer must also be in millions of dollars.

CHECK YOUR UNDERSTANDING

Evaluate.

1. $3 + (-6) + 5 + (-7) + (-10)$

2. $-2 + 7 + 10 + (-4) + (-1)$

3. Solve the equation $P = \$4 + (-\$3) + \$7$.

Answers: 1. -15; 2. 10; 3. $P = \$8$.

EXERCISES SECTION 2.2

Find the sum using a number line.

1. $3 + 5$

... 0 1 2 3 4 5 6 7 8 9 10 11 12 ...

2. $3 + (-5)$

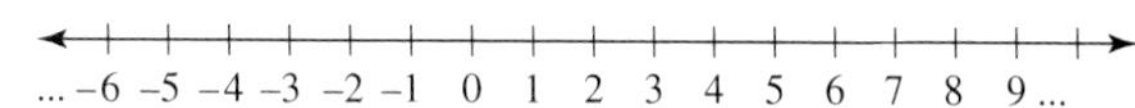

3. $-5 + 3$

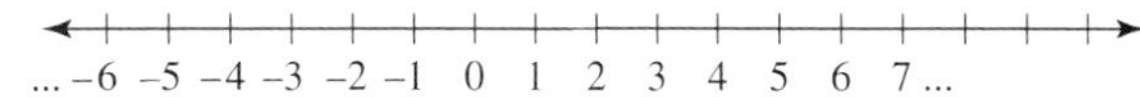

4. $-5 + (-3)$

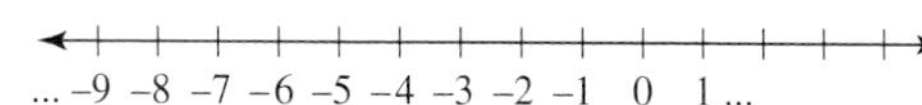

5. $0 + (-5)$

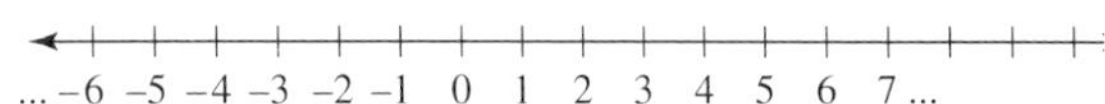

6. $0 + 3$

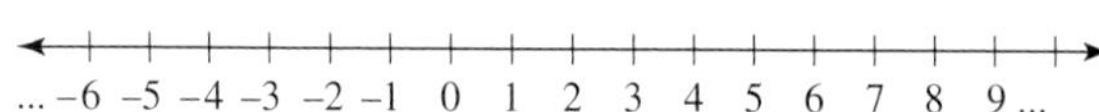

7. $10 + (-2)$

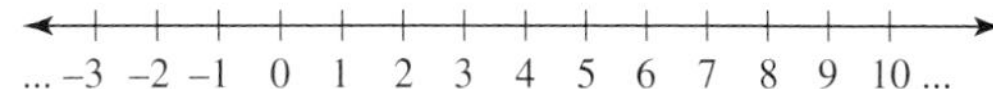

8. $-10 + 2$

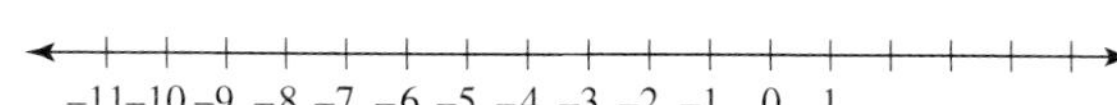

9. $-10 + (-2)$

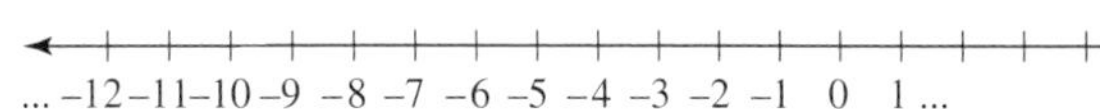

10. $-5 + 5$

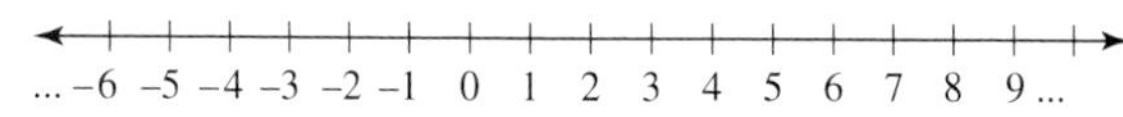

Find the sum.

11. 12 meters + 3 meters

12. $12 + (-3)$

13. $-12°C + (-3°C)$

14. $-3 + (-12)$

15. $-12 + 3$

16. $-5 + 5$

17. $-7°F + 9°F$

18. 7 pounds + 9 pounds

19. $7 + (-9)$

20. $7 + (-7)$

21. $-9 + (-7)$

22. $-9 + 9$

23. $8 + (-5)$

24. $-8 + (-5)$

25. $-8 + 5$

26. $-4 + (-9) + 8 + (-4)$

27. $8 + 2 + (-9) + (-6) + 5 + (-4)$

28. $-15°C + (-25°C)$

Solve the word problem using the five-step problem-solving strategy.

29. Phil has \$524 in his checking account. He deposits his wage check of \$926. He writes checks in the amounts \$42, \$198, and \$582. He also withdraws \$100 from the Automatic Teller Machine (ATM). After these transactions, how much money is left in Phil's account?

30. A hiker starts at an elevation of 5,600 feet above sea level. She climbs 1,538 feet up a mountain, then descends 864 feet to a mountain lake. What is the elevation of the lake?

31. In a game of Monopoly, Curtis started with $1,500. He bought a property for $135, won a beauty contest and received $14, had to pay the luxury tax of $75, bought another property for $180, passed Go and earned $200, landed in jail and paid $50 to get out, and paid a rent of $75. How much money does he still have?

Find the sum.

32. $7 + (-3)$

33. $\$3 + (-\$7)$

34. $-7 + (-3)$

35. $2 + (-6)$

36. $-6 + (-2)$

37. $-80 + 100$

38. $80 + (-100)$

39. $-80 + (-100)$

40. $0 + (-4)$

41. Johnny was playing on the hotel elevator. He started on the ninth floor. The elevator rose 5 floors, dropped 3 floors, dropped 2 floors, rose 12 floors, and dropped 7 floors. On which floor did it finally stop?

Solve the equation.

42. $y = -20 + 8$

43. $x = -200 + 100 + 500 + (-30)$

Solve the word problem using the five-step problem-solving strategy.

44. Rhonda has a balance of \$450 in her checking account. She writes a check for \$525. What is the new balance in her account?

45. Rhonda's bank account is overdrawn \$389. She makes a deposit of \$701. What is the new balance in her account?

46. The temperature in an orange grove in Citrusville, Florida, is 82°F. The weatherman predicts a dramatic fall in temperatures of 50 degrees today. What will the new temperature be?

47. Write and solve a word problem involving withdrawals and deposits into a savings account. The problem should include positive numbers (deposits) and negative numbers (withdrawals).

48. Write and solve a word problem involving temperatures. The problem should include positive numbers (above zero temperatures) and negative numbers (below zero temperatures).

49. The Stock Market Crash in 1929 resulted in huge losses in the values of stock. On October 24, 1929, the price of a share of Montgomery Ward stock was \$83 per share. Earlier in 1929 the stock had been worth \$156 a share. What was the decrease in value of the stock?

ERROR ANALYSIS

50. **Problem:** Evaluate $3 + (-4) + (-7) + 9$.

Answer:

$$\begin{aligned} & 3 + (-4) + (-7) + 9 \\ &= 7 + (-7) + 9 \\ &= 0 + 9 \\ &= 9 \end{aligned}$$

Describe the mistake made in solving this problem.

Redo the problem correctly.

REVIEW

51. What is the opposite of six?

52. What is the opposite of negative six?

53. Evaluate $15 - 8$.

54. Evaluate $15 + 8$.

Section 2.3 Subtraction of Integers

LEARNING TOOLS

CD-ROM

SSM

VIDEO

Sneak Preview

Subtraction is the opposite operation of addition. Subtraction of integers can be visualized with number lines or rewritten as addition.

After completing this section, you should be able to:

1) Subtract integers.

2) Solve equations that include subtraction of integers.

3) Solve word problems that include subtraction of integers.

INVESTIGATION Subtraction of Integers

When subtracting whole numbers, there is only one minus sign: $5 - 3 = 2$. When subtracting integers, we may have both a minus sign and one or more negative numbers.

Evaluate.

1. $6 - 2$

2. $6 + (-2)$

3. Is $6 - 2$ equal to $6 + (-2)$?

4. $8 - 5$

5. $8 + (-5)$

6. Is $8 - 5$ equal to $8 + (-5)$?

As these examples show, we can rewrite a subtraction of a positive number as addition of a negative number.

7. Rewrite $7 - 2$ as addition of a negative number. Find the sum.

8. Rewrite $-7 - 2$ as addition of a negative number. Find the sum.

Answers: 1. 4; 2. 4; 3. yes; 4. 3; 5. 3; 6. yes; 7. $7 + (-2) = 5$; 8. $-7 + (-2) = -9$.

When we rewrite subtraction as addition, we may have to evaluate the opposite of a negative number.

9. What number is the opposite of six?

10. What number is the opposite of negative six?

11. Evaluate $-(-6)$.

12. Rewrite $10 - (-6)$ as addition. Find the sum.

13. Evaluate $10 + -(-6)$.

14. Rewrite $7 - (-3)$ as addition. Find the sum.

15. Rewrite $-7 - (-3)$ as addition. Find the sum.

16. Rewrite $-7 - 3$ as addition. Find the sum.

Answers: 9. -6; 10. 6; 11. 6; 12. $10 + 6 = 16$; 13. $10 + 6 = 16$; 14. $7 + -(-3) = 10$; 15. $-7 + 3 = -4$; 16. $-7 + (-3) = -10$.

DISCUSSION

Subtraction of Integers

When we add a positive number to another number, we increase the original number by a certain amount. The sum will be found to the right of the original number on the number line. Subtraction is the opposite of addition. When we subtract a positive number from another number, we are decreasing that number by a certain amount. We move to the left on the number line.

Example 1 **Subtract $-2 - 3$.**

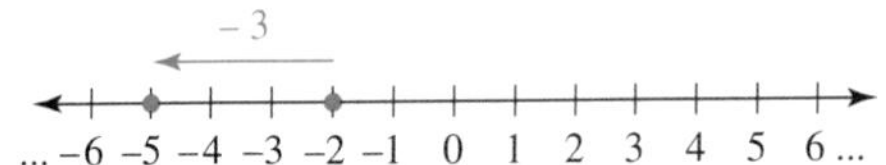

Start at −2 on the number line.
Move 3 units to the left.
$-2 - 3 = -5$

When we subtract $-2 - 3$, we get the same result as when we add $-2 + (-3)$. We can always rewrite subtraction as addition by adding the opposite of the number being subtracted.

PROCEDURE

To subtract two integers:

1. Rewrite the subtraction as addition of the opposite of the number being subtracted.
2. Complete the addition.

Example 2 **Subtract 5 − 9.**

$5 - 9$ *We are subtracting 9, so add the opposite of 9: −9.*
$= 5 + (-9)$ *Since the signs are different, subtract the smaller absolute value from the larger absolute value and attach the sign of the number with the larger absolute value.*
$= -4$

Example 3 **Subtract −30 − 80.**

$-30 - 80$ *We are subtracting 80, so add the opposite of 80: −80.*
$= -30 + (-80)$ *Since the signs are the same, add the numbers and replace the negative sign.*
$= -110$

To subtract a negative number, we rewrite the subtraction as an addition and find the opposite of the negative number being subtracted. The result is addition of a positive number.

Example 4 **Subtract −30 − (−80).**

$-30 - (-80)$
$= -30 + (-(-80))$ *Rewrite the subtraction as addition of an opposite.*
$= -30 + 80$ *The opposite of a negative number is a positive number.*
$= 50$ *Add.*

This process of rewriting subtraction as addition and evaluating the opposite of a negative number can be combined into a single step.

Example 5 **Subtract 30 − (−80).**

$30 - (-80)$
$= 30 + 80$ *Subtracting a negative number is equivalent to adding the number.*
$= 110$ *Add.*

When the variable in an equation is isolated on one side of the equation, we can solve the equation by simplifying the expression on the other side.

Example 6 **Solve the equation $m = 10 - 36 - (-12)$.**

$m = 10 - 36 - (-12)$
$= 10 + (-36) + 12$ *Rewrite subtraction as addition of the opposite.*
$= -14$ *Add from left to right.*

Example 7 **On a January morning the temperature in Moab, Utah, was 5°F and the temperature in Salt Lake City, Utah, was −6°F. Find the difference of the two temperatures.**

I know . . .
D = Difference in temperatures
Temperature in Moab = 5°F
Temperature in Salt Lake City = −6°F

Equations
Difference in temperatures = Temperature in Moab − Temperature in Salt Lake City

$$T = 5°\text{F} - (-6°\text{F})$$

Solve
$T = 5°\text{F} - (-6°\text{F})$
$= 5°\text{F} + 6°\text{F}$ *Rewrite subtraction as addition of the opposite.*
$= 11°\text{F}$ *Add.*

Check

The difference in temperatures will be the number of degrees from 0°F to 5°F added to the number of degrees from −6°F to 0°F. This requires adding 5°F to 6°F, which equals 11°F. So, this answer is reasonable.

Answer

The difference in the two temperatures is 11°F.

CHECK YOUR UNDERSTANDING

Rewrite each subtraction as addition. Add.

1. $15 - 3$

2. $-15 - 3$

3. $-15 - (-3)$

4. $-15 + 3$

5. $-15 + (-3)$

6. $15 + 3$

7. $0 - (-15)$

8. $0 - 15$

Solve the equation.

9. $S = 23 - (-16)$

10. $T = 9°\text{F} - 15°\text{F}$

Answers: 1. 12; 2. −18; 3. −12; 4. −12; 5. −18; 6. 18; 7. 15; 8. −15; 9. S = 39; 10. T = −6°F.

EXERCISES SECTION 2.3

Find the sum or difference.

1. $7 - 3$

2. $7 + (-3)$

3. $\$3 - \$7$

4. $3 - (-7)$

5. $0 - 4$

6. $0 - (-4)$

7. $2 - 6$

8. $-6 - (-2)$

9. $7 - (-3)$

10. $80 + (-100)$

11. $80 - 100$

12. $-80 - 100$

13. $15 - (-4)$

14. $-15 - (-4)$

15. $-15 + (-4)$

16. $-15 + 4$

17. $12 - 1 - 6$

18. $-12 - 1 - 6$

19. $-12 + 1 - 6$

20. $100 - 1{,}000$

21. $50 - 50 - 50$

22. $58 - (-37)$

23. $-81 - 77$

24. $-81 + 77$

25. $750 + (-266)$

26. $13{,}451 - 20{,}000$

27. $-8{,}590 - (-2{,}261)$

28. $3 - 5 - 8$

29. $-5 - (-6) + 12 - 25$

30. $0 - 11 - 20$

Solve the word problem using the five-step problem-solving strategy.

31. The temperature in Portland, Maine, was -3°C when it was 14°C in Portland, Oregon. What was the difference in temperature between Portland, Oregon, and Portland, Maine?

32. Jaime dove into a 10-foot-deep swimming pool from a 12-foot diving platform. What is the distance from the platform to the bottom of the pool?

33. Complete the checking account register.

CHECK NO.	PAYEE	DEBIT AMOUNT	CREDIT AMOUNT	BALANCE
				$532
454	Safeway	$38		
455	Bookstore	$162		
ATM	Cash Withdrawal	$45		
456	Ray J. White Properties	$475		
457	TR Video	$6		
458	Domino's Pizza	$16		
459	GM Finance	$384		
DEP	Deposit		$831	
460	Lamont's	$72		
461	TR Video	$6		
462	Valley Medical Center	$125		
463	Kmart Pharmacy	$15		
464	Albertson's	$82		
465	Bob's Pet Store	$20		
466	Ray J. White Properties	$475		

Solve the equation.

34. $y = -1 - 6$

35. $x = 5 - 27$

36. $t = 19 - 15 - (-4)$

37. $m = 19 - (-15) - (-4)$

Solve using the five-step problem-solving strategy.

38. The highest point in Asia is at the top of Mount Everest, 8,848 meters above sea level. The lowest point in Asia is the Dead Sea at 408 meters below sea level. What is the difference between the highest and the lowest points in Asia?

The Dead Sea *(Mia & Klaus Matthes/Superstock)*

39. Emma has a balance of \$450 in her checking account. If she writes a check for \$525, what is the new balance in her account?

40. Uncle Ralph has offered to help his nephew go to college by providing no-interest loans to help with expenses. Sean borrowed \$3,500 to pay for his tuition. When he sold his truck and bought a motorcycle to replace it, he paid his uncle \$1,100 of the amount he owed. The next semester he borrowed \$2,600. How much does he now owe his uncle?

41. The temperature in Lethbridge, Alberta, was $-8°F$. The temperature in Buffalo, New York, was $-12°F$. What was the difference in temperatures between Buffalo and Lethbridge?

42. Frank is standing on top of a cliff overhanging a 200-foot-deep glacial lake in Montana. The cliff is 35 feet high. How far is it from the top of the cliff to the bottom of the lake?

43. Write and solve a word problem involving integers.

44. When a hospital is at full occupancy, the number of patients that can be admitted depends on beds being emptied by patients leaving the hospital. St. Anthony's Hospital begins the day with all of its 300 beds full. During the course of the morning 48 patients were discharged, 3 died, and 31 patients were admitted. How many beds were occupied by the end of the morning?

45. Emily used a GPS to determine her elevation at the low and high points of her day hike in the Wasatch Mountains. The trailhead was at 8,143 feet elevation above sea level. She hiked up to a pass that was at 9,501 feet, descended to a valley that was at 8,462 feet, hiked up to a second pass that was at 8,859 feet elevation, and finally descended to a campsite at 7,950 feet. How many feet of her hike required her to hike uphill?

ERROR ANALYSIS

46. Problem: Evaluate $-4 + 7$.

Answer:

$$-4 + 7$$
$$= -4 + (-7)$$
$$= -11$$

Describe the mistake made in solving this problem.

Redo the problem correctly.

47. Problem: Evaluate $-7 - 8$.

Answer:

$$-7 - 8$$
$$= -7 + 8$$
$$= -1$$

Describe the mistake made in solving this problem.

Redo the problem correctly.

REVIEW

48. Describe the order of operations.

49. Evaluate: $18 \div 2 \cdot 4$

50. Evaluate: $30 - (6 + 4 \cdot 2) \div 2$

51. Evaluate: $\frac{20}{4}$

Section 2.4 Multiplication and Division of Integers

LEARNING TOOLS

CD-ROM SSM VIDEO

Sneak Preview

Multiplication and division of integers is based on the same principles as multiplication and division of whole numbers. However, the sign of the product or quotient must also be determined.

After completing this section, you should be able to:

1) Multiply integers.
2) Divide integers.
3) Evaluate expressions using the distributive property.
4) Solve equations that include multiplication or division of integers.
5) Solve word problems that involve integers.

DISCUSSION

Multiplication and Division of Integers

Multiplication and division of integers is similar to multiplication and division of whole numbers. However, we must also determine whether the product or quotient is negative or positive. The multiples of 2 show what the sign of a product or quotient must be.

$$\left.\begin{array}{l} \ldots \\ (2)(3) = 6 \\ (2)(2) = 4 \\ (2)(1) = 2 \\ (2)(0) = 0 \end{array}\right\}$$

The product of 2 and any whole number is a positive number.
Each product in the pattern is two less than the one above it.

This pattern can be continued, multiplying two by the negative integers. Again, each product is two less than the one before it.

$$\left.\begin{array}{ll} 2(-1) = -2 & \textit{Subtract 2 from 0.} \\ 2(-2) = -4 & \textit{Subtract 2 from } -2. \\ 2(-3) = -6 & \textit{Subtract 2 from } -4. \\ \ldots & \end{array}\right\}$$

The product of 2 and a negative number is always negative.

This pattern in the products shows that the product of a positive and a negative number is a negative number. We can build a similar pattern with the multiples of -2 that reveals the sign of the product of a negative number and another negative number.

$$\left.\begin{array}{l} \ldots \\ -2(3) = -6 \\ -2(2) = -4 \\ -2(1) = -2 \\ -2(0) = 0 \end{array}\right\}$$

The product of -2 and the whole numbers are all negative numbers.
Each product is two more than the one above it.

We continue the pattern, multiplying -2 by the negative integers. Again, each product is two more than the one before it.

$$\left.\begin{array}{ll} -2(-1) = 2 & \textit{Add 2 to 0.} \\ -2(-2) = 4 & \textit{Add 2 to 2.} \\ -2(-3) = 6 & \textit{Add 2 to 4.} \\ \ldots & \end{array}\right\}$$

The product of -2 and a negative number is positive.

When the second number is negative, the product is positive. This pattern shows that the product of two negative numbers is always a positive number.

PROCEDURE

To multiply or divide integers:

1. Multiply or divide the absolute value of the integers.
2. Determine the sign of the product or quotient. The sign is positive if the two integers have the same sign. The sign is negative if the two integers have opposite signs.

Example 1 **Evaluate $4(-2)$.**

$4(-2) = -8$ *The product of the absolute values is 8; the signs are different so the product is negative.*

Example 2 **Evaluate $-12 \div (-6)$.**

$-12 \div (-6) = 2$ *The quotient of the absolute values is 2; the signs are the same so the quotient is positive.*

Example 3 **Evaluate $-5 \cdot 0$.**

$-5 \cdot 0 = 0$ *The product of any number and zero is always zero.*

Example 4 **Evaluate $15 \div 0$.**

$15 \div 0$ is undefined *Division by zero is always undefined.*

Example 5 **Evaluate $(-9)(-5)$.**

$(-9)(-5) = 45$ *The product of the absolute values is 45; the signs are the same so the product is positive.*

Example 6 **Evaluate $0 \div (-13)$.**

$0 \div (-13) = 0$ *The quotient of zero divided by any number is always zero.*

Example 7 **Gita decides to automatically transfer \$200 from her checking account into her savings account each month. Calculate the amount of money that will be deducted in 2 years.**

I know . . .

m = Total money

$-\$200$ = Money per month

2 = Number of years

24 = Number of months

Equations

Total money = (Money per month)(Number of months)

$m = (-\$200)(24)$

Solve

$m = (-\$200)(24)$

$= -\$4{,}800$ *The product of a negative and a positive is a negative.*

Check

We can get a rough estimate of the total by rounding 24 to 20 and multiplying. Since $(-\$200)(20) = -\$4{,}000$ is close to our exact answer of $-\$4{,}800$, this answer is reasonable.

Answer

In 2 years, a total of \$4,800 will be deducted from her checking account.

When multiplying more than two integers, either of two methods can be used. In the first method, we just follow the order of operations and multiply the numbers two at a time from left to right.

Example 8 **Evaluate 3(−2)(−1)(7).**

$3(-2)(-1)(7)$
$= (-6)(-1)(7)$ *Multiply 3(−2).*
$= (6)(7)$ *Multiply (−6)(−1).*
$= 42$ *Multiply (6)(7).*

In the second method, we use the **commutative** and **associative properties of multiplication** to reorder and regroup the numbers before multiplying. Multiplication is done in pairs to speed up the process.

Example 9 **Solve $x = 3(-2)(-1)(7)$.**

$x = 3(-2)(-1)(7)$
$= (-6)(-7)$ *Multiply in pairs from left to right: 3(−2) = −6 and (−1)(7) = −7.*
$= 42$ *Continue multiplying.*

Example 10 **Evaluate 7(−1)(−2)(3)(−1).**

$7(-1)(-2)(3)(-1)$
$= (-7)(-6)(-1)$ *Multiply in pairs from left to right.*
$= (42)(-1)$ *Multiply from left to right.*
$= -42$ *Multiply.*

Example 11 **Solve $t = (-3)(4)(-1)(-2)(0)(-8)(4)$.**

$t = (-3)(4)(-1)(-2)(0)(-8)(4)$
$= (-12)(2)(0)(4)$ *Multiply in pairs from left to right.*
$= (-24)(0)$ *Multiply in pairs from left to right.*
$= 0$ *The product of 0 and any number is always zero.*

These examples show that the product of an even number of negative factors is always positive. The product of an odd number of negative factors is always negative. If zero is a factor, the product is zero.

Unlike multiplication, the commutative and associative properties do not apply to division. We cannot regroup and reorder division problems to make the process quicker. Instead, we must follow the order of operations.

Example 12 **Evaluate 100 ÷ (−2) ÷ (−25).**

$100 \div (-2) \div (-25)$
$= (-50) \div (-25)$ *Divide from left to right. Use parentheses to separate signs.*
$= 2$ *The quotient of the absolute values is 2. The signs are the same so the quotient is positive.*

The **distributive property,** $a(b + c) = ab + ac$ is true for integers as well as whole numbers. If we evaluate $a(b + c)$ following the order of operations, we must add inside the parentheses first. The distributive property permits us to multiply first and elminate the parentheses. Either method results in the same answer.

Example 13 **Evaluate $-3(2 - 7)$ using the distributive property and evaluate $-3(2 - 7)$ following the order of operations.**

Distributive Property

$-3(2 - 7)$
$= -3(2) - 3(-7)$ *Distribute.*
$= -6 + 21$
$= 15$

Order of Operations

$-3(2 - 7)$
$= -3(-5)$ *Subtract inside the parentheses.*
$= 15$

CHECK YOUR UNDERSTANDING

Evaluate.

1. $3(8)$

2. $-24 \div (-6)$

3. $-13 \div 0$

4. $-8(-8)$

5. $-56 \div 7$

6. $(-3)(-5)(-2)$

7. $140 \div (-2) \div (-1) \div 7$

8. $(-1)(-1)(-1)$

9. $0 \div (-15)$

10. $(18)(-2)(0)(4)$

11. $5(-2 - 7)$

12. $-2(3 - 5)$

Answers: 1. 24; 2. 4; 3. undefined; 4. 64; 5. −8; 6. −30; 7. 10; 8. −1; 9. 0; 10. 0; 11. −45; 12. 4.

EXERCISES SECTION 2.4

Evaluate.

1. $3(-2)$
2. $-8(-7)$
3. $3(8)$
4. $4(-5)$
5. $6(-20)$
6. $8(-3)$
7. $-2 \cdot 10{,}000$
8. $21 \cdot 3$
9. $(-60{,}000)(40)$
10. $(14)(-1)(-6)$
11. $30(-93)$
12. $(-12)(5)(-10)$
13. $0 \cdot (-4)$
14. $(-1)(44)$
15. $-44 \cdot 1$
16. $42 \cdot 200$
17. $17(-17)$
18. $(-1)(-2)(-3)(-4)(-5)$
19. $(-3)(-1)(4)(-9)(0)(-4)(-9)$
20. $4 \cdot 7 \cdot (-3) \cdot (-2)$

Evaluate.

21. $24 \div (-3)$
22. $-24 \div (-3)$
23. $3{,}500 \div 70$
24. $3{,}500 \div (-7)$
25. $-55 \div 11$
26. $38 \div (-2)$
27. $150{,}000 \div (-3)$
28. $-175 \div (-25)$
29. $165 \div 0$
30. $45 \div (-9)$
31. $0 \div (-6)$
32. $180 \div (-1)$
33. $225 \div (-5) \div (-1)$
34. $225 \div 5 \div (-1)$
35. $-225 \div (-5) \div (-1)$

Evaluate using the distributive property.

36. $-4(5 + 8)$
37. $-4(5 - 8)$
38. $-4(-5 - 8)$

39. $4(-5 + 8)$

40. $6(3 - 10 - 2)$

41. $-6(3 - 10 - 2)$

42. $-6(-3 - 10 - 2)$

43. $-3(5 + 8 - 3)$

44. $-3(-5 - 8 + 3)$

Solve using the five-step strategy.

45. Kendra uses automatic withdrawal at work to contribute to her 403b savings plan. She gets paid twice a month. If she takes $55 out of her take-home pay every paycheck, how much less take-home pay will she have per year?

46. A convenience store is losing an average of $15 a day to shoplifting. If this happens 365 days a year, what is the yearly loss?

47. Craig is in charge of making coffee in his office. One day he decides to slowly switch everyone over to decaffeinated coffee. He uses 10 tablespoons of coffee for each pot. Every day he uses one more teaspoon of decaffeinated coffee and one less teaspoon of caffeinated coffee. How many days will it take to switch totally over to decaffeinated coffee? (1 tablespoon = 3 teaspoons)

48. When plumbing a house that depends on a well for water, the designer must make sure that there is sufficient water pressure at every faucet, toilet, and shower. The water pressure at any outlet is equal to the pressure of the water as it enters the house minus any loss in pressure as the water travels through the house. For every 7 feet of vertical run in the water pipe, a drop of about 3 pounds per square inch of pressure occurs. Walt is plumbing a two-story vacation home in the Adirondack Mountains. The shower on the second floor is 42 feet above the holding tank for the well. What will be the loss in pressure between the holding tank and the shower head in this home?

49. An electric power company in the Northwest sent a flyer to its customers suggesting ways to conserve electricity and giving the potential savings per summer month based on an average monthly bill of $70. Shutting a hot tub off during the summer months, setting the thermostat on central air-conditioning to 78°F, removing a second refrigerator or freezer, using compact fluorescent lights, and reducing the temperature on an electric hot water heater to 115°F were suggested as energy saving measures. Calculate the total savings in electric bills for 5,000 families if 150 shut off the hot tub; 3,500 raise the thermostat on their air conditioning; 475 turn off their freezer; 2,775 lower the temperature on their electric hot water heater; and 4,200 change their light bulbs to compact fluorescent lights.

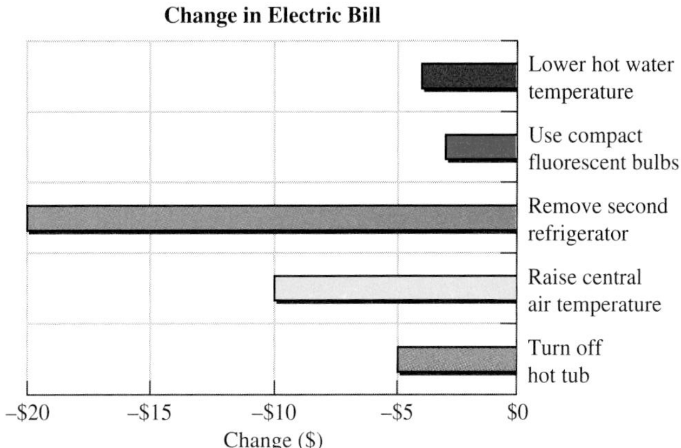

ERROR ANALYSIS

50. Problem: Evaluate $-2(-4)$.

Answer:

$$-2(-4)$$
$$= -6$$

Describe the mistake made in solving this problem.

Redo the problem correctly.

51. Problem: Evaluate $-2 - (-4)$.

Answer:

$$-2 - (-4)$$
$$= -2(4)$$
$$= -8$$

Describe the mistake made in solving this problem.

Redo the problem correctly.

REVIEW

52. Rewrite (5)(5)(5) as an exponential expression.

53. Evaluate 3^2.

54. Evaluate 4^3.

55. Evaluate $52 - 3^2 \cdot 4 + 8$.

56. Evaluate $81 \cdot (3 - 1) - 4^2$.

Section 2.5 Order of Operations

LEARNING TOOLS

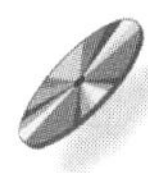
CD-ROM

SSM

VIDEO

Sneak Preview

Exponents are used to represent repeated multiplication of whole numbers. They can also be used to represent repeated multiplication of integers. The order of operations that we used for whole numbers is also applied to integers.

After completing this section, you should be able to:

1) Rewrite repeated multiplication of integers using exponents.
2) Evaluate exponential expressions.
3) Simplify exponential expressions.
4) Evaluate expressions with integers following the order of operations.

DISCUSSION

Exponents and Opposites

We can rewrite repeated multiplications of a negative number as an exponential expression. The negative number is written inside of parentheses and the exponent is outside of the parentheses. For example, $(-2)(-2)(-2)(-2)$ can be written $(-2)^4$. The value of $(-2)^4$ is 16.

If the parentheses are removed, the value of the expression changes. -2^4 could be written $-(2 \cdot 2 \cdot 2 \cdot 2)$. Order of operations requires that the exponent be evaluated first: $2^4 = 16$. Now we determine the opposite, -16. So, $-2^4 = -16$ and $(-2)^4 = 16$.

Example 1 **Evaluate -3^2.**

-3^2

$= -(3 \cdot 3)$ *Evaluate the exponent first. The base is positive 3 so $3^2 = 3 \cdot 3$.*

$= -9$

Example 2 **Evaluate $(-3)^2$.**

$(-3)^2$

$= (-3)(-3)$ *The base is -3 so $(-3)^2 = (-3)(-3)$.*

$= 9$

CHECK YOUR UNDERSTANDING

If possible, rewrite the exponential expression as a product. Evaluate.

1. 6^2 **2.** -6^2 **3.** $(-6)^2$ **4.** $(-1)^3$

5. $(-2)^0$ **6.** $(-2)^1$ **7.** $(-2)^2$ **8.** $(-2)^3$

9. $(-2)^4$ **10.** -2^2 **11.** -2^3 **12.** -2^4

Answers: 1. $6 \cdot 6 = 36$; 2. $-(6 \cdot 6) = -36$; 3. $(-6)(-6) = 36$; 4. $(-1)(-1)(-1) = -1$; 5. Any number raised to the zero power is always equal to 1; 6. -2; 7. $(-2)(-2) = 4$; 8. $(-2)(-2)(-2) = -8$; 9. $(-2)(-2)(-2)(-2) = 16$; 10. $-(2 \cdot 2) = -4$; 11. $-(2 \cdot 2 \cdot 2) = -8$; 12. $-(2 \cdot 2 \cdot 2 \cdot 2) = -16$.

DISCUSSION

The Product Rule of Exponents

The product rule of exponents applies to integers as well as to whole numbers.

To multiply exponential expressions with the same base:

In words: Add the exponents and keep the same base.

In symbols: $x^a \cdot x^b = x^{(a+b)}$

Example 3 **Simplify $(-6)^2(-6)^3$. Write the product as an exponential expression.**

$(-6)^2(-6)^3$ *Use the product rule of exponents.*

$= (-6)^{2+3}$ *Add the exponents; keep the common base.*

$= (-6)^5$ *Write the product as an exponential expression.*

Since we were asked to *simplify* in Example 1, the product was written as an exponential expression. When we are asked to *evaluate*, the product is written in place value form without exponents. We can use the product rule of exponents or we can begin by rewriting the exponential expressions as multiplications.

Example 4 **Evaluate $(-6)^2(-6)^3$. Write the product in place value form.**

$(-6)^2(-6)^3$

$= (-6)(-6)(-6)(-6)(-6)$ *Rewrite the exponential expressions as multiplications.*

$= (\mathbf{36})(-6)(-6)(-6)$ *Multiply or divide from left to right. (Or, we could multiply in pairs.)*

$= (\mathbf{-216})(-6)(-6)$

$= (\mathbf{1{,}296})(-6)$

$= -7{,}776$

When the base is a negative number and the exponent is an even number, the product is positive. When the base is a negative number and the exponent is an odd number, the product is negative.

Example 5 **Evaluate $(-2)^3(-2)^1(-2)^2$. Express the answer in place value form.**

$(-2)^3(-2)^1(-2)^2$	
$= (-2)^{3+1+2}$	*Add the exponents; keep the common base.*
$= (-2)^6$	*Since the exponent is even, the product will be positive.*
$= (-2)(-2)(-2)(-2)(-2)(-2)$	*Multiply in pairs. (Or, we could just multiply left to right.)*
$= (4)(4)(4)$	
$= (16)(4)$	
$= 64$	*The base is a negative number and the exponent is even. The final product is a positive number.*

If the bases of the exponential expressions are not the same, the only way to evaluate is to rewrite each exponential expression as repeated multiplications.

Example 6 **Evaluate $(-3)^2(4^1)(-3)^1(4^2)$. Express the answer as an integer.**

$(-3)^2(4^1)(-3)^1(4^2)$	
$= (-3)^{2+1}(4)^{2+1}$	*Use the product rule of exponents for terms with the same base.*
$= (-3)^3(4)^3$	*These bases are not the same.*
$= (-3)(-3)(-3)(4)(4)(4)$	*Rewrite exponential expressions using multiplication.*
$= (-27)(64)$	*Multiply from left to right.*
$= -1{,}728$	*The product of a negative and a positive is a negative number.*

CHECK YOUR UNDERSTANDING

Simplify. Express each product as an exponential expression.

1. $(-8)^6(-8)^3$

2. $5^8 5^3$

3. $(-2)^6(-2)^0$

4. $(-3)^9(-3)^2(-3)^7$

5. $(-2)^7(-7)^6(-2)^8$

6. $(-1)^8(-6)^8(-1)^7(-6)^9$

Evaluate. Express each product in place value form.

7. $(-2)^3(-2)^2$

8. $(-3)^2(-5)^2$

Answers: 1. $(-8)^9$; 2. 5^{11}; 3. $(-2)^6$; 4. $(-3)^{18}$; 5. $(-2)^{15}(-7)^6$; 6. $(-1)^{15}(-6)^{17}$; 7. -32; 8. 225.

DISCUSSION

The Order of Operations

The order of operations applies to integers as well as to whole numbers.

PROCEDURE

To evaluate an expression following the order of operations:

1. **Grouping:** Operations inside of grouping symbols are done first, following rules 2–4. Grouping symbols include parentheses ( ), brackets [ ], and braces { }.
2. **Exponents:** Do the operations indicated by exponents from left to right as they appear in the expression.
3. **Multiplication or Division:** Do all multiplication and divisions from left to right as they appear in the expression.
4. **Addition or Subtraction:** Do all additions and subtractions from left to right as they appear in the expression.

Absolute values | | are regarded as grouping symbols and are evaluated at the same level as parentheses ( ), brackets [ ], and braces { }. It may be helpful to rewrite subtraction as the addition of the opposite. For example, the expression $3^2 - 16 \div (-4) + 1$ may be easier to evaluate if it is first rewritten as $3^2 + (-16) \div (-4) + 1$.

The following examples are detailed and only one operation is completed per step.

Example 7 **Evaluate $3^2 - 16 \div (-4) + 1$.**

$3^2 - 16 \div (-4) + 1$	
$= 3^2 + \mathbf{(-16)} \div (-4) + 1$	*Rewrite subtraction as addition of the opposite.*
$= \mathbf{9} + (-16) \div (-4) + 1$	*Simplify exponents from left to right.*
$= 9 + \mathbf{4} + 1$	*Multiply or <u>divide</u> from left to right.*
$= \mathbf{13} + 1$	*<u>Add</u> or subtract from left to right.*
$= 14$	*<u>Add</u> or subtract from left to right.*

Example 8 **Evaluate $(-2)^3 \div (3 \cdot 6 - 2 \cdot 11)$.**

$(-2)^3 \div (3 \cdot 6 - 2 \cdot 11)$	
$= (-2)^3 \div (\mathbf{18} - \mathbf{22})$	*Work inside parentheses; <u>multiply</u> or divide from left to right.*
$= (-2)^3 \div (18 + \mathbf{(-22)})$	*Rewrite subtraction as addition of the opposite.*
$= (-2)^3 \div \mathbf{(-4)}$	*Work inside parentheses; add or <u>subtract</u> from left to right.*
$= \mathbf{-8} \div (-4)$	*Evaluate exponents from left to right.*
$= 2$	*Divide.*

Example 9 **Evaluate $-3|-6^2 - 4| \div 2(-4)$.**

$-3|-6^2 - 4| \div 2(-4)$	
$= -3|\mathbf{-36} - 4| \div 2(-4)$	*Work inside absolute value; evaluate exponents; $-6^2 = -36$.*
$= -3|-36 + \mathbf{(-4)}| \div 2(-4)$	*Rewrite subtraction as addition of the opposite.*
$= -3|\mathbf{-40}| \div 2(-4)$	*Work inside absolute value; <u>add</u> or subtract from left to right.*
$= -3 \cdot \mathbf{40} \div 2(-4)$	*Evaluate absolute value: $|-40| = 40$.*
$= \mathbf{-120} \div 2(-4)$	*<u>Multiply</u> or divide from left to right.*
$= \mathbf{-60}(-4)$	*Multiply or <u>divide</u> from left to right.*
$= 240$	*<u>Multiply</u> or divide from left to right.*

Evaluating expressions like these can be slow and even tedious. However, the skills used in this type of evaluation are important for later work with formulas and with polynomials.

Example 10 **Evaluate $-3^2 + 4^3 - |9(2) - 3(-7)|$.**

$-3^2 + 4^3 - \lvert 9(2) - 3(-7)\rvert$	
$= -3^2 + 4^3 - \lvert \mathbf{18} - 3(-7)\rvert$	*Work inside absolute value; multiply or divide from left to right.*
$= -3^2 + 4^3 - \lvert 18 + \mathbf{21}\rvert$	*Multiply or divide from left to right.*
$= -3^2 + 4^3 - \lvert \mathbf{39}\rvert$	*Add or subtract from left to right.*
$= -3^2 + 4^3 - \mathbf{39}$	*Evaluate the absolute value: $\lvert 39\rvert = 39$.*
$= \mathbf{-9} + 4^3 - 39$	*Evaluate exponents from left to right.*
$= -9 + \mathbf{64} - 39$	*Evaluate exponents from left to right.*
$= \mathbf{55} - 39$	*Add or subtract from left to right.*
$= 16$	*Add or subtract from left to right.*

When a division bar is present in an expression, the expression is evaluated as if there were parentheses around the expression on the top and parentheses around the expression on the bottom. For example, $\frac{-6 - 3^2}{-5 + 2}$ is evaluated as if it were written $\frac{(-6 - 3^2)}{(-5 + 2)}$.

Example 11 **Evaluate $\frac{-6 - 3^2}{-5 + 2}$.**

$\frac{-6 - 3^2}{-5 + 2}$	
$= \frac{-6 - \mathbf{9}}{-5 + 2}$	*Work in the top first. Evaluate exponents: $3^2 = 9$.*
$= \frac{\mathbf{-15}}{-5 + 2}$	*Add or subtract left to right: $-6 - 9 = -15$.*
$= \frac{-15}{\mathbf{-3}}$	*Now work in the bottom. Add or subtract left to right: $-5 + 2 = -3$.*
$= 5$	*Divide: $-15 \div (-3) = 5$.*

Example 12 **Evaluate $\frac{-7(-3) - 4(3)}{5(-4) + 11}$.**

$\frac{-7(-3) - 4(3)}{5(-4) + 11}$	
$= \frac{\mathbf{21} - 4(3)}{5(-4) + 11}$	*Work in the top. Multiply or divide left to right: $-7(-3) = 21$.*
$= \frac{21 - \mathbf{12}}{5(-4) + 11}$	*Multiply or divide left to right: $4(3) = 12$.*
$= \frac{\mathbf{9}}{5(-4) + 11}$	*Add or subtract left to right: $21 - 12 = 9$.*
$= \frac{9}{\mathbf{-20} + 11}$	*Now work in the bottom. Multiply or divide left to right: $5(-4) = -20$.*
$= \frac{9}{\mathbf{-9}}$	*Add or subtract left to right: $-20 + 11 = -9$.*
$= -1$	*Divide.*

When grouping symbols are nested inside each other, such as in the expression $-2[12 \div (1 + 18 \div (-6))] + 5$, the operations inside the innermost grouping symbols are evaluated first.

Example 13 **Evaluate $-2[12 \div (1 + 18 \div (-6))] + 5$.**

$-2[12 \div (1 + 18 \div (-6))] + 5$	
$= -2[12 \div (1 + \mathbf{(-3)})] + 5$	*Inside the innermost parentheses, multiply or divide left to right.*
$= -2[12 \div \mathbf{(-2)}] + 5$	*Inside the innermost parentheses, add or subtract left to right.*
$= -2[\mathbf{-6}] + 5$	*Inside the remaining brackets, multiply or divide left to right.*
$= \mathbf{12} + 5$	*Multiply or divide left to right.*
$= 17$	*Add or subtract left to right.*

CHECK YOUR UNDERSTANDING

Evaluate each expression.

1. $3 + (-20) \div 5 + 9$

2. $10 + 20 \div (-5) - 2^3$

3. $3(-6^2 - 4) \div 2(4)$

4. $3|-6^2 - 4| \div 2(4)$

5. $3^3 - (-3)^2$

6. $\dfrac{|5(-2)|}{45 \div (-9)}$

7. $\left|\dfrac{5(-2)}{45 \div -9}\right|$

8. $|-2^2 + 5 \cdot 3|$

Answers: 1. 8; 2. −2; 3. −240; 4. 240; 5. 18; 6. −2; 7. 2; 8. 11.

EXERCISES SECTION 2.5

Evaluate. Write the product in place value form.

1. 2^4

2. 3^2

3. 5^2

4. $(-2)^2$

5. -2^2

6. $(-3)^3$

7. -4^3

8. $(-1)^2$

9. 0^4

10. $(-1)^{12}$

11. $(-1)^{15}$

12. 9^0

13. $(-4)^1(-4)^2$

14. $(-2)^0$

15. -7^2

16. $(-4)^3$

Simplify. Write the product in exponential notation.

17. $4^3 \cdot 4^5$

18. $2^5 \cdot 2^3 \cdot 2^4$

19. $(-5)^3(-5)^1$

20. $(-7)^4(-7)^3(-7)^2$

21. $(-4)^3(-4)^7(-4)^0$

22. $(-6)^4(-6)^5$

23. $4^9 \cdot 4^6$

24. $(-3)^6(-3)^7(-3)^0(-3)$

25. $(9^7)(9^4)$

Simplify. Write the product in place value form.

26. $8^2 \cdot 8^0$

27. $2^3 \cdot 2^5 \cdot 2^2$

28. $(-3)^3(-3)^1(-3)^1$

Evaluate.

29. $15 + 2(6)$

30. $(15 + 2)(6)$

31. $15 + 2(-6)$

32. $(15 + 2)(-6)$

33. $15 \div (-3) + (-10) \div 2$

34. $[15 \div (-3)] + [(-10) \div 2]$

35. $4 - 12 \div 3 - 9$

36. $(4 - 12) \div (-2) + 5$

37. $|5 + 3|^2 - |3 - 5|^2$

38. $(5 - 3)^2 + 18 \div 6 + 1$

39. $-12 - 6 - 5 \cdot 2$

40. $|2 - 3|^2$

41. $12 + 16 \div (-8)(3) + 7$

42. $\dfrac{3(-5)}{9 + 6}$

43. $\left|\dfrac{3(-5)}{9 + 6}\right|$

44. $3 + 4 - 2^3$

45. $\dfrac{4^2 - 3 \cdot 2}{3 - 4 \cdot 2 + 2^2}$

46. $(-3)^3 \div (-4^2 + 7)$

Solve using the five-step problem-solving method.

47. Shotaro works for a caterer. His responsibility at a wedding reception is to serve punch and to refill the punch bowl as needed. He knows that the punch bowl holds about 80 cups of punch. He refills the punch bowl using a pitcher that holds about 15 cups of punch. As the reception begins, he has a full punch bowl. He serves 35 cups of punch and pours 2 pitchers of punch into the bowl. He serves 48 more cups of punch and pours 3 pitchers of punch into the bowl. He serves 53 more cups of punch and pours 2 pitchers of punch into the bowl. How many cups of punch are now in the punch bowl?

48. Raul and Joanne have a joint checking account and are working on their annual budget. Raul is paid \$875 twice each month by automatic deposit; Joanne is paid \$2,400 once a month. Some of their monthly bills are paid by automatic withdrawal including their house payment of \$1,007, Joanne's truck payment of \$225 and Raul's car payment of \$310, Raul's monthly athletic club membership of \$35, Joanne's child support payment of \$500, and Raul's monthly contribution to his retirement plan of \$125. What is the difference between their *annual* automatic deposits and their *annual* automatic withdrawals?

49. Grant is daydreaming about winning a million dollars. He thinks about putting it under his mattress. The first day, he plans to spend ten dollars, $-\$10$. The next day he plans to spend that amount squared, $-\$10^2$. The next day he plans to spend that amount cubed, $-\$10^3$. If he continues to increase the amount of money spent following this pattern, when will he run out of money?

50. Elena is daydreaming about winning a million dollars. She thinks about putting it under her mattress. The first day, she plans to spend two dollars, $-\$2$. The next day she plans to spend that amount squared, $-\$2^2$. The next day she plans to spend that amount cubed, $-\$2^3$. If she continues to increase the amount of money spent following this pattern, when will she run out of money?

51. The table shows student enrollment in lower division classes at Idaho State University from 1996 to 2000.

Year	1996	1997	1998	1999	2000
Enrollment	5,041	4,911	5,108	5,090	5,150

a. Draw a bar graph that shows the enrollment by year.

b. Complete the following table for increase or decrease in enrollment from the previous year. Decreases should be represented by negative numbers.

Time	1996–1997	1997–1998	1998–1999	1999–2000
Increase or decrease				

c. Draw a bar graph that shows the increase or decrease in enrollment.

ERROR ANALYSIS

52. **Problem:** Simplify $7^3 \cdot 7^4$.

Answer:

$7^3 \cdot 7^4$
$= 7^{3 \cdot 4}$
$= 7^{12}$

Describe the mistake made in solving this problem.

Redo the problem correctly.

53. **Problem:** Simplify -5^2.

Answer:

-5^2
$= (-5)(-5)$
$= 25$

Describe the mistake made in solving this problem.

Redo the problem correctly.

REVIEW

54. Calculate the area of the rectangle:

3 feet

5 feet

55. Calculate the perimeter of the triangle:

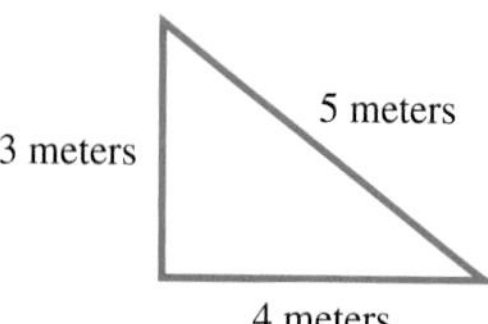

Section 2.6 Introduction to Formulas

LEARNING TOOLS

CD-ROM

SSM

VIDEO

Sneak Preview

A formula is an equation that specifies the relationship between different quantities. Formulas can be written in mathematical symbols, in words, or in a combination of words and mathematical symbols.

After completing this section, you should be able to:

1) Use a formula to solve a problem.

DISCUSSION

Using Formulas

A **formula** is an equation that contains more than one variable. It describes the relationship between different quantities. For example, the formula for the area of a rectangle, *Area* = *length* · *width,* tells the relationship between the area, length, and width of a rectangle.

Example 1 **Sean is building a new raised bed for asparagus. The length of this bed is 15 feet and the width is 5 feet. What is the area of the bed?**

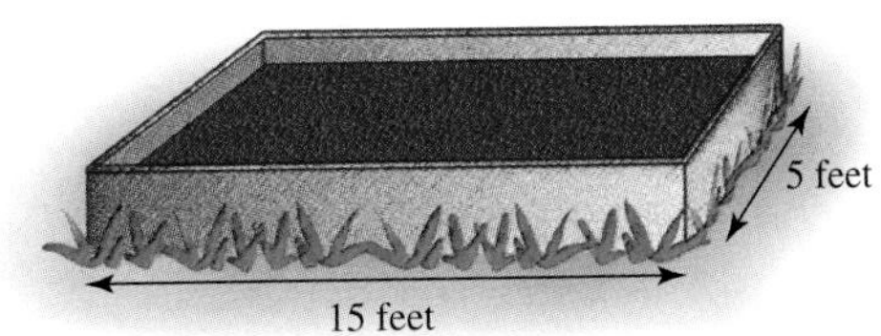

I know . . .

A = area

Length = 15 feet

Width = 5 feet

Formula: *Area* = *Length* · *Width*

Equations

Area = Length · Width

$A = (15\text{ feet})(5\text{ feet})$ *Replace length with 15 feet; replace width with 5 feet.*

Solve

$A = (15\text{ feet})(5\text{ feet})$

$= 75\text{ feet}^2$

Check

An estimate of the area is 20 feet · 5 feet = 100 feet². Since the exact answer is close to this estimate, it is reasonable.

Answer

The area of the asparagus bed is 75 square feet.

Often formulas are not written in words. Instead, variables are used to represent words. The formula *Area* = *length* · *width* could be written $A = lw$, where A is used to represent the area, l is used to represent the length, and w is used to represent the width.

Sometimes formulas contain numbers as well as variables. For example, a formula for changing degrees Celsius (°C) to degrees Fahrenheit (°F) is $F = 9C \div 5 + 32$. If we do not know why numbers such as 9, 5, and 32 are included in the formula, we cannot translate them into words in a word equation. Instead, we just write the numbers in the word equation.

Example 2 **Canada uses the Celsius temperature scale rather than the Fahrenheit temperature scale. When people who live in Buffalo, New York, hear weather broadcasts from Toronto, Ontario, the temperatures are reported in degrees Celsius. Change 20°C into its equivalent temperature in degrees Fahrenheit.**

I know . . .

F = Degrees Fahrenheit

20 = Degrees Celsius

Formula:

$F = 9C \div 5 + 32$

Equations

Degrees Fahrenheit = 9(Degrees Celsius) ÷ 5 + 32 *Word equation with numbers.*

$F = 9(20) \div 5 + 32$ *Replace "Degrees Celsius" with 20 in the math equation.*

Solve

$$
\begin{aligned}
F &= 9(20) \div 5 + 32 \\
&= 180 \div 5 + 32 \\
&= 36 + 32 \\
&= 68
\end{aligned}
$$

Check

We can check by estimating: $F = 9(20) \div 5 + 30$ *32 rounds to 30.*

$$
\begin{aligned}
&= 180 \div 5 + 30 \\
&= 66
\end{aligned}
$$

This is hardly quicker than doing the original work. Alternatively, we can check by considering whether the answer is possible and sensible. The Fahrenheit temperature certainly must be larger than the Celsius temperature because we started with a number, increased it nine times, and then only divided it five times. The proposed answer, 68°F, is larger than 20°C. Also, a temperature of 68°F is certainly a possible temperature in Toronto. The answer is reasonable.

Answer

Twenty degrees Celsius is equivalent to 68° Fahrenheit.

Example 3 **Respiratory failure is a leading reason for patient admission to intensive care units. The number of Calories needed by a patient using a ventilator is**

$$C = 1925 - 10A + 5W + 281S + 292T + 851B$$

where C = Calories needed; A = age of the patient in years; W = weight of the patient in kilograms; S = sex of the patient (male = 1, female = 0); T = presence of trauma (trauma present = 1; no trauma = 0); and B = presence of burns (burn present = 1; no burn = 0). [Source: Moore, M. C. (1997). *Nutritional Care.*]

A 55-year-old man is admitted to the intensive care unit in respiratory failure after a car accident (trauma). He has no burns. His weight is 84 kilograms. A nutritionist calculates his Calorie needs using the formula. How many Calories does he need?

I know . . .

C = Calories needed

55 = Age

84 = Weight in kilograms

1 = Sex of patient

1 = Presence of trauma

0 = Presence of burns

Formula:

$C = 1925 - 10A + 5W + 281S + 292T + 851B$

Equations

Calories needed = 1925 − 10(age) + 5(weight) + 281(sex) + 292(trauma) + 851(burns)

$$C = 1925 - 10(55) + 5(84) + 281(1) + 292(1) + 851(0)$$

Solve

$C = 1925 - 10(55) + 5(84) + 281(1) + 292(1) + 851(0)$

$= 1925 - 550 + 420 + 281 + 292 + 0$ *Multiply or divide from left to right.*

$= \mathbf{1375} + 420 + 281 + 292 + 0$ *Add or subtract from left to right.*

$= \mathbf{1795} + 281 + 292 + 0$ *Add or subtract from left to right.*

$= \mathbf{2076} + 292 + 0$ *Add or subtract from left to right.*

$= \mathbf{2368}$

Check

We can check the exact answer with an estimate.

$C = 1{,}925 - 10(55) + 5(84) + 281(1) + 292(1) + 851(0)$

$\approx 2{,}000 - 10(60) + 5(90) + 300(1) + 300(1) + 900(0)$

$= 2{,}000 - 600 + 450 + 300 + 300 + 0$

$\approx 2{,}000 - 600 + 500 + 300 + 300$

$= 2{,}500$

Instead of the usual rounding to the highest place of the largest number being added, we rounded to the hundreds. This can still be done quickly and allows us to make a more accurate estimate.

2,500 Calories is close enough to the exact answer of 2,368 Calories. The answer is reasonable.

Answer

The patient should receive 2,368 calories.

Two objects can have the same volume but have different masses. A block of iron is much heavier than a block of wood of the same size. We say that the density of iron is greater than the density of wood. Density (D) can be calculated using the formula $D = \frac{m}{V}$ where m is the mass and V is the volume of the object.

Example 4 **A hot tub contains 18 cubic meters of water that has a mass of 17,964 kilograms. Calculate the density of the water in the hot tub.**

I know . . .

$V = 18$ meters^3

$m = 17{,}964$ kilograms

$D =$ Density of the water

Formula:

$$D = \frac{m}{V}$$

Equations

$$\text{Density} = \frac{\text{Mass}}{\text{Volume}}$$

$$D = \frac{17{,}964 \text{ kilograms}}{18 \text{ meters}^3}$$

Solve

$D = 17{,}964$ kilograms $\div$ 18 meters^3

$= 998$ kilograms per cubic meter

Check

We can check division using multiplication: $18 \cdot 998 = 17{,}964$.

Answer

The water has a density of 998 kilograms per cubic meter.

CHECK YOUR UNDERSTANDING

Solve each problem using the five-step strategy.

1. The perimeter of a square can be found using the formula $P = 4s$, where s is the length of any side of the square. What is the perimeter of a square Post-it note with a side length of 3 inches?

2. The formula $A = P + Pt \div 20$ can be used to calculate the amount of money (A) in a bank account after an initial deposit of P dollars has gained interest at 5 percent for t years. (The letter P is used because another word for a deposited investment is *principal.*) How much money will be in an account after 5 years if an initial principal of $2,000 was deposited?

Answers: 1. The perimeter is 12 inches; 2. The account will have $2,500 after 5 years.

EXERCISES SECTION 2.6

Solve the problem using the five-step problem-solving strategy. The formula should be included in the *I know* step.

1. In 1865, explorers of the region around southern Lake Superior discovered ancient copper mines. Native peoples mined copper and used it to make spearheads, chisels, and other tools. One artifact that was discovered was a beating block—14 inches square and 3 inches thick—made from copper conglomerate. The volume of a cuboid can be calculated using the formula $V = lwh$ where V = volume, l = length, w = width, and h = height. Calculate the volume of this beating block.

2. Density of an object can be calculated using the formula $D = \frac{m}{v}$, where D = density, m = mass of the object, and v = volume of the object. A metal block has a mass of 1,600 grams and a volume of 200 milliliters. What is the density of this block in grams per milliliter?

3. Chemists are often interested in how much heat will be required to change a gram of a liquid into a gas. This quantity of heat is called the heat of vaporization. Heat of vaporization can be calculated experimentally using this formula: $H = h \div m$, where H is the heat of vaporization in calories per gram of liquid, h is the heat required to vaporize a sample of liquid, and m is the mass of the liquid being heated. Birgitta determines experimentally that 3,060 calories are needed to vaporize 15 grams of alcohol. What is the heat of vaporization of alcohol in calories per gram?

4. Rod and Earl are starting a small business making snowboards. They know that their cost for materials per snowboard is \$175. They have monthly costs for rent, heat, phone, and accountant fees that average \$650 a month. They want to pay themselves a salary; they decide that \$1,200 a month each is desirable. Using this information, they construct a formula for calculating their costs per month: $C = \$175x + \$650 + 2(\$1{,}200)$, where C is their total costs and x is the number of snowboards they make. If they make 58 snowboards a month, calculate their total costs.

5. Referring to Problem 4, Rod and Earl decide to project the profit of their business based on expected sales. They expect to sell all 58 snowboards they make at a price of \$279 per snowboard. Using this information, they construct a formula for calculating the profit per month: $P = R - C$, where P is the profit per month, R is the receipts from sales of snowboards, and C is the total cost of producing the snowboards (from Problem 4). Calculate the profit per month if they sell all 58 snowboards they make.

6. Marcella Hazen is a world famous authority on Italian cooking. Her recipe for fresh pasta uses the formula $E = 2N \div 3$, where E is the number of eggs used and N is the number of servings of pasta to be made. Calculate the number of eggs needed to make 36 servings of pasta.

7. The average speed of a car can be calculated using the formula $s = d \div t$ where s is the speed of the car, d is the distance the car travels, and t is the time it takes the car to travel that distance. Calculate the average speed of a car that travels 390 miles in 6 hours.

8. Mono Lake is a 700,000-year-old lake in California that has no outlet. Runoff from the Eastern Sierra Mountains washes into Mono Lake but the only loss of water is through evaporation. This results in a lake that has very high salinity. According to the California Department of Parks and Recreation, the lake is about 13 miles long east–west and 8 miles north–south. Using the formula for area, $A = lw$ where A is the area, l is the length, and w is the width, calculate the approximate area of Mono Lake.

9. The volume of a circular cylinder can be estimated using the formula $V \approx 3r^2h$ where V is the volume, r is the radius of the cylinder, and h is the height of the cylinder. The longest earthworm ever found is a South African worm that was 671 centimeters (22 feet) long and 2 centimeters across. Calculate the approximate volume of this worm.

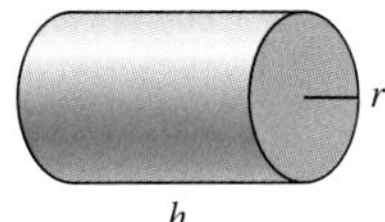

10. The volume of a circular cylinder can be estimated using the formula $V \approx 3r^2h$, where V is the volume, r is the radius of the cylinder, and h is the height of the cylinder. When the cargo ship *New Carissa* was grounded near the Oregon coast in 1999, the Coast Guard decided to tow the bow section of the ship out to sea and sink it. A special 1,100-yard-long, 5-inch radius cable was flown in from the Netherlands to connect the ship to a tugboat. What was the approximate volume of this cable?

11. Most of the gold in the world has been mined near Johannesburg in South Africa. The strip of land where the gold is located is 62 miles long and 25 miles wide and is called The Witwatersrand. Calculate the area of this strip of land using the formula $A = lw$.

12. Write your own word problem that requires the use of a formula. Solve the problem.

13. The heart of a "typical" person at rest beats 72 times per minute. The average volume of blood moved per beat is 70 milliliters. The total amount of blood in an adult is about 5 liters. The formula $t = (60{,}000V) \div (Bv)$ can be used to calculate the time it takes for the heart of a typical person at rest to move the blood in the body through the heart in seconds where t = time to move the blood through the heart in minutes, V = total volume of blood in the body in liters, B = heartbeats per minute, and v = volume of blood moved in milliliters per heartbeat. Calculate the time it takes for a typical person at rest to move the blood in the body through the heart in seconds. Round to the nearest second.

14. In a machine shop, a very small amount of metal may be removed at the end of the machining process with a grinding wheel. The result is a finish of greater smoothness than can often be obtained with a cutting tool. The surface speed of a grinding wheel can be estimated using the formula $S = 3RD \div 12$, where S is the surface speed in feet per minute, R is the revolutions of the grinding wheel per minute, and D is the diameter of the circular wheel in inches. Calculate the speed of a grinding wheel with a diameter of 9 inches that is revolving at a speed of 2,500 revolutions per minute.

15. Lumber is measured in terms of board feet. A board that is 12 inches wide, 12 inches long, and 1 inch thick contains 1 board foot of lumber. To calculate the number of board feet in a piece of lumber, the formula $B = \frac{Lwt}{12}$ can be used where B is the number of board feet, L is the length of the lumber in feet, w is the width of the lumber in inches, and t is the thickness of the lumber in inches. Calculate the board feet in a timber that is 10 inches by 10 inches by 14 feet long. Round to the nearest board foot.

16. Use the formula in Problem 26 to calculate the number of board feet in 1,200 pieces of lumber that are 2 inches by 4 inches by 18 feet long.

17. Location of studs, headers, and sills around window openings should conform to the rough opening sizes recommended by the manufacturers of the millwork around the window. For a double-hung window (single unit), the following formulas are often used:

$w = g + 6$ where w = rough opening width (in.) and g = glass width (in.)
$h = t + 10$ where h = rough opening height (in.) and t = total glass height (in.).

A double-hung window has a width of 24 inches. Each pane in the window has a height of 16 inches. Calculate the width and height of the rough frame opening.

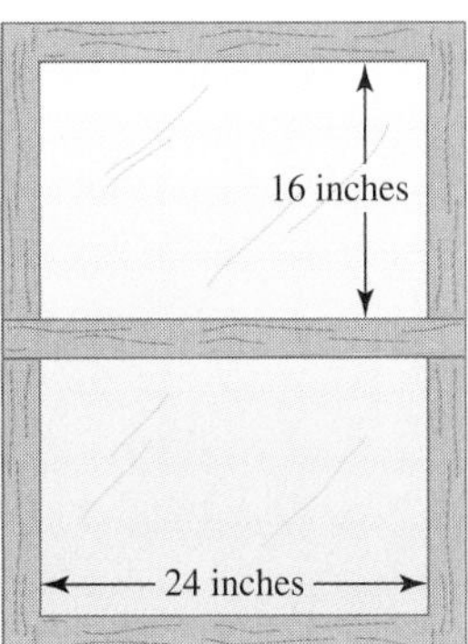

18. To estimate the area (A) of material needed to side a rectangular house, the length (L), width (W), height of the home to the eaves (C), and the height of the gable (D) need to be measured and the formula $A = 2C(L + B) + BD$ is used. Calculate the area of vinyl siding needed for a rectangular home that is 35 feet long by 26 feet wide with a height to the eaves of 9 feet and a height of the gable end of 8 feet.

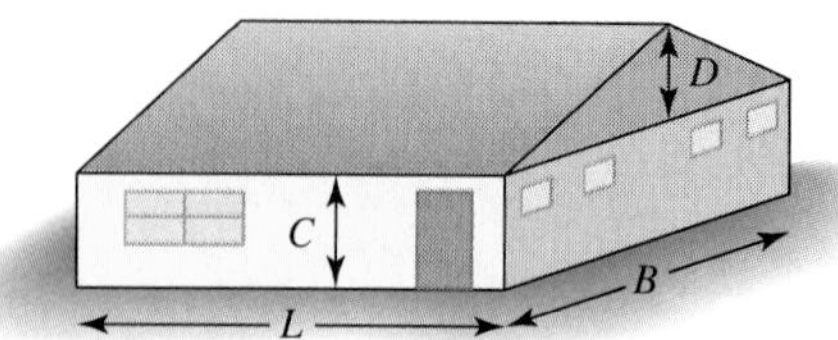

19. In rural areas, water is often supplied to homes using a pump in a well. To determine the size of pump needed, the maximum amount of water that will be used at any single time is estimated. One way to estimate maximum water use is to count the total number of fixtures that use water such as toilets, bathtubs, sinks, and dishwashers, and then use the formula $G = 60f$ where G is the maximum gallons per hour of water needed and f is the number of fixtures. Calculate the maximum water use for a two-bathroom house that has one bathtub, one shower, two toilets, three sinks, one laundry tub, one clothes washer, one dishwasher, and two garden hose faucets.

ERROR ANALYSIS

20. **Problem:** The formula $s = ut + 16t^2$ calculates the distance (s feet) an object falls under gravity in time (t seconds) with an initial speed (u feet per second). Calculate the distance a stone falls in 2 seconds when it starts with an initial velocity of 4 feet per second.

Answer:

$$\begin{aligned} s &= ut + 16t^2 \\ &= 4(2) + 16(2)^2 \\ &= 8 + 16(4) \\ &= 8 + 64 \\ &= 68 \end{aligned}$$

The stone fell 68 feet.

Describe the mistake in the answer to this problem.

Redo the problem correctly.

21. Problem: When lifting objects with a rope, the rope needs to be strong enough to raise the load without breaking. For manila hemp rope, the safe working load (S) of the rope in pounds can be determined using the formula $S = 150C^2$ where C is the circumference of the rope in inches. (Circumference is the distance around the rope.) Calculate the safe working load of a manila rope that has a circumference of 3 inches.

Answer:

$$\begin{aligned} S &= 150C^2 \\ &= 150(3)^2 \\ &= 150(6) \\ &= 900 \end{aligned}$$

The safe working load of the rope is 900 pounds.

Describe the mistake made in solving this problem.

Redo the problem correctly.

REVIEW

22. The bar graph shows the results of a World Health Organization survey.

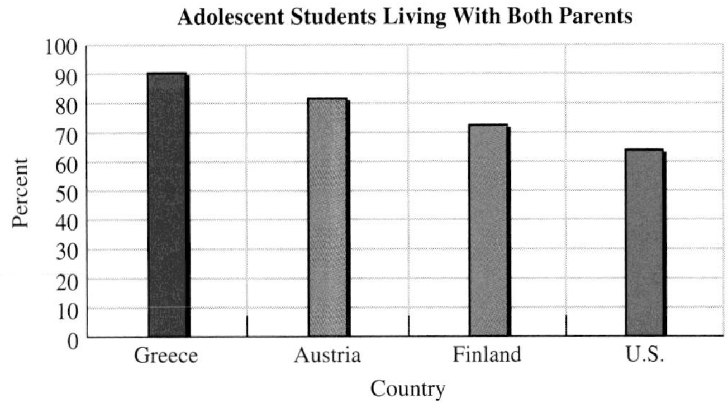

a. What percent of adolescent students in Greece lived with both parents?

b. What percent of adolescent students in the United States lived with both parents?

23. The February 27, 2000, issue of *Parade Magazine* reported the following data on what people earn per year: Fruit packer: \$9,000; Home daycare owner: \$37,000; U.S. Army private: \$13,404; Hairstylist: \$26,000; Trucker: \$65,000; Nurse's aide: \$20,000; Engineer: \$47,000; Realtor: \$65,000; Paleontologist: \$35,000; Hotel chef: \$26,000; Plumber: \$60,000; Accountant: \$26,000; Librarian: \$40,000. Present this information in a bar graph.

Section 2.7 Line Graphs

LEARNING TOOLS

Sneak Preview

Line graphs are often used to show trends in data, particularly over time. The rise and fall of a stock price over a 2-week period, or the profits of a company over a 12-month period, are examples that can be visualized on a line graph.

After completing this section, you should be able to:

1) Read and interpret line graphs.

2) Construct line graphs from a set of data.

3) Calculate the mean of a set of integers.

4) Calculate the median of a set of integers.

DISCUSSION

Arithmetic Mean

In Section 1.9, we found the arithmetic mean (or average) of a set of whole number data by dividing the sum of the data by the number of data points. We can use the same method to find the mean when there are negative numbers in the data set.

Arithmetic Mean

Arithmetic mean = (Sum of a set of numbers) ÷ (Number of members in the set)

The arithmetic mean is typically called the **average** in the media.

Example 1 **Calculate the mean of the scores 3, (−5), (9), (−3).**

Mean = (Sum of scores) ÷ (Number of scores)
= (3 + (−5) + 9 + (−3)) ÷ 4
= 4 ÷ 4
= 1

Example 2 **During her 6-week summer break, Kaye worked as a data entry clerk for a temporary agency. The table shows the number of hours she worked during the 6-week period. Calculate the average number of hours Kaye worked per week.**

Week	1	2	3	4	5	6
Hours	10	25	40	35	15	25

Mean = (Total hours worked) ÷ (Number of weeks worked)
= (10 + 25 + 40 + 35 + 15 + 25) ÷ 6
= 150 hours ÷ 6
= 25 hours

Kaye worked an average of 25 hours per week.

Example 3 **The table shows the low temperature at Bear Lake for a 1-week period in February. Calculate the average low temperature for the week.**

Day	Mon.	Tues.	Wed.	Thurs.	Fri.	Sat.	Sun.
Temperature°F	2	−1	−5	−6	−12	−4	−9

$$\begin{aligned}\text{Mean} &= (\text{Total of temperatures}) \div (\text{Number of temperatures})\\ &= (2 + (-1) + (-5) + (-6) + (-12) + (-4) + (-9))°\text{F} \div 7\\ &= -35°\text{F} \div 7\\ &= -5°\text{F}\end{aligned}$$

CHECK YOUR UNDERSTANDING

Calculate the mean.

1. 5, −4, −6, −3

2. −23, 3, −7

3. The changes in Medicare enrollment for five counties are listed below. Find the mean change in enrollment. Round to the nearest whole number.

Clearwater County: +635 recipients
Clark County: −54 recipients
Ramsey County: +627 recipients
Lewis County: −166 recipients
Boundary County: +715 recipients

Answers: 1. −2; 2. −9; 3. +351 recipients.

DISCUSSION

Median

In statistics it is often useful to find the median of a set of data. The **median** is defined as the middle point of a list of data that is arranged from lowest to highest.

PROCEDURE

To find the median of a set of data:

1. Order the data from lowest to highest.
2. Choose the middle element in this ordered list.
3. If there is no middle element, find the average of the two numbers closest to the middle of the list.

Example 4 **Find the median of the numbers 4, 9, 2, −6, 5.**

−6, 2, 4, 5, 9 *Order the set of numbers from lowest to highest.*

The median is 4. *Choose the middle element in the ordered list.*

Example 5 **Four friends decided to join a fitness and weight-loss program. The table shows the change in weight for each participant after 2 months. A weight gain is represented by a positive number; a weight loss is represented by a negative number. Find the median change in weight for the group of people.**

PARTICIPANT	CHANGE IN WEIGHT (LBS)
Agnes	−12
Agatha	−8
Amy	−10
Amber	−16

Order the numbers from lowest to highest: −16 lb, −12 lb, −10 lb, 12 lb. The median is the middle element of an ordered list. Since there is no middle element in this list of weight changes, instead find the average of the two numbers closest to the middle of the list: −12 lb and −10 lb.

$$\text{Median} = \frac{-12\text{ lb} + (-10\text{ lb})}{2}$$

$$= \frac{-22\text{ lb}}{2}$$

$$= -11\text{ lb}$$

The median change in weight is −11 pounds.

CHECK YOUR UNDERSTANDING

Find the median.

1. 4, −6, 3, 0, −7

2. $23, −$32, −$43, $12, −$15

3. 3°F, −9°F, 1°F, −7°F

4. 34, −52, −23, 23, 43, −19

Answers: 1. 0; 2. −$15; 3. −3°F; 4. 2.

DISCUSSION

Negative Numbers, Line Graphs, and Word Problems

Negative numbers can be used to represent debts, losses, falls, and subzero temperatures. On a line graph, negative numbers are located to the left of 0 on the horizontal axis and below 0 on the vertical axis. Ordered pairs can contain negative numbers and are graphed as points using the same procedure used for positive numbers.

Example 6 **The net profit and loss of a company by month is given in the table.**

MONTH	PROFIT OR LOSS
January 2000	\$4,000
February 2000	\$5,100
March 2000	−\$2,200
April 2000	−\$3,000
May 2000	\$500
June 2000	−\$750
July 2000	\$1,500

a. Present this information as a bar graph.

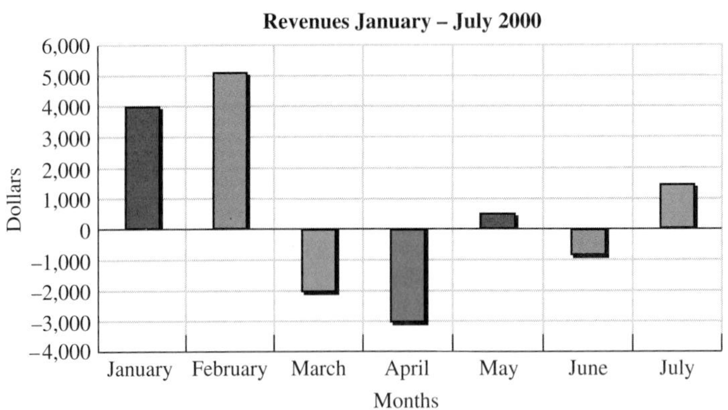

b. Present this information as a list of ordered pairs (month of the year, revenues in dollars).

(1, \$4,000) (2, \$5,100) (3, −\$2,200) (4, −\$3,000) (5, \$500)
(6, −\$750) (7, \$1,500)

c. Present this information as a line graph.

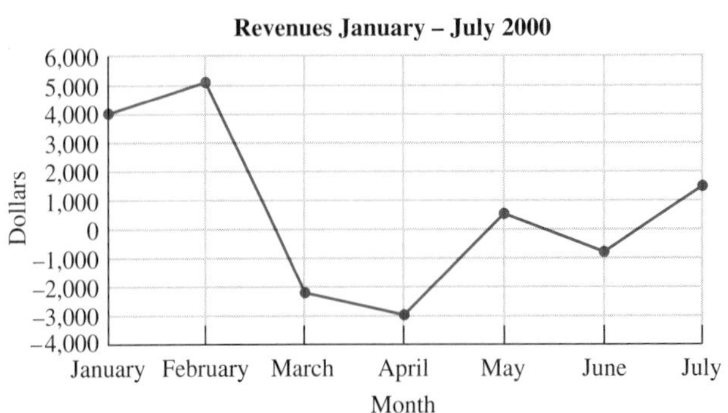

d. Calculate the net revenues of this company during this period.

I know . . .

R = Revenues

\$4,000 = January profit

\$5,100 = February profit

−\$2,200 = March loss

−\$3,000 = April loss

\$500 = May profit

−\$750 = June loss

\$1,500 = July profit

Equations

Revenues = Total profits + Total losses

$$R = \$4{,}000 + \$5{,}100 + \$500 + \$1{,}500 + (-\$2{,}200) + (-\$3{,}000) + (-\$750)$$

Solve

$R = \$11{,}100 + (-\$5{,}950)$

$= \$5{,}150$ *Since the amount is positive, this is a net profit.*

Check

Estimate total profit by rounding each profit or loss to the thousands place.

$R = (\$4{,}000 + \$5{,}000 + \$1{,}000 + \$2{,}000) + (-\$2{,}000 + (-\$3{,}000) + (-\$1{,}000))$

$= \$12{,}000 + (-\$6{,}000)$

$= \$6{,}000$ *This is close to our exact answer of \$5,150; the answer is reasonable.*

Answer

The total revenues from January to July were \$5,150.

e. Use the information to calculate the average monthly revenues over this period to the nearest dollar.

I know . . .

\$5,150 = Total revenues

7 = Number of months

A = Average revenues

Equations

$$\text{Average revenues} = \frac{\text{Total revenues}}{\text{Number of months}}$$

$$A = \frac{\$5{,}150}{7}$$

Solve

$A = \dfrac{\$5{,}150}{7}$

$= \$735\ \text{R}(5)$ *Since 5 is closer to 7 than it is to 0, the rounded quotient is \$736.*

$\approx \$736$

Check

Replacing each of the 7 months with a profit of \$736 results in total revenues of $7 \cdot \$736 = \$5{,}152$. This is very close to the total revenue of \$5,150 so the answer is reasonable.

Answer

The average revenue was \$736.

f. Calculate the median revenue over this period.

Revenues: −\$3,000, −\$2,200, −\$750, \$500, \$1,500, \$4,000, \$5,100

Median = \$500

CHECK YOUR UNDERSTANDING

1. In the English system, force is measured in pounds. Negative and positive signs are used to indicate the direction of the applied force. A couple is pushing a large hide-a-bed across the carpeted floor. They apply a forward force of 240 pounds to the hide-a-bed. However, there is friction between the bottom of the hide-a-bed and the carpet of 20 pounds. Since this friction opposes the forward motion of the hide-a-bed, it is a negative force. What is the net forward force experienced by the hide-a-bed?

Answers: 1. 220 pounds.

EXERCISES SECTION 2.7

Calculate the mean (average).

1. 4, −6, −7, 5

2. 12, −5, −6, −3, −8

3. 7, −3, 5, −8, −12, −1

4. 29, −34, −49, −57, −14

Find the median.

5. 4, 9, −3, −7, −3, 5, 12

6. −20, −41, −52, 30, 32, −71, −14

7. 3, −17, −31, 12

8. 12, −37, 21, −41, 53, 21, −17, −8

9. When there are not enough U.S. citizens to fill needed jobs, H1-B visas are used to bring foreign high-skilled employees to the United States. The following table shows the number of H1-B visas granted per year.

Year	1992	1993	1994	1995	1996	1997	1998	1999	2000
Visas	48,745	61,591	60,279	54,178	55,141	65,000	65,000	115,000	115,000

a. Construct a line graph for this data.

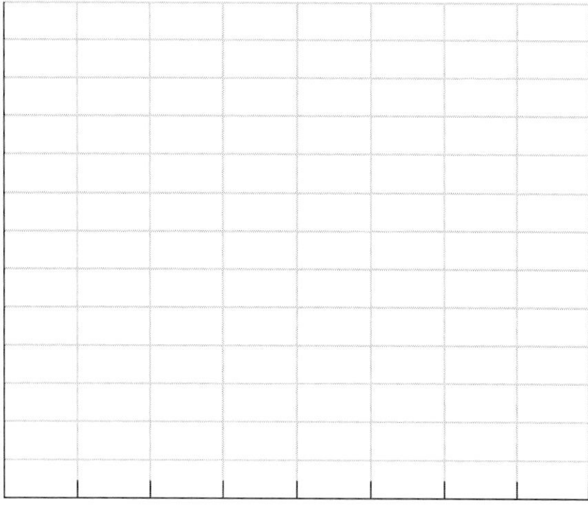

b. What is the median number of visas granted?

c. What is the average number of visas granted? Round to the nearest whole number.

10. Alex's checking account had an average balance of $420 over a 10-day period. The line graph shows the amount Alex's account balance was above or below the average value.

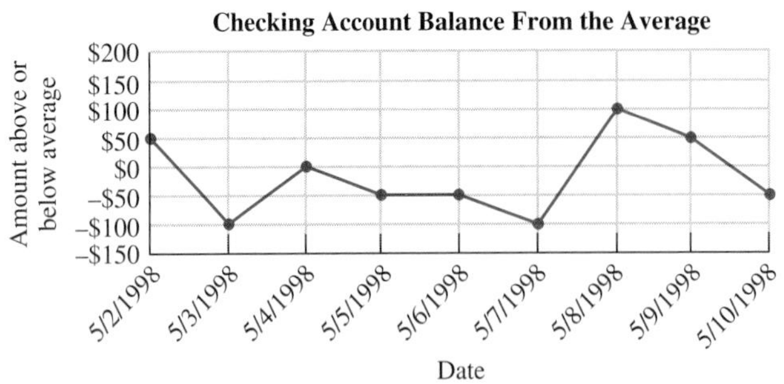

a. What is the maximum amount of money Alex had in his checking account?

b. What is the difference between the maximum and minimum amounts in Alex's checking account?

c. What is the median amount Alex has in his checking account?

11. The chart below lists the difference in the reported high temperature from the normal high temperature at Rochester Airport.

DATE	DIFFERENCE FROM NORMAL	DATE	DIFFERENCE FROM NORMAL	DATE	DIFFERENCE FROM NORMAL
April 25	−6°F	April 26	−6°F	April 27	−1°F
April 28	+3°F	April 29	+12°F	April 30	+17°F
May 1	+18°F	May 2	+10°F	May 3	+13°F
May 4	+11°F	May 5	+14°F	May 6	+16°F
May 7	+16°F	May 8	+12°F	May 9	+6°F

a. Calculate the average difference of the actual high temperature from the normal high temperature.

b. Calculate the median difference in high temperature from normal.

12. Tolerance is the amount an actual measurement can vary from the desired measurement. For example, the desired weight of a bag of cement may be 320 ounces, but the actual measurement may be up to 5 ounces heavier or 5 ounces lighter without the bag being rejected as overweight or underweight by quality control inspectors. The chart shows the amount in ounces that inspected bags of cement varied from 320 ounces. Calculate the average difference this shipment of cement bags varies from the desired weight.

BAG NO.	VARIANCE FROM 320 OUNCES	BAG NO.	VARIANCE FROM 320 OUNCES	BAG NO.	VARIANCE FROM 320 OUNCES
1	+4 ounces	2	+2 ounces	3	−1 ounce
4	0 ounces	5	−3 ounces	6	−2 ounces
7	+3 ounces	8	+2 ounces	9	0 ounces
10	−2 ounces	11	+4 ounces	12	+5 ounces

13. The double line graph shows the daily high and low temperatures measured in degrees Celsius in New York for a week in January.

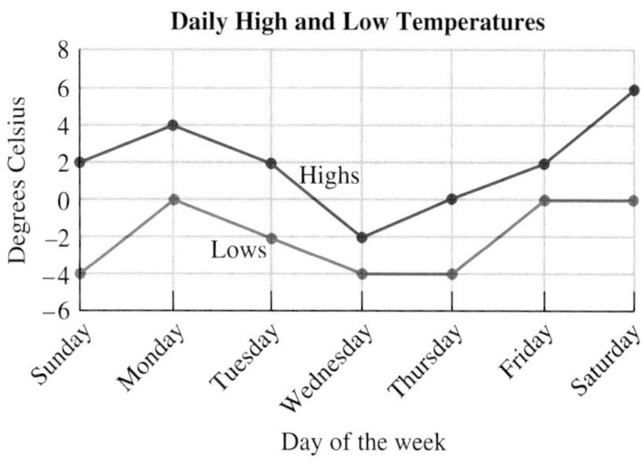

a. What is the difference between the high temperature and the low temperature recorded on Wednesday?

b. What is the average daily high temperature over the week?

c. What is the difference between the highest temperature and the lowest temperature recorded this week?

d. What is the average temperature difference between the daily high temperatures and the daily low temperatures?

e. What is the median daily low temperature?

14. As shown in the table, the expense for an average traveler for a single night's lodging and three meals has more than doubled in some cities since the 1980s and only increased slightly in other areas. All costs are in U.S. dollars.

LOCATION	COST IN 1988	COST IN 1995	COST IN 1999
Athens, Greece	\$163	\$307	\$289
Hong Kong	\$229	\$433	\$362
Mexico City	\$118	\$222	\$260
Moscow	\$139	\$376	\$391
Paris	\$293	\$396	\$359
Toronto	\$116	\$161	\$196
Tokyo	\$352	\$461	\$383

a. What was the change in cost from 1988 to 1999 in Athens?

b. What was the change in cost from 1995 to 1999 in Tokyo?

c. Construct a line graph that compares the changes in costs in Moscow and in Paris.

d. Calculate the average cost for a single night's lodging and three meals in 1999 for the seven cities.

e. Calculate the median cost for a single night's lodging and three meals in 1999 for the seven cities.

ERROR ANALYSIS

15. The table shows the number of hunting licenses sold in the state of Washington by year.

Year	1995	1996	1997	1998	1999
Licenses	267,000	258,000	235,000	229,000	237,000

Problem: What is the mean number of hunting licenses sold between 1995 and 1999?

Answer: The mean value is the middle value of 235,000 licenses.

Describe the mistake made in solving this problem.

Redo the problem correctly.

16. During the winter of 2001, drought conditions caused snowpacks at the top of Idaho ski resorts to be less than one half of the average depth. The line graph shows the depth of snow at the summits of five ski areas.

Problem: Estimate the median depth of snow at these ski areas.

Answer: The middle value of the data set is the depth at Silver Mountain, 64 inches.

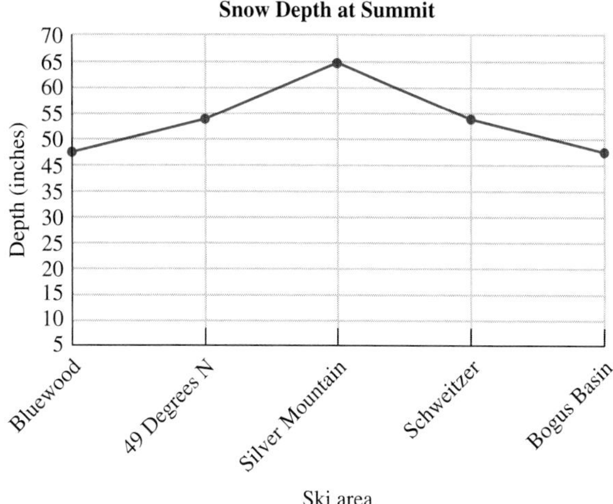

Describe the mistake made in solving this problem.

Redo the problem correctly.

REVIEW

17. What is a product? Give an example.

18. What is a factor? Give an example.

19. Describe the set of natural numbers using set notation.

20. Ellipses can be used in set notation. For example, the set of whole numbers can be written {0, 1, 2, 3, . . .). What is the meaning of the ellipses?

21. Explain the difference between a finite set and an infinite set. Give an example of each.

READ AND REFLECT

Like all mathematical concepts, our ideas and rules about negative numbers have a history. The Babylonians and Egyptians used mathematics primarily for counting and measurement. They were interested in geometric problems such as those involved in constructing the pyramids, measuring land, or measuring volumes of liquids. Since negative numbers are not essential for any of these purposes, the Babylonians and Egyptians had no use for, or conception of, negative numbers.

The Chinese wanted to solve problems that had two solutions. Using the language of algebra, we describe these problems as having two variables. Solving this kind of problem can require negative numbers. Because they needed them, the Chinese developed the concept of negative numbers and symbols to represent negative numbers. All numbers were represented by particular arrangements of colored rods. Red rods represented positive numbers; black rods represented negative numbers.

The Greeks thought about the possibility of negative numbers but rejected their existence for two reasons. First, they did most of their mathematics based on geometric models. Just as you don't count with negative numbers, you don't measure lengths, areas, or volumes with negative numbers. Second, Greek mathematicians were striving to discover the perfect mathematics system. Negative numbers were thought to be imperfect and so were unacceptable.

The Greek "negative" attitude about negative numbers persisted as Europeans began to contribute to the development of mathematics. Michael Stifel (1487–1567) called negative numbers "numeri absurdi" and Gerolamo Cardano called them "numeri ficti." However, in the absence of negative numbers, mathematicians were frustrated. After all, what is $4 - 10$ if negative numbers do not exist?

Eventually scientists and mathematicians began to accept that negative numbers could exist. Everyone did not start believing in or using negative numbers at the same time. Some mathematicians were cautious and chose to wait until prominent mathematicians of the time decided that negative numbers were legitimate. For example, when Simon Stevin (1548–1620), an influential Dutch engineer, publicly decided that negative numbers were acceptable, others who respected his opinion were convinced to use them as well.

Perhaps you see now that mathematics is a work in progress. Mathematicians argue constantly about new and old ideas. Some mathematicians are more influential and have "louder voices" than others. Sometimes it takes years for a new discovery to be accepted.

Today, ideas about negative numbers are taught beginning in elementary school, usually in the context of bank balances, above- and below-zero temperatures, and number lines. However, when a concept is difficult to represent using physical objects such as multiplication of two negative numbers, it may take longer to understand. Mathematics on the individual level, just as from a historical perspective, is not learned or accepted overnight.

RESPOND

1. Do you remember when you first learned about negative numbers? If so, describe what you remember. If not, describe when you first remember using them.

2. Describe a nonmathematical concept or fact that took some time to be accepted by the scientific community or the public.

Section	Terms and Concepts/Procedures
2.1	positive numbers; negative numbers; implied plus sign; integers; opposite; absolute value; reference point **1.** Identify the opposite of an integer. **2.** Identify the absolute value of an integer. **3.** Solve an equation that includes an absolute value. **4.** Round an integer to a specified place value.
2.2	sum; axiom; commutative property of addition; associative property of addition; term; like term **1.** Add integers using a number line. **2.** Add integers. **3.** Solve an equation that includes integers.
2.3	**1.** Rewrite subtraction as addition of the opposite. **2.** Subtract integers.
2.4	product; quotient; distributive property **1.** Multiply integers. **2.** Divide integers.
2.5	base; exponent; place value form; product rule of exponents; even number; odd number **1.** Evaluate an exponential expression. **2.** Simplify an exponential expression. **3.** Use the order of operations to evaluate or simplify an expression.
2.6	formula; variable; solution **1.** Use a formula to solve a word problem.
2.7	arithmetic mean; average; median; line graph; data **1.** Read and interpret line graphs. **2.** Construct line graphs. **3.** Calculate the mean. **4.** Calculate the median.

REVIEW EXERCISES CHAPTER 2

SECTION 2.1 INTRODUCTION TO INTEGERS

1. Represent a 30-point drop in NASDAQ prices as an integer.

2. From 1990 to 1996, the population of Norfolk, Virginia, decreased by 27,820 people. Represent this information as an integer.

3. Represent an increase in cholesterol of 30 points as an integer.

4. Graph on a number line 0, −7, 4, −3, and −1.

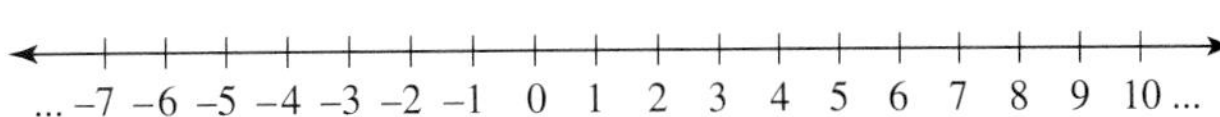

Put <, >, or = between the numbers to make a true statement.

5. −4 7

6. 6 −9

7. 5 −5

8. −6 −5

9. What is the opposite of −9?

10. What is the opposite of 40?

11. What is the opposite of a height of 700 feet?

Evaluate.

12. $-(-7)$

13. $-(+3)$

14. $-(-(-5))$

Evaluate.

15. $|-6|$

16. $-|-5|$

17. $|-5| + |6|$

Solve. Check the solution(s).

18. $|x| = 8$

19. $|y| = 0$

20. $|z| = -7$

21. $|m| = 1$

Round to the indicated place value.

22. 56 to the nearest ten

23. −673 to the nearest hundred

24. −6,399 to the nearest thousand

25. −67 to the nearest thousand

26. −4 to the nearest ten

27. −45,702 to the nearest thousand

SECTION 2.2 ADDITION OF INTEGERS

Evaluate.

28. $-3 + 9$

29. $-5 + (-9) - 7$

30. $3 + (-7) + (-7) + 5$

31. $-3 + 7 + (-5)$

32. $-3 + (-5)$

SECTION 2.3 SUBTRACTION OF INTEGERS

Evaluate.

33. $-7 - (-8)$

34. $-3 - 8 - 2$

35. $4 + 7 - 8 + (-8)$

36. $-5 + 8 - (-7) + 3$

Solve the equation.

37. $x = -9 + 7 - (-4)$

38. $m = -52 - 27 - (-36)$

Solve using the five-step problem-solving strategy.

39. The population of Lansing, Michigan, was 127,321 in 1990. In 1996 the population was 125,736. What was the change in population from 1990 to 1996?

40. The temperature in Honolulu, Hawaii, was 28°C. At the same time the temperature in Chicago, Illinois, was −5°C. What was the temperature difference between Chicago and Honolulu?

SECTION 2.4 MULTIPLICATION AND DIVISION OF INTEGERS

Evaluate.

41. $-4(3)$

42. $-3(-2)(-8)(-3)$

43. $-6 \div 3$

44. $-72 \div (-9)$

45. $\frac{-65}{0}$

46. $\frac{30}{-5}$

Evaluate. Use either the order of operations or the distributive property.

47. $5(2 - 9)$

48. $-4(3 - 2)$

49. $-3(5 - 8 + 2)$

50. $\frac{5 - 10}{5}$

51. $\frac{12 - 12}{-12}$

SECTION 2.5 ORDER OF OPERATIONS

Evaluate.

52. 3^2

53. $(-6)^2$

54. -7^2

55. $(-3)^0$

Simplify. Write the product as an exponential expression.

56. $3^4 \cdot 3^7$

57. $(-5)^3(-5)^9(-5)^6$

58. $-7 \cdot (-7)^0$

Evaluate.

59. $(-6)^2 \cdot 4 \div (-3)$

60. $\dfrac{20 + (-5)}{-3}$

61. $-3(-1)(3)(4)$

62. $\dfrac{2(-6)}{3(-1)}$

63. $-2(3 + 6)^2$

64. $(-3)^2 \cdot 4 \div (-6)$

Solve the equation.

65. $y = |-4 + 5| - |-3 - 5|$

66. $t = (-3 - 5)^2 - (7 - (-2))$

67. $x = -2^2 + 4^2 - 7(2 - 4) \div (-5 + 3)$

68. $n = \dfrac{-4(-3) + 3}{-7 + 2}$

SECTION 2.6 INTRODUCTION TO FORMULAS

69. Use the formula $C = 5(F - 32) \div 9$ to convert 68°F to degrees Celsius.

Solve the problem using the five-step problem-solving strategy.

70. The Capri Motel has a pool that measures 20 feet by 15 feet. Use the formula $A = lw$ where A is the area, l is the length and w is the width, to calculate the area of the pool.

71. Density (D) can be calculated using the formula $D = \frac{m}{V}$ where m is the mass and V is the volume of the object. A metal block has a mass of 1,800 grams and a volume of 300 milliliters. What is the density of the block in grams per milliliter?

72. The electrical power (P) in a DC circuit equals the product of the voltage drop (V) and the current (I). This can be calculated using the formula $P = VI$. A physics student observes a voltage drop of 6 volts in a DC circuit with a current of 15 amperes (amps). Calculate the electrical power that is used in this circuit. When voltage is measured in volts and current is measured in amperes, power is measured in watts.

SECTION 2.7 LINE GRAPHS

73. The line graph shows average corporate profits over an 11-year period.

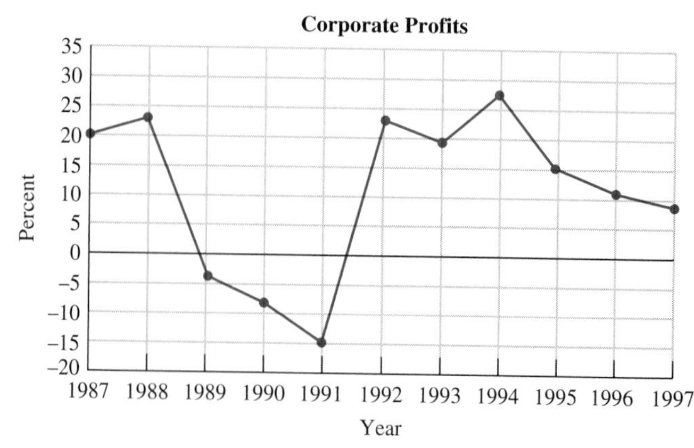

a. Which year showed the highest percent of profits?

b. What is the difference between the percent of profits in 1995 and 1991?

c. What is the difference in the percent of profits in 1987 and 1991?

d. Find the average percent profit over the 11-year period. Round the answer to the nearest percent.

e. Find the median of the percent profits over the 11-year period.

74. Ten samples of a cement were prepared. The tensile strength of each sample was tested at a different time interval. The table shows the results of these tests.

Age (weeks)	1	2	3	4	5	6	7	8	9	10
Tensile strength (lb per in.^{2})	120	150	175	195	210	223	233	242	250	255

a. Construct a line graph showing the relation between the age of the cement and its tensile strength.

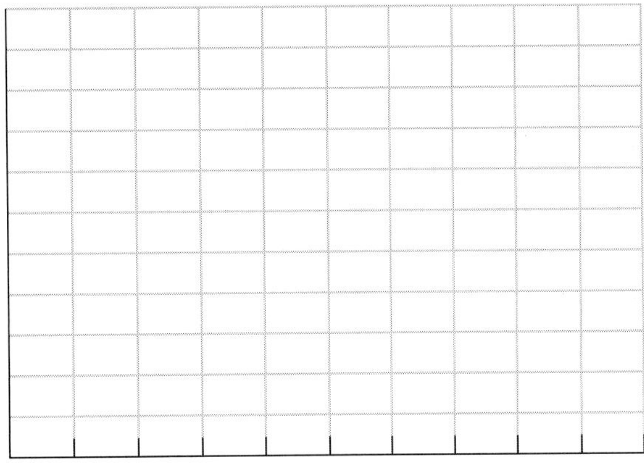

b. Calculate the mean tensile strength of the cement over the 10 weeks. Round the answer to the nearest pound per square inch.

c. Find the median tensile strength. Round the answer to the nearest pound per square inch.

TEST CHAPTER 2

1. Represent 10 degrees below zero as an intege

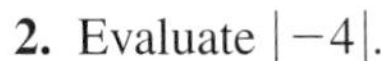

2. Evaluate $|-4|$.

3. Graph the integers on a number line: 4, -7, 0, -3, and -1.

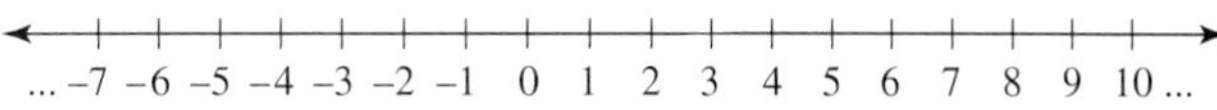

4. Evaluate $-(-8)$.

5. Put $<$, $>$, or $=$ between the numbers to make a true statement: $-34 \quad -62$.

6. What is the opposite of -5?

7. What is the opposite of a loss of 50 points?

8. Evaluate $-(-(-45))$.

9. Evaluate $-|-7|$.

10. Solve and check the equation $|x| = 8$.

11. Round -87 to the nearest ten.

12. The temperature in San Jose was 75°F at noon. Over the next 5 hours it decreased 2 degrees, decreased 1 degree, increased 4 degrees, decreased 6 degrees, and decreased 8 degrees. What was the temperature at 5 P.M.?

13. Evaluate $8 + (-5)$.

14. Evaluate $\dfrac{5(-4) - 10}{2(5)}$

15. Solve the equation $T = -12°F - (-6°F)$.

16. Evaluate $(-16)^0$.

17. Evaluate $-4(5 - 8 + 4)$.

18. Solve the equation $x = 3 + 8 - 5 - 9$.

19. Evaluate $-9 + 5 + (-7)$.

20. Solve the equation $|y| = -5$.

21. Solve the equation $y = -28 \div 7$.

22. Solve the equation $x = -3 + 9$.

23. Round -45 to the nearest hundred.

24. Evaluate $(-8)^2$.

25. Solve the equation $t = -12 - 8$.

26. Solve using the five-step problem-solving strategy: The highest point in Europe is at the summit of El'brus, 5,642 meters above sea level. The lowest point in Europe is at the Caspian Sea, 28 meters below sea level. What is the difference in elevation between the highest and lowest points in Europe?

27. Evaluate 7^0.

28. Evaluate $-7(-2)(5)$.

29. Evaluate $(7 - 21) \div (-7)$.

30. Solve the equation $x = \dfrac{-8(9)}{-2(-3)}$.

31. Simplify. Write the answer as an exponential expression: $5^7 \cdot 5^8 \cdot 5^2$.

32. Solve the equation $x = -20 \div (-5)$.

33. Find the median: 34, 27, 23, 34, -51, -26.

34. Evaluate $-6(2 - 9)$.

35. Evaluate $-4^2 + 6^2$.

36. Solve the equation $x = (-5 - 7) \div (-4)$.

37. Evaluate $-3(4) - 5(-2)$.

38. Evaluate using the distributive property: $(14 - 7) \div (-7)$.

39. Evaluate $|3 - (-2)^2|$.

40. Evaluate $3(2^3)(-4^2)$.

41. Solve the equation $A = -6 - 7(-5)$.

42. Use the formula $C = 5(F - 32) \div 9$ to convert 21°F into degrees Celsius. Round the answer to the nearest degree.

43. The double line graph shows the highest and lowest surface points on each of the seven continents.

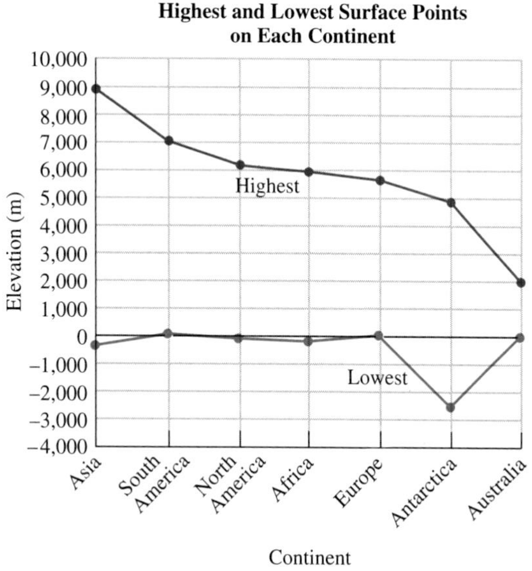

a. Which continent has the highest surface point?

b. Which continent has the lowest surface point?

c. Which continent has the greatest difference between its highest and lowest surface points?

d. Estimate the difference between the highest and lowest surface points of Antarctica.

e. Estimate the average highest surface point on the seven continents.

f. Find the median highest surface point on the seven continents.

44. Evaluate $|-6 + 4|^2 - |-3^2 - 4|$.

45. Solve the equation $s = (3 - 2)^2 - (-3 - 4)^2$.

3

Fractions

Because we use the English system of measurement, fractions surround us in our daily lives. We buy three-eighths-inch plywood and measure sugar in fractions of a cup. Fractions are not limited to English units of measure, however. We consider one half of a population or two thirds of a book. We calculate distance per hour and miles per gallon.

In this chapter, we will learn the arithmetic of fractions. We will solve word problems, continuing our work with the five-step problem-solving strategy. Later in the text, we will work with ratios, rates, and percents, important applications of fractions in daily life.

CHAPTER OUTLINE

Section 3.1 Multiples and Factors

LEARNING TOOLS

CD-ROM SSM VIDEO

Sneak Preview

In this chapter we begin doing arithmetic with fractions. To make that work easier, we will first learn more about divisors, multiples, and factors.

After completing this section, you should be able to:

1) Determine if a number is a multiple of a given number.
2) Use divisibility laws to determine if a number is a divisor of a given number.
3) Determine whether a number is prime, composite, or neither.
4) Determine the factors of a number.

DISCUSSION

Multiples, Factors, and Divisors

The natural numbers are the set {1, 2, 3, 4, 5, . . .}. A **multiple** of a number is the product of that number and a natural number. For example, 12 is a multiple of 6 since $6 \cdot 2 = 12$. The first five multiples of 6 are 6, 12, 18, 24, and 30 since $6 \cdot 1 = 6$, $6 \cdot 2 = 12$, $6 \cdot 3 = 18$, $6 \cdot 4 = 24$, and $6 \cdot 5 = 30$. We can use division to determine if a number is a multiple of a given number.

PROCEDURE

To determine if a number N is a multiple of a given number n:

1. Divide N by n.
2. If the remainder is 0, N is a multiple of n.

Example 1 **Is 60 a multiple of 8?**

$60 \div 8 = 7 \text{ R}(4)$

Since there is a remainder other than 0, 60 is not a multiple of 8.

Example 2 **Is 75 a multiple of 5?**

$75 \div 5 = 15$

Since the remainder is 0, 75 is a multiple of 5.

CHECK YOUR UNDERSTANDING

1. List the first five multiples of 4.
2. List the first five multiples of 7.
3. Is 14 a multiple of 3?
4. Is 108 a multiple of 9?

Answers: 1. 4, 8, 12, 16, 20; 2. 7, 14, 21, 28, 35; 3. No. $14 \div 3 = 4 \text{ R}(2)$; 4. Yes. $108 \div 9 = 12$.

When we multiply two whole numbers, the result is called a **product.** The numbers being multiplied are called **factors.** Every whole number greater than 1 has at least two factors, the number itself and 1. The factors of a number are always greater than or equal to 1 and less than or equal to the number itself.

Example 3 **List the factors of 2.**

The factors of 2 are 1 and 2.

Example 4 **List the factors of 6.**

The factors of 6 are 1, 2, 3, and 6.

The factors of a number are also its **divisors.** A divisor divides a number without remainder. For example, 3 is a divisor of 15 since $15 \div 3 = 5$ without remainder. However, 4 is not a divisor of 15 since $15 \div 4 = 3$ R(3). Because 0 can never be a divisor, it can never be a factor. Since factors have to be greater than 0, zero has no factors.

> **PROPERTY**
>
> **A natural number n is a divisor or a factor of another whole number N, if n divides N exactly without remainder.**

To find all the factors of a whole number, N, we look for pairs of numbers whose product is the number. We use division to determine which natural numbers {1, 2, 3, . . . } are factors. This will also reveal factor pairs.

Example 5 **List the factors of 36.**

Begin with 1. $36 \div 1 = 36$ so 1 and 36 are factors.
Try 2. $36 \div 2 = 18$ so 2 and 18 are factors.
Try 3. $36 \div 3 = 12$ so 3 and 12 are factors.
Try 4. $36 \div 4 = 9$ so 4 and 9 are factors.
Try 5. $36 \div 5 = 7$ R(1) so 5 is not a factor.
Try 6. $36 \div 6 = 6$ so 6 is a factor.
Try 7. $36 \div 7 = 5$ R(1) so 7 is not a factor.
Try 8. $36 \div 8 = 4$ R(4) so 8 is not a factor.
Try 9. $36 \div 9 = 4$ so 9 and 4 are factors.

Each pair of factors is called a factor pair. For example, 2 and 18 are a factor pair.

But wait, we already know that 9 and 4 are factors from checking if 4 is a factor. In fact, we do not need to check if 9 or any numbers greater than 9 are factors. *Once the square of a possible factor is greater than or equal to the number we are factoring, we have found all the possible factors.* In this example, we look for a number whose square is greater than or equal to the number we are factoring, 36. Since $6^2 = 36$, we need not look any higher than 6 for more factors.

The factors of 36 are 1, 2, 3, 4, 6, 9, 12, 18, and 36.

Example 6 **List the factors of 50.**

Begin with 1. $50 \div 1 = 50$ so 1 and 50 are factors.
Try 2. $50 \div 2 = 25$ so 2 and 25 are factors.
Try 3. $50 \div 3 = 16$ R(3) so 3 is not a factor.
Try 4. $50 \div 4 = 12$ R(2) so 4 is not a factor.
Try 5. $50 \div 5 = 10$ so 5 and 10 are factors.
Try 6. $50 \div 6 = 8$ R(2) so 6 is not a factor.
Try 7. $50 \div 7 = 7$ R(1) so 7 is not a factor.
Try 8. $50 \div 8 = 6$ R(2) so 8 is not a factor.

Since 8^2 is 64, which is greater than 50, we can stop looking for additional factors.

The factors of 50 are 1, 2, 5, 10, 25, and 50.

Every whole number greater than 1 has at least two factors. If a number that is greater than 1 has exactly two factors, one and itself, it is called a **prime number.** Any number that is greater than 1 that is not prime is called a **composite number.** Numbers that are less than or equal to 1 are neither prime nor composite. The set of prime numbers can be written: Primes = {2, 3, 5, 7, 11, 13, 17, 19, 23, 29, 31, . . .}. The ellipses . . . show that the set continues. There is no largest prime number.

PROPERTY

Prime and Composite Numbers

1. A prime number is any whole number greater than 1 that has only 1 and itself as factors.
2. Any whole number greater than 1 that is not prime must have more than two distinct factors and is called composite.
3. The numbers 0 and 1 are neither prime nor composite.

We can identify factors or divisors using the multiplication facts of the times table, long division, or a calculator. We can also use the following **divisibility rules.**

The Divisibility Rules

Two: A number is divisible by 2 if the number is even. A number is even if it ends in the digit 0, 2, 4, 6, or 8. Otherwise, the number is odd.

Three: A number is divisible by 3 if the sum of the digits of the number is divisible by 3.

Four: A number is divisible by 4 if the number formed by the last two digits is divisible by 4.

Five: A number is divisible by 5 if the last digit is 5 or 0.

Six: A number is divisible by 6 if it is divisible by 2 and it is also divisible by 3.

Nine: A number is divisible by 9 if the sum of its digits is divisible by 9.

Ten: A number is divisible by 10 if the last digit is 0.

Example 7 **Use the divisibility rules to determine if 356,121 is divisible by 2 or 3. Check by division.**

Is 2 a divisor? The last digit of 356,121 is an odd number so it is not divisible by 2.

Check

356,121 ÷ 2 = 178,060 R(1) Since there is a remainder, 2 is not a divisor.

Is 3 a divisor? The sum of the digits of 356,121 is 3 + 5 + 6 + 1 + 2 + 1 = 18. 18 is divisible by 3 since 18 ÷ 3 = 6 with no remainder. Since the sum of its digits is divisible by three, 356,121 is divisible by 3.

Check

356,121 ÷ 3 = 118,707 Since there is no remainder, 3 is a divisor.

Example 8 **Determine if 456,937,736 is divisible by 2 or 4.**

Is 2 a divisor? The last digit of 456,937,736 is an even number so 456,937,736 is divisible by 2.

Is 4 a divisor? The last two digits of 456,937,736 form the number 36. Since 36 is divisible by four, 456,937,736 is divisible by 4.

Example 9 **Determine if 1,008 is divisible by 6 or 9.**

Is 6 a divisor? Since the last digit of 1,008 is even, 1,008 is divisible by 2. The sum of the digits of 1,008 is $1 + 0 + 0 + 8 = 9$. Since 9 is divisible by 3, 1,008 is also divisible by 3. Since 1,008 is divisible by 2 and 3, it is also divisible by 6.

Is 9 a divisor? The sum of the digits of 1,008 is $1 + 0 + 0 + 8 = 9$. Since 9 is divisible by 9 the number 1,008 is also divisible by 9.

Example 10 **Determine if 127 is a prime or composite number.**

1 and itself are factors of 127.

Try 2. 127 is not an even number, so 2 is not a factor.

Try 3. The sum of the digits is $1 + 2 + 7 = 10$. $10 \div 3 = 3$ R(1), so 3 is not a factor.

Try 4. The last two digits are 27. $27 \div 4 = 6$ R(3), so 4 is not a factor.

Try 5. The last digit of 127 is not 0 or 5, so 5 is not a factor.

Try 6. Since 127 is not divisible by 2 or 3, it is not divisible by 6.

Try 7. No divisibility rule so divide. $127 \div 7 = 18$ R(1), so 7 is not a factor.

Try 8. No divisibility rule so divide. $127 \div 8 = 15$ R(7), so 8 is not a factor.

Try 9. The sum of the digits is 10. $10 \div 9 = 1$ R(1), so 9 is not a factor.

Try 10. The last digit is not 0 so 10 is not a factor.

Try 11. No divisibility rule so divide. $127 \div 11 = 11$ R(6), so 11 is not a factor.

Try 12. No divisibility rule so divide. $127 \div 12 = 10$ R(7), so 12 is not a factor.

Since 12^2 is 144 and 144 is greater than 127, we do not have to check any more numbers. *The only factors of 127 are 1 and 127. 127 is a prime number.*

Example 11 **Which of the numbers 2, 3, 4, 5, 6, 9, and 10 are factors of 703?**

2: 703 is odd so is not divisible by 2. 2 is not a factor.

3: The sum of the digits $= 7 + 0 + 3 = 10$, which is not divisible by 3. 3 is not a factor.

4: The last two digits form the number, 03 or 3, which is not divisible by 4. 4 is not a factor.

5: The number does not end in a 0 or a 5. 5 is not a factor.

6: The number is not divisible by 2 or by 3. 6 is not a factor.

9: The sum of the digits is 10, which is not divisible by 9. 9 is not a factor.

10: The number does not end in a 0. 10 is not a factor.

The numbers 2, 3, 4, 5, 6, 9, and 10 are not factors of 703. This does not necessarily show that 703 is prime. Other numbers could be factors of 703. In fact, $703 = 19 \cdot 37$ so 19 and 37 are factors of 703. So, 703 is composite.

CHECK YOUR UNDERSTANDING

Use the divisibility rules to determine if:

1. 3 is a factor of 51

2. 3 is a factor of 840

3. 6 is a factor of 31,503

4. 9 is a factor of 315,324

5. 5 is a factor of 51

6. 2 is a factor of 840

7. 4 is a factor of 728

8. 10 is a factor of 315,322,500

List the factors of each number and identify it as prime, composite, or neither.

9. 17

10. 45

11. 22

12. 23

13. 1

14. 37

15. 0

16. 16

Answers: 1. yes; 2. yes; 3. no; 4. yes; 5. no; 6. yes; 7. yes; 8. yes; 9. 1, 17; prime; 10. 1, 3, 5, 9, 15, 45; composite; 11. 1, 2, 11, 22; composite; 12. 1, 23; prime; 13. 1; neither prime nor composite; 14. 1, 37; prime; 15. no factors; neither prime nor composite since it is less than 1; 16. 1, 2, 4, 8, 16; composite.

EXERCISES SECTION 3.1

List five multiples of the number.

1. 2 **2.** −2 **3.** 32 **4.** 1

5. Find a number that is a multiple of 2, a multiple of 3, and is greater than 15.

6. Find a number that is a multiple of 9, a multiple of 5, and is greater than 50.

7. Find a number that is a multiple of 3, a multiple of 4, a multiple of 10, and is greater than 40.

8. Tim says that 4 is a *factor* of 12. Rhonda disagrees, claiming that 4 is a *divisor* of 12. Both Tim and Rhonda are correct. Explain why 4 is *both* a factor and a divisor of 12.

List the factors of the number.

9. 24 **10.** 100 **11.** 31

12. 64 **13.** 77 **14.** 12

15. List the first 20 prime numbers.

16. List the first 20 composite numbers.

17. List the first 20 natural numbers.

18. No prime numbers are negative. Explain why.

19. Zero is not a prime number. Explain why.

Complete the tables using the divisibility rules.

	NUMBER	EVEN OR ODD NUMBER?	DIVISIBLE BY 2? (YES OR NO)	SUM OF THE DIGITS?	DIVISIBLE BY 3? (YES OR NO)
20.	8				
21.	12				
22.	65				
23.	1,257				
24.	100,002				
25.	540				

	NUMBER	DIVISIBLE BY 2? (YES OR NO)	SUM OF THE DIGITS?	DIVISIBLE BY 3? (YES OR NO)	DIVISIBLE BY 9? (YES OR NO)
26.	54				
27.	5,499				
28.	92,430				
29.	105				

	NUMBER	DIVISIBLE BY? (YES OR NO)						
		2	3	4	5	6	9	10
30.	924							
31.	18							
32.	24							
33.	9,240							
34.	40,815							
35.	12,016							
36.	36							
37.	564							

38. During the Idaho gold rush in the mid-1800s, panning for gold from river gravels was labor intensive. The gold was so fine and light that $1 of gold was equal to about 100,000 flecks of gold. Use the divisibility rules to explain if 100,000 flecks of gold could be divided equally among five people.

39. On the third day of the Civil War battle of Gettysburg, the Confederate Army sent three divisions—13,000 men—against the center of the Union Army line. The Union guns opened fire at the right of the Confederate advancing line. An officer wrote that as many as ten men at a time were destroyed by a single bursting shell. Use the divisibility rules to explain if each division could have had the same number of soldiers.

Gettysburg battlefield *(© 2001 Beth Strauss, Inter Productions)*

40. An office machine that functions as a printer, fax machine, copy machine, and scanner has a sale price of $399. If only whole-dollar amounts are used (no cents), use the divisibility rules to explain if this price could be divided into nine equal payments.

41. Clay is driving a distance of 1,542 miles. If only whole-mile divisions are allowed (no fractions of a mile), use the divisibility rules to explain if this distance could be divided into four equal parts.

42. Patti is a teacher's aide at a Head Start preschool. She needs to divide a bag of 96 jelly beans equally into baskets. Use the divisibility rules to explain if these candies can be divided equally into six baskets.

43. A circle has a total angle measure of 360 degrees. Use the divisibility rules to explain if this measure can be divided into three equal parts.

44. Find the smallest whole number that is a multiple of both 6 and 15.

45. Find the smallest whole number that is a multiple of both 20 and 30.

46. Find the smallest whole number that is a multiple of both 8 and 12.

47. Find the smallest whole number that is a multiple of both 10 and 8.

48. Find the smallest whole number that is a multiple of both 6 and 9.

ERROR ANALYSIS

49. Problem: Use the divisibility rules to explain if 741 is divisible by 6.

Answer: $7 + 4 + 1 = 12$

12 is divisible by 6, so 741 is divisible by 6.

Describe the mistake made in solving this problem.

Redo the problem correctly.

50. Problem: List the factors of 12.

Answer: 2, 3, 4, 6

Describe the mistake made in solving this problem.

Redo the problem correctly.

51. Problem: List the first ten prime numbers.

Answer: 1, 2, 3, 5, 7, 11, 13, 17, 19, 23

Describe the mistake made in solving this problem.

Redo the problem correctly.

REVIEW

52. Describe the set of natural numbers using set notation.

53. Identify the divisor, dividend, and quotient in $15 \div 3 = 5$.

54. Identify the factors and the product in $6 \cdot 4 = 24$.

Section 3.2 Introduction to Fractions

LEARNING TOOLS

CD-ROM SSM VIDEO

Sneak Preview

Fractions represent part of a whole. There are an infinite number of fractions on the number line between each pair of whole numbers.

After completing this section, you should be able to:

1) Identify a fraction as proper or improper.

2) Identify the numerator and denominator of a fraction.

3) Describe the set of rational numbers.

4) Graph a fraction on the number line.

5) Write a fraction with a given denominator that is equivalent to another fraction.

6) Simplify a fraction to an equivalent fraction in lowest terms.

7) Write the prime factorization of a composite number.

DISCUSSION

Numerators, Denominators, Proper and Improper Fractions

This rectangle is divided into four equal parts. One part is shaded. This shaded part is a **fraction** of the whole rectangle.

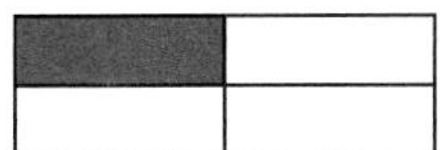

To describe this fraction mathematically, we compare the number of shaded parts to the total number of parts in the whole rectangle.

$$\text{Fraction of a whole} = \frac{\text{Number of parts that are shaded}}{\text{Total number of equal parts in the whole}} = \frac{1}{4}$$

We read this fraction as "one fourth" or "one quarter." The number on the top line of the fraction is called the **numerator** and the number on the bottom of the fraction is the **denominator.**

All fractions can be represented by partially shaded geometric figures. For example, to represent the fraction $\frac{3}{8}$ using a circle, divide a circle into eight equal parts. Shade three of the parts. The denominator of the fraction is 8; the numerator is 3.

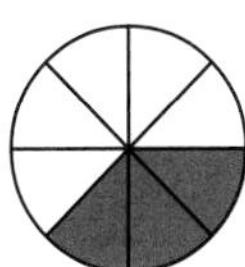

To represent the fraction $\frac{3}{3}$ using a triangle, divide the triangle into three equal parts. Since the numerator is 3, all of the parts should be shaded.

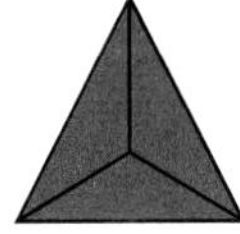

We can also think of fractions in terms of part of a whole collection or set. A class of 30 students may include 7 students who speak two languages. We can represent the fraction of the students in the class who speak two languages as $\frac{7}{30}$. The number of students in the class is the entire collection and is represented by the denominator 30. The number of students who speak two languages is the numerator 7.

Example 1 **Carolina Biological Supply conducts environmental science courses for secondary school teachers. Using nitrate test kits, the classes have tested water all over Maryland for nitrogen. A significant cause of high nitrogen levels is groundwater runoff from farms, home landscapes, and golf courses that have been fertilized. The highest level measured was 15 parts per million from a well in a residential area near Baltimore. Express 15 parts per million as a fraction.**

The entire collection of parts reported is the denominator: one million or 1,000,000. The parts of the million that are nitrogen, 15, is the numerator. 15 parts per million written as a fraction is $\frac{15}{1{,}000{,}000}$.

The fractions we have seen so far all refer to part of one whole figure or collection. They are all **proper fractions.** In a proper fraction, the numerator is always less than the denominator. We can also write fractions that describe more than one shaded figure or collection. These are called **improper fractions.** In an improper fraction, the numerator is always greater than or equal to the denominator.

Example 2 **Represent the improper fraction $\frac{5}{4}$ using shaded rectangles.**

 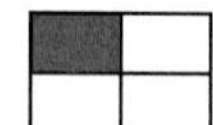

Since the denominator is 4, each figure will be divided into four parts. A total of five parts need to be shaded so we will need to use two rectangles. Completely shade one rectangle before shading any parts in the next rectangle.

Improper fractions can also be used to represent any natural number {1, 2, 3, . . .}.

Example 3 **Write 4 as a fraction.**

The number 4 represents four whole figures such as four circles. Each circle has not been divided, so each has one part. The denominator of the fraction must be 1. Since there are four of these wholes, the numerator is 4. So, 4 can be written as $\frac{4}{1}$.

The set of fractions with numerators and denominators that are integers is called the **set of rational numbers.** All integers are rational numbers because they can be written as improper fractions with a denominator of 1.

CHECK YOUR UNDERSTANDING

Represent each fraction with a shaded figure(s) and then identify as proper or improper.

1. $\frac{2}{3}$
2. $\frac{8}{3}$
3. $\frac{9}{9}$
4. $\frac{7}{6}$
5. $\frac{2}{1}$

Answers: 1.  *proper* 2. 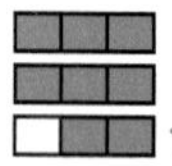 *improper* 3. 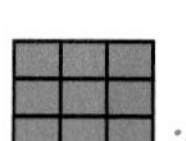 *improper* 4. 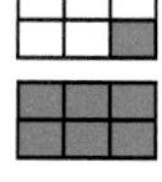 *improper* 5. 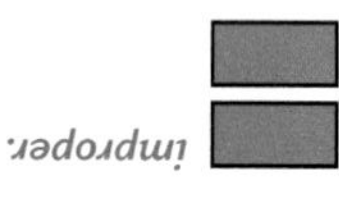 *improper*

DISCUSSION

Graphing Fractions on a Number Line

When we think of fractions only as shaded parts of a figure, we are limited to positive numbers. However, we can divide the distance between any integer and 0 on a number line into the required number of parts. This means that both positive and negative fractions can be graphed on a number line.

Example 4 **Graph $\frac{3}{4}$ on a number line.**

This is a proper fraction so it lies between 0 and 1. Since the denominator of the fraction is 4, we draw three lines between 0 and 1 to divide this interval into four equal parts.

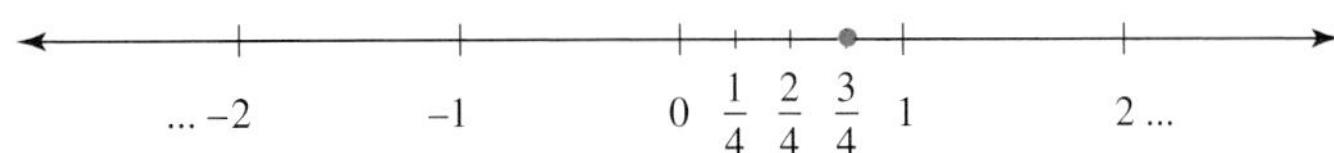

Example 5 **Graph $-\frac{1}{3}$ on a number line.**

This fraction is between 0 and −1. Since the denominator of the fraction is 3, we divide the number line between 0 and −1 into 3 equal parts. We do this by drawing two vertical lines between 0 and −1.

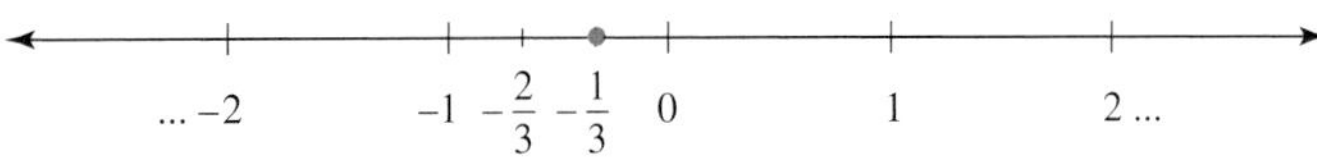

CHECK YOUR UNDERSTANDING

Graph each fraction on its own number line.

1. $\frac{3}{5}$

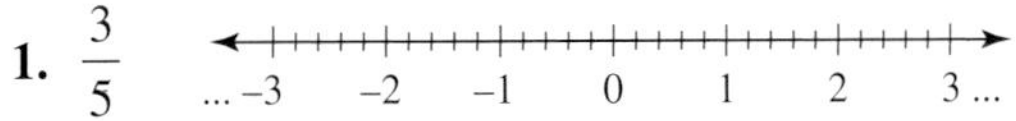

2. $-\frac{3}{5}$ ... −3 −2 −1 0 1 2 3 ...

3. $-\frac{10}{5}$ ... −3 −2 −1 0 1 2 3 ...

4. $-\frac{13}{5}$ ... −3 −2 −1 0 1 2 3 ...

5. $\frac{0}{5}$ ... −3 −2 −1 0 1 2 3 ...

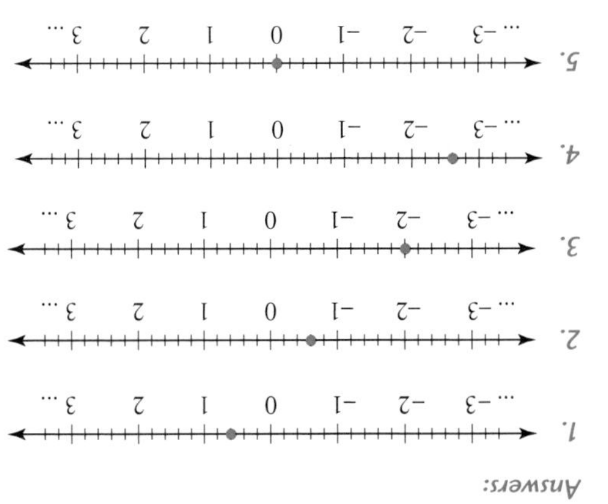

DISCUSSION

Negative Fractions, Opposites, and Absolute Values

A fraction that is negative can be written with the − sign in any one of three different places: in front of the entire fraction, next to the numerator, or next to the denominator.

Example 6 **Rewrite negative $\frac{3}{4}$ using a − sign.**

Negative $\frac{3}{4}$ can be written as $-\frac{3}{4}$, $\frac{-3}{4}$, or $\frac{3}{-4}$.

The − sign can be written next to the numerator or the denominator because the fraction bar between the numerator and the denominator also acts as a division sign. The quotient of a negative number and a positive number (− sign in the numerator) and the quotient of a positive number and a negative number (− sign in the denominator) are both negative numbers.

We graph and evaluate the opposite or absolute value of a fraction just as we did for whole numbers.

Example 7 **Graph the opposite of $\frac{4}{5}$ on a number line.**

... −3 −2 −1 0 1 2 3 ...

$-\frac{4}{5}$ is found $\frac{4}{5}$ of a unit to the left of 0 on a number line.

Example 8 **Evaluate the opposite of negative $\frac{4}{5}$.**

The opposite of negative $\frac{4}{5}$ is written $-\left(-\frac{4}{5}\right)$ and is equal to positive $\frac{4}{5}$.

Example 9 **Evaluate $\left|\frac{11}{8}\right|$.**

$\left|\frac{11}{8}\right| = \frac{11}{8}$

Example 10 **What is the absolute value of $-\frac{11}{8}$?**

$\left|-\frac{11}{8}\right|$ is equal to $\frac{11}{8}$

To determine whether a negative fraction is proper or improper, look at its absolute value. If the absolute value is proper, the original fraction is proper. If the absolute value is improper, the original fraction is improper. For example, $-\frac{11}{8}$ is an improper fraction because $\left|-\frac{11}{8}\right| = \frac{11}{8}$ is improper.

CHECK YOUR UNDERSTANDING

Evaluate each expression.

1. $\left|-\frac{3}{4}\right|$

2. $\left|\frac{3}{4}\right|$

3. $-\left(-\frac{3}{4}\right)$

4. $-\left(-\left(-\frac{3}{4}\right)\right)$

Identify each fraction as proper or improper.

5. $-\frac{5}{7}$

6. $\frac{-6}{1}$

7. $-\frac{7}{7}$

8. $\frac{2}{-3}$

Answers: 1. $\frac{3}{4}$; 2. $\frac{3}{4}$; 3. $\frac{3}{4}$; 4. $-\frac{3}{4}$; 5. proper; 6. improper; 7. improper; 8. proper

DISCUSSION

Equivalent Fractions

Fractions that represent the same part of a whole but have different denominators are called **equivalent fractions.** For example, $\frac{1}{2}$ and $\frac{2}{4}$ are represented by the same shaded area in a figure:

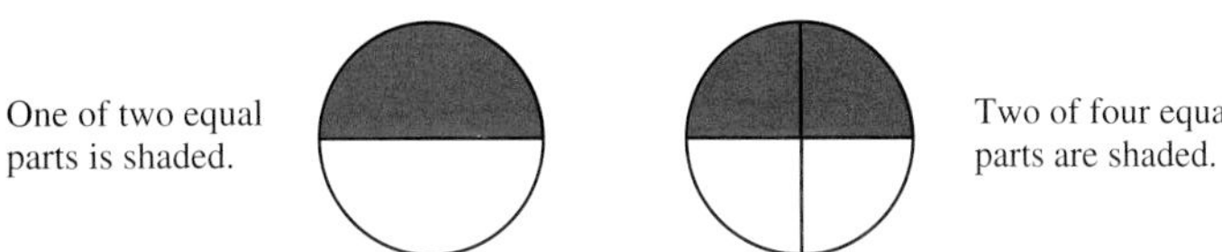

We build equivalent fractions by multiplying or dividing the numerator and the denominator by the same nonzero number.

PROPERTY

In words: Multiplying or dividing both the numerator and the denominator by the same nonzero number results in an equivalent fraction.

In symbols: If a, b, and c are numbers with $b \neq 0$ and $c \neq 0$,

$$\frac{a}{b} = \frac{ac}{bc} \quad \text{and} \quad \frac{a}{b} = \frac{a \div c}{b \div c}$$

The value of the fraction, the shaded area, does not change even though the numerator and denominator change.

Example 11 **Write a fraction with a denominator of 12 that is equivalent to $\frac{2}{3}$.**

$\frac{2}{3}$

$= \frac{2 \cdot \mathbf{4}}{3 \cdot \mathbf{4}}$ *To change the original denominator of 3 to 12, multiply by 4. We must also multiply the numerator by 4.*

$= \frac{8}{12}$

The fraction $\frac{4}{4}$ can be rewritten as $4 \div 4$. Since $4 \div 4 = 1$, $\frac{4}{4}$ also equals 1. When we multiply both the numerator and denominator by 4, it is equivalent to multiplying the whole fraction by $\frac{4}{4}$ or 1. Multiplying by 1 does not change the value of the fraction. When we multiply $\frac{2}{3}$ by $\frac{4}{4}$, the product, $\frac{8}{12}$, has the same value as $\frac{2}{3}$. Multiplying the numerator and denominator by the same number is called **building an equivalent fraction.**

Example 12 **Three fourths of the freshman students in a mathematics class work more than ten hours a week at a job. About 6,000 freshman students are enrolled on campus.**

a. Build a fraction with a denominator of 6,000 equivalent to $\frac{3}{4}$.

$\frac{3}{4}$

$= \frac{3 \cdot \mathbf{1{,}500}}{4 \cdot \mathbf{1{,}500}}$ *To change the original denominator to 6,000, multiply by 1,500. The numerator must also be multiplied by 1,500.*

$= \frac{4{,}500}{6{,}000}$

b. If the number of students in this mathematics class that work is representative of all freshman students, use this fraction to predict the total number of freshman students that work more than ten hours a week at a job.

The total number of freshmen that work more than ten hours a week at a job is about 4,500 students.

We can also build equivalent fractions for negative numbers.

Example 13 **Build a fraction with a denominator of 42 that is equivalent to $-\frac{2}{7}$.**

$-\frac{2}{7}$

$= -\frac{2 \cdot 6}{7 \cdot 6}$ *To change the original denominator of 7 to 42, multiply by 6. The numerator must also be multiplied by 6.*

$= -\frac{12}{42}$ *The − sign can be written in the numerator, in front of the entire fraction, or in the denominator.*

Dividing the numerator and denominator by the same number is called **simplifying the fraction.** Simplifying a fraction is the opposite process of building a fraction.

Example 14 **Write a fraction with a denominator of 5 that is equivalent to $-\frac{9}{15}$.**

$-\frac{9}{15}$

$= -\frac{9 \div 3}{15 \div 3}$ *To change the original denominator of 15 to 5, divide by 3. The numerator must also be divided by 3.*

$= -\frac{3}{5}$

$-\frac{3}{5}$ is equivalent to $-\frac{9}{15}$.

CHECK YOUR UNDERSTANDING

Write a fraction with a denominator of 4 that is equivalent to each of the following fractions.

1. $\frac{1}{2}$ **2.** $\frac{25}{20}$ **3.** $\frac{-36}{8}$ **4.** $\frac{1}{1}$

Answers: 1. $\frac{2}{4}$; 2. $\frac{5}{4}$; 3. $-\frac{18}{4}$; 4. $\frac{4}{4}$.

DISCUSSION

Simplifying Fractions to Lowest Terms

When a fraction is written in its simplest form, it is written in **lowest terms.** We use common factors to simplify a fraction to lowest terms. When a fraction is in lowest terms, the numerator and denominator have no common factors other than 1.

PROPERTY

The *greatest common factor* of two or more numbers is the largest factor that all the numbers have in common.

Example 15 **Identify the greatest common factor of 40 and 72.**

The factors of 40 are 1, 2, 4, 5, 8, 10, 20, and 40.

The factors of 72 are 1, 2, 3, 4, 6, 8, 9, 12, 18, 24, 48, and 72.

Factors that are shared by both 40 and 72 are called common factors. The common factors of 40 and 72 are 1, 2, 4, and 8. Since 8 is the largest of the common factors, it is the **greatest common factor.**

PROCEDURE

To simplify a fraction to lowest terms using the greatest common factor:

1. Identify the greatest common factor of the numerator and denominator.
2. Divide the numerator and denominator by the greatest common factor.

Example 16 **Simplify $\frac{40}{72}$.**

$\frac{40}{72}$ *The greatest common factor of 40 and 72 is 8.*

$= \frac{40 \div \mathbf{8}}{72 \div \mathbf{8}}$ *Divide the numerator and denominator by the greatest common factor, 8.*

$= \frac{5}{9}$

Example 17 **Simplify $\frac{12}{18}$ to lowest terms.**

The factors of 12 are 1, 2, 3, 4, 6, and 12. The factors of 18 are 1, 2, 3, 6, 9, and 18. The greatest common factor is 6.

$\frac{12}{18}$

$= \frac{12 \div \mathbf{6}}{18 \div \mathbf{6}}$ *Divide the numerator and denominator by the greatest common factor, 6.*

$= \frac{2}{3}$ *The only common factor of 2 and 3 is 1. This fraction is in lowest terms.*

To simplify, it is not necessary to find the greatest common factor. However, we must continue to divide the fraction by common factors until the only common factor of the numerator and the denominator is 1.

PROCEDURE

To simplify a fraction to lowest terms using any common factor:

1. Identify a common factor of the numerator and denominator.
2. Divide the numerator and denominator by the common factor.
3. Repeat this process until the only common factor shared by the numerator and denominator is 1.

Example 18 **Simplify $\frac{1,440}{450}$ to lowest terms.**

Finding the greatest common factor of 1,440 and 450 by listing factors is a lengthy process. Instead we can look for *any* factor of both numbers. Recall that a number is divisible by 10 if it ends in a zero. Since both 1,440 and 450 end in zeros, they are both divisible by 10.

$$\frac{1,440}{450}$$

$$= \frac{1,440 \div \mathbf{10}}{450 \div \mathbf{10}}$$

$$= \frac{144}{45}$$ *Although simpler than before, this fraction is still not in lowest terms.*

Now find a common factor of 144 and 45.

144: The sum of the digits $1 + 4 + 4 = 9$, which is divisible by 9.
45: The sum of the digits $4 + 5 = 9$, which is divisible by 9.

Using the divisibility law for 9, both 144 and 45 are divisible by 9. Use 9 as the common factor.

$$\frac{144}{45}$$

$$= \frac{144 \div \mathbf{9}}{45 \div \mathbf{9}}$$

$$= \frac{16}{5}$$ *Since 16 and 5 have only 1 as a common factor, the fraction is now in lowest terms.*

Example 19 **42 out of 88 people who filled out a survey were women. Simplify $\frac{42}{88}$ to lowest terms.**

$$\frac{42}{88}$$ *42 and 88 are even so they are both divisible by 2.*

$$= \frac{42 \div \mathbf{2}}{88 \div \mathbf{2}}$$ *Divide both the numerator and the denominator by 2.*

$$= \frac{21}{44}$$ *21 and 44 have no common factors other than 1. This fraction is now in lowest terms.*

Example 20 **Simplify $-\frac{630}{720}$ to lowest terms.**

$$-\frac{630}{720}$$ *630 and 720 both end in zero so they are both divisible by 10.*

$$= -\frac{630 \div \mathbf{10}}{720 \div \mathbf{10}}$$ *Negative numerators and denominators have the same common factors as positive numerators and denominators.*

$$= -\frac{63}{72}$$ *63 and 72 are both divisible by 9:*
6 + 3 = 9, which is divisible by 9.
7 + 2 = 9, which is divisible by 9.

$$= -\frac{63 \div \mathbf{9}}{72 \div \mathbf{9}}$$

$$= -\frac{7}{8}$$ *7 and 8 have only 1 as a common factor. This fraction is in lowest terms.*

CHECK YOUR UNDERSTANDING

Simplify the following fractions to lowest terms.

1. $\frac{16}{18}$ **2.** $-\frac{121}{77}$ **3.** $\frac{360}{80}$ **4.** $\frac{5}{20}$

Answers: 1. $\frac{8}{9}$; 2. $-\frac{11}{7}$; 3. $\frac{9}{2}$; 4. $\frac{1}{4}$.

Every composite number has at least three factors. Every composite number can also be written as the product of two or more prime numbers. This product of prime numbers is called its **prime factorization.**

The Fundamental Theorem of Arithmetic

Every composite number can be written as a unique product of prime factors.

This is called its *prime factorization.*

Example 21 **Find the prime factorization of 24.**

24

$= 4 \cdot 6$ *4 and 6 are composite, so factor both numbers.*

$= 2 \cdot 2 \cdot 2 \cdot 3$ *2 and 3 are prime numbers, so this is the prime factorization.*

The initial choice of factors does not affect the prime factorization. Instead of $24 = 4 \cdot 6$, $24 = 12 \cdot 2$ or $24 = 8 \cdot 3$ can be used. The prime factorization will not change.

Example 22 **Find the prime factorization of 30.**

30

$= 3 \cdot 10$ *3 is prime but 10 is composite. Factor 10.*

$= 3 \cdot 2 \cdot 5$ *3, 2, and 5 are prime so this is the prime factorization.*

$= 2 \cdot 3 \cdot 5$ *Arrange the prime factors from smallest to largest.*

Sometimes it is more useful to write a prime factorization as the product of exponential expressions.

Example 23 **Write the prime factorization of 100 as the product of exponential expressions.**

100

$= 10 \cdot 10$ *10 is composite. Factor 10.*

$= 2 \cdot 5 \cdot 2 \cdot 5$ *2 and 5 are prime, so this is the prime factorization.*

$= 2^2 \cdot 5^2$ *Order the prime factors from smallest to largest and write using exponents.*

Example 24 **Write the prime factorization of 324 as the product of exponential expressions.**

324

$= 3 \cdot 108$ *3 is prime, 108 is composite.*

$= 3 \cdot 12 \cdot 9$ *12 and 9 are composite.*

$= 3 \cdot 3 \cdot 4 \cdot 3 \cdot 3$ *4 is composite.*

$= 3 \cdot 3 \cdot 2 \cdot 2 \cdot 3 \cdot 3$ *This is a product of primes.*

$= 2^2 \cdot 3^4$ *Order from smallest to largest and write using exponents.*

CHECK YOUR UNDERSTANDING

Write the prime factorization as the product of exponential expressions.

1. 20 **2.** 108 **3.** 51 **4.** 132

Answers: 1. $2^2 \cdot 5$; 2. $2^2 \cdot 3^3$; 3. $3 \cdot 17$; 4. $2^2 \cdot 3 \cdot 11$.

Prime factorizations can be used to simplify fractions to lowest terms. The prime factorization is written separately for the numerator and the denominator of a fraction. The fraction is then simplified by dividing the numerator and the denominator by the common prime factors.

PROCEDURE

To simplify a fraction to lowest terms using prime factorization:

1. Find the prime factorization of the numerator and denominator. Write the factorization without exponents.
2. Divide the numerator by the common prime factors in the denominator.
3. Multiply the remaining factors to rewrite the fraction.

Example 25 **Simplify the fraction $\frac{90}{126}$ to lowest terms using prime factorization.**

Find the prime factorizations for the numerator and the denominator.

90	126
$= 10 \cdot 9$	$= 6 \cdot 21$
$= \mathbf{2 \cdot 5 \cdot 3 \cdot 3}$	$= \mathbf{2 \cdot 3 \cdot 3 \cdot 7}$

$$\frac{90}{126}$$

$$= \frac{\mathbf{2 \cdot 3 \cdot 3} \cdot 5}{\mathbf{2 \cdot 3 \cdot 3} \cdot 7}$$ *Replace 90 and 126 with their prime factorizations.*

$$= 1 \cdot 1 \cdot 1 \cdot \frac{5}{7}$$ *Simplify common factors in the numerator and the denominator to 1.*

$$= \frac{5}{7}$$

Example 26 **Simplify the fraction $-\frac{6}{24}$ to lowest terms using prime factorization.**

$$-\frac{6}{24}$$

$$= -\frac{\mathbf{2 \cdot 3}}{2 \cdot 2 \cdot \mathbf{2 \cdot 3}}$$ *Write the prime factorizations for 6 and 24.*

$$= -\frac{\mathbf{2}}{\mathbf{2}} \cdot \frac{\mathbf{3}}{\mathbf{3}} \cdot \frac{1}{2 \cdot 2}$$ *Simplify the fraction. Divide the numerator and the denominator by the common factors 2 and 3.*

$$= -1 \cdot 1 \cdot \frac{1}{4}$$

$$= -\frac{1}{4}$$

CHECK YOUR UNDERSTANDING

Simplify the fraction to lowest terms using prime factorization.

1. $\dfrac{72}{117}$

2. $-\dfrac{168}{120}$

3. $\dfrac{153}{102}$

4. $-\dfrac{20}{100}$

Answers: 1. $\dfrac{8}{13}$; 2. $-\dfrac{7}{5}$; 3. $\dfrac{3}{2}$; 4. $-\dfrac{1}{5}$.

EXERCISES SECTION 3.2

1. Graph $-\frac{3}{4}$ on a number line.

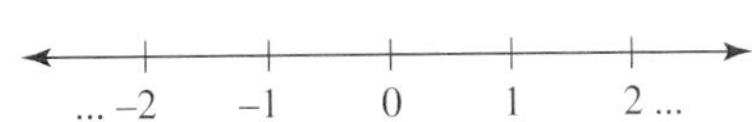

2. Graph $\frac{4}{3}$ on a number line.

3. Use rectangles with shading to represent the improper fraction $\frac{11}{3}$.

4. Explain the difference between a proper fraction and an improper fraction.

5. Explain how to rewrite any whole number as an improper fraction.

Evaluate.

6. $\left|\frac{3}{4}\right|$

7. $\left|-\frac{3}{4}\right|$

8. $\left|\frac{-3}{4}\right|$

9. $\left|\frac{3}{-4}\right|$

Write a >, <, or = symbol between the numbers to make a true statement.

10. $\frac{1}{6} \quad \frac{3}{6}$

11. $\frac{1}{6} \quad -\frac{1}{6}$

12. $\frac{2}{6} \quad \frac{1}{3}$

13. $1 \quad \frac{1}{3}$

14. $-1 \quad \frac{1}{3}$ **15.** $0 \quad \frac{0}{3}$ **16.** $\frac{5}{5} \quad \frac{2}{2}$ **17.** $\frac{99}{100} \quad \frac{999}{1,000}$

Build an equivalent fraction that has a denominator of 48.

18. $\frac{5}{6}$ **19.** $\frac{1}{8}$ **20.** $-\frac{5}{12}$

21. $-\frac{3}{4}$ **22.** $\frac{0}{3}$ **23.** $-\frac{7}{24}$

Build an equivalent fraction that has a denominator of 72.

24. $\frac{5}{6}$ **25.** $\frac{1}{8}$

26. $-\frac{5}{12}$ **27.** $-\frac{3}{4}$

28. A system administrator of an Internet service provider estimates that $\frac{7}{10}$ of the subscribers log on once a day.

a. Build a fraction with a denominator of 1,000 that is equivalent to $\frac{7}{10}$.

b. According to this equivalent fraction, if there are 1,000 subscribers to the Internet service, how many of them log on once a day?

29. Since 1978, the American Society of Newspaper Editors has tracked the race/ethnicity of people employed in the newsrooms of daily newspapers. In 2000, the annual survey revealed that $\frac{13}{100}$ of reporters were Hispanic, Native American, Asian, or Black.

a. Build a fraction with a denominator of 5,000 that is equivalent to $\frac{13}{100}$.

b. According to this equivalent fraction, of 5,000 randomly selected reporters in 2000, how many were Hispanic, Native American, Asian, or Black?

Write the prime factorization of the number as the product of exponential expressions.

30. 50

31. 17

32. 63

33. 64

34. 98

35. 120

Simplify to lowest terms. Some fractions may already be in lowest terms.

36. $\frac{12}{15}$

37. $-\frac{12}{18}$

38. $\frac{5}{20}$

39. $\frac{21}{63}$

40. $-\frac{3}{24}$

41. $\frac{6}{42}$

42. $\frac{4}{11}$

43. $\frac{45}{9}$

44. $\frac{9}{30}$

45. $\frac{72}{120}$

46. $-\frac{72}{124}$

47. $\frac{126}{72}$

48. $\frac{80}{100}$

49. $\frac{8}{100}$

50. $\frac{33}{100}$

51. Major League baseballs must be constructed following precise specifications. The coverings of the ball are sewn together with exactly 108 stitches using exactly 88 inches of thread. The interior layers of red and black rubber are wound with three different layers of wool yarn. The first layer is 121 yards of four-ply gray yarn, the second layer is 45 yards of three-ply white yarn, and the third layer is 53 yards of three-ply gray yarn. What fraction of the total length of yarn is four-ply gray yarn? (The fraction must be in lowest terms.)

52. An ombudsman is an employee of a media organization whose job is to investigate questionable journalistic conduct and to recommend action. The first newspapers to use an ombudsman were the *Louisville Times* and *Courier-Journal* in 1967. In 1993, about 30 of the more than 1,600 daily newspapers in the United States employed an ombudsman. What fraction of the newspapers employed an ombudsman? (The fraction must be in lowest terms.)

53. Cocoa trees are evergreen trees that grow best in locations that are within 20 degrees of the equator. Cocoa beans are roasted and used to make chocolate. West Africa currently produces about 50,000 tons of cocoa beans per year. The total amount of cocoa beans grown worldwide is about 75,000 tons. What fraction of the world's cocoa bean crop is grown in West Africa? (The fraction must be in lowest terms.)

54. Tracy has both men and women in her refresher first aid course for flight attendants. Of a total enrollment of 210, 150 are men. What fraction of the students are men? (The fraction must be in lowest terms.)

55. Excessive amounts of nitrate in drinking water may harm infants. In the human stomach, nitrate is chemically changed to nitrite. Most adults will convert only about $\frac{1}{20}$ of the nitrate in their stomach to nitrite. However, infants have less stomach acid, resulting in more conversion of nitrate to nitrite. Nitrite can combine irreversibly with hemoglobin in the blood, preventing the hemoglobin from doing its essential task of carrying oxygen around the body. The U.S. Environmental Protection Agency (EPA) has set a limit of 10 parts per million of nitrogen as nitrate allowed in public drinking water. Express this limit as a fraction written in lowest terms.

ERROR ANALYSIS

56. **Problem:** Simplify $\frac{48}{60}$.

Answer:

$$\frac{48}{60}$$
$$= \frac{48 \div 3}{60 \div 3}$$
$$= \frac{16}{20}$$

Describe the mistake made in solving this problem.

Redo the problem correctly.

57. **Problem:** Build a fraction with a denominator of 24 that is equivalent to $-\frac{5}{8}$.

Answer:

$$-\frac{5}{8}$$
$$= -\frac{5 \cdot 4}{8 \cdot 4}$$
$$= -\frac{20}{24}$$

Describe the mistake in the answer to this problem.

Redo the problem correctly.

REVIEW

58. Rewrite $3 \cdot 5$ as addition.

59. State the associative property of multiplication.

60. Explain how to determine if the product of two integers is a negative or a positive number.

61. State the commutative property of multiplication.

Section 3.3 Multiplication of Fractions

LEARNING TOOLS

Sneak Preview

Multiplication is the operation used to find a fraction of a fraction. If we need to know $\frac{1}{2}$ of a number, we will multiply the number by $\frac{1}{2}$.

After completing this section, you should be able to:

1) Multiply fractions.

2) Solve word problems that include multiplication of a fraction.

INVESTIGATION Finding a Fraction of a Fraction

A fraction can represent part of a set such as this set of 16 cubes.

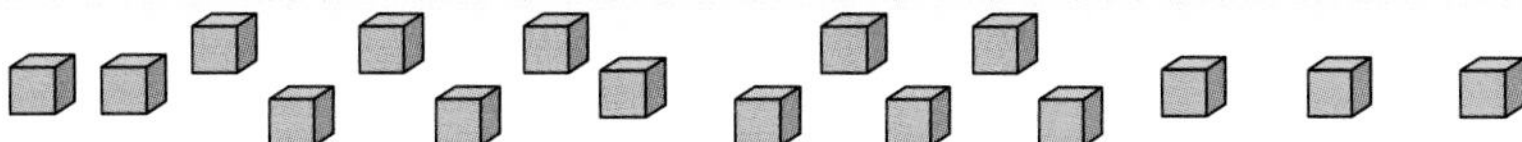

If we need only $\frac{1}{2}$ of the cubes, we group the cubes into two circles. Each circle represents $\frac{1}{2}$ of the set and contains 8 cubes. We write $\frac{1}{2} \cdot 16 = 8$.

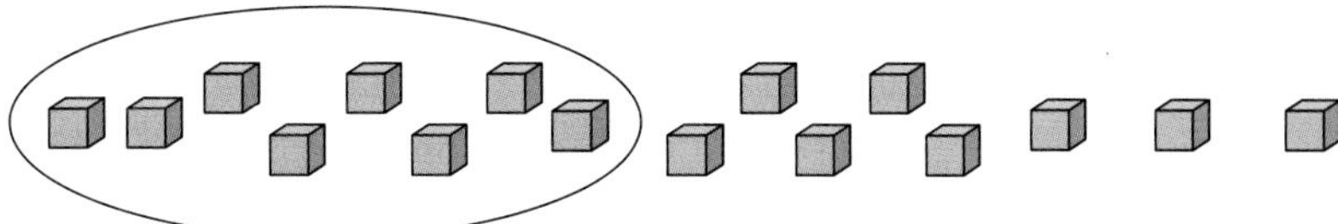

We can use the same method to find other products of a fraction and 16.

1. Group the cubes in four circles. Each circle should have the same number of cubes. Count the number of cubes in a single circle. $\frac{1}{4} \cdot 16 =$ ____________

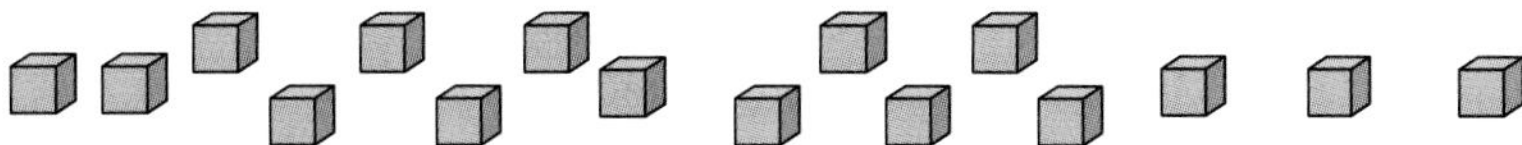

2. Use the same method of circling cubes to find the product: $\frac{1}{8} \cdot 16 =$ ____________

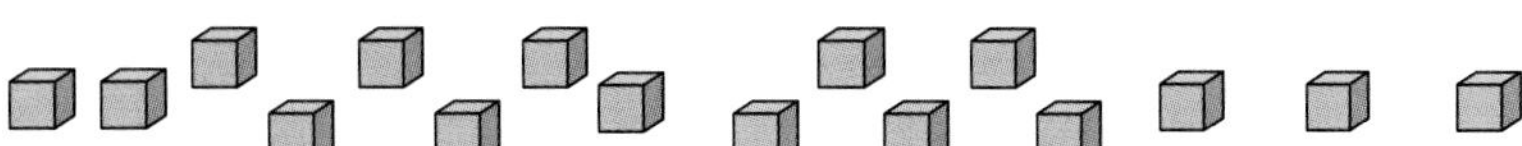

3. Draw a picture of circled cubes that shows $\frac{1}{3} \cdot 12 = 4$.

4. Draw a picture of circled cubes that shows $\frac{1}{6} \cdot 12 = 2$.

Answers: 1. 4; 2. 2.

We can also find a fraction of a fraction of an integer such as $\frac{1}{4} \cdot \frac{1}{2} \cdot 16$.

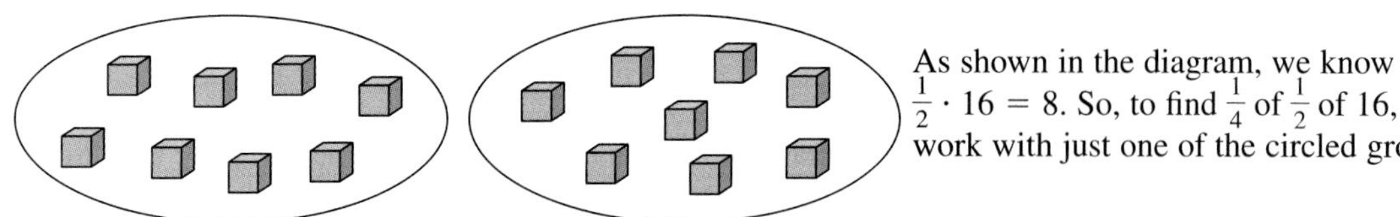

As shown in the diagram, we know that $\frac{1}{2} \cdot 16 = 8$. So, to find $\frac{1}{4}$ of $\frac{1}{2}$ of 16, we work with just one of the circled groups.

5. Group the cubes in the left-hand circle into four smaller rectangles. Each rectangle should have the same number of cubes. Count the number of cubes in a single rectangle. $\frac{1}{4} \cdot \frac{1}{2} \cdot 16 =$ ______________

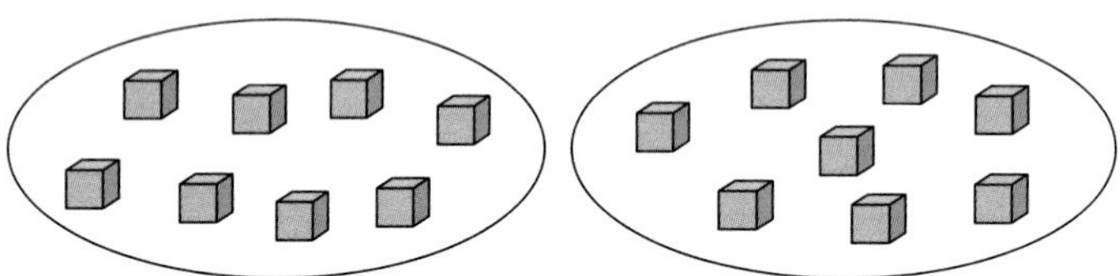

6. Draw 16 small squares below. Use circles and rectangles to find $\frac{1}{2} \cdot \frac{1}{4} \cdot 16$.

7. We found out in Question 5 that $\frac{1}{4} \cdot \frac{1}{2} \cdot 16 = 2$. Is this the same product that we found in Question 6 for $\frac{1}{2} \cdot \frac{1}{4} \cdot 16$?

8. According to these examples, does the commutative property of multiplication apply to fractions? Explain.

9. Draw 24 small squares below. Use circles and rectangles to find $\frac{1}{3} \cdot \frac{1}{2} \cdot 24$.

10. Draw 24 small squares below. Divide them into groups to find $\frac{1}{4} \cdot \frac{1}{6} \cdot 24$.

Answers: 5. 2 cubes; 6. 2; 7. Yes; 8. Yes, because the commutative property of multiplication says that we can change the order of multiplication and not change the product. In these examples, we changed the order in which we multiplied the fractions and we still found the same product; 9. 4; 10. 1.

We need a method that is quicker than counting cubes to find the product of fractions. The following examples show an alternate way to calculate the products of fractions.

EXAMPLE A:

$$\frac{1}{2} \cdot \frac{1}{4} \cdot \frac{16}{1}$$

$$= \frac{1 \cdot 1 \cdot 16}{2 \cdot 4 \cdot 1}$$

$$= \frac{16}{8}$$

$$= 2$$

EXAMPLE B:

$$\frac{2}{3} \cdot \frac{1}{4}$$

$$= \frac{2 \cdot 1}{3 \cdot 4}$$

$$= \frac{2}{12}$$

$$= \frac{1}{6}$$

11. Describe what was done in Example A to find the product of $\frac{1}{2} \cdot \frac{1}{4} \cdot \frac{16}{1}$.

12. Describe what was done in Example B to find the product of $\frac{2}{3} \cdot \frac{1}{4}$.

Use this faster procedure to find each product. The product must be in lowest terms.

13. $\frac{1}{6} \cdot \frac{1}{3}$

14. $\frac{1}{8} \cdot \frac{3}{5}$

15. $\frac{2}{3} \cdot \frac{6}{7}$

16. $\frac{5}{9} \cdot \frac{1}{8}$

17. $\frac{0}{5} \cdot \frac{1}{2}$

18. $\frac{2}{9} \cdot \frac{6}{7}$

19. $\frac{2}{3} \cdot 14$

20. $\frac{1}{8} \cdot \frac{5}{6} \cdot \frac{3}{10}$

Answers: 13. $\frac{1}{18}$; 14. $\frac{3}{40}$; 15. $\frac{4}{7}$; 16. $\frac{5}{72}$; 17. 0; 18. $\frac{4}{21}$; 19. $\frac{28}{3}$; 20. $\frac{1}{32}$.

DISCUSSION

Multiplication of Fractions

Finding a fraction of a fraction is equivalent to multiplying two fractions. To multiply two fractions, find the product of the numerators and the product of the denominators. Simplify the resulting fraction to lowest terms.

PROCEDURE

To multiply two fractions:

In words: Multiply the numerators, multiply the denominators, then simplify the resulting fraction.

In symbols: If a, b, c, and d are numbers and $b \neq 0$ and $d \neq 0$,

$$\frac{a}{b} \cdot \frac{c}{d} = \frac{ac}{bd}$$

Example 1 **Evaluate $\frac{3}{5} \cdot \frac{2}{3}$.**

$\frac{3}{5} \cdot \frac{2}{3}$

$= \frac{3 \cdot 2}{5 \cdot 3}$ *Multiply the numerators. Multiply the denominators.*

$= \frac{6}{15}$ *A common factor of 6 and 15 is 3. Divide numerator and denominator by 3.*

$= \frac{2}{5}$ *The product is in lowest terms.*

The fraction was simplified to lowest terms *after* multiplication of the numerators and denominators. Simplifying can also be done *before* multiplication.

$\frac{3}{5} \cdot \frac{2}{3}$

$= \frac{\mathbf{3} \cdot 2}{5 \cdot \mathbf{3}}$ *Before we multiply, we divide the numerator and the denominator by 3.*

$= \frac{2}{5}$

When multiplying a fraction and an integer, the integer can be rewritten as a fraction by writing it over a denominator of 1. For example, $4 = \frac{4}{1}$.

Example 2 **Evaluate $\frac{7}{10} \cdot 5$.**

$\frac{7}{10} \cdot 5$

$= \frac{7}{10} \cdot \frac{\mathbf{5}}{\mathbf{1}}$ *Rewrite the whole number as a fraction.*

$= \frac{35}{10}$ *Multiply the numerators. Multiply the denominators.*

$= \frac{7}{2}$ *Simplify to lowest terms. Divide numerator and denominator by 5.*

Again, we can choose to simplify before we multiply.

$$\frac{7}{10} \cdot 5$$

$$= \frac{7}{10} \cdot \frac{\mathbf{5}}{\mathbf{1}}$$ *Rewrite the whole number as a fraction.*

$$= \frac{7}{\mathbf{10}} \cdot \frac{\mathbf{5}}{1}$$ *The numerator and denominator have a common factor, 5.*

$$= \frac{7 \cdot \mathbf{1}}{\mathbf{2} \cdot 1}$$ *Simplify before multiplying by dividing numerator and denominator by 5.*

$$= \frac{7}{2}$$

Example 3 **Find the product of $-\frac{4}{5}$ and $\frac{20}{12}$.**

$$-\frac{4}{5} \cdot \frac{20}{12}$$

$$= -\frac{4 \cdot 20}{5 \cdot 12}$$ *Multiply the numerators. Multiply the denominators.*

$$= -\frac{80}{60}$$

$$= -\frac{8}{6}$$ *Simplify to lowest terms. Both 80 and 60 are divisible by 10.*

$$= -\frac{4}{3}$$ *Simplify further. Both 8 and 6 are divisible by 2.*

If we simplify before multiplying:

$$-\frac{4}{5} \cdot \frac{20}{12}$$

$$= -\frac{\mathbf{4} \cdot 20}{5 \cdot \mathbf{12}}$$ *4 and 12 have a common factor of 4. Divide by 4.*

$$= -\frac{1 \cdot \mathbf{20}}{\mathbf{5} \cdot 3}$$ *20 and 5 have a common factor of 5. Divide by 5.*

$$= -\frac{1 \cdot 4}{1 \cdot 3}$$

$$= -\frac{4}{3}$$

Fractions also appear in word problems. In the following example, a fraction is used to describe the characteristics of a group of people. By multiplying the fraction by the number of people in the group, we can make predictions about the group.

Example 4 **A recent study claims that one in three deaths in the United States is the result of cigarette smoking. A city records 270 deaths in one month. How many of these deaths would the study claim could be from smoking cigarettes?**

I know . . .

D = Deaths from smoking

$\frac{1}{3}$ = Fraction of deaths resulting from smoking

270 = Total deaths

Equations

Deaths from smoking = Fraction of deaths · Total deaths

$$D = \frac{1}{3} \cdot 270$$

Solve

$$\begin{aligned} D &= \frac{1}{3} \cdot 270 \\ &= \frac{1}{\mathbf{3}} \cdot \frac{\mathbf{270}}{1} \\ &= \frac{1 \cdot \mathbf{90}}{\mathbf{1} \cdot 1} \\ &= 90 \end{aligned}$$

Check

Since 90 is one third of the total deaths, $3 \cdot 90$ should equal the total deaths. Since $3 \cdot 90 = 270$, the total deaths, the answer is reasonable.

Answer

The study indicates that 90 of the 270 deaths could be the result of smoking cigarettes.

Multiplication of fractions also appears in formulas. A **pyramid** is a solid figure that has a base and that rises to a point above the center of the base. We usually think of pyramids with square, rectangular, or circular bases. Most ancient Egyptian pyramids had a square base. However, the base of a pyramid can be any shape. The volume of a pyramid can be calculated using the formula *Volume* = $\frac{1}{3}$*(area of base)(height).*

PROPERTY

Volume of a pyramid

In words: Volume = $\frac{1}{3}$(area of the base)(height)

In symbols: $V = \frac{1}{3}Ah$

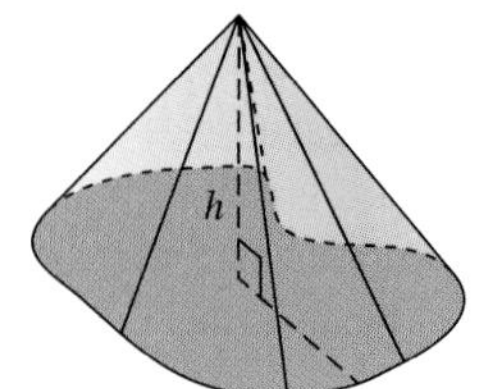

Example 5 **The Great Pyramid of Khufu or Cheops (2680 B.C.E.) is one of the Seven Wonders of the Ancient World and the largest pyramid ever constructed. The area of its base is 566,280 square feet. It is 482 feet high. Calculate the volume of this pyramid.**

© *Philip H. Coblentz/World Travel Images, Inc.*

I know . . .

V = Volume

566,280 square feet = Area of the base

482 feet = Height

Equations

$$\text{Volume} = \frac{1}{3}\text{ (area of base)(height)}$$

$$V = \frac{1}{3}(566{,}280 \text{ feet}^2)(482 \text{ feet})$$

Solve

$$V = \frac{1}{3}(566{,}280 \text{ feet}^2)(482 \text{ feet})$$

$$= \frac{1 \cdot 566{,}280 \cdot 482 \cdot \text{feet}^2 \cdot \text{feet}}{3}$$

$$= \frac{272{,}946{,}960 \text{ feet}^3}{3}$$

$$= 90{,}982{,}320 \text{ feet}^3$$

Check

Estimate the volume by rounding the measurements.

$$V = \frac{1 \cdot 600{,}000 \text{ feet}^2 \cdot 500 \text{ feet}}{3}$$

$$= \frac{300{,}000{,}000}{3}\text{ feet}^3$$

$$= 100{,}000{,}000 \text{ feet}^3$$

This is reasonably close to our exact answer of 90,982,320 feet^3.

Answer

The volume of the Great Pyramid of Cheops is 90,982,320 cubic feet.

A **trapezoid** is a geometric figure with four sides. Exactly two of the sides are parallel and called the bases. The other two sides are called legs. The height of the trapezoid is the shortest distance between the two parallel sides. The area of a trapezoid can be calculated using the formula Area $= \frac{1}{2}$(height)(base 1 + base 2).

PROPERTY

Area of a trapezoid

In words: Area $= \frac{1}{2}$(height)(base 1 + base 2)

In symbols: $A = \frac{1}{2}h(b_1 + b_2)$

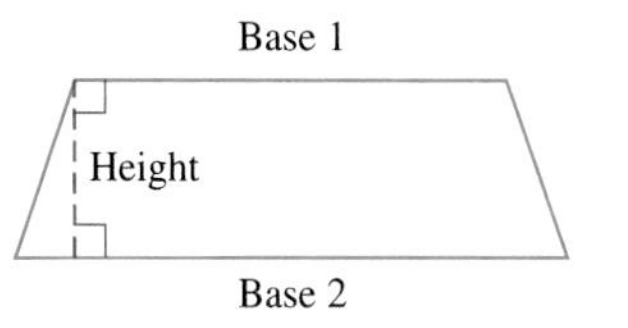

Example 6 **A trapezoid has a height of $\frac{3}{4}$ inch. One base of the trapezoid is 3 inches long; the other base is 5 inches long. Calculate the area of the trapezoid.**

$$\begin{aligned}\text{Area} &= \frac{1}{2}(\text{height})(\text{base 1} + \text{base 2}) \\ &= \frac{1}{2}\left(\frac{3}{4}\text{ inch}\right)(3\text{ inches} + 5\text{ inches}) \\ &= \frac{1}{2}\left(\frac{3}{4}\text{ inch}\right)(8\text{ inches}) \\ &= \frac{1 \cdot 3 \cdot 8}{2 \cdot 4}\text{ inches}^2 \\ &= \frac{24}{8}\text{ inches}^2 \\ &= 3\text{ inches}^2\end{aligned}$$

3 in.

$\frac{3}{4}$ in.

5 in.

The area of the trapezoid is 3 square inches.

CHECK YOUR UNDERSTANDING

Multiply. Simplify each product to lowest terms.

1. $\frac{4}{5} \cdot \frac{3}{7}$

2. $\left(\frac{3}{8}\text{ mile}\right)\left(\frac{2}{9}\text{ mile}\right)$

3. $\left(-\frac{6}{5}\right)(20)$

4. $\frac{1}{2}(12\text{ meter})(3\text{ meter} + 2\text{ meter})$

5. $\frac{1}{2}(8\text{ meter}^2)(5\text{ meters})$

Answers: 1. $\frac{12}{35}$; 2. $\frac{1}{12}$ mile^2; 3. -24; 4. 30 meter^2; 5. 20 meters^3.

EXERCISES SECTION 3.3

All answers that are fractions should be simplified to lowest terms.

Evaluate.

1. $\frac{2}{3} \cdot \frac{1}{8}$

2. $\frac{2}{9}\left(-\frac{1}{4}\right)$

3. $\frac{2}{3} \cdot \frac{5}{6}$

4. $\frac{10}{3} \cdot \frac{9}{2}$

5. $\left(\frac{5}{3}\right)\left(-\frac{12}{5}\right)$

6. $\frac{0}{6} \cdot \frac{3}{8}$

7. $\frac{5}{6} \cdot \frac{3}{10}$

8. $-\frac{5}{6}\left(-\frac{8}{15}\right)$

9. $-\frac{2}{9} \cdot \frac{3}{14}$

10. $\frac{15}{28} \cdot \frac{20}{33}$

11. $\frac{5}{8} \cdot \frac{2}{3}$

12. $\frac{1}{2}(10)$

13. $\frac{1}{8} \cdot \frac{1}{8}$

14. $\left(\frac{1}{4}\right)^2$

15. $\left(-\frac{1}{4}\right)^2$

16. $-\left(\frac{1}{4}\right)^2$

17. $\frac{1}{6} \cdot \frac{6}{7}$

18. $\left(\frac{4}{5}\right)\left(-\frac{3}{2}\right)$

19. $\left(-\frac{4}{15}\right)\left(-\frac{3}{2}\right)$

20. $\frac{1}{2} \cdot \frac{4}{4}$

21. $\left(\frac{3}{2}\right)\left(\frac{8}{9}\right)$

22. $\left(\frac{4}{5}\right)\left(\frac{7}{2}\right)$

23. $\left(-\frac{1}{2}\right)(6)$

24. $\left(-6\right)\left(\frac{3}{20}\right)$

25. $\left(\frac{-10}{3}\right)\left(-\frac{21}{20}\right)$

26. $\frac{1}{2} \cdot \frac{3}{4} \cdot \frac{8}{9}$

27. $\left(\frac{2}{3}\right)^3$

28. $\left(-\frac{2}{3}\right)^3$

29. $-\left(\frac{2}{3}\right)^3$

30. $\left(\frac{5}{5}\right)\left(\frac{9}{9}\right)\left(-\frac{4}{4}\right)$

31. $\left(\frac{8}{9}\right)\left(\frac{2}{3}\right)\left(\frac{15}{16}\right)$

32. $\left(\frac{129}{126}\right)\left(\frac{80}{99}\right)$

33. $\left(\frac{15}{28}\right)^3$

Solve using the five-step problem-solving strategy.

34. Toenails grow $\frac{5}{1,000}$ of an inch per week. Fingernails grow four times faster. How much does a fingernail grow per week?

35. In baseball, the pitcher's mound is 60 feet 6 inches from home plate. What is $\frac{1}{3}$ of this distance?

36. Marjorie is making 32 angel costumes for a Christmas play. Each costume requires $\frac{5}{8}$ yard of white fabric. How much fabric will Marjorie need to make all the costumes?

37. A bushel of dried peas weighs about 60 pounds. What is the weight of $\frac{1}{4}$ of a bushel of dried peas?

38. Jennifer has a garden with an area of 16 square yards. She plants $\frac{1}{2}$ of the garden with flowers and $\frac{1}{2}$ of the garden with vegetables. If she wants $\frac{1}{4}$ of the vegetable garden to be planted with tomatoes, how much area should she reserve for tomatoes?

39. When determining the eligibility of a buyer for a home loan, mortgage companies typically allow a borrower to spend $\frac{2}{5}$ of *monthly* take-home pay on house payments. If Quincy has an *annual* take-home pay of $24,000, how much will he be allowed to spend on his house payments per month?

40. Gasoline is a mixture of different carbon compounds including one called iso-octane. The octane rating of gasoline measures the fraction of the gasoline that is iso-octane. Higher octane gasoline can prevent engine knocking. 90-octane gasoline is $\frac{9}{10}$ iso-octane. Bethany purchases 20 gallons of 90-octane gasoline. How many of those gallons are iso-octane?

41. In 1996, farm production in Washington State generated $5,770 million. Fruits, nuts, and berries account for $\frac{1}{20}$ of farm production. What was the value of these crops?

42. To graduate with a teaching major in geology requires 128 credits including completion of the core requirement. The core curriculum requirements are 52 credits. What fraction of the credits completed are core curriculum requirements?

43. Mono Lake is an ancient lake in California with no outlet. Salts and minerals that wash into the lake from the streams of the Eastern Sierra Nevada mountains never leave. Consequently, the lake is much saltier than seawater. Snowy plovers nest along the windswept alkali flats of Mono Lake's eastern shore. This colony of birds is about $\frac{1}{10}$ of the California population of this species. The total California population is 4,000 snowy plovers. How many birds are in the Mono Lake colony?

44. A librarian is prioritizing requests from faculty for new books for the library. Only $\frac{2}{5}$ of the new selections may be books that were published more than 3 years ago. If the total number of new books will be 125, how many of them may be books that were published more than 3 years ago?

45. A survey of preschoolers entering kindergarten in a school district reported that $\frac{2}{3}$ of the children could recognize the letters of the English alphabet. If the survey was of 600 children, how many of them could recognize their letters?

46. An ice cube measures $\frac{1}{2}$ inch high, $\frac{3}{4}$ inch wide, and $\frac{7}{8}$ inch long.

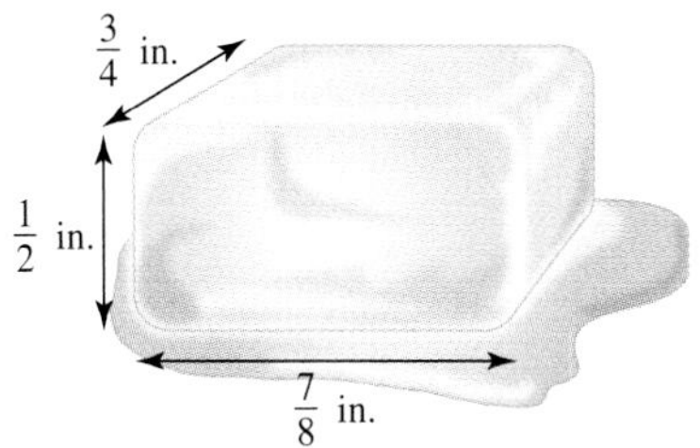

a. What is the volume of this ice cube?

b. Why is the name *ice cube* inappropriate for this object?

47. About 1,960 romance novels were published in 1998. About $\frac{3}{10}$ of these novels were historical romances, set in a time prior to World War II. How many of the novels were historical romances?

48. The bar graph shows the number of words in the title of the motion picture that won the Oscar for Best Picture.

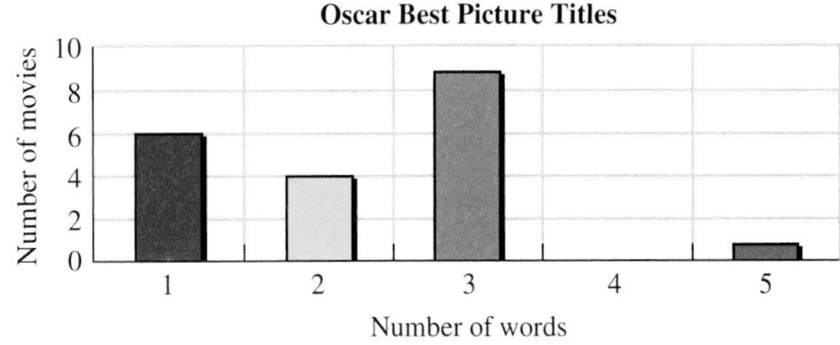

a. Explain why this graph shows that $\frac{1}{20}$ of the titles had 5 words.

b. Explain why this graph shows that $\frac{1}{5}$ of the titles had 2 words.

49. A buyer for a discount chain can obtain copies of designer lamps for about $\frac{1}{20}$ of the cost of the original lamps. The price of the designer lamp is \$340. What is the price of a copy?

50. What is the area of a stamp that measures $\frac{3}{4}$ inch by $\frac{7}{8}$ inch?

51. The area of a triangle can be calculated using the formula Area $= \frac{1}{2}$(base)(height). What is the area of a triangle with a base of $\frac{5}{8}$ inch and a height of $\frac{7}{8}$ inch?

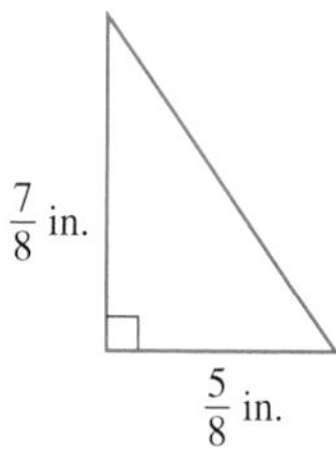

52. The area of a trapezoid can be calculated using the formula Area $= \frac{1}{2}$(height)(base 1 + base 2). A trapezoid has a height of $\frac{7}{8}$ inch. One base of the trapezoid is 2 inches long; the other base is 4 inches long. Calculate the area of the trapezoid.

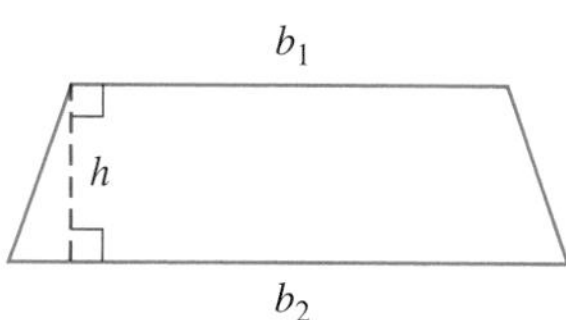

53. The side window of Kim's truck shell is trapezoidal in shape. Calculate the area of this window.

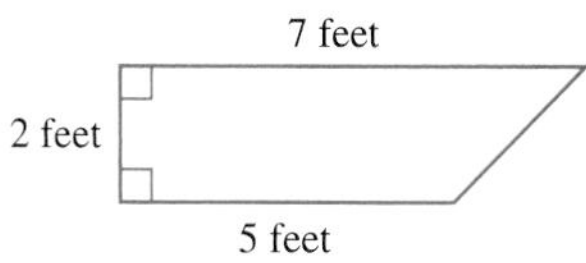

54. As shown in the diagram, a classroom table that has a laminate surface is trapezoidal in shape. (See Exercise 52 for formula.)

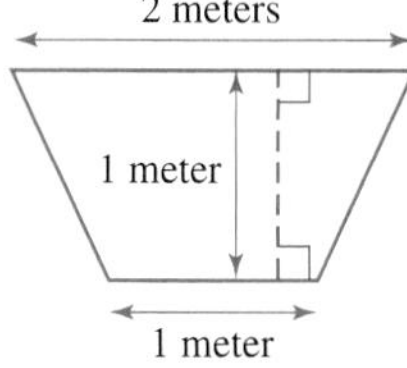

a. Calculate the area of the surface of the table.

b. Calculate the area of laminate surface in a classroom of 30 tables.

55. The formula for finding the volume of a pyramid is Volume $= \frac{1}{3}$(area of base)(height). A pyramid has a square base with an area of 196 square feet. The height of the pyramid is 60 feet. What is the volume of the pyramid?

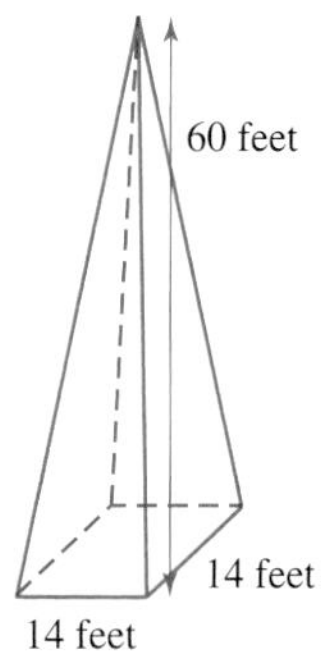

56. Wood chips often are stored at river ports in piles that are pyramids with circular bases. The area of a circular pyramid of wood chips is 6,000 square feet. The height is 40 feet. The formula for finding the volume of a pyramid is Volume $= \frac{1}{3}$(area of base)(height).

a. Calculate the volume of wood chips in this pile.

b. What is another name commonly used to describe a circular pyramid?

Port of Wilma, Whitman County, Washington

ERROR ANALYSIS

57. Problem: Multiply $\frac{3}{10} \cdot \frac{7}{10}$.

Answer:

$$\frac{3}{10} \cdot \frac{7}{10}$$
$$= \frac{3 \cdot 7}{10}$$
$$= \frac{21}{10}$$

Describe the mistake made in solving this problem.

Redo the problem correctly.

58. Problem: Multiply $\frac{5}{9} \cdot 2$.

Answer:

$$\frac{5}{9} \cdot 2$$
$$= \frac{5}{9} \cdot \frac{1}{2}$$
$$= \frac{5}{18}$$

Describe the mistake made in solving this problem.

Redo the problem correctly.

REVIEW

59. Identify the dividend, quotient, and divisor in $65 \div 5 = 13$.

60. Explain how to check a quotient using multiplication.

61. Explain why division by zero is undefined.

62. List the factors of 75.

Section 3.4 Division of Fractions

LEARNING TOOLS

CD-ROM SSM VIDEO

Sneak Preview

Division is the opposite operation of multiplication. Division is the operation we use to find how many fractions are contained in a number.

After completing this section, you should be able to:

1) State the reciprocal of a number.
2) Divide fractions.
3) Solve word problems that involve division of a fraction.

DISCUSSION

Reciprocals

When a fraction is turned upside down (inverted), the resulting fraction is called its **reciprocal.** In symbols, we say that the **reciprocal** of any fraction $\frac{a}{b}$ is $\frac{b}{a}$ $(a \neq 0; b \neq 0)$.

Example 1 **State the reciprocal of the fractions $\frac{4}{5}$ and $\frac{5}{9}$.**

The reciprocal of $\frac{4}{5}$ is $\frac{5}{4}$.

The reciprocal of $\frac{5}{9}$ is $\frac{9}{5}$.

When a fraction is negative, the negative sign can be assigned to either the whole fraction, the numerator or the denominator. The reciprocal of a negative fraction is also a negative fraction.

Example 2 **State the reciprocal of these fractions: $\frac{-5}{7}$, $\frac{8}{-3}$, and $-\frac{7}{4}$.**

The reciprocal of $\frac{-5}{7}$ is $\frac{7}{-5}$.

The reciprocal of $\frac{8}{-3}$ is $\frac{-3}{8}$.

The reciprocal of $-\frac{7}{4}$ is $-\frac{4}{7}$.

To find the reciprocal of an integer, we first rewrite the integer as a fraction with a denominator of 1.

Example 3 **State the reciprocal of 4, −5, and 0.**

4 can be written as $\frac{4}{1}$. The reciprocal of 4 is $\frac{1}{4}$.

−5 can be written as $-\frac{5}{1}$. The reciprocal of −5 is $-\frac{1}{5}$.

0 can be written as $\frac{0}{1}$. However, 0 has no reciprocal because $\frac{1}{0}$ is undefined.

The product of a number and its reciprocal is always 1. The reciprocal of a number is also called its **multiplicative inverse.**

Example 4 **Multiply $-\frac{1}{3}$ by its multiplicative inverse.**

The multiplicative inverse of $-\frac{1}{3}$ is its reciprocal, $-\frac{3}{1}$.

$$\left(-\frac{1}{3}\right)\left(-\frac{3}{1}\right)$$
$$= \frac{-3}{-3}$$
$$= 1$$

Example 5 **Find the product of $\frac{3}{5}$ and its reciprocal.**

The reciprocal of $\frac{3}{5}$ is $\frac{5}{3}$.

$$\frac{3}{5} \cdot \frac{5}{3}$$
$$= \frac{3 \cdot 5}{5 \cdot 3} \quad \text{Multiply numerators; multiply denominators.}$$
$$= \frac{15}{15} \quad \text{Simplify to lowest terms.}$$
$$= 1$$

CHECK YOUR UNDERSTANDING

State the reciprocal of each number.

1. $\frac{7}{12}$ 2. $\frac{1}{6}$ **3.** $-\frac{6}{7}$ **4.** -1 **5.** 0 **6.** 9

Answers: 1. $\frac{12}{7}$; 2. $\frac{6}{1}$; 3. $-\frac{7}{6}$; 4. -1; 5. 0 has no reciprocal; 6. $\frac{1}{9}$.

INVESTIGATION How Many Fractions Fit in a Fraction?

1. Identify the *dividend,* the *divisor,* and the *quotient* in $12 \div 6 = 2$.

2. $12 \div 6 = 2$ can be translated into words as: "If 12 objects are divided into 6 groups, there will be 2 objects in each group." Translate $12 \div 4 = 3$ into words.

Answers: 1. The dividend is 12, the divisor is 6, and the quotient is 2; 2. If 12 objects are divided into 4 groups, there will be 3 objects in each group.

When a rectangle is divided into eight parts, each part is $\frac{1}{8}$ of the rectangle.

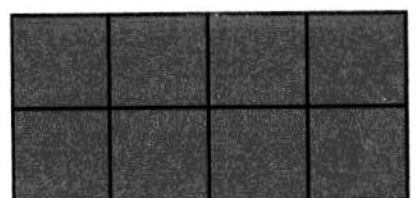

There are **8** one-eighths in this rectangle. In mathematical symbols, we write $1 \div \frac{1}{8} = \mathbf{8}$.

We can also determine how many one-eighths are found in $\frac{1}{2}$ of the rectangle.

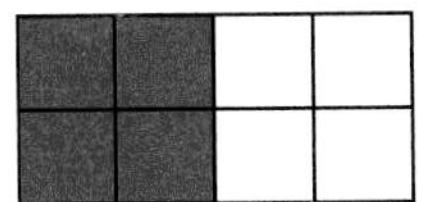

There are **4** one-eighths in $\frac{1}{2}$ of this rectangle. In mathematical symbols, we write $\frac{1}{2} \div \frac{1}{8} = \mathbf{4}$.

Dividing a dividend by $\frac{1}{8}$ determines how many one-eighths are contained in the dividend. Dividing any dividend by a fraction determines how many of the fractions are contained in the dividend.

Find each quotient by counting shaded areas.

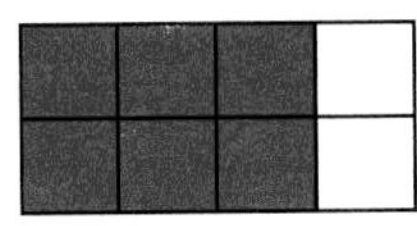

3. $\frac{3}{4} \div \frac{1}{8} = ?$ (How many one-eighths are shaded?)

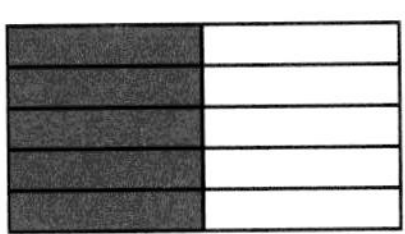

4. $\frac{1}{2} \div \frac{1}{10} = ?$ (How many one-tenths are shaded?)

Answers: 3. 6; 4. 5.

Calculating quotients by dividing and shading rectangles and counting shaded areas is time-consuming. Instead, we can use the following procedure to find these quotients.

EXAMPLE A:

$$\frac{3}{4} \div \frac{1}{8} = \frac{3}{4} \cdot \frac{8}{1} = \frac{3 \cdot 8}{4 \cdot 1} = \frac{24}{4} = 6$$

EXAMPLE B:

$$\frac{1}{2} \div \frac{1}{10} = \frac{1}{2} \cdot \frac{10}{1} = \frac{1 \cdot 10}{2 \cdot 1} = \frac{10}{2} = 5$$

5. Describe the procedure used to find the quotients in the examples.

Use this procedure to divide. Simplify to lowest terms.

6. $\frac{1}{2} \div \frac{1}{4}$

7. $\frac{3}{4} \div \frac{5}{7}$

8. $\frac{3}{4} \div -\frac{1}{5}$

9. $-6 \div \frac{1}{8}$

10. $0 \div \frac{3}{4}$

11. $-\frac{15}{6} \div \frac{10}{3}$

12. $\frac{15}{2} \div \frac{2}{15}$

13. $\frac{4}{5} \div \frac{0}{7}$

14. $\frac{5}{6} \div 2$

Answers: 6. 2; 7. $\frac{21}{20}$; 8. $-\frac{15}{4}$; 9. -48; 10. 0; 11. $-\frac{3}{4}$; 12. $\frac{225}{4}$; 13. undefined; 14. $\frac{5}{12}$.

Solve the word problem using the five-step problem-solving strategy.

15. Bill is preparing homemade sausage for freezing by wrapping each package in plastic wrap and then in white freezer paper. Each package of sausage requires $\frac{2}{3}$ foot of plastic wrap. Bill has a roll of plastic wrap that is 250 feet long. How many packages of sausage will this roll of plastic wrap cover?

I know . . .

Equations

Solve

Check

Answer

STOP

Answer: 15. This roll will cover 375 packages.

DISCUSSION

Division

Cutting a 10-meter rope into two parts can be thought of either as a division or as a multiplication.

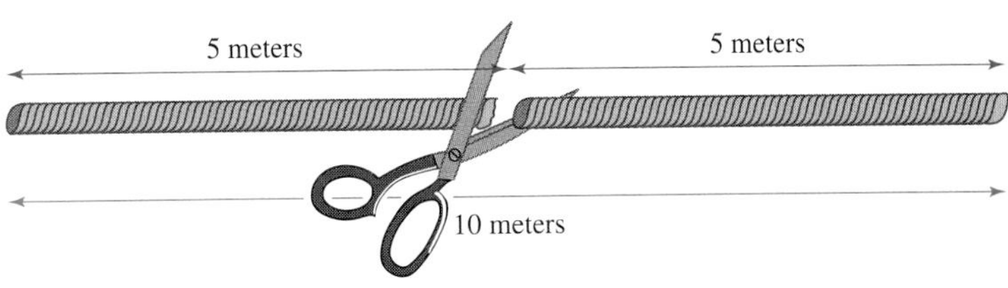

We can divide the length of the rope into two parts: 10 meters ÷ 2 = 5 meters. Or, we can find one half of the length of the rope by multiplying: 10 meters $\cdot \frac{1}{2}$ = 5 meters. Dividing by 2 gives the same result as multiplying by $\frac{1}{2}$. Since 2 and $\frac{1}{2}$ are reciprocals, this example shows that *dividing by a number gives the same result as multiplying by its reciprocal.*

PROCEDURE

To divide a number by a fraction:

In words: Multiply the number by the reciprocal of the fraction.

In symbols: If a, b, c, and d are numbers and $b \neq 0$, $c \neq 0$, and $d \neq 0$,

$$\frac{a}{b} \div \frac{c}{d} = \frac{a}{b} \cdot \frac{d}{c}$$

Example 6 **Evaluate $\frac{4}{5} \div \frac{5}{3}$.**

$$\frac{4}{5} \div \frac{5}{3}$$

$$= \frac{4}{5} \cdot \frac{3}{5}$$ *Rewrite division as multiplication by the reciprocal.*

$$= \frac{12}{25}$$ *Multiply numerators; multiply denominators. This fraction is in lowest terms.*

The same rules for determining the sign of the quotient of two integers (Section 2.4) apply to division of fractions. The quotient of two numbers with the same signs is a positive number; the quotient of two numbers with different signs is a negative number.

Example 7 **Evaluate $-\frac{5}{16} \div \left(-\frac{7}{4}\right)$.**

$$-\frac{5}{16} \div \left(-\frac{7}{4}\right)$$

$$= -\frac{5}{16} \cdot \left(-\frac{4}{7}\right)$$ *Rewrite division as multiplication by the reciprocal.*

$$= \frac{5 \cdot \mathbf{4}}{\mathbf{16} \cdot 7}$$ *Simplify before multiplying. A common factor of 4 and 16 is 4.*

$$= \frac{5 \cdot \mathbf{1}}{\mathbf{4} \cdot 7}$$

$$= \frac{5}{28}$$

When dividing a fraction by an integer, the integer is rewritten as a fraction with a denominator of 1. The division is then rewritten as multiplication by the reciprocal.

Example 8 **Find the quotient of $\frac{7}{8}$ and 3.**

$$\frac{7}{8} \div 3$$

$$= \frac{7}{8} \div \frac{\mathbf{3}}{\mathbf{1}} \quad \textit{Write the integer as a fraction with a denominator of 1.}$$

$$= \frac{7}{8} \cdot \frac{\mathbf{1}}{\mathbf{3}} \quad \textit{Rewrite division as multiplication by the reciprocal.}$$

$$= \frac{7}{24}$$

Division of fractions is required to solve some word problems. When dividing a measurement by a measurement, divide the numbers first and then divide the units.

Example 9 **A bookshelf requires boards that are each $\frac{3}{4}$ foot long. How many of these boards can be cut from a board that is 6 feet long? (Disregard any length of board lost in sawdust during cutting.)**

I know . . .

B = Number of boards

$\frac{3}{4}$ ft = Length per board

6 ft = Total length

Equations

Number of boards = Total length ÷ Length per board

$$B = 6 \text{ ft} \div \frac{3}{4} \text{ ft}$$

Solve

$$B = 6 \text{ ft} \div \frac{3}{4} \text{ ft}$$

$$= \frac{6}{1} \text{ ft} \div \frac{3}{4} \text{ ft} \quad \textit{Rewrite the integer as a fraction with a denominator of 1.}$$

$$= \left(\frac{6}{1} \div \frac{3}{4}\right)(\text{ft} \div \text{ft}) \quad \textit{Divide the numbers; divide the units.}$$

$$= \left(\frac{6}{1} \div \frac{3}{4}\right)(1)$$

$$= \left(\frac{6}{1} \cdot \frac{4}{3}\right) \quad \textit{Simplify before multiplying.}$$

$$= \frac{2 \cdot 4}{1 \cdot 1}$$

$$= \frac{8}{1}$$

$$= 8$$

Check

$\frac{3}{4}$ ft is almost 1 ft; we can estimate that B = 6 feet ÷ 1 foot = 6. Since the boards are a little shorter than 1 foot in length, it is reasonable to think that we could get two more boards than our estimate.

Answer

Eight $\frac{3}{4}$-foot boards can be cut from a board that is 6 feet long.

CHECK YOUR UNDERSTANDING

Find the quotients.

1. $\frac{3}{7} \div \frac{2}{5}$

2. $\frac{6}{5} \div \frac{3}{10}$

3. $\frac{3}{4} \div (-2)$

Answers: 1. $\frac{15}{14}$; 2. 4; 3. $-\frac{3}{8}$.

EXERCISES SECTION 3.4

Write the reciprocal of the number.

1. $\frac{1}{3}$

2. 3

3. $\frac{4}{5}$

4. $-\frac{4}{5}$

5. $\frac{1}{8}$

6. 0

7. -50

8. $\frac{-6}{7}$

9. What is the product of any number except 0 and its reciprocal?

10. Explain why zero does not have a reciprocal.

Find the quotient.

11. $\frac{2}{3} \div \frac{1}{8}$

12. $\frac{2}{9} \div \left(-\frac{1}{4}\right)$

13. $\frac{2}{3} \div \frac{5}{6}$

14. $\frac{0}{6} \div \frac{3}{8}$

15. $\frac{5}{6} \div \frac{3}{10}$

16. $-\frac{5}{6} \div \left(-\frac{8}{15}\right)$

17. $-\frac{2}{9} \div \frac{3}{14}$

18. $\frac{15}{28} \div \frac{20}{21}$

19. $\frac{5}{8} \div \frac{2}{3}$

20. $\frac{6}{5} \div \left(-\frac{3}{10}\right)$

21. $\frac{1}{8} \div 4$

22. $8 \div \frac{1}{8}$

23. $\frac{1}{8} \div 8$

24. $\frac{5}{6} \div \frac{6}{5}$

25. $\frac{1}{6} \div \frac{6}{7}$

26. $\frac{4}{5} \div \left(-\frac{8}{15}\right)$

27. $-\frac{4}{15} \div \left(-\frac{2}{3}\right)$

28. $\frac{1}{2} \div \frac{4}{4}$

29. $\frac{3}{8} \div \frac{1}{2}$

30. $\frac{6}{25} \div \left(-\frac{8}{15}\right)$

31. $\frac{6}{7} \div 0$

32. $\frac{1}{8} \div \frac{1}{8}$

33. $\frac{5}{6} \div \frac{12}{7}$

34. $\frac{-2}{3} \div \frac{4}{9}$

Solve using the five-step problem-solving strategy.

35. Wade grows organic herbs commercially. The herbs are shipped in bulk with their stems in water to the distributor. The distributor puts the herbs in $\frac{1}{4}$-ounce packages and ships them to grocery stores to be sold for \$1.95 each. How many packages can be packed from 3 *pounds* of herbs?

36. If a snail moves $\frac{7}{8}$ of an inch in one minute, how long will it take the snail to move 14 inches?

37. A bookstore has allocated 15 feet 4 inches of shelf space for the textbooks for a math class. Each textbook is $\frac{7}{8}$ inch thick. How many textbooks will fit on this shelf?

38. The U.S. Coast Guard determines ferry seating on benches using the assumption that the average ferry traveler occupies $\frac{3}{2}$ feet of bench. How many people should be able to sit on a bench that is 30 feet long?

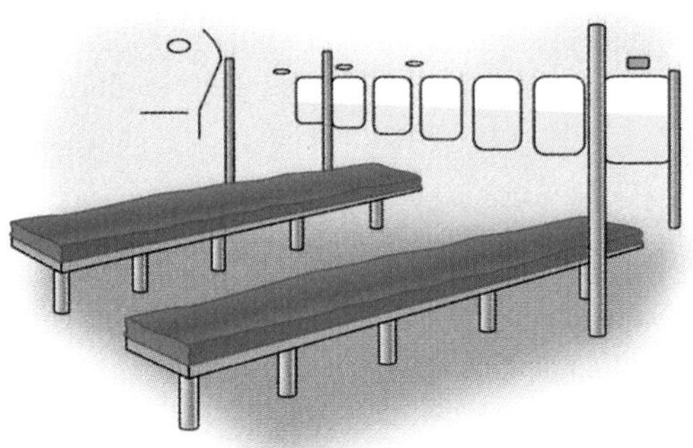

39. When geology students go in the field to observe minerals, they may take along a small box of identified samples for comparison. One such sample box is 5 inches long by 3 inches wide by $\frac{5}{16}$ inches high. The sample boxes are stored in a geology storeroom. The height between the shelves is 12 inches. How many sample boxes will fit in a stack on such a shelf?

Evaluate.

40. $\frac{3}{4} \cdot \frac{1}{2}$

41. $\frac{3}{4} \div \frac{1}{2}$

42. $\frac{1}{2} \div \frac{3}{4}$

43. $\frac{20}{3} \cdot \frac{9}{10}$

44. $\frac{20}{3} \div \frac{9}{10}$

45. $\frac{1}{5} \cdot \frac{1}{5}$

46. $\frac{1}{5} \div \frac{1}{5}$

47. $-\frac{14}{15} \div \frac{15}{14}$

48. $-\frac{14}{15} \cdot \frac{15}{14}$

Solve using the five-step strategy. Either multiplication or division may be needed.

49. According to a U.S. Department of Justice Bureau of Statistics Bulletin (NCJ-151651), "the number of inmates held in local jails on June 30, 1994 reached a record high of 490,442. Local jails, which are operated by counties and municipalities and administered by local government agencies, housed about a third of all persons incarcerated in the U.S. at midyear 1994; the other two-thirds were in State or Federal prisons." Calculate how many inmates were held in state or federal prisons.

50. Kylie buys hamburger in bulk at a warehouse, divides the hamburger into $\frac{3}{4}$-pound packages, and freezes them for later use. If she has 15 pounds of hamburger, how many packages will she make?

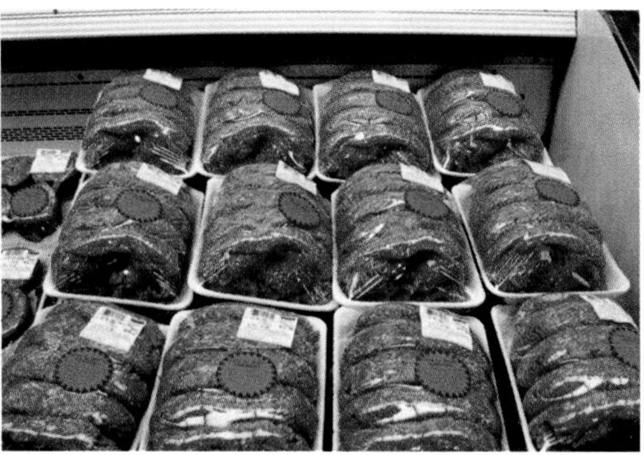

51. New tires can be manufactured with a maximum of $\frac{3}{20}$ recycled rubber. If a tire weighs 30 pounds, how much of that weight can be from recycled rubber?

52. Connor is planning a 60-mile relay race to raise money for charity. Each person will run $\frac{2}{3}$ of a mile. How many runners will he need?

53. Connor is planning a 60-mile relay race to raise money for charity. Each person will run $\frac{3}{4}$ of a mile. How many runners will he need?

54. Connor is planning a 60-mile relay race to raise money for charity. He has 130 runners. If each person is going to run the same distance, what distance should that be?

ERROR ANALYSIS

55. Problem: Divide $\frac{5}{11} \div \frac{3}{7}$.

Answer:

$$\frac{5}{11} \div \frac{3}{7}$$
$$= \frac{11}{5} \cdot \frac{3}{7}$$
$$= \frac{33}{35}$$

Describe the mistake made in solving this problem.

Redo the problem correctly.

56. Problem: Divide $\frac{4}{9} \div \frac{0}{6}$.

Answer:

$$\frac{4}{9} \div \frac{0}{6}$$
$$= \frac{4}{9} \cdot \frac{6}{0}$$
$$= \frac{24}{0}$$
$$= 0$$

Describe the mistake made in solving this problem.

Redo the problem correctly.

REVIEW

57. Rewrite $18 - 5 = 13$ as addition of an opposite.

58. Rewrite $20 - (-8) = 28$ as addition of an opposite.

59. Evaluate $12 - (-4)$.

60. Evaluate $-12 - (-4)$.

Section 3.5 Addition and Subtraction of Fractions

LEARNING TOOLS

CD-ROM SSM VIDEO

Sneak Preview

When we add fractions, the fractions must be counting parts that are the same size. This means that we can only add or subtract fractions that have the same denominator.

After completing this section, you should be able to:

1) Add or subtract fractions with the same (common) denominator.

2) Find a common multiple for any two integers.

3) Change a fraction into an equivalent fraction with a given denominator.

4) Add or subtract fractions with different denominators.

5) Calculate the perimeter of a straight-sided figure.

6) Solve word problems that require the addition or subtraction of fractions.

DISCUSSION

Adding and Subtracting Fractions with a Common Denominator

Adding and subtracting fractions with the same denominator is similar to adding and subtracting like units. We know that 3 feet + 4 feet = 7 feet. In fractions, the "like unit" is found in the denominator. For example, $\frac{3}{10} + \frac{4}{10} = \frac{7}{10}$ can be rewritten 3 tenths + 4 tenths = 7 tenths. The like unit is the common denominator, tenths. So, to add or subtract fractions with the same (common) denominator, add or subtract the numerators and keep the common denominator.

PROCEDURE

To add or subtract fractions with a common denominator:

In words: Add or subtract the numerators. Keep the common denominator.

In symbols: $\frac{a}{c} + \frac{b}{c} = \frac{a+b}{c}$ $\quad \frac{a}{c} - \frac{b}{c} = \frac{a-b}{c}$

where a, b, and c are integers and $c \neq 0$.

Example 1 Evaluate $\frac{5}{8} + \frac{2}{8}$.

$\frac{5}{8} + \frac{2}{8}$

$= \frac{5+2}{8}$ *Add the numerators; the denominator remains the same.*

$= \frac{7}{8}$ *This fraction is in lowest terms.*

A − sign to the left of a fraction can be assigned to the numerator before adding or subtracting. The procedures for adding and subtracting integers are used to combine the numerators.

Example 2 **Add $\frac{3}{5} + \left(-\frac{4}{5}\right)$.**

$$\frac{3}{5} + \left(-\frac{4}{5}\right)$$

$$= \frac{3}{5} + \left(\frac{-4}{5}\right)$$ *Assign the − sign to the numerator.*

$$= \frac{3 + (-4)}{5}$$ *Add the numerators; keep the common denominator.*

$$= \frac{-1}{5}$$

$$= -\frac{1}{5}$$ *The − sign can be placed in the numerator, in the denominator, or to the left of the fraction.*

Measurements that are fractions can be combined provided that the units of measure are the same. Add or subtract the numbers. The unit of measurement is unchanged.

Example 3 **Find the difference: $\frac{5}{8}$ inch − $\frac{3}{8}$ inch.**

$$\frac{5}{8}\text{ inch} - \frac{3}{8}\text{ inch}$$

$$= \left(\frac{5}{8} - \frac{3}{8}\right)\text{inch}$$ *Subtract the fractions; keep the common measurement unit (inch).*

$$= \frac{5 - 3}{8}\text{ inch}$$ *Subtract the numerators; keep the common denominator.*

$$= \frac{2}{8}\text{ inch}$$

$$= \frac{1}{4}\text{ inch}$$ *Simplify to lowest terms.*

Example 4 **Simplify $\frac{7}{9} + \frac{4}{9} - \left(-\frac{5}{9}\right)$.**

$$\frac{7}{9} + \frac{4}{9} - \left(-\frac{5}{9}\right)$$

$$= \frac{7 + 4 + 5}{9}$$ *Rewrite subtraction as addition of the opposite. The opposite of a negative number is a positive number.*

$$= \frac{16}{9}$$ *Add the numerators; keep the common denominator.*

Example 5 **A 15-square-mile area of fruit orchards was converted into a housing development with streets following the rectangular grid of the old orchard roads. The blocks in the development are $\frac{3}{8}$ mile long and $\frac{1}{8}$ mile wide. Tanner jogs around the block where he lives three times a week. Calculate the perimeter of a block to determine the length of each jog.**

I know . . .

P = Perimeter of a block

$\frac{3}{8}$ mile = Length of block

$\frac{1}{8}$ mile = Width of block

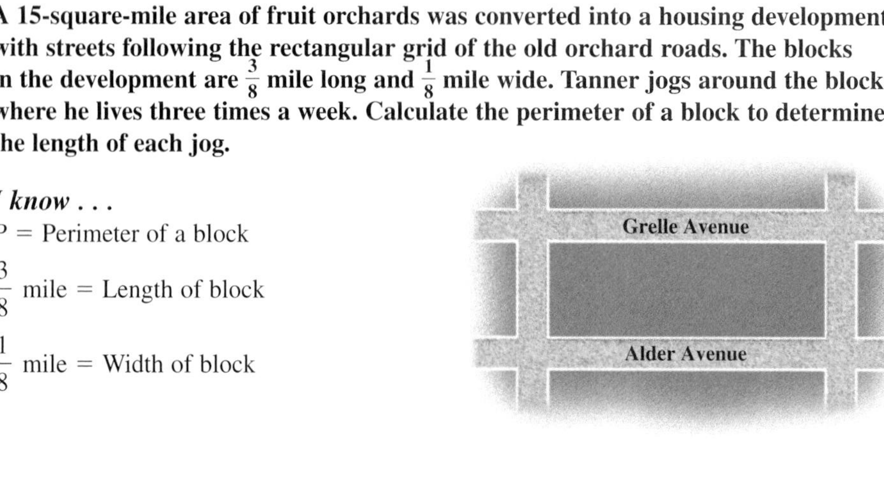

Equations

Perimeter = 2(length) + 2(width)

$$P = 2\left(\frac{3}{8}\text{ mile}\right) + 2\left(\frac{1}{8}\text{ mile}\right)$$

Solve

$$P = 2\left(\frac{3}{8}\text{ mile}\right) + 2\left(\frac{1}{8}\text{ mile}\right)$$

$$= \frac{6}{8}\text{ mile} + \frac{2}{8}\text{ mile} \qquad \textit{Multiply or divide from left to right.}$$

$$= \frac{6+2}{8}\text{ mile} \qquad \textit{Add or subtract from left to right.}$$

$$= \frac{8}{8}\text{ mile}$$

$$= 1\text{ mile}$$

Check

Each fraction is less than $\frac{1}{2}$ so it seems reasonable that a sum of four lengths would be somewhere around 1 mile.

Answer

The perimeter of each block is 1 mile.

CHECK YOUR UNDERSTANDING

Add or subtract. Simplify each sum or difference to lowest terms.

1. $\frac{3}{4}$ foot + $\frac{1}{4}$ foot

2. $-\frac{5}{8} + \frac{7}{8}$

3. $\frac{2}{3} - \frac{7}{3}$

Answers: 1. 1 foot; 2. $\frac{1}{4}$; 3. $-\frac{5}{3}$.

DISCUSSION

Adding and Subtracting Fractions with Different Denominators

Fractions can only be added or subtracted if they have the same denominator. To add or subtract fractions with different denominators, we must first build equivalent fractions that have the same denominator. When two fractions have the same denominator, we say they share a **common denominator.**

Example 6 **Evaluate $\frac{1}{4} + \frac{1}{3}$.**

To build equivalent fractions, we first choose a common denominator. A common denominator is a multiple of both of the original denominators.

The multiples of 3 are 3, 6, 9, 12, 15, 18, 21, 24, 27, 30, 36, . . .

The multiples of 4 are 4, 8, 12, 16, 20, 24, 28, 32, 36, . . .

The set of **common multiples** of 3 and 4 is {12, 24, 36, . . .}. Any of these multiples can be used as the common denominator for the equivalent fractions. If 36 is chosen, each original fraction is multiplied to build an equivalent fraction with a denominator of 36.

$$\frac{1}{4} = \frac{1}{4} \cdot \frac{\mathbf{9}}{\mathbf{9}} = \frac{9}{36}$$ *Multiply the numerator and denominator of $\frac{1}{4}$ by 9.*

$$\frac{1}{3} = \frac{1}{3} \cdot \frac{\mathbf{12}}{\mathbf{12}} = \frac{12}{36}$$ *Multiply the numerator and denominator of $\frac{1}{3}$ by 12.*

Now the fractions have common denominators and can be added.

$$\frac{1}{4} + \frac{1}{3}$$

$$= \frac{9}{36} + \frac{12}{36}$$

$$= \frac{9 + 12}{36}$$ *Add the numerators; keep the denominator.*

$$= \frac{21}{36}$$

$$= \frac{7}{12}$$ *Simplify to lowest terms.*

PROCEDURE

To add or subtract fractions with different denominators:

1. Find a common denominator.
2. Build equivalent fractions with this denominator for each original fraction.
3. Add or subtract the numerators. Keep the common denominator.
4. If possible, simplify to lowest terms.

Any common multiple can be used as the common denominator. The product of the two denominators is always a common multiple and is quick to find.

Example 7 **Evaluate $\frac{5}{6} - \frac{3}{4}$.**

$$\frac{5}{6} - \frac{3}{4}$$ *The product of the two denominators is 24. Use 24 as the common denominator.*

$$= \frac{5 \cdot \mathbf{4}}{6 \cdot \mathbf{4}} - \frac{3 \cdot \mathbf{6}}{4 \cdot \mathbf{6}}$$ *Build equivalent fractions with a denominator of 24.*

$$= \frac{20}{24} - \frac{18}{24}$$

$$= \frac{20 - 18}{24}$$ *Subtract the numerators. Keep the common denominator.*

$$= \frac{2}{24}$$ *Simplify to lowest terms.*

$$= \frac{1}{12}$$

Example 8 **The sizes of softwood lumber are standardized for convenience in ordering. However, the actual width and thickness of a board are less than this standard size. A board with a standard size of 1 inch thickness is actually $\frac{3}{4}$ inch if it is dry and $\frac{25}{32}$ inch thick if it is green (a moisture content over 19 percent). Find the difference in thickness of a dry board and a green board.**

I know . . .

D = Difference

$\frac{3}{4}$ inch = Thickness of dry board

$\frac{25}{32}$ inch = Thickness of green board

Equations

Difference = Thickness of green board − Thickness of dry board

$$D = \frac{25}{32} \text{ inch} - \frac{3}{4} \text{ inch}$$

Solve

$$D = \frac{25}{32} \text{ inch} - \frac{3}{4} \text{ inch}$$

$$= \left(\frac{25}{32} - \frac{3}{4}\right) \text{inch}$$ *A common denominator of 32 and 4 is 32; since 32 is a multiple of 32 and 4.*

$$= \left(\frac{25}{32} - \frac{3 \cdot \mathbf{8}}{4 \cdot \mathbf{8}}\right) \text{inch}$$ *Multiply the numerator and denominator by 8 to build an equivalent fraction with a denominator of 32.*

$$= \left(\frac{25}{32} - \frac{24}{32}\right) \text{inch}$$

$$= \frac{25 - 24}{32} \text{ inch}$$ *Subtract the numerators; keep the same denominator.*

$$= \frac{1}{32} \text{ inch}$$

Check

The dry board is $\frac{3}{4}$ inch thick. We know that it should be just slightly thinner than the wet board because it has lost moisture during drying. A difference of $\frac{1}{32}$ inch seems reasonable.

Answer

The difference in thickness is $\frac{1}{32}$ inch.

When adding fractions, we can use any common multiple of the two denominators as the common denominator. However, we can minimize the amount of simplifying to lowest terms that needs to be done if we choose the smallest possible common multiple. The **least common multiple** (**LCM**) is the smallest of the common multiples. One way to find the LCM is to create a list of multiples of each denominator and choose the smallest number found on both lists. Another method for finding the LCM uses the **prime factorization** of each denominator.

PROCEDURE

To find the least common multiple (LCM) of a set of numbers using prime factorization:

1. Write the prime factorization of each number.
2. For each different prime that appears in any of the prime factorizations, choose the prime with the highest exponent.
3. The least common multiple is the product of the chosen primes.

Example 9 **Evaluate** $\frac{3}{40} + \frac{5}{12}$.

To select a common denominator, find the least common multiple of 40 and 12.

$$40 = 2^3 \cdot 5$$
$$12 = 2^2 \cdot 3$$

1. Write the prime factorization for each denominator.

$$\text{LCM} = 2^3 \cdot 3 \cdot 5$$
$$= 8 \cdot 3 \cdot 5$$

2. For each different prime that appears in any of the prime factorizations, choose the prime with the highest exponent.

$$= 120$$

3. The least common multiple is the product of the chosen primes.

Now add, using the equivalent fractions with the LCM as their denominators.

$$\frac{3}{40} + \frac{5}{12}$$
$$= \frac{3}{40} \cdot \frac{\mathbf{3}}{\mathbf{3}} + \frac{5}{12} \cdot \frac{\mathbf{10}}{\mathbf{10}}$$ *Build equivalent fractions with a common denominator of 120.*
$$= \frac{9}{120} + \frac{50}{120}$$
$$= \frac{59}{120}$$ *Add numerators; keep the common denominator.*

This same process can also be used to add or subtract three or more fractions.

Example 10 **Evaluate** $\frac{3}{14} + \frac{1}{10} - \frac{2}{35}$.

To select a common denominator, find the least common multiple of 14, 10, and 35.

$$14 = 2 \cdot 7$$
$$10 = 2 \cdot 5$$
$$35 = 5 \cdot 7$$

1. Write the prime factorization for each denominator.

2. For each different prime that appears in any of the prime factorizations, choose the prime with the highest exponent.

$$\text{LCM} = 2 \cdot 5 \cdot 7$$
$$= 70$$

3. The least common multiple is the product of the chosen primes.

Rewrite each fraction as an equivalent fraction with a denominator of 70.

$$\frac{3}{14} + \frac{1}{10} - \frac{2}{35}$$
$$= \frac{3}{14} \cdot \frac{\mathbf{5}}{\mathbf{5}} + \frac{1}{10} \cdot \frac{\mathbf{7}}{\mathbf{7}} - \frac{2}{35} \cdot \frac{\mathbf{2}}{\mathbf{2}}$$
$$= \frac{15}{70} + \frac{7}{70} - \frac{4}{70}$$ *Build fractions with the LCM as the common denominator.*
$$= \frac{15 + 7 - 4}{70}$$
$$= \frac{18}{70}$$ *Add and subtract the numerators from left to right; keep the denominator.*
$$= \frac{9}{35}$$ *Simplify to lowest terms.*

Instead of using prime factorizations to find least common multiples, we can use any common multiple or simply multiply the denominators together to find a common denominator. The choice depends on whether we prefer to do the work at the beginning in finding the least common multiple or to do the work at the end in simplifying the final answer to lowest terms. However, since least common multiples that include variables are also used in later work with algebraic fractions, learning to do prime factorizations is an important skill.

CHECK YOUR UNDERSTANDING

Use prime factorization to find the least common multiple.

1. 12 and 18

2. 30, 22, and 32

3. 3, 16, and 36

Add or subtract.

4. $\frac{2}{15} + \frac{1}{24}$

5. $\frac{5}{12} + \frac{7}{18}$

6. $\frac{1}{7} - \frac{19}{21}$

Answers: 1. 36; 2. 5,280; 3. 144; 4. $\frac{7}{40}$; 5. $\frac{29}{36}$; 6. $-\frac{16}{21}$.

EXERCISES SECTION 3.5

Find the sum or difference. Simplify to lowest terms.

1. $\frac{1}{4} + \frac{1}{4}$

2. $\frac{1}{9} + \frac{5}{9}$

3. $-\frac{11}{20} - \frac{3}{20}$

4. $\frac{11}{20} - \frac{3}{20}$

5. $\frac{11}{20} + \frac{3}{20}$

6. $-\frac{11}{20} + \frac{3}{20}$

7. $\frac{5}{6} - \frac{7}{6}$

8. $-\frac{5}{6} - \frac{7}{6}$

9. $-\frac{5}{6} + \frac{7}{6}$

10. $-\frac{1}{5} - \frac{3}{5}$

11. $-\frac{1}{5} - \left(-\frac{3}{5}\right)$

12. $-\frac{5}{8} - \frac{3}{8} - \frac{1}{8}$

Solve using the five-step problem-solving strategy.

13. Some lenses used in eyeglasses begin as uniform plastic "blanks" that are about $\frac{3}{4}$ inch thick. These blanks are ground and polished to the exact requirements needed for a particular pair of glasses. Finished eyeglass lenses are typically reground to a thickness of $\frac{1}{4}$ inch. What thickness of plastic is ground away to make the typical finished lens?

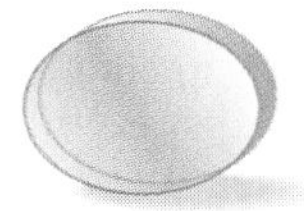

14. A standard compact disc is $\frac{7}{1000}$ inch thick. Charles took five CDs out of his CD wallet and stacked them on his dresser. What is the height of this stack?

15. The jelly dispensers at a local restaurant hold individual servings of jelly. Janel spreads two containers of jelly on her toast. Each container holds $\frac{5}{8}$ ounce of jelly. What is the total amount of jelly that Janel put on her toast?

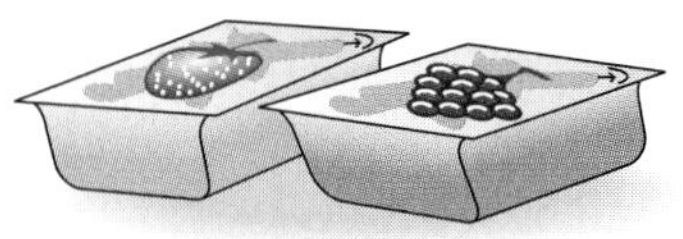

16. Florence needs to modify a sewing pattern for a skirt so that it is slightly larger. The standard size for a seam is $\frac{5}{8}$ inch wide. (A seam is made placing two pieces of material together. A line of stitching is made using a sewing machine or by hand.) If the seam is reduced to a width of $\frac{3}{8}$ inch, what is the effect on the skirt?

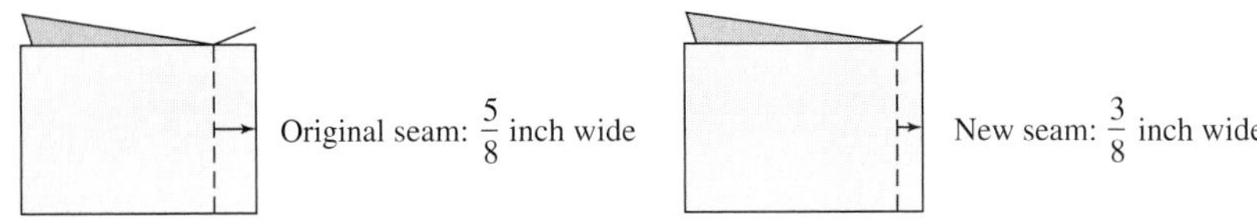

Rewrite the given fraction as an equivalent fraction with the given denominator.

17. $\frac{2}{3}$; denominator 18

18. $\frac{5}{6}$; denominator 18

19. $\frac{5}{8}$; denominator 72

20. $\frac{2}{5}$; denominator 250

21. $\frac{1}{8}$; denominator 56

22. $\frac{4}{7}$; denominator 28

23. $\frac{3}{4}$; denominator 48

24. $\frac{1}{12}$; denominator 36

Find the least common multiple.

25. 8; 42

26. 95; 60

27. 72; 120

28. 32; 48; 7

29. 33; 4; 18

30. 120; 160

31. 8; 60

32. 9; 15

33. 17; 19

Find the sum or difference. Simplify to lowest terms.

34. $\frac{1}{5} + \frac{2}{3}$

35. $\frac{3}{8} + \frac{2}{9}$

36. $\frac{3}{14} - \frac{2}{3}$

37. $\frac{1}{5} - \left(-\frac{2}{15}\right)$

38. $\frac{0}{12} + \frac{2}{7}$

39. $\frac{5}{8} - \frac{2}{7}$

40. $\frac{9}{10} - \frac{3}{4}$

41. $\frac{5}{6} + \frac{2}{9}$

42. $\frac{1}{3} + \frac{2}{9} + \frac{1}{6}$

43. $\frac{1}{14} - \frac{3}{8}$

44. $-\frac{1}{7} + \frac{4}{9}$

45. $\frac{1}{10} + \frac{4}{5}$

46. $\frac{1}{10} + \frac{1}{100} + \frac{1}{1,000}$

47. $\frac{99}{100} + \frac{1}{1,000}$

48. $\frac{1}{10} + \frac{29}{100}$

Solve the problem using the five-step problem-solving strategy.

49. What is the total thickness of a floor constructed from one piece of $\frac{5}{8}$-inch plywood and one piece of $\frac{1}{4}$-inch-thick vinyl tile?

50. Yves mixes $\frac{1}{3}$ cup of lemon juice and $\frac{3}{4}$ cup of soy sauce to make a marinade for some fish. What is the total volume of the marinade?

51. Chie is building a sheet metal enclosure to protect her mailbox from vandalism. She wants to bolt together two pieces of metal that are each $\frac{9}{16}$ inch thick. A $\frac{1}{8}$-inch lockwasher and a $\frac{1}{4}$-inch nut will be used to hold the bolt in place. What is the minimum length of bolt needed to fasten the metal together?

52. The distance from a hotel to a convention center is $\frac{3}{4}$ kilometer. It is an additional $\frac{1}{8}$ kilometer to the subway entrance. What is the total distance from the hotel to the entrance to the subway?

53. Mike is framing a photograph with three different mats: the outer mat is dark green and has a visible width of $\frac{9}{16}$ inch, the middle dark blue mat has a visible width of $\frac{5}{8}$ inch, and the innermost light green mat has a visible width of $\frac{1}{4}$ inch. What is the distance between the picture frame and the edge of the photograph?

Each of the following problems requires the addition, subtraction, multiplication, or division of fractions. Identify the required operation.

54. Eric has two cans of oil. One can is $\frac{1}{4}$ full of oil. The other can is $\frac{1}{2}$ full of oil. Eric pours both cans into his leaf shredder. Eric calculates that he has poured $\frac{3}{4}$ of a can of oil into the shredder. Is Eric adding, subtracting, multiplying, or dividing the fractions in this problem?

55. Jen cut a cheesecake into ten pieces and brought it into the office to share with her colleagues. At 10 A.M., there is $\frac{7}{10}$ of a cheesecake remaining. Four people come into the office at 10:30 A.M. and each of them takes a piece. Jen calculates that $\frac{3}{10}$ of a cheesecake is left. Is Jen adding, subtracting, multiplying, or dividing the fractions in this problem?

56. Deidre wants to spend $\frac{1}{3}$ of her tax refund of \$840 on new clothes. She calculates that $\frac{1}{3}$ of the refund is \$280. Is Deidre adding, subtracting, multiplying, or dividing \$840 and $\frac{1}{3}$?

57. Write a word problem that requires the addition of $\frac{1}{3}$ and $\frac{1}{2}$. Solve this word problem using the five-step problem-solving strategy.

ERROR ANALYSIS

58. **Problem:** Evaluate $\frac{2}{9} + \frac{9}{10}$.

Answer:

$$\frac{2}{9} + \frac{9}{10}$$

$$= \frac{2+9}{9+10}$$

$$= \frac{11}{19}$$

Describe the mistake made in solving this problem.

Redo the problem correctly.

59. **Problem:** Evaluate $\frac{3}{5} + \frac{2}{3}$.

Answer:

$$\frac{3}{5} + \frac{2}{3}$$
$$= \frac{3 + 2}{5 + 3}$$
$$= \frac{\cancel{3} + 2}{5 + \cancel{3}}$$
$$= \frac{2}{5}$$

Describe the mistake made in solving this problem.

Redo the problem correctly.

REVIEW

60. Explain the difference between a proper fraction and an improper fraction.

61. Draw shaded rectangles to represent the fraction $\frac{7}{4}$.

62. Round 936 to the nearest ten.

63. Round -936 to the nearest ten.

Section 3.6 Mixed Numbers

LEARNING TOOLS

CD-ROM SSM VIDEO

Sneak Preview

A mixed number is an integer and a proper fraction. Mixed numbers are important in the English system of measurement. Every mixed number can be rewritten as an equivalent improper fraction.

After completing this section, you should be able to:

1) Graph a mixed number on a number line.

2) Write an improper fraction that is equivalent to a given mixed number.

3) Write a mixed number that is equivalent to a given improper fraction.

4) Round a mixed number to the nearest integer.

5) Multiply or divide mixed numbers.

6) Simplify a product of two fractions before multiplying.

DISCUSSION

Mixed Numbers and Improper Fractions

Fractions are either **proper** or **improper.** In a proper fraction the numerator is less than the denominator. In an improper fraction, the numerator is greater than or equal to the denominator. $\frac{5}{3}$ is an example of an improper fraction.

$\frac{5}{3}$

0 1 2 ...

We can also write $\frac{5}{3}$ as $\frac{3}{3} + \frac{2}{3}$. Since $\frac{3}{3} = 1$, $\frac{5}{3} = 1 + \frac{2}{3}$. If we omit the plus sign, $\frac{5}{3} = 1\frac{2}{3}$. This is an example of a **mixed number.** A mixed number is the sum of an integer and a proper fraction. The plus sign is always **implied.**

1 $\frac{2}{3}$

0 1 2 ...

When a mixed number is greater than zero, it is the sum of a positive integer and a positive fraction: $\frac{8}{3} = 2 + \frac{2}{3}$. When a mixed number is less than zero, it is the sum of a negative integer and a negative fraction: $-\frac{8}{3} = -2 - \frac{2}{3}$.

Example 1 **Graph $-2\frac{2}{3}$ on a number line.**

$$-2\frac{2}{3} = -2 + \left(-\frac{2}{3}\right)$$

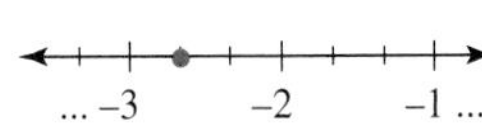

Any improper fraction can be rewritten as an equivalent mixed number or integer.

> **PROCEDURE**
>
> **To rewrite an improper fraction as a mixed number or integer:**
>
> 1. Work with the absolute value of the improper fraction.
> 2. Divide the numerator by the denominator.
> 3. If there is no remainder, the improper fraction simplifies to an integer. If there is a remainder, write the integer and then write the fraction, $\dfrac{\text{Remainder}}{\text{Divisor}}$. Simplify the fraction to lowest terms.
> 4. If the original improper fraction was negative, replace the negative sign on the mixed number.

Example 2 **Rewrite $\dfrac{23}{7}$ as a mixed number.**

$\dfrac{23}{7}$

$= 23 \div 7$ *Divide the numerator by the denominator.*

$= 3\ \text{R}(2)$

$= 3\dfrac{2}{7}$ *The quotient 3 is the integer part of the mixed number. The numerator of the fraction is the remainder 2. The denominator of the fraction is the divisor 7.*

Example 3 **Rewrite $-\dfrac{38}{5}$ as a mixed number.**

$\left|-\dfrac{38}{5}\right|$

$= \dfrac{38}{5}$ *Work with the absolute value of the improper fraction.*

$= 7\ \text{R}(3)$ *Divide the numerator by the denominator.*

$= 7\dfrac{3}{5}$ *The quotient 7 is the integer part of the mixed number. The numerator of the fraction is the remainder 3 and the denominator of the fraction is the divisor 5.*

Replace the negative sign on the mixed number: $-\frac{38}{5} = -7\frac{3}{5}$.

To divide or multiply mixed numbers, each mixed number must first be rewritten as an equivalent improper fraction.

> **PROCEDURE**
>
> **To rewrite a mixed number as an improper fraction:**
>
> 1. Work with the absolute value of the mixed number.
> 2. Multiply the denominator of the fraction by the integer of the mixed number.
> 3. Add the numerator of the fraction to this product.
> 4. Write this sum over the denominator.
> 5. If the mixed number is negative, replace the negative sign on the improper fraction.

Example 4 **Rewrite $3\frac{5}{8}$ as an improper fraction.**

$$3\frac{5}{8}$$

$$= \frac{3 \cdot 8 + 5}{8}$$ *Multiply the denominator of the fraction by the integer of the mixed number.*

$$= \frac{24 + 5}{8}$$ *Add this product to the numerator of the fraction. Keep the same denominator.*

$$= \frac{29}{8}$$

Example 5 **Rewrite $-2\frac{3}{5}$ as an improper fraction.**

$$\left|-2\frac{3}{5}\right|$$

$$= 2\frac{3}{5}$$ *Work with the absolute value of the mixed number.*

$$= \frac{2 \cdot 5 + 3}{5}$$ *Multiply the denominator of the fraction by the integer of the mixed number.*

$$= \frac{10 + 3}{5}$$ *Add this product to the numerator of the fraction. Keep the same denominator.*

$$= \frac{13}{5}$$

Replace the negative sign on the mixed number: $-2\frac{3}{5} = -\frac{13}{5}$.

CHECK YOUR UNDERSTANDING

Rewrite each improper fraction as a mixed number or integer.

1. $\frac{17}{3}$ **2.** $\frac{8}{2}$ **3.** $-\frac{7}{4}$ **4.** $-\frac{35}{4}$

Rewrite each mixed number as an improper fraction.

5. $-9\frac{2}{5}$ **6.** $1\frac{1}{4}$ **7.** $6\frac{3}{8}$ **8.** $-3\frac{5}{16}$

Answers: 1. $5\frac{2}{3}$; 2. 4; 3. $-1\frac{3}{4}$; 4. $-8\frac{3}{4}$; 5. $-\frac{47}{5}$; 6. $\frac{5}{4}$; 7. $\frac{51}{8}$; 8. $-\frac{53}{16}$.

DISCUSSION

Rounding Mixed Numbers

Mixed numbers can be rounded to the nearest integer by choosing the closest integer to the mixed number on a number line.

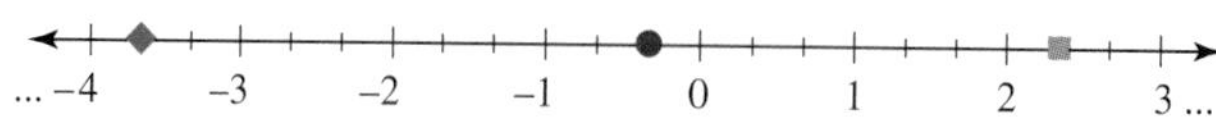

- ◆ $-3\frac{2}{3}$ is closer to -4 than it is to -3. So $-3\frac{2}{3}$ rounds to -4.
- ● $-\frac{1}{3}$ is closer to 0 than it is to -1. So $-\frac{1}{3}$ rounds to 0.
- ■ $2\frac{1}{3}$ is closer to 2 than it is to 3. So, $2\frac{1}{3}$ rounds to 2.

To round without graphing, use absolute values and pay close attention to the fraction part of the mixed number.

PROCEDURE

To round a mixed number to the nearest integer:

1. Work with the absolute value of the mixed number.
2. Look at the fraction:
   If the numerator is less than half the denominator, round to the whole number of the absolute value of the mixed number.
   If the numerator is greater than or equal to half the denominator, add 1 to the whole number of the absolute value of the mixed number.
3. If the mixed number is negative, replace the negative sign on part 2.

Example 6 **Round $4\frac{5}{6}$ to the nearest integer.**

In the fraction $\frac{5}{6}$, the denominator is 6. One half of the denominator is 3. The numerator 5 is greater than 3, so add 1 to the integer 4 and omit the fraction: $4\frac{5}{6}$ rounds to 5.

Example 7 **Round $7\frac{1}{4}$ feet to the nearest foot.**

In the fraction $\frac{1}{4}$, the denominator is 4. One half of the denominator is 2. The numerator 1 is less than 2, so leave the integer the same and omit the fraction: $7\frac{1}{4}$ feet rounds to 7 feet.

Example 8 **Round $-4\frac{3}{5}$ to the nearest integer.**

Work with the absolute value: $\left|-4\frac{3}{5}\right| = 4\frac{3}{5}$.
Look at the fraction. The denominator is 5. One half of the denominator is $2\frac{1}{2}$. The numerator 3 is greater than $2\frac{1}{2}$, so add 1 to the integer 4 and omit the fraction: $4\frac{3}{5}$ rounds to 5. Finally, replace the negative sign: $-4\frac{3}{5}$ rounds to -5.

CHECK YOUR UNDERSTANDING

Round each mixed number to the nearest integer.

1. $4\frac{1}{3}$ **2.** $1\frac{4}{7}$ **3.** $-2\frac{3}{8}$ **4.** $-9\frac{1}{2}$

Answers: 1. 4; 2. 2; 3. −2; 4. −10.

DISCUSSION

Multiplication and Division of Mixed Numbers

To multiply or divide mixed numbers, we first rewrite each mixed number as an equivalent improper fraction.

PROCEDURE

To multiply or divide mixed numbers:

1. Rewrite each mixed number as an equivalent improper fraction.
2. Multiply or divide the improper fractions.
3. Rewrite the quotient as an equivalent mixed number.

Example 9 **Calculate the product of $3\frac{3}{5}$ and $4\frac{4}{9}$.**

$3\frac{3}{5} \cdot 4\frac{4}{9}$

$= \frac{18}{5} \cdot \frac{40}{9}$ *Rewrite the mixed numbers as improper fractions.*

$= \frac{720}{45}$ *Multiply numerators; multiply denominators.*

$= \frac{2 \cdot 2 \cdot 2 \cdot 2 \cdot \mathbf{3} \cdot \mathbf{3} \cdot \mathbf{5}}{\mathbf{3} \cdot \mathbf{3} \cdot \mathbf{5}}$ *Simplify by writing a prime factorization of the numerator and the denominator. Divide common factors.*

$= 2 \cdot 2 \cdot 2 \cdot 2$

$= 16$

In Example 9, we simplified the product to lowest terms *after* multiplying. As shown in the next example, we can also choose to simplify *before* multiplying by first dividing the numerator and denominator by common factors.

Example 10 **Find the product of $3\frac{3}{5}$ and $4\frac{4}{9}$.**

$3\frac{3}{5} \cdot 4\frac{4}{9}$

$= \frac{18}{5} \cdot \frac{40}{9}$ *Rewrite the mixed numbers as improper fractions.*

$= \frac{\mathbf{2} \cdot 40}{5 \cdot \mathbf{1}}$ *Simplify before multiplying. 18 and 9 are divisible by 9.*

$= \frac{2 \cdot \mathbf{8}}{\mathbf{1} \cdot 1}$ *Simplify before multiplying. 40 and 5 are divisible by 5.*

$= 16$

To find a fraction of an original amount, we multiply the original amount by the fraction.

Example 11 **A full plastic jug contains $1\frac{4}{5}$ gallons of shampoo. When the jug is only $\frac{2}{3}$ full, how much shampoo does it contain?**

I know . . .

s = Amount of shampoo

$1\frac{4}{5}$ gallons = Amount in full jug

$\frac{2}{3}$ = Fraction of jug filled

Equations

Amount of shampoo = (Fraction of jug filled)(Amount in full jug)

$$s = \frac{2}{3} \cdot 1\frac{4}{5} \text{ gallons}$$

Solve

$$\begin{aligned} s &= \frac{2}{3} \cdot 1\frac{4}{5} \text{ gallons} \\ &= \frac{2}{3} \cdot \frac{9}{5} \text{ gallons} \\ &= \frac{2 \cdot 3}{1 \cdot 5} \text{ gallons} \\ &= \frac{6}{5} \text{ gallons} \\ &= 1\frac{1}{5} \text{ gallons} \end{aligned}$$

Check

$1\frac{4}{5}$ gallons is approximately equal to 2 gallons. $\frac{2}{3} \cdot 2$ gallons $= \frac{4}{3}$ or $1\frac{1}{3}$ gallons, which is reasonably close to our exact answer of $1\frac{1}{5}$ gallons.

Answer

The jug will contain $1\frac{1}{5}$ gallons of shampoo when it is two thirds full.

When we divide by a mixed number, we must first rewrite the mixed number as an improper fraction. We then rewrite the division as multiplication by the reciprocal of the improper fraction. We can choose to simplify before or after multiplying.

Example 12 **Evaluate $-3\frac{1}{2} \div 2\frac{1}{4}$. Write the quotient as a mixed number.**

$$-3\frac{1}{2} \div 2\frac{1}{4}$$

$= \frac{-7}{2} \div \frac{9}{4}$ *Rewrite mixed numbers as improper fractions.*

$= \frac{-7}{2} \cdot \frac{4}{9}$ *Rewrite division by a fraction as multiplication by its reciprocal.*

$= -\frac{7 \cdot 2}{1 \cdot 9}$ *Simplify before multiplying. 4 and 2 are divisible by 2.*

$= -\frac{14}{9}$

$= -1\frac{5}{9}$ *Rewrite the improper fraction as a mixed number.*

Example 13 **Evaluate $5\frac{1}{2} \div 8$.**

$$5\frac{1}{2} \div 8$$

$$= \frac{11}{2} \div \frac{8}{1} \quad \textit{Rewrite the mixed number as an improper fraction.}$$

$$= \frac{11}{2} \cdot \frac{1}{8} \quad \textit{Rewrite division by a fraction as multiplication by its reciprocal.}$$

$$= \frac{11}{16}$$

When we divide a positive fraction by another positive fraction, we are asking how many of the divisors are contained in the dividend. The quotient of $\frac{1}{2} \div \frac{3}{7}$ is $\frac{7}{6}$. We expect that the quotient should be greater than 1 since $\frac{3}{7} < \frac{1}{2}$. More than one $\frac{3}{7}$ can "fit" in $\frac{1}{2}$. In contrast, the quotient of $\frac{1}{2} \div \frac{7}{8}$ is $\frac{4}{7}$. The quotient is less than 1 because $\frac{7}{8} > \frac{1}{2}$. Not even one $\frac{7}{8}$ can be contained in a $\frac{1}{2}$; it is "too big to fit." This gives us a quick way to check the reasonability of the quotient of two positive fractions.

Example 14 **Predict whether the quotient of $\frac{5}{6} \div \frac{1}{4}$ will be greater than or less than 1.**

Since $\frac{1}{4} < \frac{5}{6}$, the quotient should be greater than 1.

CHECK YOUR UNDERSTANDING

Predict whether the quotient will be less than or greater than one.

1. $3 \div 4$

2. $4 \div 3$

3. $1\frac{1}{3} \div \frac{3}{4}$

4. $\frac{3}{4} \div 1\frac{1}{3}$

5. $5\frac{1}{2} \div 1\frac{3}{4}$

6. $7\frac{1}{8} \div 9$

Find the product or quotient. If possible, simplify before multiplying.

7. $\left(2\frac{2}{5} \text{ feet}\right)\left(2\frac{1}{2} \text{ feet}\right)$

8. $5\frac{1}{2} \div 1\frac{3}{4}$

9. $-1\frac{7}{8} \cdot 3\frac{1}{3}$

Answers: 1. less than 1; 2. greater than 1; 3. greater than 1; 4. less than 1; 5. greater than 1; 6. less than 1; 7. 6 feet²; 8. $3\frac{1}{7}$; 9. $-6\frac{1}{4}$.

EXERCISES SECTION 3.6

Rewrite the mixed number as an improper fraction.

1. $3\frac{1}{8}$

2. $5\frac{3}{8}$

3. $7\frac{2}{9}$

4. $-5\frac{3}{8}$

5. $2\frac{7}{16}$

6. $-4\frac{2}{9}$

Graph the mixed number on a number line.

7. $3\frac{1}{8}$

... −6 −5 −4 −3 −2 −1 0 1 2 3 4 5 6 7 8 9 ...

8. $5\frac{3}{8}$

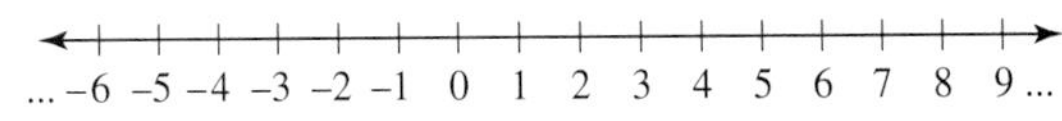

9. $-5\frac{3}{8}$

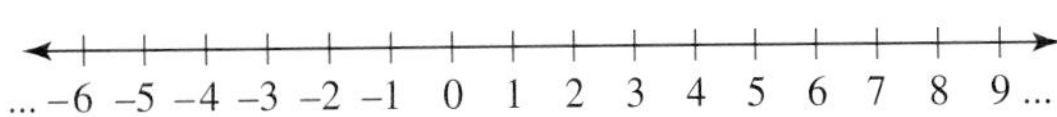

10. $2\frac{7}{16}$

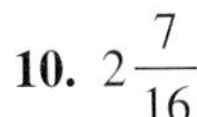

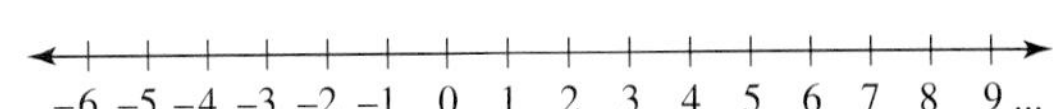

11. $3\frac{1}{4}$

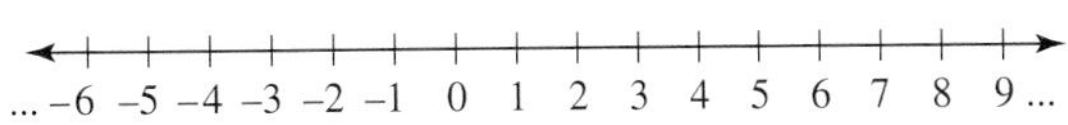

12. $-1\frac{1}{3}$

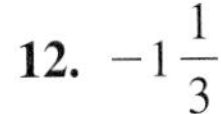

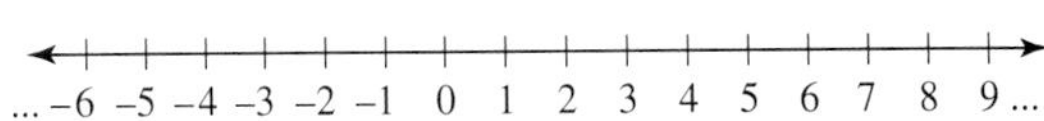

Rewrite the improper fraction as a mixed number.

13. $\frac{13}{6}$

14. $\frac{7}{2}$

15. $-\frac{10}{3}$

16. $\frac{15}{1}$

17. $-\frac{250}{3}$

18. $\frac{243}{2}$

Graph the improper fraction on a number line.

19. $\frac{13}{6}$

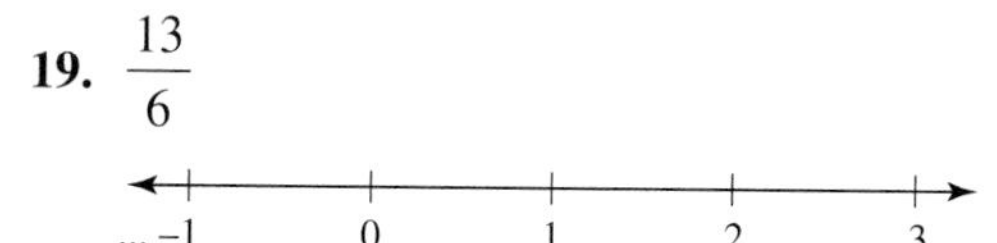

20. $\frac{7}{2}$

... 2 3 4 5 ...

21. $-\frac{10}{3}$

... −4 −3 −2 −1 0 1 ...

22. $\frac{14}{3}$

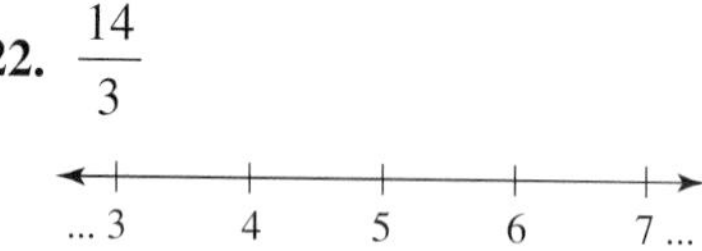

23. $-\frac{5}{4}$

... −2 −1 0 1 2 ...

24. $-\frac{3}{2}$

... −3 −2 −1 0 1 ...

Round the mixed number or improper fraction to the nearest integer.

25. $3\frac{1}{4}$

26. $-3\frac{1}{4}$

27. $\frac{13}{4}$

28. $-\frac{13}{4}$

29. $-14\frac{3}{8}$

30. $-14\frac{1}{2}$

Solve using the five-step problem-solving method.

31. John is providing food for a Forest Service crew that is doing trail maintenance on the Appalachian Trail. All of the food is brought in on pack mules. A chili recipe calls for two $14\frac{1}{2}$-ounce cans of tomatoes and two 19-ounce cans of red beans as well as spices, chopped onion, and chopped garlic, which weigh about $\frac{1}{4}$ pound. John needs to triple the recipe to feed the crew. Estimate the weight of the ingredients for this chili in ounces.

32. Tree roots can eventually push up and crack a sidewalk from below. City crews surveyed several blocks for sidewalk damage. The table below shows the length of tree-damaged sidewalk for five different city blocks. Assuming that similar damage has occurred throughout the 50-block neighborhood, estimate the total length of damaged sidewalk that will have to be replaced.

BLOCK	LENGTH OF DAMAGED SIDEWALK
1400 10th Avenue	$53\frac{1}{4}$ feet
1500 10th Avenue	$11\frac{1}{4}$ feet
1600 10th Avenue	$27\frac{1}{2}$ feet
1400 9th Avenue	$20\frac{3}{4}$ feet
1500 9th Avenue	28 feet

33. Without doing the actual multiplication,

a. Predict which product is greater: $\frac{1}{4} \cdot 3\frac{1}{8}$ or $\frac{1}{2} \cdot 3\frac{1}{8}$.

b. Explain why you think so.

c. Check your prediction by calculating the products.

Evaluate. Write the product as an integer or as a mixed number.

34. $\frac{1}{4} \cdot 3\frac{1}{8}$

35. $\left(\frac{1}{2}\right)\left(3\frac{1}{8}\right)$

36. $-3\frac{1}{8} \cdot 4$

37. $-\frac{1}{2} \cdot \left(-8\frac{1}{4}\right)$

38. $\frac{1}{4} \cdot 8\frac{1}{4}$

39. $3\frac{1}{8} \cdot \left(-4\frac{3}{5}\right)$

40. Without doing the actual division,

a. Predict which quotient is greater: $3\frac{1}{8} \div \frac{1}{4}$ or $3\frac{1}{8} \div \frac{1}{2}$.

b. Explain why you think so.

c. Check your prediction by calculating both quotients.

Evaluate. Write the quotient as an integer or as a mixed number.

41. $6\frac{1}{2} \div \frac{1}{4}$

42. $6\frac{1}{2} \div \frac{1}{8}$

43. $6\frac{1}{2} \div \left(-\frac{1}{2}\right)$

44. $10\frac{1}{3} \div 2\frac{1}{9}$

45. $-100\frac{1}{4} \div \left(-20\frac{1}{2}\right)$

46. $3\frac{5}{6} \div \left(-1\frac{3}{4}\right)$

Solve using the five-step problem-solving strategy. The answer should be written as an integer or a mixed number, not an improper fraction.

47. A recipe for a chili cheese dip calls for one 15-ounce can of chili, 1 cup of chopped onion, $\frac{1}{2}$ teaspoon of cayenne pepper, and $1\frac{1}{2}$ cups of processed cheese cut into cubes. Fergus is going to triple the recipe. How many cups of processed cheese will he need?

48. Jacie is cutting cotton strips that are $2\frac{1}{4}$ inches wide for a quilt. The cotton fabric she is using is $44\frac{1}{2}$ inches wide.
a. How many strips can she cut from a width of fabric?

b. What is the width of the leftover fabric?

49. A recipe for carrot soup requires $5\frac{1}{3}$ cups of chicken broth and serves eight people. Ben needs to make $\frac{1}{4}$ of a recipe so he can make just enough for himself and his girlfriend. How many cups of chicken broth will Ben need?

50. The cover of a textbook measures $7\frac{1}{4}$ inches by $10\frac{1}{4}$ inches. Each textbook is $1\frac{3}{4}$ inches thick. If the textbook is stacked in piles of 25 in the bookstore, what is the height of each pile of textbooks?

51. A standard piece of graph paper is $8\frac{1}{2}$ inches by 11 inches. Each square on the graph paper measures $\frac{1}{4}$ inch by $\frac{1}{4}$ inch. How many squares are on a sheet of this graph paper? (The squares cover the paper; there are no margins.)

52. A catalog advertises that a daypack has a volume of 1,590 cubic inches. The dimensions of the daypack are reported as 17 inches by 13 inches by $5\frac{1}{4}$ inches.

a. Calculate the difference of the advertised volume to the volume calculated using the length, width, and height of the daypack.

b. What could account for this difference?

53. The secretary of a biology department is ordering microscope slide cover glasses. Each box contains 50 cover glasses. The shipping weight of each box is $\frac{3}{4}$ pound. What is the total shipping weight for an order of 100 boxes?

54. A plumber has been subcontracted to do all the plumbing work for a new hotel. The shower in each bathroom requires a length of $\frac{1}{2}$-inch-diameter copper pipe that is $45\frac{3}{4}$ inches long. If copper pipe is sold in lengths of 20 feet, how many pieces of pipe can be cut from each length?

55. A grocery store has allocated 5 feet 4 inches of shelf space at the end of an aisle for a special breakfast cereal promotion. Each box of cereal is $7\frac{3}{4}$ inches across. How many boxes of cereal will fit across this shelf?

56. Jeff walks $4\frac{2}{5}$ miles in 1 hour. If he continues at the same pace, how far will he walk in $2\frac{1}{2}$ hours?

57. Sue is buying new carpet for her living room. The room measures $16\frac{2}{3}$ feet by $22\frac{1}{2}$ feet. What is the area of the living room?

58. The lower layer of roof covering is called the roof sheathing. Formulas can be used to estimate the area of different styles of roof and the amount of roof sheathing needed. The area of a gable roof can be calculated using the formula Area = 2(length)(width) as shown in the diagram. Find the area of a gable roof when the length is 30 feet and the width is $18\frac{1}{2}$ feet (18 feet 6 inches).

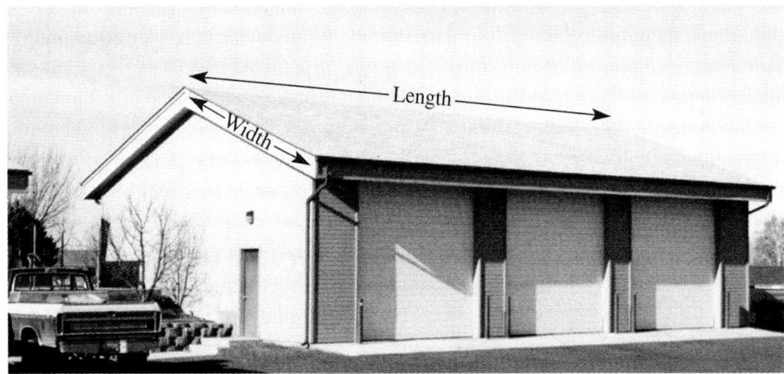

Area = 2 (length) (width)

59. The lower layer of roof covering is called the roof sheathing. Formulas can be used to estimate the area of different styles of roof and the amount of roof sheathing needed. The area of a hip roof on a square building can be calculated using the formula $A = 4\left(\frac{1}{2}\right)$(length)(width) as shown on the photo.

Area = 4 ($\frac{1}{2}$) (length) (width)

a. Find the area of a hip roof when the width is $13\frac{2}{3}$ feet (13 feet 8 inches) and the length is 26 feet.

b. The formula for the area of a triangle is $A = \left(\frac{1}{2}\right)$(base)(height). Explain why the formula in this problem calculates the total area of a hip roof of a square building.

Evaluate. Express the answer as an improper fraction.

60. $3\frac{1}{10}$ meters $\div \frac{3}{4}$

61. $3\frac{1}{10} \cdot \frac{3}{4}$

62. $-\frac{3}{4} \div 3\frac{1}{10}$

63. $\left(-\frac{3}{4}\right)\left(3\frac{1}{10}\right)$

64. $5\frac{2}{7}$ inches $\div 1\frac{1}{6}$

65. $\left(3\frac{4}{5}\right)\left(6\frac{2}{9}\right)\left(-\frac{1}{4}\right)$

ERROR ANALYSIS

66. **Problem:** Multiply $9\frac{1}{8} \cdot 10\frac{1}{2}$.

Answer: $90\frac{1}{16}$

Describe the mistake made in solving this problem.

Redo the problem correctly.

67. **Problem:** Evaluate $-7\frac{1}{5} \cdot 4\frac{1}{2}$.

Answer:

$$\left(-7\frac{1}{5}\right)\left(4\frac{1}{2}\right)$$
$$= \left(-\frac{34}{5}\right)\left(\frac{9}{2}\right)$$
$$= -\frac{306}{10}$$
$$= -\frac{153}{5}$$
$$= -30\frac{3}{5}$$

Describe the mistake made in solving this problem.

Redo the problem correctly.

REVIEW

68. Explain how to determine the sign of the product of two integers.

69. Describe the order of operations. Include an example.

70. When is carrying necessary in addition? Include an example.

71. When is borrowing necessary in subtraction? Include an example.

Section 3.7 Arithmetic with Mixed Numbers

LEARNING TOOLS

CD-ROM SSM VIDEO

Sneak Preview

Mixed numbers are common in the English system of measurement. To use formulas with these measurements, we need to be able to add, subtract, multiply, and divide mixed numbers. As with integers, we will follow the order of operations.

After completing this section, you should be able to:

1) Add or subtract mixed numbers.

2) Evaluate expressions with mixed numbers following the order of operations.

3) Solve word problems that include mixed numbers.

DISCUSSION

Addition and Subtraction of Mixed Numbers

Two different methods can be used to add and subtract mixed numbers. In the first method, we change the mixed numbers to improper fractions before adding or subtracting.

Example 1 **Evaluate $9\frac{1}{4} - 5\frac{3}{4}$. Write the difference as a mixed number.**

$9\frac{1}{4} - 5\frac{3}{4}$

$= \frac{37}{4} - \frac{23}{4}$ *Rewrite the mixed numbers as improper fractions.*

$= \frac{37 - 23}{4}$ *Subtract the numerators; keep the common denominator.*

$= \frac{14}{4}$

$= 3\frac{2}{4}$ *Rewrite the improper fraction as a mixed number.*

$= 3\frac{1}{2}$ *Simplify the fraction part of the mixed number to lowest terms.*

PROCEDURE

Method 1

To add and subtract mixed numbers:

1. Rewrite the mixed numbers as improper fractions.
2. Combine the improper fractions using a common denominator.
3. Rewrite the sum or difference as a mixed number, if necessary.

Example 2 **Add $2\frac{1}{3} + 3\frac{3}{4}$. Write the sum as a mixed number.**

$2\frac{1}{3} + 3\frac{3}{4}$

$= \frac{7}{3} + \frac{15}{4}$ *Rewrite the mixed numbers as improper fractions.*

$= \frac{7 \cdot \mathbf{4}}{3 \cdot \mathbf{4}} + \frac{15 \cdot \mathbf{3}}{4 \cdot \mathbf{3}}$

$= \frac{28 + 45}{12}$ *Build equivalent fractions with a common denominator of 12.*

$= \frac{73}{12}$ *Add the numerators; keep the common denominator.*

$= 6\frac{1}{12}$ *Rewrite the improper fraction as a mixed number.*

We can also add or subtract mixed numbers by combining the integer parts and then combining the fractional parts of the mixed numbers. We do not change the mixed number to an improper fraction in this method.

PROCEDURE

Method 2

To add or subtract mixed numbers:

1. Rewrite each mixed number as the addition of an integer and a proper fraction.
2. Add the integers; add the proper fractions.
3. If the sum of the fractions is greater than 1, rewrite the fraction as a mixed number and add the integers.
4. If the integer and the fraction have the same sign, rewrite as a mixed number.
5. If the integer and the fraction do not have the same sign, borrow 1 or -1 from the integer. Combine the fraction with the 1 or -1. Now rewrite as a mixed number.

Example 3 **Evaluate $2\frac{1}{3} + 3\frac{3}{4}$.**

$2\frac{1}{3} + 3\frac{3}{4}$

$= \left(2 + \frac{1}{3}\right) + \left(3 + \frac{3}{4}\right)$ *Rewrite the mixed numbers including the implied plus signs.*

$= (2 + 3) + \left(\frac{1}{3} + \frac{3}{4}\right)$ *Regroup the integers and the fractions (associative property of addition).*

$= 5 + \left(\frac{1}{3} + \frac{3}{4}\right)$ *Add the integers: 2 + 3 = 5.*

$= 5 + \left(\frac{1 \cdot \mathbf{4}}{3 \cdot \mathbf{4}} + \frac{3 \cdot \mathbf{3}}{4 \cdot \mathbf{3}}\right)$ *Build equivalent fractions with a common denominator 12.*

$= 5 + \frac{4 + 9}{12}$

$= 5 + \frac{13}{12}$ *Add the numerators; keep the common denominator.*

$= 5 + 1\frac{1}{12}$ *The sum of the fractions is greater than 1. Rewrite the improper fraction as a mixed number.*

$= 6\frac{1}{12}$ *Add the integers: 5 + 1 = 6. The final sum is a mixed number.*

Example 4 **Evaluate $3\frac{1}{2} - 5\frac{4}{7}$.**

$$3\frac{1}{2} - 5\frac{4}{7}$$

$= 3\frac{1}{2} + \left(-5\frac{4}{7}\right)$ *Rewrite the subtraction as addition of the opposite.*

$= 3 + \frac{1}{2} + (-5) + \left(-\frac{4}{7}\right)$ *Rewrite the mixed numbers as sums.*

$= (3 - 5) + \left(\frac{1}{2} - \frac{4}{7}\right)$ *Regroup the addition to combine the integers and combine the fractions.*

$= -2 + \left(\frac{7}{14} - \frac{8}{14}\right)$ *Build equivalent fractions with the common denominator 14.*

$= -2 + \left(-\frac{1}{14}\right)$ *Add the fractions.*

$= -2\frac{1}{14}$ *Rewrite the mixed number.*

After adding or subtracting, the integer and the fraction may have different signs. If this happens, the fraction must borrow enough value from the integer to become the same sign as the integer. If the integer is positive, borrow 1. If the integer is negative, borrow -1. Combine the fraction with the borrowed number. The fraction and integer should now have the same sign and can be rewritten as a mixed number.

Example 5 **Evaluate $10\frac{5}{7} - 3\frac{6}{7}$.**

$$10\frac{5}{7} - 3\frac{6}{7}$$

$= 10\frac{5}{7} + \left(-3\frac{6}{7}\right)$ *Rewrite subtraction as addition of the opposite.*

$= 10 + \frac{5}{7} + (-3) + \left(-\frac{6}{7}\right)$

$= (10 + (-3)) + \left(\frac{5}{7} + \left(-\frac{6}{7}\right)\right)$ *Regroup.*

$= 7 + \left(-\frac{1}{7}\right)$ *The fraction is negative and the integer is positive.*

$= 6 + \left(1 - \frac{1}{7}\right)$ *Borrow 1 from the integer.*

$= 6 + \left(\frac{7}{7} - \frac{1}{7}\right)$ *Change the borrowed 1 to an equivalent fraction with a common denominator of 7. Combine the fractions.*

$= 6 + \frac{6}{7}$

$= 6\frac{6}{7}$ *The integer and fraction now have the same sign. Rewrite as a mixed number.*

In this situation, it may be quicker to rewrite both numbers as improper fractions and combine the improper fractions. The sum or difference can then be rewritten as a mixed number, if necessary.

Example 6 **Evaluate $-4\frac{1}{2} + 2\frac{3}{4}$.**

$$-4\frac{1}{2} + 2\frac{3}{4}$$
$$= -4 - \frac{1}{2} + 2 + \frac{3}{4}$$
$$= (-4 + 2) + \left(-\frac{1}{2} + \frac{3}{4}\right)$$
$$= -2 + \left(-\frac{2}{4} + \frac{3}{4}\right)$$
$$= -2 + \frac{1}{4}$$
$$= -1 + \left(-1 + \frac{1}{4}\right)$$
$$= -1 + \left(-\frac{4}{4} + \frac{1}{4}\right)$$
$$= -1 + \left(-\frac{3}{4}\right)$$
$$= -1\frac{3}{4}$$

Either method for combining mixed numbers can be used to solve word problems. When mixed numbers are included in a word problem, the answer is also usually expressed as a mixed number.

Example 7 **Ed needs a board that is $12\frac{5}{8}$ inches long to mend his fence. He uses a 20-inch board from his garage, cutting it with a saw blade that is $\frac{1}{8}$ inch wide. How long is the remaining piece of board? (A length of board equal to the width of the saw blade will be lost as sawdust during cutting.)**

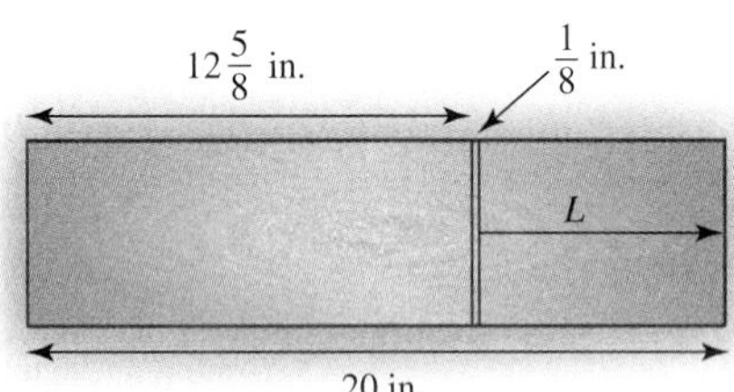

I know . . .

L = Remaining length

$12\frac{5}{8}$ inches = Board length

20 inches = Original length

$\frac{1}{8}$ inch = Saw blade width

Equations

Remaining length = Original length − Board length − Saw blade width

$$L = 20 \text{ inches} - 12\frac{5}{8} \text{ inches} - \frac{1}{8} \text{ inch}$$

Solve

$$L = 20 \text{ inches} - 12\frac{5}{8} \text{ inches} - \frac{1}{8} \text{ inch}$$

$$= \left(20 - 12\frac{5}{8} - \frac{1}{8}\right) \text{ inches}$$

$$= \left(20 - \frac{101}{8} - \frac{1}{8}\right) \text{ inches}$$

$$= \left(\frac{160}{8} - \frac{101}{8} - \frac{1}{8}\right) \text{ inches}$$

$$= \frac{58}{8} \text{ inches}$$

$$= 7\frac{2}{8} \text{ inches}$$

$$= 7\frac{1}{4} \text{ inches}$$

Check

The difference is about 20 inches − 13 inches = 7 inches. This is reasonably close to the exact answer of $7\frac{1}{4}$ inches.

Answer

The remaining board is $7\frac{1}{4}$ inches long.

CHECK YOUR UNDERSTANDING

Evaluate.

1. $3\frac{4}{7} - 2\frac{1}{7}$

2. $6\frac{4}{7} - 2\frac{5}{7}$

3. $6\frac{4}{7} + 2\frac{5}{7}$

4. $2\frac{1}{5} + 4\frac{3}{8}$

5. $7\frac{1}{3} - 3\frac{3}{4}$

6. 4 inches $- 2\frac{5}{6}$ inches

Answers: 1. $1\frac{3}{7}$; 2. $3\frac{6}{7}$; 3. $9\frac{2}{7}$; 4. $6\frac{23}{40}$; 5. $3\frac{7}{12}$; 6. $1\frac{1}{6}$ inches.

DISCUSSION

Order of Operations and Formulas

Order of operations specifies the order in which we must do different arithmetic operations as well as the purpose of parentheses, brackets, and braces. The order in which arithmetic operations are performed is the same for fractions and mixed numbers as it is for whole numbers (Section 1.9) and integers (Section 2.5.)

Example 8 **Evaluate $\frac{1}{4} + 3\frac{1}{2} \div 8$.**

$\frac{1}{4} + 3\frac{1}{2} \div 8$

$= \frac{1}{4} + \frac{7}{2} \div \frac{8}{1}$ *Rewrite the mixed number as an improper fraction. Write 8 as a fraction with a denominator of 1.*

$= \frac{1}{4} + \frac{7}{2} \cdot \frac{1}{8}$ *Rewrite division as multiplication by the reciprocal.*

$= \frac{1}{4} + \frac{7}{16}$ *Multiply numerators; multiply denominators.*

$= \frac{1}{4} \cdot \frac{\mathbf{4}}{\mathbf{4}} + \frac{7}{16}$ *To add, use a common denominator of 16. Build an equivalent fraction with this denominator.*

$= \frac{4}{16} + \frac{7}{16}$

$= \frac{11}{16}$

As with integers, the absolute value of a mixed number is its distance from zero on the number line.

Example 9 **Evaluate $\left|\frac{2}{3}\right| - \left(1\frac{1}{2}\right)^2$.**

$\left|\frac{2}{3}\right| - \left(1\frac{1}{2}\right)^2$

$= \frac{2}{3} - \left(\frac{3}{2}\right)^2$ *Evaluate the absolute value. Rewrite the mixed number as an improper fraction.*

$= \frac{2}{3} - \frac{9}{4}$ *Evaluate the exponential expression.*

$= \frac{2 \cdot \mathbf{4}}{3 \cdot \mathbf{4}} - \frac{9 \cdot \mathbf{3}}{4 \cdot \mathbf{3}}$ *Build equivalent fractions with a common denominator of 12.*

$= \frac{8 - 27}{12}$

$= -\frac{19}{12}$ *Subtract numerators; keep the common denominator.*

$= -1\frac{7}{12}$ *Rewrite the improper fraction as a mixed number.*

In the next example, we simplify before multiplying to keep the numbers smaller.

Example 10 **Evaluate** $-\frac{2}{3}\left(2\frac{2}{5}\right)^2 - \left(-\frac{2}{3}\right)^2 \cdot 2\frac{2}{5}$.

$$-\frac{2}{3}\left(2\frac{2}{5}\right)^2 - \left(-\frac{2}{3}\right)^2 \cdot 2\frac{2}{5}$$

$$= -\frac{2}{3}\left(\frac{12}{5}\right)^2 - \left(-\frac{2}{3}\right)^2 \cdot \frac{12}{5}$$ *Rewrite the mixed numbers as improper fractions.*

$$= -\frac{2}{\mathbf{3}} \cdot \frac{\mathbf{144}}{25} - \frac{4}{\mathbf{9}} \cdot \frac{\mathbf{12}}{5}$$ *Evaluate the exponents.*

$$= -\frac{2}{\mathbf{1}} \cdot \frac{\mathbf{48}}{25} - \frac{4}{\mathbf{3}} \cdot \frac{\mathbf{4}}{5}$$ *Simplify to lowest terms before multiplying.*

$$= -\frac{96}{25} - \frac{16}{15}$$

$$= -\frac{288}{75} - \frac{80}{75}$$ *The LCM of 25 and 15 is 75. Write equivalent fractions with a denominator of 75.*

$$= -\frac{368}{75}$$ *Combine the numerators and keep the common denominator.*

$$= -4\frac{68}{75}$$ *Rewrite as a mixed number.*

When evaluating formulas, the order of operations is always followed. Formulas are often used to find the perimeter or area of a geometric figure such as a triangle or a rectangle.

A **triangle** has three sides and three angles.

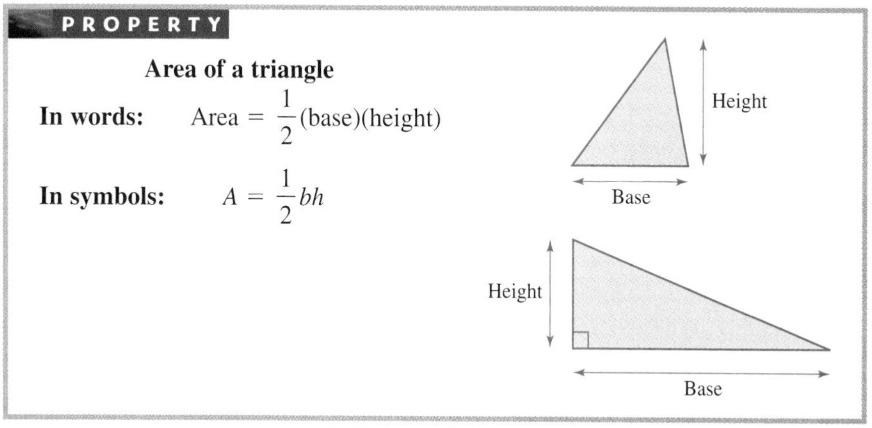

PROPERTY

Area of a triangle

In words: Area $= \frac{1}{2}$(base)(height)

In symbols: $A = \frac{1}{2}bh$

A **rectangle** has four sides and has right angles between adjacent sides. The opposite sides of a rectangle are parallel. The area of a rectangle is equal to the product of its length and width.

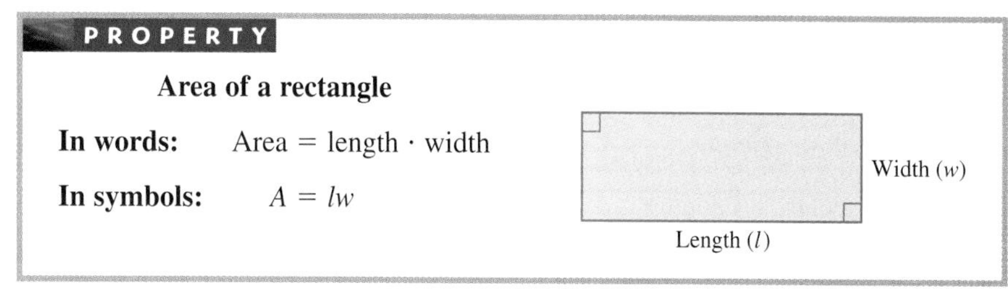

PROPERTY

Area of a rectangle

In words: Area = length · width

In symbols: $A = lw$

Example 11 **Nathan wants to spread a $\frac{1}{4}$ foot thick layer of gravel in a triangular area in his backyard that measures about $7\frac{1}{4}$ feet by $9\frac{1}{2}$ feet by 12 feet as shown. About what volume of gravel will he need in cubic feet? Round the answer to the next largest cubic foot.**

12 ft

$7\frac{1}{4}$ ft

$9\frac{1}{2}$ ft

I know . . .

V = Volume of gravel

$\frac{1}{4}$ foot = Depth of gravel

$7\frac{1}{4}$ feet $= \frac{29}{4}$ feet = Height of triangle

$9\frac{1}{2}$ feet $= \frac{19}{2}$ feet = Base of triangle

Conversion factor: 12 inches = 1 foot

Formula: Area of triangle $= \frac{1}{2}$base · height

Formula: Volume = Area · depth

Equations

Volume of gravel = Area of triangle · Depth of gravel

$$V = \left(\frac{1}{2} \cdot \frac{19}{2}\text{ feet} \cdot \frac{29}{4}\text{ feet}\right)\left(\frac{1}{4}\text{ foot}\right)$$

Solve

$$\begin{aligned} V &= \left(\frac{1}{2} \cdot \frac{19}{2}\text{ feet} \cdot \frac{29}{4}\text{ feet}\right)\left(\frac{1}{4}\text{ foot}\right) \\ &= \left(\frac{1}{2} \cdot \frac{19}{2} \cdot \frac{29}{4} \cdot \frac{1}{4}\right)(\text{feet} \cdot \text{feet} \cdot \text{foot}) \\ &= \frac{551}{64}\text{ feet}^3 \\ &= 8\frac{39}{64}\text{ feet}^3 \end{aligned}$$

Rounding to the next cubic foot, $V = 9$ feet^3.

Check

Round the measurements so that the base is 10 feet and the height is 7 feet. The area of the triangle is then $\frac{1}{2} \cdot 10$ feet $\cdot$ 7 feet = 35 feet^2. Multiply the estimated area by the depth: 35 feet$^2 \cdot \frac{1}{4}$ feet $= \frac{35}{4}$ feet^3. Since $\frac{35}{4}$ feet$^3 = 8\frac{3}{4}$ feet^3, this is reasonably close to the exact answer.

Answer

Nathan needs about 9 cubic feet of gravel.

Example 12 **The back window in Kruin's pickup truck is trapezoidal in shape. He needs to replace the glass in the window. If the top of the window is $40\frac{1}{2}$ inches, the bottom of the window is 48 inches, and the height of the window is 36 inches, what is the area of the window?**

$40\frac{1}{2}$ inches

36 inches

48 inches

I know . . .

A = Area of window

Formula: Area $= \frac{1}{2}$(height)(base 1 + base 2)

36 inches = Height

$40\frac{1}{2}$ inches = Length of top edge = Base 1

48 inches = Length of bottom edge = Base 2

Equations

$$\text{Area} = \frac{1}{2}(\text{height})(\text{base 1} + \text{base 2})$$

$$A = \frac{1}{2} \cdot 36 \text{ inches} \cdot \left(40\frac{1}{2} \text{ inches} + 48 \text{ inches}\right)$$

Solve

$$\begin{aligned} A &= \frac{1}{2}(36 \text{ inches})\left(40\frac{1}{2} \text{ inches} + 48 \text{ inches}\right) \\ &= \frac{1}{2}(36 \text{ inches})\left(88\frac{1}{2} \text{ inches}\right) \\ &= \frac{1}{2}(36 \text{ inches})\left(\frac{177}{2} \text{ inches}\right) \\ &= \frac{1 \cdot 36 \cdot 177}{2 \cdot 2} \text{ inches} \cdot \text{inches} \\ &= \frac{6{,}372}{4} \text{ inches}^2 \\ &= 1{,}593 \text{ inches}^2 \end{aligned}$$

Check

Rounding the measurements to a height of 40 inches, a top length of 40 inches, and a bottom length of 50 inches, the area formula becomes:

$$\begin{aligned} A &= \frac{1}{2}(40)(40 + 50) \\ &= \frac{1}{2}(40)(90) \\ &= 20(90) \\ &= 1{,}800 \end{aligned}$$

1,800 square inches is close to our exact answer. We expect that the estimate should be somewhat larger than the exact answer because both the top length and the bottom length were rounded up to the nearest ten.

Answer

The area of the window is 1,593 square inches.

CHECK YOUR UNDERSTANDING

Evaluate these expressions.

1. $\left(3 - \frac{2}{5}\right) \div \left(4 + \frac{5}{2}\right)$

2. $\frac{4}{5} \cdot \left(\frac{3}{2}\right)^2 - |-2|$

3. $\left(\frac{3}{4}\right)^2 - 4\left(\frac{5}{8}\right)\left(1\frac{1}{5}\right)$

4. $\left(\frac{1}{2}\left(4 - \frac{4}{3}\right)^2 + \frac{5}{6}\right)$

Answers: 1. $\frac{2}{5}$; 2. $-\frac{1}{5}$; 3. $-2\frac{7}{16}$; 4. $4\frac{7}{18} = \frac{79}{18}$.

EXERCISES SECTION 3.7

Find the sum or difference. Write the answer as a mixed number.

1. $6\frac{3}{4} + 5\frac{1}{9}$

2. $6\frac{3}{4}$ yards $- 5\frac{1}{9}$ yards

3. $-6\frac{3}{4} + 5\frac{1}{9}$

4. $-6\frac{3}{4} - 5\frac{1}{9}$

5. $55\frac{3}{4} + 61\frac{1}{16}$

6. $13 - 81\frac{7}{10}$

7. $3\frac{9}{10} + 8\frac{99}{100} + 7\frac{999}{1{,}000}$

8. 99 inches $- 6\frac{9}{10}$ inches

9. $7\frac{3}{5} - 5\frac{2}{9}$

10. $-6\frac{1}{7} + 3\frac{1}{4}$

11. $-6\frac{1}{7} - \left(-3\frac{1}{4}\right)$

12. $2\frac{1}{4}$ feet + $5\frac{1}{6}$ feet

13. $44\frac{1}{2} - 100$

14. $13\frac{1}{2} + 12\frac{3}{4} - 8\frac{1}{5} + \frac{1}{2}$

15. $-3\frac{2}{3} - \frac{4}{7} + 3\frac{6}{7} - 1\frac{2}{3}$

16. $2\frac{3}{4} - 1\frac{7}{8}$

Solve the problem following the five-step problem-solving strategy. Write the answer as a mixed number.

17. A wildlife biology department has an opportunity to purchase surplus museum cases for displaying animal specimens. The ceiling in the room in which the cases will be put is 11 feet high. The cases come in three heights: $46\frac{1}{4}$ inches, $35\frac{1}{2}$ inches, and $42\frac{5}{8}$ inches. If one case of each size is used to make a stack, what will be the distance of the top of the stack from the ceiling?

18. A roll of wallpaper is 5 yards long. Karen is cutting strips of wallpaper to cover a wall that is $96\frac{5}{8}$ inches high. In addition, she wants to add $3\frac{1}{4}$ inches to each strip for matching the pattern and an additional $1\frac{1}{2}$ inches to each strip for trimming at the top and the bottom. How many strips will she be able to cut from each roll?

19. The sum of the three angles in any triangle always equals 180 degrees. A triangle has one angle that measures 90 degrees and another angle that measures $22\frac{1}{4}$ degrees. What is the measure of the third angle?

20. The volume of a teaspoon is $\frac{17}{100}$ liquid ounce. The volume of a tablespoon is $\frac{1}{2}$ liquid ounce. If a recipe calls for 1 tablespoon + 1 teaspoon of vanilla extract, what is the volume of the vanilla in liquid ounces?

21. Jim is a long-haul trucker who has been told by his doctor to drink more liquids when he drives. He is keeping track of how much he drinks. On one day, he drank a thermos of coffee ($1\frac{1}{10}$ quart), $2\frac{1}{3}$ quarts of bottled water, and $\frac{1}{2}$ quart of orange juice. What volume of liquid did he drink that day?

22. Pat bought 4 gallons of white semigloss paint to paint the woodwork in her house. She used $\frac{1}{3}$ gallon to paint the dining room woodwork. Comparing the amount of woodwork in the living room to what she had painted in the dining room, Pat estimates that she will need $1\frac{1}{2}$ gallons to paint the living room. After finishing the living room, how much paint will be left for painting the remaining woodwork?

23. A food bank encourages gardeners to "grow a row" of vegetables to donate to the food bank. During August, the food bank receives huge amounts of tomatoes from local gardeners. On one day alone the bank received five different donations of tomatoes: $1\frac{1}{4}$ bushels, $2\frac{1}{8}$ bushels, $3\frac{1}{2}$ bushels, $\frac{2}{3}$ bushel, and 4 bushels. How many bushels of tomatoes were donated that day?

Evaluate.

24. $\frac{1}{2} + 3\frac{1}{4} \cdot 2 - 6$

25. $\frac{1}{2} + 3\frac{1}{4} \cdot 2\frac{1}{3} - 6$

26. $\dfrac{3\frac{1}{4} - 2\frac{3}{4}}{12\frac{2}{3} - 9\frac{1}{3}}$

27. $\dfrac{8\frac{1}{4} - \left(-2\frac{3}{4}\right)}{12\frac{2}{3} - 9}$

28. $6 - 2\frac{1}{3} \div \frac{1}{2} + \left(5\frac{1}{2}\right)^2$

29. $15 - 12\frac{1}{8} \div \frac{1}{2} - \left(\frac{1}{2}\right)^2$

30. $3\frac{1}{4} \cdot 2 - 1 \div \frac{1}{4}$

31. $3\frac{1}{4} \cdot 6 - 10 \div 2\frac{1}{4}$

32. $5\frac{1}{4} + \frac{3}{4} \div 2 - \frac{1}{3}$

33. $\left(5\frac{1}{4} + \frac{3}{4}\right) \div 2 - \frac{1}{3}$

34. $3\frac{1}{2} - \left|-4\frac{1}{4} \div 3\frac{1}{3}\right|$

35. $\left|5\frac{2}{7} - 7\frac{2}{3}\right|^2$

Solve using the five-step problem-solving strategy.

36. The total number of gallons of paint required to paint the exterior of a house can be found using the formula

$$\text{Gallons of paint} = \left(\frac{\text{Area to be painted}}{\text{Coverage of paint}}\right)(\text{Number of coats of paint}).$$

Anil's house has an exterior area of 1,280 square feet. The coverage of the paint she is using is 350 square feet per gallon. She is going to put two coats of paint on the house.

a. Calculate the exact number of gallons of paint needed using this formula. Write the answer as a mixed number.

b. How many gallons of paint should Anil actually buy? Explain.

37. The *total rise* of a staircase is the vertical distance between the finished floors on either end of the staircase. The *unit rise* of the staircase is the vertical distance between each tread. The *unit run* of the staircase is the horizontal distance between each tread. The unit run can be calculated using this formula:

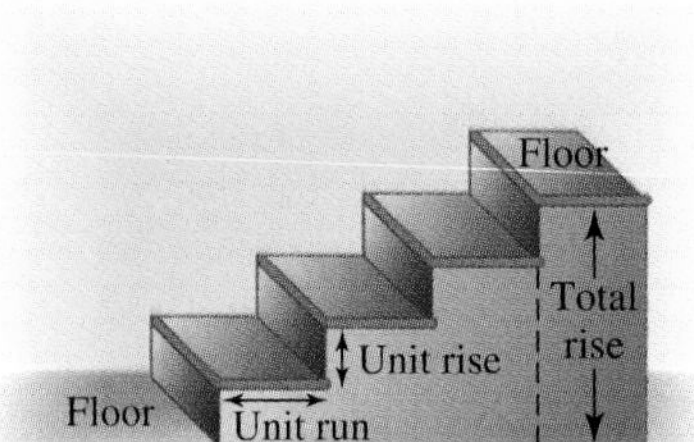

Unit run $= 17\frac{1}{2}$ inches $-$ Unit rise.

The *total run* of the staircase is equal to the product of the total number of treads and the unit run.

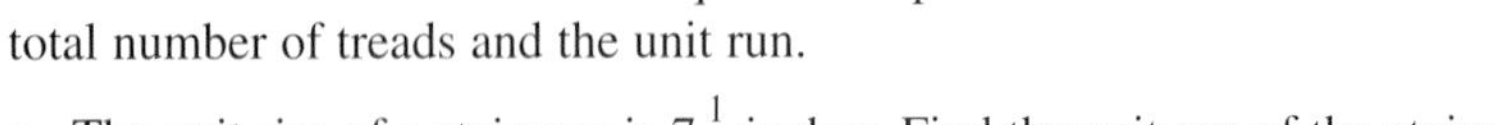

a. The unit rise of a staircase is $7\frac{1}{8}$ inches. Find the unit run of the staircase.

b. There are 8 treads on this staircase, not including the finished floors on either end of the staircase. Draw a diagram of this staircase.

c. Find the total run of the staircase.

d. Find the total rise of the staircase.

38. Standard cotton fabric is $2\frac{2}{3}$ feet wide. Jenna is buy remnants of fabric from a Mills End store to make decorative pillows. She buys one piece of fabric that is $1\frac{1}{2}$ yards long, another piece that is $2\frac{1}{4}$ yards long, and a third piece that is $\frac{5}{8}$ yard long. What is the total area of the fabric she has purchased in square feet?

39. *Heating degree-days* are the number of degrees the daily average temperature is below 65°F. This measure is used by the heating industry to estimate energy needs and times of peak demand. A day with an average temperature of 40°F has 25 heating degree-days; a day with an average temperature of 80°F has none. We can compare the normal degree-days in different parts of the country to compare energy used for heating. Chicago has $4\frac{1}{2}$ times more annual heating degree-days than New Orleans. Assuming that heating costs the same in both cities, if a house costs $325 a year to heat in New Orleans, what will it cost to heat the same house in Chicago?

40. At a lumber yard, the standard measurements of a cedar post are 3 inches by 4 inches by 6 feet. Actually, the post measures $2\frac{5}{8}$ inches by $3\frac{5}{8}$ inches by 6 feet. Calculate the volume of this post in cubic inches.

41. An interior decorator wants to use large triangular silk scarves for a special window treatment. Each edge of the scarf will measure 3 feet. The height of the scarf will be $2\frac{7}{10}$ feet. Calculate the area of silk in three such scarves.

42. Any rectangle can be divided into two triangles by connecting two opposite corners with a straight line (the diagonal). A rectangle measures $2\frac{1}{4}$ meters by $5\frac{1}{8}$ meters.

a. Calculate the area of this rectangle.

b. If this rectangle is divided into two triangles by a diagonal, calculate the area of one of these triangles.

c. Show that the area of the rectangle is equal to (2)(area of one triangle).

ERROR ANALYSIS

43. Problem: Subtract $25\frac{1}{7} - 4\frac{3}{7}$.

Answer:

$$25\frac{1}{7} - 4\frac{3}{7}$$
$$= 21\frac{1-3}{7}$$
$$= 21\frac{2}{7}$$

Describe the mistake made in solving this problem.

Redo the problem correctly.

44. Problem: Evaluate $1\frac{1}{2} + 3\frac{1}{2} \div \frac{1}{4}$.

Answer:

$$1\frac{1}{2} + 3\frac{1}{2} \div \frac{1}{4}$$
$$= 5 \div \frac{1}{4}$$
$$= \frac{5}{4}$$
$$= 1\frac{1}{4}$$

Describe the mistake made in solving this problem.

Redo the problem correctly.

READ AND REFLECT

Mathematics students often ask, "When am I ever going to use this?" Some mathematical concepts and skills are directly useful in life outside of school. An obvious application is money and finance. Unless you understand and can apply concepts such as percentages, interest rates, exponential growth, and compounding, you have to trust the advice of others. Since many who make recommendations about purchases and investments stand to benefit if you follow their advice, ignorance makes you vulnerable to exploitation. For example, when Sean went to a car dealership to buy a truck, he had many decisions to make that involved mathematics. He had to consider whether to lease or buy. He had to decide how much of his limited savings to use for the down payment. He had to choose between a dealer rebate or a reduced interest rate on a loan. Since he knows very little practical mathematics, he had no alternative but to trust the salesperson and her calculator.

Similar knowledge is required to make thoughtful decisions about credit cards, mortgages, insurance, investments, and other loans. People who have limited mathematical understanding cannot compare different alternatives. Their decisions may not be in their own best interest and may be profitable for those who earn their living on commissions or loan interest.

Like money, probabilities and statistics surround us. Political polls, the performance of the stock market, the odds of winning a lottery, and the probability of surviving a cancer diagnosis are all examples. Training in mathematics helps you to ask the right questions about the statistics you hear and read so that you can disregard misleading information. If you know the mathematics behind percentages, ratios, medians, averages, and probabilities, you may avoid being manipulated by advertising. You will have a better sense of what questions need to be asked about reported statistics or probabilities to make thoughtful decisions.

Estimation is perhaps the most widely used skill learned in mathematics. Estimation is the mathematical procedure that is typically used for calculating about how much time it will take to drive to the airport, the approximate sale price of an item, or the cost of a tank of gas. Checking exact solutions with estimation can reveal unreasonable solutions before mistakes are made. Estimation used in this way can alert us to a potentially fatal dose of medication, a ridiculous order for carpet, or an outrageous calculation of sales tax.

In addition to these direct applications of mathematics, some of the value of completing a mathematics course is indirect. For example, in this course you analyze completed examples. From this study, you develop a procedure for working similar problems. You may seldom or even never use this procedure outside the classroom. However, the process of developing the procedure can help you improve your ability to think critically in other situations.

RESPOND

1. Describe a situation where you often do estimates.

2. Describe a financial situation other than the ones listed in the reading in which knowledge of mathematics is important.

3. Describe another class you are taking or have already completed that had subject matter that you felt would be directly useful to you in your life outside of school. Include examples of what you felt would be useful to you and why you think it will be useful.

4. What mathematics will be most useful to you in your future daily life or chosen profession?

Section	Terms and Concepts/Procedures
3.1	multiple; prime number; composite number; factor; divisor 1. Use the divisibility laws to identify divisors of a number. 2. Identify the factors of a number.
3.2	fraction; numerator; denominator; proper fraction; improper fraction; rational numbers; equivalent fractions; lowest terms; prime factorization 1. Build an equivalent fraction with a given denominator. 2. Simplify a fraction to lowest terms. 3. Write the prime factorization of a number.
3.3	product; element of a set; pyramid; trapezoid 1. Multiply fractions. 2. Evaluate a formula that includes fractions.
3.4	quotient; dividend; divisor; reciprocal; multiplicative inverse 1. Divide fractions.
3.5	common denominator; set of common multiples of two numbers; least common multiple 1. Add or subtract fractions that have the same denominator. 2. Find a common denominator of two or more fractions. 3. Add or subtract fractions that have different denominators.
3.6	mixed number; implied plus sign 1. Graph a mixed number on a number line. 2. Rewrite a mixed number as an improper fraction. 3. Rewrite an improper fraction as a mixed number. 4. Multiply mixed numbers. 5. Divide mixed numbers.
3.7	triangle; rectangle; trapezoid; order of operations 1. Add or subtract mixed numbers. 2. Evaluate expressions that include mixed numbers.

REVIEW EXERCISES CHAPTER 3

SECTION 3.1 MULTIPLES AND FACTORS

1. List the first five multiples of 7.

2. Find a number that is a multiple of 4, a multiple of 5, and a multiple of 15.

3. List the factors of 40.

4. List the factors of 29.

5. Identify the composite numbers in this set: {0, 1, 2, 6, 12, 13, 15, 19, 20, 21, 22, 23}.

6. Use the divisibility rules to determine if 38,532 is divisible by:

a. 2 **b.** 3 **c.** 4

d. 9 **e.** 10

Determine if the number is prime or composite.

7. 23

8. 45

9. 51

10. 31

SECTION 3.2 INTRODUCTION TO FRACTIONS

11. Graph $\frac{7}{10}$ on a number line.

12. Graph $-\frac{4}{5}$ on a number line.

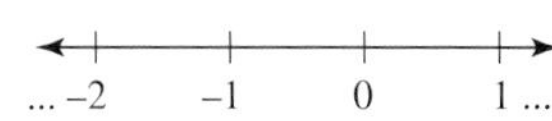

13. List the improper fractions in this set: $\left\{\frac{4}{5}, \frac{9}{2}, \frac{8}{8}, \frac{11}{10}, \frac{6}{7}, \frac{0}{4}, \frac{5}{5}\right\}$.

14. State the opposite of $\frac{-7}{4}$.

15. Evaluate $\left|-\frac{6}{7}\right|$.

Place a <, >, or = sign between the numbers to make a true statement.

16. $\frac{3}{4} \quad \frac{-5}{6}$

17. $\frac{-5}{8} \quad -\frac{7}{8}$

18. Build a fraction equivalent to $\frac{2}{3}$ that has a denominator of 9.

19. Build a fraction equivalent to $-\frac{3}{7}$ that has a denominator of 21.

Simplify to lowest terms.

20. $\frac{12}{48}$

21. $-\frac{126}{39}$

22. $-\frac{320}{800}$

23. $\frac{56}{12}$

24. Ten of the 28 students in Ben's math class are women. What fraction of the class are women?

25. What fraction of an hour is 35 minutes?

26. The U.S. Army pays its pilots a $12,000 bonus. The maximum bonus for pilots allowed by law is $25,000. What fraction of the maximum bonus do army pilots receive?

27. Write the prime factorization of 35.

28. Write the prime factorization of 71.

29. Write the prime factorization of 500.

30. Write the prime factorization of 96.

SECTION 3.3 MULTIPLICATION OF FRACTIONS

31. Evaluate $\frac{11}{12} \cdot \frac{6}{5}$.

32. Evaluate $\frac{-24}{5} \cdot \frac{7}{6} \cdot 10$.

33. Rita spends $\frac{2}{5}$ of a typical 10-hour work day in meetings. How many hours does she spend in meetings?

34. Bill pays $\frac{7}{25}$ of his salary to the IRS in the form of income tax. If he pays \$525 in tax per month, what is his monthly salary?

SECTION 3.4 DIVISION OF FRACTIONS

Write the reciprocal of the number.

35. -8

36. $\frac{-9}{5}$

37. $2\frac{1}{3}$

38. $\frac{1}{4}$

Find the quotient.

39. $\frac{12}{5} \div \frac{20}{7}$

40. $\frac{-14}{5} \div \frac{7}{10}$

41. $\frac{22}{15} \div \left(-\frac{6}{5}\right)$

SECTION 3.5 ADDITION AND SUBTRACTION OF FRACTIONS

Evaluate.

42. $\frac{7}{5} + \frac{3}{5}$

43. $\frac{3}{7} - \frac{5}{7}$

44. $\frac{8}{9} - \left(-\frac{7}{9}\right)$

45. $\frac{5}{8} + \frac{4}{5}$

46. $\frac{8}{5} - \frac{9}{11}$

47. $\frac{5}{12} - \frac{8}{9} - \left(-\frac{3}{4}\right)$

48. Find the least common multiple of 24 and 90.

49. Find the least common multiple of 8, 36, and 42.

Solve using the five-step problem-solving strategy.

50. The exterior wall of Shelly's house is made of $\frac{5}{8}$-inch drywall, $\frac{9}{4}$-inch insulation, $\frac{3}{4}$-inch plywood sheathing, $\frac{1}{8}$-inch house wrap, and $\frac{7}{16}$-inch siding. What is the total thickness of the exterior wall?

SECTION 3.6 MIXED NUMBERS

Rewrite the improper fraction as a mixed number or integer.

51. $\frac{18}{7}$

52. $\frac{-62}{30}$

53. $\frac{20}{6}$

Rewrite the mixed number or integer as an improper fraction.

54. $3\frac{2}{7}$

55. $-6\frac{3}{8}$

Round the mixed number or improper fraction to the nearest integer.

56. $\frac{21}{4}$

57. $-3\frac{5}{7}$

58. $-34\frac{2}{9}$

59. $\frac{-54}{12}$

Solve using the five-step problem-solving strategy.

60. Calculate the volume of a rectangular backpack that is $32\frac{1}{4}$ inches tall, $14\frac{1}{2}$ inches wide, and has a depth of $8\frac{3}{4}$ inches.

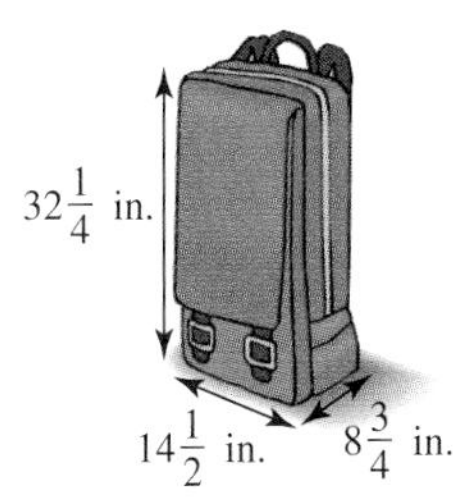

61. Find the volume of a rectangular sleeping bag that has dimensions of $32\frac{1}{4}$ inches by $77\frac{5}{8}$ inches and a loft of $3\frac{1}{2}$ inches.

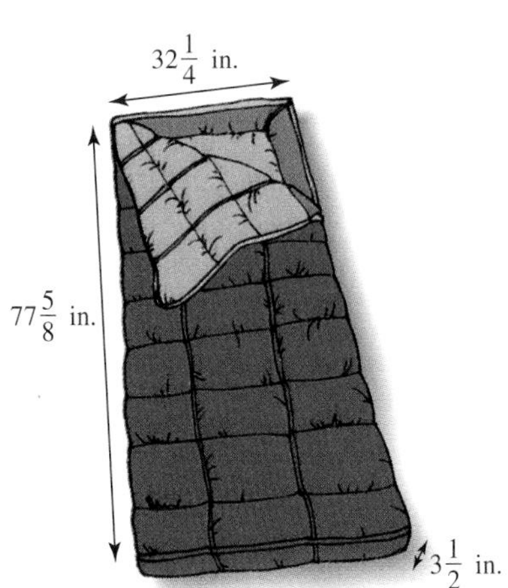

62. A rectangular sink has dimensions of $32\frac{1}{4}$ inches by $21\frac{5}{8}$ inches by $7\frac{5}{6}$ inches. What is the volume of the sink?

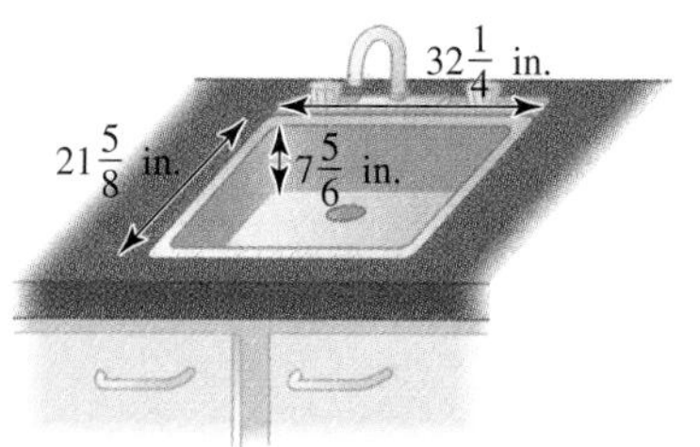

Evaluate.

63. $3\frac{2}{7} \cdot 6\frac{3}{5}$

64. $17\frac{1}{4} \div 3\frac{4}{5}$

65. $2\frac{3}{5} \cdot 5\frac{3}{8} \div \left(-1\frac{1}{4}\right)$

SECTION 3.7 ARITHMETIC WITH MIXED NUMBERS

Evaluate.

66. $2\frac{6}{7} + \frac{4}{5} + 3\frac{7}{8}$

67. $-5 - \frac{4}{5}$

68. $-7\frac{3}{4} - 3\frac{4}{5}$

69. $3\frac{4}{5} - \frac{7}{8} - 2\frac{3}{4}$

70. $-3\frac{1}{4}\left(2\frac{9}{10} - 5\frac{3}{5}\right)$

71. $\left(3 - \frac{3}{5}\right)^2 - \frac{4}{9}$

72. Calculate the area of the trapezoid.

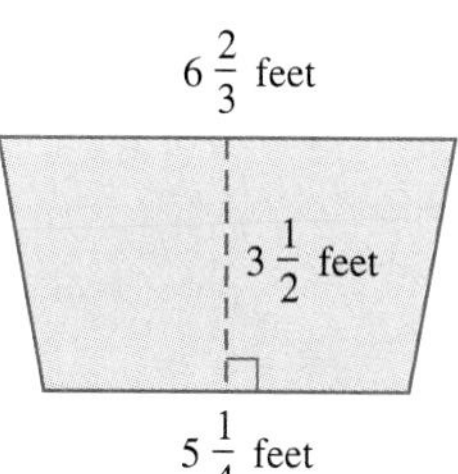

73. A triangle has a base that measures $5\frac{2}{3}$ inches and a height of $7\frac{1}{2}$ inches. Calculate the area of the triangle.

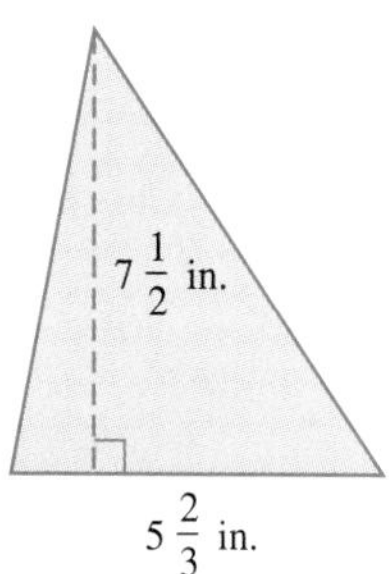

74. A green (more than 19 percent moisture) standard two-by-four pine board actually has a thickness of $1\frac{9}{16}$ inches and a width of $3\frac{9}{16}$ inches. A dry standard two-by-four pine board actually has a thickness of $1\frac{1}{2}$ inches and a width of $3\frac{1}{2}$ inches.

a. Calculate the difference in thickness of a green and a dry two-by-four pine board.

b. Calculate the difference in width of a green and a dry two-by-four pine board.

c. Find the difference in volume between a green and a dry two-by-four board that is 8 feet long. (The length must be changed into inches before calculating the volume.)

Solve using the five-step problem-solving strategy.

75. Wilma has a garbage disposal with a $\frac{1}{3}$-horsepower motor. Betty's garbage disposal has a $\frac{5}{8}$-horsepower motor. What is the difference in the size of the motors?

1. List the first five multiples of 8.

2. Five of the nine births at St. Joseph's Hospital were boys. What fraction of the births were girls?

3. Divide $\frac{18}{7} \div \frac{12}{5}$.

4. List the factors of 36.

5. Determine using divisibility rules if 54,872 is divisible by:
a. 2 **b.** 3 **c.** 4 **d.** 5 **e.** 9

6. Graph $-\frac{2}{3}$ on a number line.

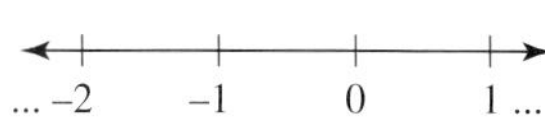

7. List the proper fractions in this set: $\left\{\frac{4}{5}, \frac{9}{2}, \frac{8}{8}, \frac{11}{10}, \frac{6}{7}, \frac{0}{4}, \frac{5}{5}\right\}$.

8. Evaluate $\left|-\frac{4}{9}\right|$.

9. Place a <, >, or = sign between the numbers to make a true statement: $-\frac{3}{4} \quad -\frac{3}{7}$.

10. Build a fraction equivalent to $\frac{2}{7}$ that has a denominator of 21.

11. List the factors of 73.

12. Simplify to lowest terms: $-\frac{132}{144}$

13. What fraction of a gallon is a quart?

14. The population of Alburgh was 336 in 1999. Only $\frac{3}{16}$ of the population was under 21. How many people were under 21?

15. Tom's lawnmower has a $6\frac{1}{2}$-horsepower gas m[illegible]er has a $5\frac{3}{4}$-horsepower gas motor. What is the difference in the power of the motors?

16. Write the prime factorization of 53.

17. Round to the nearest integer: $\frac{37}{4}$.

18. Evaluate $2\frac{1}{4}\left(\frac{9}{10} - 5\frac{3}{4}\right)$.

19. A mail-order catalogue charges $\frac{3}{100}$ of the total sales charge for packaging. If Sallee buys $200 worth of items from the catalogue, how much is charged for packaging?

20. Find the least common multiple of 36 and 24.

21. Multiply $\frac{-18}{5} \cdot \frac{7}{12} \cdot 2$.

22. Write the reciprocal of $\frac{-4}{7}$.

23. Write the prime factorization of 27.

24. Identify the prime numbers in this set: $\{0, 1, 7, 11, 15, 21, 31, 51\}$.

25. Graph $\frac{6}{5}$ on a number line.

26. Evaluate $\frac{2}{5} + \frac{9}{5}$.

27. Evaluate $\frac{4}{7} - \frac{8}{7}$.

28. Evaluate $\frac{5}{3} + \frac{7}{12}$.

29. Evaluate $\frac{8}{5} - \left(-\frac{3}{7}\right)$.

30. Write $\frac{12}{100}$ as a fraction in lowest terms.

31. Find the least common multiple of 6, 22, and 33.

32. Rewrite as a mixed number or integer: $\frac{-12}{7}$.

33. Rewrite as an improper fraction: $5\frac{8}{9}$.

34. Estimate the area of a rectangular courtyard that measures $42\frac{1}{2}$ yards by $19\frac{4}{7}$ yards.

35. Evaluate $4\frac{1}{4} \cdot 3\frac{3}{5}$.

36. Write the reciprocal: -34.

37. Evaluate $11\frac{1}{4} \div 2\frac{4}{5}$.

38. Evaluate $2\frac{4}{9} + \frac{3}{5} + \left(-3\frac{2}{3}\right)$.

39. Evaluate $\left(5 - \frac{4}{5}\right) - \left(\frac{4}{5}\right)^2$.

40. Evaluate $\frac{-8}{5} \div \frac{12}{25}$.

41. Evaluate $\frac{5}{12} \cdot \frac{8}{15}$.

42. Write the prime factorization of 510.

43. Calculate the area of the trapezoidal window.

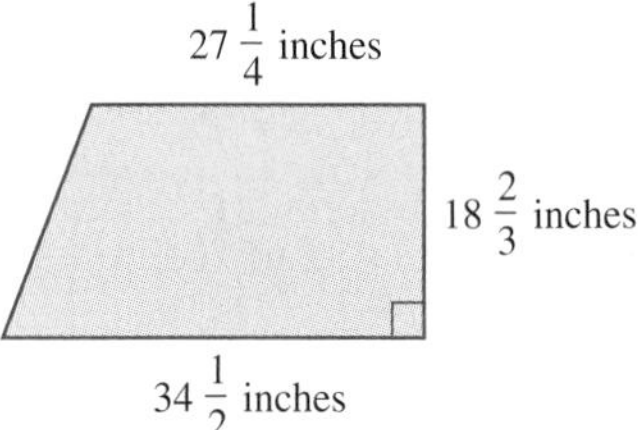

4

Decimals

As soon as we learn about pennies, dimes, and dollars, we are learning to work with decimal numbers. Decimal numbers are the numbers used in metric measurements and are used to represent irrational numbers. One important irrational number introduced in this chapter is pi, π.

In this chapter, we will learn the arithmetic of decimals. Our knowledge of multiples of ten and exponents will be used to learn scientific notation. We will solve word problems and extend our skills in converting measurements. We will learn about square roots and the Pythagorean theorem.

CHAPTER OUTLINE

Section 4.1 Introduction to Decimal Numbers

LEARNING TOOLS

CD-ROM SSM VIDEO

Sneak Preview

Decimal numbers are the numbers of our monetary system and the metric system. They extend the place value system of integers to the right of the decimal point.

After completing this section, you should be able to:

1) Write the word name of a decimal number.

2) Change a decimal number to an equivalent fraction or mixed number.

3) Compare the size of decimal numbers.

4) Round a decimal number to a specified place value.

5) Add or subtract decimal numbers.

6) Estimate a sum or difference with decimal numbers.

7) Solve word problems that include decimal numbers.

DISCUSSION

Decimals, Place Value, and Equivalent Fractions

Decimal numbers extend the place value system of integers to the right of the decimal point. A decimal point is used to separate the integer part from the fractional part of the number. Each place value to the right is ten times smaller than the previous place value.

Decimal point (between ONES and TENTHS)

THOUSANDS	HUNDREDS	TENS	ONES	TENTHS	HUNDREDTHS	THOUSANDTHS
1,000	100	10	1	$\frac{1}{10}$	$\frac{1}{100}$	$\frac{1}{1,000}$
10^3	10^2	10^1	10^0	$\frac{1}{10^1}$	$\frac{1}{10^2}$	$\frac{1}{10^3}$

Place value is used to write word names for decimals. When writing word names, the word "and" is reserved for the decimal point.

Example 1 **Write the word name for each decimal number.**

2.854 *two and eight hundred fifty-four thousandths*

−9.82 *negative nine and eighty-two hundredths*

0.3 *three tenths*

0.0071 *seventy-one ten-thousandths*

CHECK YOUR UNDERSTANDING

Write the word name for each decimal.

1. 3.97 **2.** 0.5 **3.** −2.915 **4.** 0.03

Answers: 1. three and ninety-seven hundredths; 2. five tenths; 3. negative two and nine hundred fifteen thousandths; 4. three hundredths.

We can compare the value of decimals by graphing the decimals on a number line. The smaller the decimal number, the further to the left it is found on the number line.

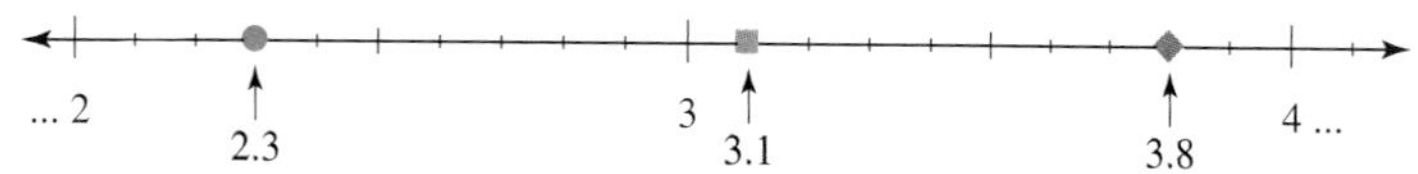

Since 2.3 lies to the left of 3.1, which lies to the left of 3.8, $2.3 < 3.1 < 3.8$.

As with integers, a zero indicates that a place value is empty. Adding zeros at the end of a decimal to the right of the decimal point does not change the value of the decimal. For example, $126.5 = 126.50$. However, adding zeros at the end of a number to the left of the decimal point does change its value; for example, $126.5 \neq 1{,}260.5$.

Sometimes zeros are used to show that an instrument or machine can measure to a certain place value. If a scale can determine the mass of an object to the thousandths place, the mass may be reported as 50.000 grams rather than just 50 grams. Here zeros are being used to indicate the **precision** of a measurement.

CHECK YOUR UNDERSTANDING

Place a $<$, $>$, or $=$ between the numbers to make a true statement.

1. 3.467 3.458 **2.** 3.999 3.99 **3.** −0.8 −0.6

Answer the questions.

4. Which is the smaller dose of medication: 0.134 grams or 0.5134 grams?

5. Which is the bigger wire: a wire with a diameter of 1.25 millimeters or a wire with a diameter of 1.255 millimeters?

Answers: 1. $>$; 2. $>$; 3. $<$; 4. 0.134 grams; 5. 1.255 millimeters

DISCUSSION

Rational Numbers

A **rational number** is any number that can be written as a fraction or as a mixed number. **Terminating decimals** and **repeating decimals** are rational numbers. A terminating decimal has a finite number of place values after the decimal point. For example, 0.2 has one place value after the decimal point; −2.94578 has five place values after the point. Both of these numbers can be rewritten as fractions and are rational numbers.

Repeating decimals are decimal numbers that have a neverending repeating pattern of digits. For example, if we divide 1 by 3, the quotient is 0.3333..., and the threes never stop. To indicate that the three repeats, we place a bar above the repeating digit or pattern of digits: $1 \div 3 = 0.333\ldots = 0.\overline{3}$. Similarly, if we divide 1 by 11, the quotient is 0.090909... and can be written $0.\overline{09}$. Repeating decimals can be written as fractions and are rational numbers.

Some decimal numbers continue on forever, never stopping and never repeating. These are conveniently called **nonrepeating, nonterminating decimals.** These numbers cannot be written as fractions. They are elements of the set of **irrational numbers.** If we measure the circumference of any circle (the distance around) and divide it by the diameter of that circle (the distance across through the center), the quotient is an irrational number called **pi.** The first few digits of pi are 3.14159265359..., but it is a number that goes on forever without repeating in a pattern.

All the rational numbers combined with all the irrational numbers form the set of numbers called the **real numbers.**

PROCEDURE

To rewrite a terminating decimal as a fraction:

1. Write the number without the decimal point as the numerator of the fraction.
2. Identify the place value of the last digit to the right. This place value determines the denominator of the fraction.
3. Simplify the fraction to lowest terms.

Example 2 **Rewrite 0.43 as a fraction.**

The last digit is in the hundredths place. The denominator will be 100.

$$0.43 = \frac{43}{100}$$

Example 3 **Rewrite −15.711 as a fraction.**

The last digit is in the thousandths place. The denominator will be 1,000.

$$-15.711 = -\frac{15{,}711}{1{,}000}$$

PROCEDURE

To rewrite a finite decimal as a mixed number:

1. Write the number in front of the decimal point as the integer of the mixed number.
2. Write the number after the decimal point as the numerator of the fraction with the denominator determined by the place value of the last digit to the right.
3. Simplify the fraction to lowest terms.

Example 4 **Rewrite −15.711 as a mixed number.**

The last digit is in the thousandths place. The denominator of the fraction is 1,000.

$$-15.711 = -15\frac{711}{1{,}000}$$

CHECK YOUR UNDERSTANDING

Identify the decimal as a rational number or an irrational number.

1. $0.\overline{2}$ **2.** 0.21678276687... **3.** $0.\overline{512}$ **4.** 0.512

Rewrite as an equivalent proper fraction or mixed number. Simplify to lowest terms.

5. 0.7 **6.** −0.5 **7.** 2.3 **8.** −10.1

Answers: 1. rational; 2. irrational; 3. rational; 4. rational; 5. $\frac{7}{10}$; *6.* $-\frac{1}{2}$; *7.* $2\frac{3}{10}$; *8.* $-10\frac{1}{10}$.

DISCUSSION

Rounding Decimals

Rounding approximates a number to a particular place value. We round decimal numbers in a similar way as we round integers. However, there is one additional step.

PROCEDURE

To round a decimal number:

1. If the number is negative, work with the absolute value of the number.
2. Identify the place value to which you are rounding. This is the rounding digit.
3. If the digit immediately to the right of the rounding digit is less than 5, the rounding digit is not changed. If this digit is greater than or equal to 5, increase the rounding digit by 1.
4. If the rounding digit is to the left of the decimal point, replace all digits to the right of the rounding digit by 0 and stop at the decimal point. If the rounding digit is to the right of the decimal point, omit all of the digits to the right of the rounding digit.
5. If the original number was negative, replace the negative sign.

Example 5 **Round 5.0378 to the hundredths place.**

The rounding digit is 3. The digit immediately following is 7, which is greater than or equal to 5. So, increase the rounding digit by 1 to 4. Since the rounding digit is to the right of the decimal point, omit all the digits after the rounding digit.

5.0378 rounded to the hundredths place is 5.04.

Example 6 **Round −9.4921 to the nearest thousandth.**

The rounding digit is 2. The digit immediately following is 1, which is less than 5. So, keep the rounding digit. Since the rounding digit is to the right of the decimal point, omit all the digits after the rounding digit.

−9.4921 rounded to the nearest thousandth is −9.492.

Example 7 **Round 36.0489 to the tenths place.**

The rounding digit is 0. The digit immediately to the right of the rounding digit is 4. Since $4 < 5$, the rounding digit is not changed. Since the rounding digit is to the right of the decimal point, omit all digits to the right of the decimal point.

36.0489 rounded to the tenths place is 36.0. (Do not remove the final 0 even though 36 is equivalent to 36.0. The 0 shows that the number was rounded to the tenths place, not the ones place.)

CHECK YOUR UNDERSTANDING

Round each decimal number to the indicated place value.

1. 99.0271; tenths

2. 99.0271; hundredths

3. 99.0271; tens

Answers: 1. 99.0; 2. 99.03; 3. 100.

DISCUSSION

Addition and Subtraction of Decimal Numbers

The procedure used for addition and subtraction of integers is used for decimal numbers. However, the decimal points must be aligned to ensure that digits in the same place value are combined.

PROCEDURE

To add or subtract decimal numbers:

1. All the numbers must have the same number of digits to the right of the decimal point. If necessary, write additional zeros at the end of the numbers.
2. Write the numbers in a vertical column, aligning the decimal points.
3. Add or subtract following the procedure for adding or subtracting integers.
4. Place the decimal point in the sum or difference directly below its position in the numbers being added or subtracted.

Sometimes one of the numbers being added does not have the same number of digits to the right of the decimal point. Because the value of the number is not changed, we can write additional zeros at the right end of the number.

Example 8 **Evaluate 4.027 + 3.2 + 8.513.**

$$\begin{array}{r} {\scriptstyle 1} \\ 4.027 \\ 3.200 \\ +8.513 \\ \hline 15.740 \end{array}$$

Two zeros are written at the end of 3.2 so that it has three digits following the decimal point.

Just as with integers, we can combine measurements that have the same units.

Example 9 **Evaluate 5.04 grams − 3.29 grams.**

$$\begin{aligned} &5.04 \text{ grams} - 3.29 \text{ grams} \\ &= (5.04 - 3.29) \text{ grams} \\ &= 1.75 \text{ grams} \end{aligned}$$

$$\begin{array}{r} \scriptstyle 4\ 9\ 14 \\ \not{5}.\not{0}\not{4} \\ -3.29 \\ \hline 1.75 \end{array}$$

Order of operations applies to decimal numbers. When we add or subtract three or more decimal numbers, we work from left to right.

Example 10 **Evaluate −0.926 + (−2.1) − 8.57.**

$$\begin{aligned} &-0.926 + (-2.1) - 8.57 \\ &= -3.026 - 8.57 \quad \textit{Add or subtract left to right.} \\ &= -11.596 \end{aligned}$$

Since digital equipment usually uses decimal numbers, applications involving technology also use decimal numbers.

Example 11 **Charlotte has a digital indoor/outdoor thermometer mounted in her kitchen window. The outdoor temperature reads −3.8°F. The indoor temperature reads 67.5°F. To the nearest degree Fahrenheit, what is the difference between the indoor and outdoor temperatures?**

I know . . .

d = Difference in temperature

67.5°F = Indoor temperature

−3.8°F = Outdoor temperature

Equations

Difference in temperature = Indoor temperature − Outdoor temperature

$$d = 67.5°\text{F} - (-3.8°\text{F})$$

Solve

$$\begin{aligned} d &= 67.5°\text{F} - (-3.8°\text{F}) \\ &= 67.5°\text{F} + 3.8°\text{F} \\ &= 71.3°\text{F} \\ &\approx 71°\text{F (rounded to the nearest degree)} \end{aligned}$$

$$\begin{array}{r} \scriptstyle 1\ 1 \\ 67.5 \\ +\ 3.8 \\ \hline 71.3 \end{array}$$

Check

To find the difference between the two temperatures, we can add the total number of degrees of separation. 67.5°F is about 68°F and −3.8°F is about −4°F. 68°F + 4°F = 72°F, which is reasonably close to our exact answer.

Answer

The indoor and outdoor temperatures differ by 71°F.

CHECK YOUR UNDERSTANDING

1. Evaluate $3.4 + 5.87$.

2. Evaluate 4.67 grams − 3.89 grams.

3. Evaluate $-2.45 + 3.2 - (-0.971)$.

4. At the beginning of a trip, the odometer in Gracie's car read 45,872.3 miles. At the end of the trip the odometer read 47,209.8 miles. To the nearest mile, how many miles did Gracie drive on her trip?

Answers: 1. 9.27; 2. 0.78 grams; 3. 1.721; 4. 1,338 miles.

DISCUSSION

Estimating Sums and Differences

The same procedure used to estimate sums and differences of integers can be used to estimate the sums and differences of decimals.

PROCEDURE

To estimate a sum or difference:

1. Round each number to the largest place value that occurs in any of the numbers.
2. Add or subtract the rounded numbers.

Example 12 **Estimate the sum of \$2.30, \$0.03, \$0.56, and \$0.92.**

The largest place value in any of the numbers is the ones place.

$\$2.30 + \$0.03 + \$0.56 + \$0.92$

$\approx \$2 + \$0 + \$1 + \$1$ *Round each number to the ones place.*

$= \$4$

Example 13 **Estimate the total rainfall per year at Lake of the Ozarks.**

RAINFALL PER MONTH IN INCHES AT LAKE OF THE OZARKS, MISSOURI

JAN	FEB	MARCH	APRIL	MAY	JUNE	JULY	AUG	SEPT	OCT	NOV	DEC
1.82	1.79	3.06	3.67	4.69	4.54	2.40	3.58	4.23	2.89	2.27	1.88

I know . . .
r = Estimated total rainfall

RAINFALL PER MONTH IN INCHES AT LAKE OF THE OZARKS, MISSOURI (ROUNDED TO THE ONES PLACE)

JAN	FEB	MARCH	APRIL	MAY	JUNE	JULY	AUG	SEPT	OCT	NOV	DEC
2	2	3	4	5	5	2	4	4	3	2	2

Equations
Estimated total rainfall = Sum of rounded rainfall for all 12 months
$$r = (2 + 2 + 3 + 4 + 5 + 5 + 2 + 4 + 4 + 3 + 2 + 2) \text{ inches}$$

Solve
$r = (2 + 2 + 3 + 4 + 5 + 5 + 2 + 4 + 4 + 3 + 2 + 2)$ inches
$= 38$ inches

Check
The highest rainfall is 5 inches per month. If there was this much rain every month, the total would be 5 inches $\cdot$ 12 = 60 inches. The lowest rainfall is 2 inches per month. If there was this little rain every month, the total would be 2 inches $\cdot$ 12 = 24 inches. The exact answer should fall between these two extremes of 24 inches and 60 inches. Since 38 inches is between these two extremes, it is a reasonable answer.

Answer
The estimated total annual rainfall at Lake of the Ozarks is 38 inches.

CHECK YOUR UNDERSTANDING

1. Estimate the total cost of these groceries: bread (\$2.39), chicken (\$4.85), cereal (\$3.20), gum (\$0.65), soda (\$0.89), and milk (\$1.59).

2. In 1991, the Duke versus Kansas NCAA basketball championship earned a 22.7 point Nielsen television rating. In 1996, the Kentucky versus Syracuse game earned an 18.3 point Nielsen television rating. Estimate the difference in television ratings of the two games by rounding to the nearest whole number.

3. Estimate the difference between 0.0582 seconds and 0.024 seconds.

4. Estimate the perimeter of a rectangle with a length of 3.092 meters and a width of 1.963 meters.

Answers: 1. \$14; 2. 5 points; 3. 0.04 second; 4. 10 meters.

EXERCISES SECTION 4.1

Identify the place value of 4 in the number.

1. 25.14 **2.** −401.69 **3.** 0.0004

Identify the place value of 8 in the number.

4. 8,000.01 **5.** −0.85 **6.** 0.08

Write the word name for the number.

7. Maurizio is writing a check for tuition for $1,459.09. What word name will he write on the check?

8. Write "fifty-three and two hundred thirty-nine ten-thousandths" as a decimal number.

Rewrite the decimal as a fraction in lowest terms.

9. 0.75 **10.** −0.0008 **11.** 0.25

12. 0.125 **13.** −0.60

14. A college teacher suggested that her students memorize the fractions that are equivalent to 0.125, 0.625, 0.20, 0.25, 0.5, and 0.75. Rewrite each of these decimals as an equivalent fraction.

Write the decimal as an *improper* fraction in lowest terms.

15. 1.56 **16.** −1.32 **17.** 100.35

Rewrite the decimal as a *mixed number* in lowest terms.

18. −1.32

19. −64.002

20. 100.35

Use a greater than (>) or less than (<) symbol to write a true statement that compares the two account balances.

	BANK OF AMERICA	STATEMENT	FIRST SECURITY BANK
21.	\$121.35		\$121.30
22.	\$121.90		\$121.99
23.	−\$121.00		\$121.00

Insert a > or < between the numbers to make a true statement.

24. 0.09 0.009

25. −0.09 −0.009

26. 0.09 −0.009

27. 0.99 0.099

28. 0.1 −0.9

29. 3.01 3.10

Round each measurement to the hundredths place.

30. 48.055 grams

31. −48.055 degrees

32. 43.7008 meters

33. \$1,066.999

Round each measurement to the tenths place.

34. \$92.51

35. 71.33 feet

36. 88.99 gallons

37. 0.01 liters

Evaluate.

38. \$27.86 + \$11.55

39. −\$27.96 + \$11.55

40. 77.16 grams − 35.81 grams

41. 100.010 milliliters + 0.99 milliliters

42. $0.24 - 0.024$

43. $903 + 0.55$

44. $\$1,672,070.99 - \$3,663.45$

45. $13.4 + 2.4 - 3.1 + 10.01$

46. $15 - 0.02 + 8 - 0.03$

47. $5.3 - 1.9 - 4.6 - 9$

Round the measurement to the nearest *tenth.*

48. 52.905 meters

49. 56.991 grams

Round the measurement to the nearest *hundredth.*

50. 52.905 meters

51. 56.991 grams

Round the measurement to the nearest *ten.*

52. 52.905 meters

53. 56.991 grams

Estimate the sum or difference.

54. $111.92 + 396.05$

55. $11.92 + (-0.05)$

56. $11.92 + 96.05$

57. $-18 + 0.03$

Solve the word problem using the five-step strategy.

58. Bud is keeping a mental estimate of his charges as he fills his shopping cart on the Internet at Amazon.com. He has two CDs at $14.99 each, a book at $6.34, a book at $17.50, a poster at $8.50, and a book at $23.95. Estimate the sum of his charges.

59. Kenny is mailing a letter that he wants only delivered to his mother and he needs a return receipt showing to whom and when the letter was delivered. The postage on the letter is $0.95. Use the information in the table to calculate the total cost of the letter.

Fee for certified mail (additional to required postage)	$1.10
Delivery restricted to one person	$1.75
Return receipts showing delivery to whom and when	$1.10
Return receipts showing delivery to whom, when, and the delivery address	$1.50

60. The annual operating budgets in millions of dollars of museums in the Puget Sound Area are listed in the table. What is the total annual operating budget of these museums?

Museum	Budget	Museum	Budget
Northwest Trek Wildlife Park	\$2.1	Pacific Science Center	\$10.2
Washington Park Arboretum	\$1.8	Museum of Flight	\$6.9
Point Defiance Zoo & Aquarium	\$6.2	The Seattle Aquarium	\$3.9
Washington State Historical Society	\$3.6	Woodland Park Zoo	\$13.4
The Burke Museum of Natural History and Culture	\$2.2	Tacoma Art Museum	\$2.0
Museum of History & Industry	\$3.6	Seattle Art Museum	\$9.7
Henry Art Gallery, Faye G. Allen Center for the Visual Arts	\$2.1	The Children's Museum	\$1.5
The Whatcom Museum of History & Art	\$1.5	Frye Art Museum	\$1.5
Wing Luke Asian Museum	\$0.8	Bellevue Art Museum	\$1.2
Nordic Heritage Museum	\$0.5	Bloedel Reserve	\$0.9
Center on Contemporary Art	\$0.3	Center for Wooden Boats	\$0.4
Museum of Northwest Art	\$0.3	Northwest Railway Museum	\$0.3
Skagit County Historical Museum	\$0.2		

61. In 1997, the average consumption in gallons of ice cream eaten per person in the United States was 5.4 gallons. In France, the average consumption was 1.9 gallons. How many more gallons, on average, do Americans eat of ice cream per year than the French?

© *Adam Smith/Superstock*

62. According to the U.S. State Department, the leading U.S. weapons importers in fiscal year 1997 were (in millions) Egypt ($1,001.0), South Korea ($905.7), Spain ($839.5), Saudi Arabia ($758.2), Japan ($670.5), Taiwan ($626.5), the United Kingdom ($616.9), and Israel ($557.2). What is the total dollar amount of U.S. weapons imported by these countries in 1997?

63. Potassium is needed for proper nerve and muscle action in the human body. The normal range of potassium is 3.5 to 5.5 milliequivalents per liter. Kaitlin's grandmother has a potassium level of 6.01 milliequivalents. How far above the normal range is this potassium reading?

64. Calcium is the most abundant mineral in the body, involved in bone metabolism, protein absorption, fat transfer, muscular contraction, transmission of nerve impulses, blood clotting, and cardiac function. The normal adult range is from 8.5 to 10.3 milliequivalents per deciliter. The optimal adult reading is 9.4 milliequivalents per deciliter. What is the difference between the high end of the normal adult range and the optimal reading?

65. In 1995, American consumers spent \$7.1709 billion on books for adults and \$2.3266 billion on books for children. How much more money was spent on books for adults?

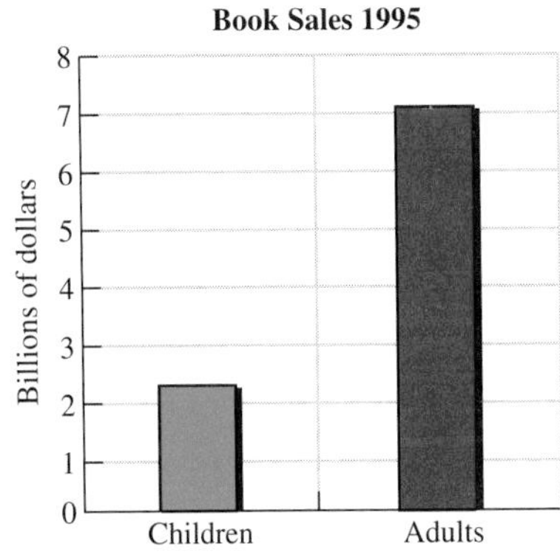

66. In 1992, the top U.S. market for book sales was the Los Angeles–Long Beach, California, region with sales of \$383.902 million. New York, New York, was second with sales of \$359.716 million. How much less was spent on books in the New York region?

67. In 1996, Canadian cable television had 7.867 million subscribers. This was an increase from 1995 when there were 7.791 million subscribers. What was the increase?

ERROR ANALYSIS

68. Problem: Round 34.827 to the nearest tenth.

Answer: The number in the tenths place is 3. The number to its right is 4. Since 4 is less than 5, leave the number in the tenths place alone and replace the other digits with 0.
34.827 rounds to 30.000.

Describe the mistake made in solving this problem.

Redo the problem correctly.

69. Problem: Evaluate 3.12 + 14.23.

Answer:

$$\begin{array}{r} 3.12 \\ +14.23 \\ \hline 45.43 \end{array}$$

Describe the mistake made in solving this problem.

Redo the problem correctly.

REVIEW

70. Multiply 24 · 36.

71. Find the product of 45 and 8.

72. Estimate the product of 2,367 and 371.

Section 4.2 Multiplication of Decimal Numbers

LEARNING TOOLS

CD-ROM SSM VIDEO

Sneak Preview

Multiplying wages per hour by the hours worked in a week may involve multiplication of decimal numbers. Buying more than one of anything often requires multiplying decimal numbers.

After completing this section, you should be able to:

1) Multiply decimal numbers.

2) Estimate the product of decimal numbers.

3) Solve word problems that include decimal numbers.

DISCUSSION

Multiplication of Decimals

When we multiply two whole numbers, we are doing repeated additions. For example, $3 \cdot 4 = 3 + 3 + 3 + 3$. The product and sum are both 12. When we multiply a decimal by a whole number, we are doing repeated additions of the decimal. For example, $3.14 \cdot 2 = 3.14 + 3.14$. The product and sum are both 6.28. We can modify the procedure used to multiply integers to multiply decimal numbers. By counting the digits that are to the right of the decimal point in each factor, we can correctly place the decimal point in the product.

PROCEDURE

To multiply two decimals:

1. For this step, ignore the decimal points. Use the multiplication procedure to find the product of the resulting integers.
2. Count the total number of digits to the right of the decimal point in both of the original factors.
3. Place the decimal point in the product from Step 1 so that the product has the same number of digits to the right of the decimal point as determined in Step 2. If necessary, write more zeros at the end of the number to allow correct placement of the decimal point.

Example 1 **Find the product of 5.8 and 2.5.**

5.8	*5.8 has 1 digit to the right of the decimal point*	58	*The product*
×2.5	*2.5 has 1 digit to the right of the decimal point*	×25	*of 58 and 25,*
	Product will have 2 digits to the right of the decimal point	290	*ignoring the*
		1160	*decimal points,*
		1450	*is 1450.*

We place the decimal point so that the product has two digits to the right of the decimal point: 14.50. The product of 5.8 and 2.5 is 14.50.

Decimals can be positive or negative. The rules for signs of products of integers are also used for the sign of the product of two decimals (see Section 2.4).

Example 2 **Evaluate (4.5)(−1.69).**

1.69 has 2 digits to the right of the decimal point	169
4.5 has 1 digit to the right of the decimal point	× 45
Product will have 3 digits to the right of the decimal point	845
	6760
	7605

We place the decimal point so that the product has three digits to the right of the decimal point: 7.605. The product of a positive number and a negative number is negative. So, $(4.5)(-1.69) = -7.605$.

Products of decimals can be checked using estimation. The same rules for estimating products of integers applies to estimating products of decimals.

PROCEDURE

To estimate the product of decimals:

1. Round each number to its own highest place value.
2. Multiply the rounded numbers.

Example 3 **Estimate the product of 4.5 and −1.69.**

4.5 rounds to 5 and −1.69 rounds to −2.

$4.5(-1.69)$

$\approx 5(-2)$

$= -10$ *This is close to the exact answer of −7.605 found in Example 2.*

U.S. money is based on decimal numbers. Applications that involve money will require decimal arithmetic including multiplication.

Example 4 **A warehouse is operated by a port district as a small business incubator. Reduced rents help small businesses survive low profits at start-up. Port commissioners charge $0.22 per square foot rent for the first year of operation. A paperboard recycling operation wishes to rent 40,000 square feet of the warehouse for a year. Calculate the rent for one year.**

I know . . .

r = Total rent

$0.22 = Price per square foot

40,000 = Number of square feet rented

Equations

Total rent = Number of square feet rented · Price per square foot

$r = (40{,}000)(\$0.22)$

Solve

$r = (40{,}000)(\$0.22)$

40,000 has no digits to the right of the decimal point	40000
0.22 has 2 digits to the right of the decimal point	× 22
Product will have 2 digits to the right of the decimal point	80000
	800000
	880000

The decimal point is placed so that the product has two digits to the right of the decimal point: 8800.00.

Check

\$0.22 is about one quarter of a dollar. If the rent were \$1 per square foot, the total cost would be \$40,000. If this is divided by four, we should be close to the exact answer. \$40,000 ÷ 4 = \$10,000, which is reasonably close to the exact answer.

Answer

The rent for the warehouse will be \$8,800 per year.

When earning or spending money, we often make decisions or purchases based on estimates.

Example 5 **Jose earns \$9.35 per hour as a teaching assistant at a preschool. He receives a paycheck every two weeks. If he works 37.5 hours each week, estimate his gross pay per paycheck.**

I know . . .

p = Estimate of pay

2 = Weeks

$37.5 \approx 40$ = Estimated hours per week

$\$9.35 \approx \$9$ = Estimated pay per hour

Equations

Estimate of pay = (Estimated hours per week)(Estimated pay per hour) (Weeks)

$$p = (40)(\$9)(2)$$

Solve

$$\begin{aligned} p &= (40)(\$9)(2) \\ &= (\$360)(2) \\ &= \$720 \end{aligned}$$

Check

This estimate can be checked estimating still further. Use \$10 an hour for pay instead of \$9 an hour. If Jose earned \$10 an hour and worked 80 hours in two weeks, he would earn \$800, reasonably close to this estimate.

Answer

Jose's gross pay per paycheck will be about \$720.

CHECK YOUR UNDERSTANDING

1. Multiply (3.7 meters)(3.14).

2. Multiply (3.14)(0.2 centimeters)(0.2 centimeters).

3. Estimate the product of (3.14)(0.69 meter)(0.69 meter)(1.52 meters).

Answers: 1. 11.618 meters; 2. 0.1256 centimeters^2; 3. 2.94 meters^3.

Just as with integers, we follow the order of operations with decimal numbers. If a problem involves both addition or subtraction and multiplication, we complete the multiplication before doing the addition or subtraction. We always work from left to right.

Example 6 **Annie has scraped the old paint off a shed and will now prime the bare wood with a stain blocking primer and then paint with a good latex exterior paint. She needs 6 gallons of stain-blocking primer and 4 gallons of paint. The primer costs $18.95 a gallon. The paint costs $15.95 a gallon. What is the total cost for the primer and the exterior paint?**

I know . . .

c = Total cost

6 = Gallons primer

4 = Gallons paint

$18.95 = Cost per gallon of primer

$15.95 = Cost per gallon of paint

Equations

Total cost = (Gallons primer)(Cost primer) + (Gallons paint)(Cost paint)

c = (6)($18.95) + (4)($15.95)

Solve

c = (6)($18.95) + (4)($15.95)

= $113.70 + $63.80 *Multiply from left to right.*

= $177.50 *Add.*

Check

Round the values in the equation: $c \approx$ (6)($20) + (4)($20).

Solving, c = $120 + $80 = $200. This is close to our exact answer of $177.50. The answer is reasonable.

Answer

The total cost of primer and exterior paint is $177.50.

CHECK YOUR UNDERSTANDING

1. 2($1.60) + 4($9.24)

2. (10.5)(6.1) − (3.2)(9)

3. 12(6.00 inches) + 15(2.5 inches) − 8(1.75 inches)

Answers: 1. $40.16; 2. 35.25; 3. 95.5 inches.

EXERCISES SECTION 4.2

Evaluate.

1. (31)(64)

2. (3.1)(6.4)

3. $(-0.31)(6.4)$

4. (0.31)(0.64)

5. (0.031)(0.064)

6. $(-31)(0.064)$

7. $(2.35)(-1.7)$

8. $(-0.75)(-8.00)$

9. (99)(0.10)

10. 34.6(2.9)

11. $-7.08(8.3)$

12. $-3.2(-0.0065)$

Estimate the product.

13. 23.5(9.3)

14. $-0.7(-0.87)$

15. 2.487(3.9654)

16. $-0.008(0.06)$

17. $12.9(3.2)(-0.005)$

18. $-0.03(0.0067)$

19. 5.2(12.5)

20. $-13.678(0.058)$

21. 0.00789(0.0369)

Use the five-step problem-solving strategy.

22. According to the American Egg Board, one large egg has 215 milligrams of cholesterol and 1.5 grams of saturated fat. The total fat in an egg is 4.5 grams. How many total grams of fat are there in a dozen eggs?

23. Scruples Salon repairs acrylic nails for $3.50 per nail. Jill needs seven nails repaired. How much will this cost?

24. Kyle's pickup gets 17.2 miles per gallon on his commute to school. He has 25 gallons of gas in his truck. How many miles can he drive?

25. A real estate broker posts property tax rates on his Web page so that potential buyers can calculate their property tax bills. In Princeton Township, the rate is $2.02 per $100.00 of assessed property evaluation. How much would the property taxes be on a house with an assessed value of $124,000?

26. One inch is equivalent to 2.54 centimeters. Knowing that 12 inches equals one foot, how many centimeters are there in a foot?

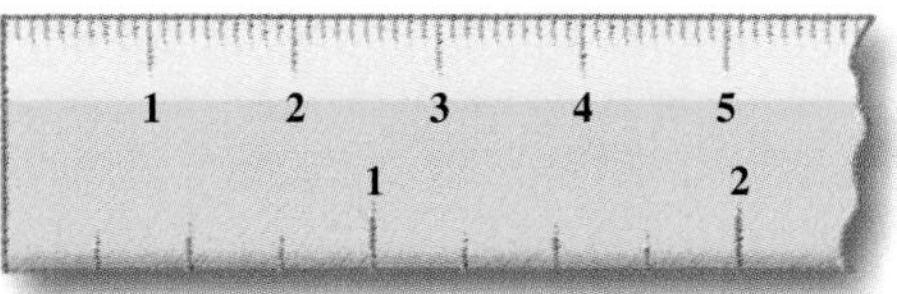

27. In August 2001, the U.S. Department of Agriculture sold 7,522.5 tons of sugar to ethanol producers. J. R. Simplot Company in Idaho purchased 7,500 tons; International Ingredient Corporation in Texas purchased 22.5 tons. There are 2,000 pounds in a ton. How many pounds of sugar did the U.S. Department of Agriculture sell?

28. Detasseling corn to prevent improper pollination is often done by teenagers. Minimum ages for detasseling vary from 10 in South Dakota to 14 in Iowa. In Nebraska, lawmakers kept the minimum age at 12 and limit detasselers under 18 to 9 hours per day and 48 hours per week. A beginning detassler makes about $5.15 an hour. How much will a beginning detassler make for a 48-hour week of work?

29. "Retail beef prices are reflecting the summer grilling season and edging down after the past several months of hefty increases. Last year, petite sirloin steak—the most popular cut of beef at Rosauers—was selling for $3.49 a pound, according to meat department manager Dale Riener. Last week that price was up to $4.69 a pound. This week, it's back down to $3.98 a pound." (Source: *Lewiston Morning Tribune*, 8/01/01.) Calculate the cost of buying 2.5 pounds of petite sirloin steak at each of the three different prices.

30. Wyeth-Ayerst, a pharmaceutical company, charges \$5.29 for a single dose of vaccine for a flu shot. A clinic orders 5,357 doses. What will be the charge for this order? (Source: *USA Today,* 7/30/01).

31. The *Portland Oregonian* charges \$1.50 per line per day for garage sale classified ads. What is the cost of a garage sale ad with six lines that runs for a total of 5 days?

32. A fencing and decking supplier charges \$15.50 per linear foot for the materials for a white vinyl privacy fence that is 6 feet high. If the supplier also delivers the materials and sets the posts, the charge is \$20.50 per linear foot. If the fence is completely installed, the charge is \$26.99 per foot. Wendi needs 125 feet of fence.

a. Calculate the cost of buying the materials for the fence.

b. Calculate the cost of buying the materials and having the supplier also deliver the materials and set the posts.

c. What is the difference in cost of buying the materials for the fence and having the fence completely installed?

33. One serving of Veracruz-style red snapper includes 1 fish fillet, $\frac{1}{2}$ cup of salsa, and 1 lime wedge. This contains 3.2 grams of fat, 28.2 grams of protein, and 14.5 grams of carbohydrate. Malik ate 2 servings. How many grams of carbohydrate did he eat? (Source: *Cooking Light,* August 2001.)

34. Jeno's Crisp 'n' Tasty Pizza cost $0.79 each; Good Day Mac 'n' Cheese cost $0.33 each; Hunt's Manwich Sauce cost $1.50 each; a package of hamburger buns costs $0.69 each. If Tom buys 5 pizzas, 15 Mac 'n' Cheese, 8 Manwich Sauces, and a package of hamburger buns, what is his total cost?

35. There are about 2.2 pounds in a kilogram. An antidrug operation led to seizures of 116,188 kilograms of cocaine, 748 kilograms of crack, and 872,777 kilograms of marijuana. What is the weight of these seizures in pounds? (Source: *TIME,* July 30, 2001.)

36. A catalog sells cotton long sleeve T-shirts for \$6.50. Barb is the coordinator of a century ride (100 miles) for a bike club and needs 75 shirts for printing. Calculate the cost of the shirts.

37. Stephanie has found a used car to buy that costs \$16,000. If she pays \$1,000 in cash and borrows the rest in a 4-year loan at 5 percent interest, her monthly car payment will be \$345.55. If she pays \$4,000 in cash and borrows the rest in a 4-year loan at 5 percent interest, her monthly car payment will be \$276.35. How much less will Stephanie spend on the car if she pays \$4,000 down in cash?

38. In 2000, an average American ate 82.1 pounds of chicken and 125.1 pounds of red meat per year. In 1980, an average American ate 46.6 pounds of chicken and 136.7 pounds of red meat per year. For a family of three, calculate the increase in annual consumption of chicken from 1980 to 2000.

39. Older toilets use a minimum flush of 3.5 gallons per flush. Ultra low-flush toilets use 1.6 gallons per flush. Assuming that a person uses the toilet at home an average of four times a day and that 35,000 households have an average of 3.2 people, calculate the minimum savings in water if all these households change to ultra low-flush toilets.

40. Write your own word problem that requires the multiplication of two decimal numbers. Solve this problem.

ERROR ANALYSIS

41. Problem: Evaluate 3.52(1.2).

Answer:

$$\begin{array}{r} 3.52 \\ \times 1.2 \\ \hline 704 \\ 352 \\ \hline 10.56 \end{array}$$

Describe the mistake in the answer to this problem.

Redo the problem correctly.

REVIEW

42. Use long division to calculate $864 \div 2$. Check the solution.

43. Explain how to check the quotient in a long division problem when the quotient has an integer and a remainder.

44. Use long division to calculate $864 \div 20$. Check the solution.

45. Use long division to calculate $864 \div 200$. Check the solution.

Section 4.3 Division of Decimal Numbers

LEARNING TOOLS

Sneak Preview

We divide decimals the same way we divide integers, with the addition of one extra step to place the decimal point in the quotient. The procedures used to simplify expressions and solve equations that contain integers can also be extended to decimal numbers.

After completing this section, you should be able to:

1) Divide a decimal by an integer or by another decimal.
2) Estimate a quotient of decimals.
3) Evaluate expressions that contain decimals following the order of operations.
4) Solve word problems that include decimals.

INVESTIGATION Division of Decimals

The long division procedure for integers can also be used to divide decimal numbers. However, we need to know where to put the decimal point in the quotient. The first case to consider is when the divisor is an integer.

EXAMPLE A: Calculate the quotient of 864 ÷ 2.

We are asking, "How many twos are contained in 864?"

$$\begin{array}{r} 432 \\ 2\overline{)864} \end{array}$$

The quotient is **432.** This is reasonable since $400 \cdot 2 = 800$.

1. Calculate the quotient of 216 ÷ 2 using long division. Explain why your answer is reasonable.

EXAMPLE B: Calculate the quotient of 86.4 ÷ 2.

We are asking, "How many twos are contained in 86.4?"

$$\begin{array}{r} 43.2 \\ 2\overline{)86.4} \end{array}$$

The quotient is 43.2. This is reasonable since $40 \cdot 2 = 80$.

2. Think about the quotient of 21.6 ÷ 2. About how many twos are contained in 21.6?

EXAMPLE C: Calculate the quotient of 0.864 ÷ 2.

We are asking, "How many twos are contained in 0.864?"

$$\begin{array}{r} 0.432 \\ 2\overline{)0.864} \end{array}$$

The quotient is 0.432. This is reasonable. Since 2 is larger than 0.864, there should be less than one 2 in 0.864.

3. Think about the quotient 0.216 ÷ 2. Is there more than one, exactly one, or less than one 2 contained in 0.216?

4. Using these examples, describe where the decimal point is placed in the quotient of a decimal number and an integer.

Evaluate. Show intermediate steps.

5. 52.8 ÷ 3

6. 5.28 ÷ 3

7. 0.528 ÷ 3

Answers: 5. 17.6; 6. 1.76; 7. 0.176.

The second case to consider is when the divisor is a *decimal number.*

EXAMPLE D: Calculate the quotient of 864 ÷ 0.2.

We are asking, "How many two-tenths are contained in 864?"

$$0.2\overline{)864.0}\quad \text{quotient: } 4320.$$

The quotient is 4,320. This is a reasonable quotient. There are 5 two-tenths in 1. Since we are asking how many two-tenths are contained in 864, the answer should be about 5 · 860 or 4,300.

8. Think about the quotient of 216 and 0.2. About how many two-tenths are contained in 216? Explain your thinking.

EXAMPLE E: Calculate the quotient of 86.4 ÷ 0.2.

We are asking, "How many two-tenths are contained in 86.4?"

$$0.2\overline{)86.4}\quad \text{quotient: } 432.$$

The quotient is 432.

9. Think about the quotient of 21.6 and 0.2. About how many two-tenths are contained in 21.6? Explain your thinking.

EXAMPLE F: Calculate the quotient of 0.864 ÷ 0.002.

We are asking, "How many two-thousandths are contained in 0.864?"

$$0.002\overline{)0.864}\quad \text{quotient: } 432$$

The quotient is 432.

10. Think about the quotient of 0.216 and 0.002. Is there more than one, exactly one, or less than one two-thousandths contained in 0.216?

11. When dividing a decimal number by a decimal number, where is the decimal point placed in the quotient?

Calculate the quotients. Mentally check the quotient for reasonability.

12. $62.05 \div 5$

13. $62.05 \div 0.5$

14. $0.6205 \div 0.05$

15. $0.3572 \div 0.4$

16. $0.001 \div 0.5$

17. $1 \div 0.5$

Answers: 12. 12.41; 13. 124.1; 14. 12.41; 15. 0.893; 16. 0.002; 17. 2.

DISCUSSION

Division of Decimals

The same basic long division procedure used to divide integers is used to divide decimals by a *whole number.* Instead of stopping at the decimal point and writing a remainder, we continue the process of dividing and bringing down numbers from the dividend. If necessary, we write additional zeros at the end of the dividend to allow continued division. If the divisor is a *whole number,* the decimal point in the quotient is placed directly above the decimal point in the dividend.

Example 1 **Evaluate 4.29 ÷ 3**

```
   1.43
 3)4.29
  -3
   1 2
  -1 2
     09
     -9
      0
```

Example 2 **Evaluate 45.3 ÷ 15**

```
     3.02
 15)45.30
   -45
     0 3
     -0
      30
     -30
       0
```

An additional zero is written at the end of 45.3 to allow the division to be completed.

A quotient either will terminate or will repeat. The repetition may not appear immediately, and the repetition may be more than one digit.

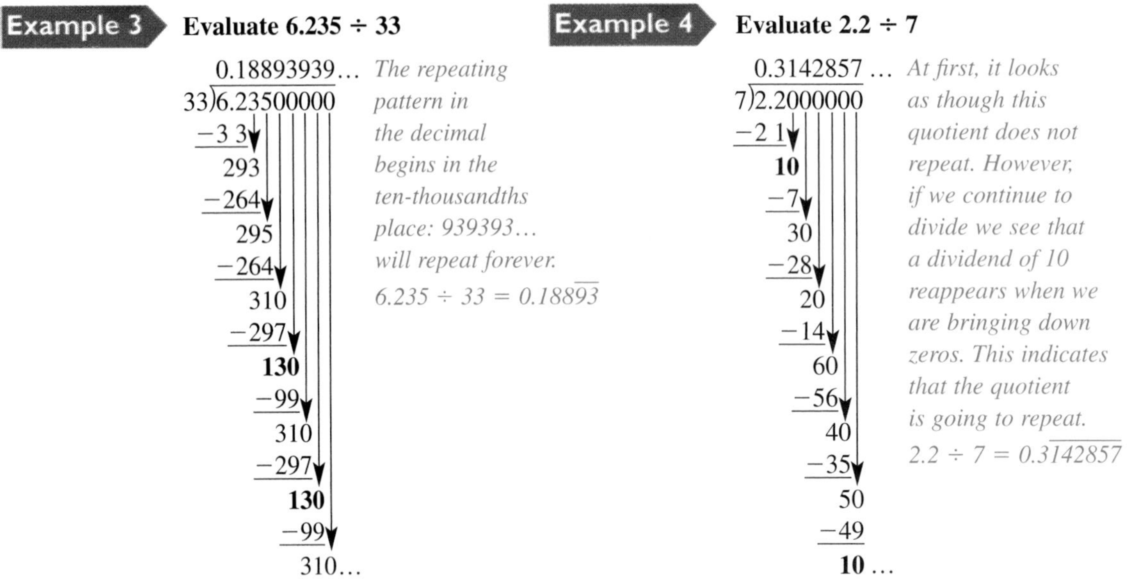

The long division procedure must be modified if the divisor is a *decimal.* The divisor must be multiplied by the power of ten that makes the divisor a whole number. Since the divisor is multiplied by a power of ten, the dividend must also be multiplied by the same power of ten.

A practical way to turn the divisor into a whole number is to count the number of digits to the right of the decimal point in the divisor and move the decimal point this many places to the right. Now move the decimal point the same number of places to the right in the dividend. Divide, remembering to place the decimal point in the quotient directly above the new position of the decimal point in the dividend.

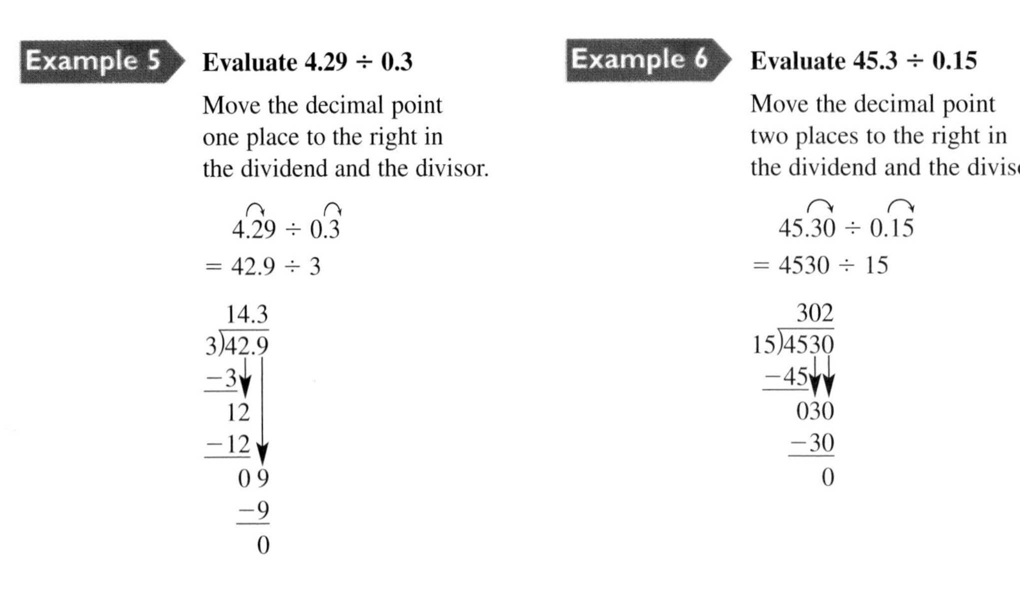

CHECK YOUR UNDERSTANDING

Divide.

1. $2.34 \div 1.2$

2. $-3.7 \div 0.03$

3. $2.34 \div 0.6$

Answers: 1. 1.95; 2. $-123.\overline{3}$*; 3. 3.9.*

DISCUSSION

Estimation of Quotients

In everyday life it is often more convenient to estimate a quotient rather than calculate an exact quotient. Quotients can be estimated by rounding each number to its own highest place value before dividing.

PROCEDURE

To estimate a quotient:

1. Round each number to its own highest place value.
2. Divide the rounded numbers.
3. Round this quotient, if necessary.

Example 7 **Estimate the quotient: 14.67 ÷ 3.81**

Round 14.67 to the tens place: 10. Round 3.81 to the ones place: 4

```
    2.5
 4)10.0
  -8 ↓
   2 0
  -2 0
     0
```

An extra zero is written after the decimal point in 10 to allow the division to be completed. The decimal point is placed in the quotient directly above the decimal point in the dividend.

An estimate of the quotient of $14.67 \div 3.81$ is 2.5.

Example 8 **Estimate the quotient: 784.963 ÷ 0.00753**

Round 784.963 to the hundreds place: 800.

Round 0.00753 to the thousandths place: 0.008.

The estimated quotient is 800 ÷ 0.008. To divide by the decimal, move the decimal point three places to the right in the dividend and the divisor.

$$\begin{aligned} &800 \div 0.008 \\ &= 800{,}000 \div 8 \quad \textit{This is whole number division.} \\ &= 100{,}000 \end{aligned}$$

The estimated quotient of 784.963 ÷ 0.00753 is 100,000.

Example 9 **Estimate the quotient of 15.784 and −0.6854**

15.784 rounds to 20.

−0.6854 rounds to −0.7.

$$\begin{aligned} &15.784 \div (-0.6854) \\ &\approx 20 \div (-0.7) \quad \textit{Move the decimal point in each number to the right one place value.} \\ &= 200 \div (-7) \\ &= 28.571\ldots \\ &\approx 30 \quad \textit{Round.} \end{aligned}$$

The final estimate can be rounded to any place value that seems reasonable. Either 29 or 30 are reasonable answers.

CHECK YOUR UNDERSTANDING

Estimate each quotient.

1. 27.3 ÷ 0.039

2. 0.065 ÷ 0.052

3. 0.072 ÷ 1.2

Answers: 1. 750; 2. 1.4; 3. 0.07

DISCUSSION

Order of Operations

We evaluate the operations in an expression that contains decimals following the order of operations.

PROCEDURE

To evaluate a numerical expression following the order of operations:

1. **Grouping:** Operations inside of grouping symbols are done first, following rules 2–4. Grouping symbols include parentheses ( ), brackets [ ], braces { }, absolute values | |, and fraction bars —.
2. **Exponents:** Evaluate all exponents and square roots from left to right.
3. **Multiplication or Division:** Do all multiplication and divisions from left to right.
4. **Addition or Subtraction:** Do all additions and subtractions from left to right.

Example 10 **Evaluate (20)(4.5) + (0.5)(9.80)(4.5)².**

$(20)(4.5) + (0.5)(9.80)(4.5)^2$

$= (20)(4.5) + (0.5)(9.80)\mathbf{(4.5)(4.5)}$ *Evaluate exponents; rewrite as multiplication.*

$= (20)(4.5) + (0.5)(9.80)(20.25)$

$= \mathbf{90} + (0.5)(9.80)(20.25)$ *Multiply or divide left to right.*

$= 90 + (\mathbf{4.9})(20.25)$ *Multiply or divide left to right.*

$= 90 + \mathbf{99.225}$ *Multiply or divide left to right.*

$= 189.225$ *Add or subtract left to right.*

Example 11 **A mole is a unit in chemistry that is used to measure a quantity of atoms or molecules. A mole of carbon dioxide molecules (CO_2) contains one mole of carbon atoms and two moles of oxygen atoms. A mole of carbon atoms has a mass of 12.011 grams. A mole of oxygen atoms has a mass of 15.994 grams. What is the mass of one mole of carbon dioxide?**

I know . . .

m = Mass of carbon dioxide

12.011 grams = Mass of carbon

15.994 grams = Mass of oxygen

Equations

Mass of carbon dioxide = Mass of carbon + 2 · Mass of oxygen

m = 12.011 grams + 2(15.994 grams)

Solve

m = 12.011 grams + 2(15.994 grams)

= 12.011 grams + 31.988 grams

= 43.999 grams

Check

A mole of carbon is about 10 grams; a mole of oxygen is about 20 grams.

The mass of carbon dioxide should be around $10 + 2 \cdot 20 = 50$, reasonably close to the exact answer.

Answer

The mass of one mole of carbon dioxide is 43.999 grams.

Example 12 **A Grohe faucet, imported from Germany, has a safety temperature shut-off valve to prevent scalding. Since it is made in a country that uses the metric system, the temperature settings are in degrees Celsius. A family with small children wants to set the safety temperature at 120°F. The formula for converting Fahrenheit temperatures into Celsius is $C = (F - 32) \div 1.8$. What is the equivalent temperature in degrees Celsius? Round the answer to the nearest degree.**

I know . . .

C = Degrees Celsius

120 = Degrees Fahrenheit

Formula: $C = (F - 32) \div 1.8$

Equations

Degrees Celsius = (Degrees Fahrenheit − 32) ÷ 1.8

$C = (120 - 32) \div 1.8$

Solve

$C = (120 - 32) \div 1.8$

$= 88 \div 1.8$

$= 48.888\ldots$

≈ 49

Check

To estimate the subtraction, round each number to the hundreds place. The formula becomes

$$C \approx (100 - 0) \div 2$$
$$= 100 \div 2$$
$$= 50°\text{C}, \text{ which is reasonably close to the exact answer}$$

Answer

The faucet temperature should be set at 49°C.

Example 13 **Evaluate $\dfrac{3(1.2) - (0.3)^2}{1.2(0.3) + (-2.1)}$. Round the answer to the nearest tenth.**

$= \dfrac{3(1.2) - (0.3)^2}{1.2(0.3) + (-2.1)}$

$= \dfrac{3(1.2) - 0.09}{1.2(0.3) + (-2.1)}$ — *Evaluate the exponential expression in the numerator.*

$= \dfrac{3.6 - 0.09}{0.36 - 2.1}$ — *Multiply in the numerator and in the denominator.*

$= \dfrac{3.51}{-1.74}$ — *Do subtraction in the numerator; do subtraction in the denominator.*

$= -2.01724\ldots$ — *Divide the numerator by the denominator.*

≈ -2.0 — *Round the quotient to the nearest tenth.*

CHECK YOUR UNDERSTANDING

Evaluate.

1. $15.5 \div 3.1 + 6(4.1 - 0.9)$

2. (55.847 grams)(2) + (15.994 grams)(3)

3. $\dfrac{(0.240)(0.144)}{0.240 + 0.144}$

4. $(\$0.45)(80) + \$22.50$

Answers: 1. 24.2; 2. 159.676 grams; 3. 0.09; 4. $58.50.

EXERCISES SECTION 4.3

Evaluate using long division.

1. $12.84 \div 6$

2. $12.84 \div 0.6$

3. $-12.84 \div 0.06$

4. $12.84 \div 0.006$

5. $-0.008 \div 0.2$

6. $0.008 \div (-2)$

7. $0.008 \div 0.0002$

8. $7.62 \div 0$

9. $0 \div 7.62$

10. $12.936 \div 5.6$

11. $24.095 \div (-7.9)$

12. $0.5976 \div 8.3$

13. $0.099 \div 0.01$

14. $(-48.28) \div (-7.1)$

15. $116.55 \div 0.21$

16. $1.62 \div 81$

17. $205.32 \div 708$

18. $59.1 \div 0$

Estimate the quotient.

19. $5.2 \div 2.3$

20. $-55.3 \div 3.4$

21. $456.879 \div 23.2$

22. $75.3 \div 38.3$

23. $-67.2 \div 3.8$

24. $-678.23 \div 1.2$

25. $8.1 \div 32.09$

Solve the problem using the five-step strategy.

26. When John Glenn returned to space on the space shuttle, he orbited the Earth 134 times, traveling 3.6 million miles. What was the distance of each orbit? Round to the nearest mile.

27. In 1997, the Girl Scouts sold more than 190 million boxes of cookies, generating about $370 million in operating revenues. What amount of operating revenues was generated for each box of cookies sold? Round to the nearest cent.

28. On January 1, 1999, the European Commission set irrevocable conversion rates between the new euro currency and currencies from the 11 participating nations. One euro is equal to 1.95583 German marks, 6.55957 French francs, 1,936.27 Italian lire, 166.386 Spanish pesetas, 2.20371 Dutch guilders, 40.3399 Belgian francs, 13.7603 Austrian schillings, 200.482 Portuguese escudos, 5.94573 Finnish markka, 0.787564 Irish pounds, and 40.3399 Luxembourg francs. Hans has 5,000 German marks in his checking account. What is the value of his checking account in euros? Round to the nearest hundredth of a euro.

29. One advantage of the euro currency is that tourists can travel in Europe without converting their money as they cross national borders. In September 2001, the euro was equal to about $0.9059 dollar. If a tourist enters France with $750, how much money does she have in euros? Round to the nearest hundredth of a euro.

30. The price of a regular season adult lift ticket at Vail, Colorado, in 2001 was $59 per day. If a skier made nine runs (using the ski lift nine times), what price did she pay per ride? Round to the nearest cent.

31. In 1992, the U.S. Agriculture Department's Agricultural Stabilization and Conservation Service brought 900 of its employees to Washington, D.C., for an award ceremony. The total cost of the event came from airfare ($426,790), hotel cost and receptions ($24,000), plaques ($7,978), and certificates, badges, signs, banners and agendas ($13,656). What was the cost per employee for the event? Round to the closest dollar. (Source: *Seattle Post-Intelligencer,* July 6, 1992.)

32. On the San Diego transit system, a Ready Pass is accepted for unlimited rides within a calendar month on San Diego Transit and San Diego Trolley as well as other public transportation. In 2001, a Ready Pass that includes all local and urban buses costs $54. If Marti rides the bus to work and back home 19 days a month, how much does she pay per day for her transportation? Round to the nearest cent.

33. Write your own word problem that requires the division of two decimal numbers. Solve your problem.

Evaluate following the order of operations.

34. $14.89 + 12.04 \div 2$

35. $14.89 - 12.04 \div 2$

36. $(-14.89 - 12.04) \div 2$

37. (0.5)(3.14)(8.9 centimeters)^2

38. $\dfrac{(0.54)(0.0826)(288)}{(0.967)}$

39. $(4.3)(8.1) + (0.5)(9.8)(8.1)^2$

40. (0.5)(3.14)(4.0 feet)^2 + (4.0)(7.5 feet)^2

41. (6.72 meter)^2 + (1.01 meter)^2

Solve using the five-step problem-solving strategy.

42. The classified ads section of a local newspaper charges \$7.50 for a four-line ad that runs 7 days for an item that is priced between \$101 and \$1,500. For items priced between \$1,501 and \$3,500, the charge is \$11.50. Wyatt placed one ad to sell a waterbed for \$275 and placed another ad to sell a 1989 Chevy S-10 pickup for \$1,700. It took 2 weeks to sell the waterbed and 1 week to sell the pickup. What was the total cost of the two ads?

43. An oil change at a local rapid oil change costs \$23.50, including tax. For newer cars, most manufacturers recommend an oil change every 5,000 miles instead of every 3,000 miles. Carl generally drives a car 90,000 miles before trading it in for a new car. How much would Carl save over the time he owns a car if he changed the oil every 5,000 miles instead of every 3,000 miles?

44. An executive for a large aeronautical company exercised a stock option and bought 10,000 shares of stock in the company at \$23.63 each. He sold them at \$50.06 each 30 days later. How much was his profit?

45. Daphne's monthly natural gas bill is a basic charge of \$3.28 plus an additional charge of \$0.35387 per therm of natural gas she uses to heat her home, heat her hot water, and run her gas stove. (A therm is a unit used to measure natural gas. One therm equals the heating capacity of approximately 100,000 matches.) In December, she used 47 therms. What was her gas bill (excluding any taxes)?

46. A shipping warehouse heats its facility with a natural gas furnace and is rated a "large general service" company by the gas company. The monthly rate is \$0.37025 per therm for the first 200 therms of natural gas, \$0.35387 per therm for the next 800 therms used, and \$0.26165 per therm for all additional therms. The warehouse uses 1,240 therms in January. Calculate the heating bill.

47. Electricity is relatively cheap in the northwestern United States because the power is generated from hydroelectric dams. A regional utility company charges residential customers \$0.04026 per kilowatt for the first 600 kilowatts, \$0.04790 for the next 700 kilowatts, and \$0.05436 for all additional kilowatts. Mike lives in Portland and uses 1,508 kilowatts in March. Calculate his electrical bill.

48. In 2001, a combined admission ticket to the San Diego Zoo and the Wild Animal Park was \$47.85 for an adult and \$29.45 for a child. What was the cost for two mothers and six children to buy tickets?

49. In 1938, the minimum wage was instituted at a rate of \$0.25 per hour. In 1999, the minimum wage was \$5.15. For someone working 40 hours a week for minimum wage, what is the difference in gross wages per week received in 1938 and 1999?

50. When comparing wages in different years, the value of the wage should be adjusted for inflation. In 1954, the minimum wage was \$0.75 per hour; adjusted for inflation to 1996 dollars, it was equal to \$4.37 per hour. In 1996, the minimum wage was \$4.75. What was the difference in pay for a 40-hour week for a worker earning minimum wage in 1954 (in adjusted dollars) and 1996?

51. Manhattan Island is 13.4 miles long and 2.3 miles across at its widest point. Its area is 22.5 square miles. Show that Manhattan Island is not rectangular in shape.

Manhattan, circa 1963 (*© Charles E. Rotkin/CORBIS*)

52. The table lists the estimated income from endorsements between June 1998 and May 1999 for ten star athletes. Calculate the average of these ten incomes. (Source: *The Wall Street Journal Almanac*, 1999.)

ATHLETE	ENDORSEMENT INCOME (IN MILLIONS OF DOLLARS)
Michael Jordan	\$42
Tiger Woods	\$30
Shaquille O'Neal	\$28
Arnold Palmer	\$23
Jack Nicklaus	\$19
Grant Hill	\$15
Cal Ripkin, Jr.	\$9.75
Dale Earnhardt	\$5.8
Ken Griffey, Jr.	\$6.5
Jeff Gordon	\$5.5

53. In late 1999, farm prices for wheat, soybeans, and corn fell far below previous prices. Determine the difference in the amount a farmer received for 750 bushels of wheat in 1999 and for 750 bushels of wheat at the highest price in the 1990s. Answers will vary, depending on the estimated value from the graph.

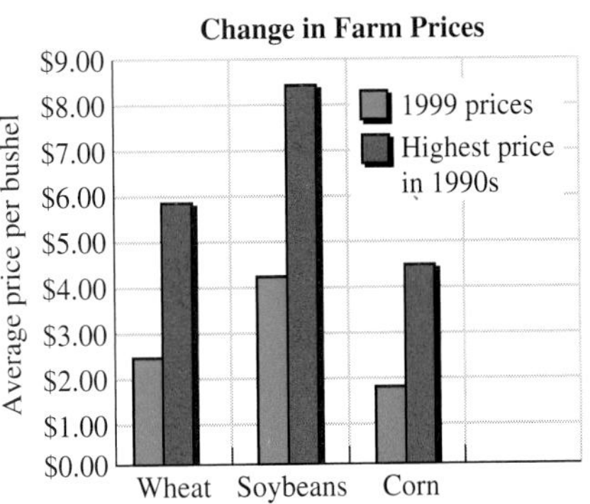

ERROR ANALYSIS

54. Problem: Divide 3.4 ÷ 0.2.

Answer:

$$\begin{array}{r} 1.7 \\ 0.2\overline{)3.4} \\ \underline{34} \end{array}$$

Describe the mistake made in solving this problem.

Redo the problem correctly.

REVIEW

55. Write 100,000 as an exponential expression with a base of 10.

56. Write 10^4 in place value form.

57. Evaluate 625,000 ÷ 100.

58. Evaluate (625,000)(100).

Section 4.4 Powers of Ten and Scientific Notation

LEARNING TOOLS

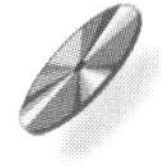

CD-ROM SSM VIDEO

Sneak Preview

Integers that are powers of ten can be multiplied or divided by adding or removing zeros. A similar process can be done with decimals. Large and small numbers can be expressed as products of powers of ten in a notation called "scientific notation."

After completing this section, you should be able to:

1) Write positive numbers and measurements in scientific notation.

2) Change numbers and measurements in scientific notation to place value notation.

DISCUSSION

Multiplying and Dividing by Powers of Ten

Decimal numbers are written in a base ten system. Moving to the right, each place value decreases by a power of ten. We can multiply or divide by a power of ten by moving the decimal point the appropriate number of place values. When we *multiply* by a power of ten, the decimal point moves to the right. Division is the opposite operation of multiplication. When we *divide* by a power of ten, the decimal point moves to the left.

PROCEDURE

To multiply a decimal number by 10^n, where n is a positive integer:

Move the decimal point n places to the *right.*

To divide a decimal number by 10^n, where n is a positive integer:

Move the decimal point n places to the *left.*

Example 1 **Multiply 0.045 by 10.**

$10 = 10^1$; move the decimal point one place to the right.

$(0.045)(10) = 0.45$

Example 2 **Multiply 2.679 by 1,000.**

$1{,}000 = 10^3$; move the decimal point three places to the right.

$(2.679)(1{,}000) = 2{,}679$

Example 3 **Multiply 4.782 by 100.**

$100 = 10^2$; move the decimal point two places to the right.

$(4.782)(100) = 478.2$

Example 4 **Divide 5.8 by 10^3.**

Move the decimal point three places to the left. Write zeros to the left of the number to allow the correct placement of the decimal point.

$5.8 \div 10^3 = 0.0058$

Example 5 **Divide 45.96 by 10^5.**

Move the decimal point five places to the left. Write zeros to the left of the number to allow the correct placement of the decimal point.

$45.96 \div 10^5 = 0.0004596$

CHECK YOUR UNDERSTANDING

Evaluate.

1. $(67.3)(10^5)$

2. $(0.00356)(10^7)$

3. $567.3 \div 10^3$

4. $34.2 \div 10^4$

5. $(87.2)(10^2)$

6. $(7)(10^5)$

Answers: 1. 6,730,000; 2. 35,600; 3. 0.5673; 4. 0.00342; 5. 8,720; 6. 700,000.

DISCUSSION

Negative Exponents

The powers of ten that are greater than or equal to 1 follow a pattern:

$$1{,}000 = 10^3$$
$$100 = 10^2$$
$$10 = 10^1$$
$$1 = 10^0$$

Moving to the left from the decimal point, each place value is ten times larger than the previous place value. The exponent of the power of ten increases by 1. The number of zeros in the decimal number also increases by 1.

The pattern continues as we move to numbers that are less than 1.

$$0.1 = \frac{1}{10} = \frac{1}{10^1} = 10^{-1}$$
$$0.01 = \frac{1}{100} = \frac{1}{10^2} = 10^{-2}$$
$$0.001 = \frac{1}{1{,}000} = \frac{1}{10^3} = 10^{-3}$$
$$0.0001 = \frac{1}{10{,}000} = \frac{1}{10^4} = 10^{-4}$$

Moving to the right from the decimal point, each place value is ten times smaller than the previous place value. In fraction notation, the power of ten is found in the denominator. When we write a negative exponent, it is a quick way to write a fraction whose numerator is 1 and whose denominator is a power of ten. Such fractions have a fractional value between 0 and 1.

All negative numbers are found to the left of 0 on the number line. As we see in this pattern, however, a negative exponent does not tell us that a number is negative or positive.

PROPERTY

The meaning of a negative exponent

If n is a positive integer and $a \neq 0$, then $a^{-n} = \frac{1}{a^n}$.

Example 6 **Rewrite $\frac{1}{2^4}$ using a negative exponent.**

$$\frac{1}{2^4} = 2^{-4}$$

Example 7 **Rewrite $\frac{1}{10^3}$ using a negative exponent.**

$$\frac{1}{10^3} = 10^{-3}$$

Example 8 **Rewrite $\frac{1}{10{,}000}$ using a negative exponent.**

$$\frac{1}{10{,}000}$$
$$= \frac{1}{10^4}$$
$$= 10^{-4}$$

CHECK YOUR UNDERSTANDING

Rewrite using a negative exponent.

1. $\frac{1}{100{,}000}$
2. $\frac{1}{10^3}$
3. $\frac{1}{10}$
4. $\frac{1}{100}$

Answers: 1. 10^{-5}; 2. 10^{-3}; 3. 10^{-1}; 4. 10^{-2}.

DISCUSSION

Scientific Notation

Scientists often work with very large or very small numbers. For example, light travels approximately 5,865,696,000,000 miles in 1 year. The wavelength of red light is approximately 0.000000072 centimeters. To write very large and very small numbers more efficiently, **scientific notation** is used. A number written in scientific notation is written as the product of a number between 1 and 10 and a power of ten. Instead of using parentheses ( ) or a dot · to show multiplication, the × symbol is used.

PROPERTY

A number is in **scientific notation** when it is of the form
$a \times 10^n$ where $1 \leq a < 10$ and n is an integer.

Example 9 **Write 300 meters in scientific notation.**

300 meters
$= 3 \times 100$ meters
$= 3 \times 10^2$ meters   *$100 = 10^2$*

Example 10 **The speed of light in a vacuum is about 30,000,000,000 centimeters per second. Write this speed in scientific notation.**

30,000,000,000 centimeters per second
$= 3 \times 10{,}000{,}000{,}000$ centimeters per second
$= 3 \times 10^{10}$ centimeters per second   *$10{,}000{,}000{,}000 = 10^{10}$*

Example 11 **The flow rate of a glacier is 0.0005 kilometers per hour. Write this measurement in scientific notation.**

0.0005 kilometers per hour
$= 5 \times 0.0001$ kilometers per hour
$= 5 \times 10^{-4}$ kilometers per hour   *$0.0001 = 10^{-4}$*

We can move directly from a positive decimal number to scientific notation by counting the number of places we move the decimal point. If the original number is between 0 and 1, the exponent will be a negative number. If the original number is greater than 1, the exponent will be a positive number.

PROCEDURE

To write a positive number in scientific notation:

1. Place the decimal point immediately following the first nonzero digit to form a number a where $1 \leq a < 10$.
2. Count the number of decimal places, n, that the decimal point was moved in Step 1.
3. If the decimal point was moved to the left in Step 1, the number in scientific notation is $a \times 10^n$.
4. If the decimal point was moved to the right in Step 1, the number in scientific notation is $a \times 10^{-n}$.

Scientific notation is very convenient for writing very small or very large measurements.

Example 12 **The mass of a proton is about 0.000000000000000000000001673 grams. Write this measurement in scientific notation.**

0.000000000000000000000001673   *Place the decimal point here. $0 < 1.673 < 10$*

The decimal point moved 24 places to the right so the exponent is -24.
0.000000000000000000000001673 grams $= 1.673 \times 10^{-24}$ grams

Example 13 **The amount of coal in the world that is considered to be recoverable totals about 13,800,000,000,000 tons. Write this measurement in scientific notation.**

13,800,000,000,000 tons   *Place the decimal point here. $0 < 1.38 < 10$*

The decimal point moved 13 places to the left so the exponent is 13.
13,800,000,000,000 tons $= 1.38 \times 10^{13}$ tons

CHECK YOUR UNDERSTANDING

Write the measurement in scientific notation.

1. 0.03527 gram
2. The ionization constant of acetic acid at 25°C is 0.000018.
3. The length of a year is 365.2422 days.
4. The end of the Mesozoic era and the probable extinction of the dinosaurs happened 65,000,000 years ago.

Answers: 1. 3.527×10^{-2} grams; 2. 1.8×10^{-5}; 3. 3.652422×10^{2} days; 4. 6.5×10^{7} years.

We can also change measurement in scientific notation to place value notation. The exponent tells us how many place values to move the decimal point to the right or left. Each time we move the decimal point one place value, we are changing the value of the number by a power of ten.

Example 14 **The average distance from the Earth to the Sun is 1.496×10^{11} meters. Write this distance as a decimal number.**

Multiplying by 10^{11} is equivalent to moving the decimal point 11 places to the right.

1.496×10^{11} meters $= 149{,}600{,}000{,}000$ meters

Example 15 **The metric system measures energy in joules. Another unit of energy is the food Calorie. One Calorie is equivalent to 4.185×10^{-3} joules. Write this measurement as a decimal number.**

Multiplying by 10^{-3} is equivalent to moving the decimal point three places to the left.

4.185×10^{-3} joules $= 0.004185$ joules

CHECK YOUR UNDERSTANDING

Write the measurement in place value notation.

1. The diameter of Jupiter is 1.428×10^{5} kilometers.
2. The estimated total volume of the two large ice sheets that cover Antarctica is 2.4×10^{7} cubic kilometers.
3. The estimated insured loss due to Hurricane Andrew in August 1992 was $\$1.55 \times 10^{10}$.

Answers: 1. 142,800 km; 2. 24,000,000 km^{3}; 3. \$15,500,000,000.

EXERCISES SECTION 4.4

Evaluate.

1. (40.64)(10)

2. (−40.64)(100)

3. (40.64)(0.1)

4. (−40.64)(−0.01)

5. (−40.64)(0.001)

6. (40.64)(0.0001)

7. 55.992 ÷ 10

8. (−55.992) ÷ (−100)

9. (−55.992) ÷ (0.1)

Solve the problem. Use the five-step problem-solving strategy.

10. At the request of the American West Steamboat Company, the Port of Clarkston spent $30,476 to extend its dock 100 feet. How much did this extension cost per foot?

Port of Clarkston, Clarkston, Washington

11. An executive for a large aeronautical company exercised a stock option and bought 10,000 shares at $23.63 each. What was the cost of this purchase?

12. The human heart pumps 1.3 gallons of blood each minute. How many gallons of blood are pumped in 10 minutes?

13. A laboratory mouse weighs about 0.79 ounce. If a laboratory orders 1,000 mice, what is their total weight?

14. Sandy has a navel orange that contains about 65 calories. She peels it and divides it into ten sections. How many calories are there in each section?

15. An 8-ounce California avocado contains 305 calories and 30 grams of fat. If it is chopped into 100 nearly equal pieces and put out for a salad bar, how many calories does each piece of avocado contain?

Change each measurement into scientific notation.

16. Speed of light: 669,600,000 miles an hour

17. Weekly receipts of the Broadway show *The Lion King:* $829,156

18. Approximate human population of the Earth: 5,100,000,000 people

19. Amount the three television networks paid to broadcast National Football League games until 2005: $17,600,000,000

20. Altitude of a communications satellite above the Earth: 42,240 kilometers

21. Approximate age of the solar system: 4,500,000,000 years

22. Mass of the Sun: 1,989,100,000,000,000,000,000,000,000,000 kilograms

23. Mass of the Earth: 5,974,200,000,000,000,000,000,000 kilograms

24. Mass of an electron: 0.000000000000000000000000000000911 kilogram

25. Mass of a human being: 75.3 kilograms

26. Duration of a lightning flash: 0.002 second

27. The wavelength of green visible light: 0.0000006 meter

Rewrite in place value notation.

28. 3.45×10^5

29. 4.6×10^3

30. 1.2×10^7

31. 5×10^1

32. 3×10^{-3}

33. 7.13×10^{-4}

34. 2.9×10^0

35. 3.39×10^{-2}

36. 8.99×10^8

37. When rewriting a number that is between 0 and 1 in scientific notation, will the exponent be negative or positive?

38. When rewriting a number that is greater than 1 in scientific notation, will the exponent be negative or positive?

Rewrite each number using a negative exponent.

39. $\frac{1}{1{,}000}$

40. $\frac{1}{10^5}$

41. $\frac{1}{1{,}000{,}000}$

42. 0.01

43. 0.00001

44. 0.0001

45. $\frac{1}{10}$

46. $\frac{1}{100}$

47. 0.1

ERROR ANALYSIS

48. **Problem:** Write in scientific notation: 3,700,000,000.

Answer: 3.7×10^{-9}

Describe the mistake made in solving this problem.

Redo the problem correctly.

49. **Problem:** Convert to scientific notation: 0.00786.

Answer: 7.86×10^{-2}

Describe the mistake made in solving this problem.

Redo the problem correctly.

REVIEW

50. State the commutative property of multiplication.

51. Evaluate: $\frac{5}{5}$.

52. Simplify: $\frac{3\text{ miles}}{3\text{ miles}}$.

53. Give an example of a conversion factor.

54. What is the product of 1 and any number?

Section 4.5 Unit Analysis

LEARNING TOOLS

CD-ROM SSM VIDEO

Sneak Preview

Unit analysis is an efficient method of unit conversion. Especially appropriate for work in the sciences, unit analysis is based on repeated multiplication by 1.

After completing this section, you should be able to:

1) Convert measurements with one or more units using unit analysis.

2) Solve word problems that require conversion of units.

DISCUSSION

Unit Analysis

Unit analysis is the method used in science to convert from one measurement to another. Some conversions involve changing from a unit in the English system to the metric system of measurement, or from one English unit to another. In science, conversions usually involve changing from one metric unit to another.

Frequently used conversion factors are listed in Table 4.1. The symbol ≈ means "approximately equal to" and indicates that the conversion factor has been rounded. The prefixes for metric units are listed in Table 4.2. The metric-to-metric conversions can be remembered by learning the meaning of these prefixes (see Section 1.8).

TABLE 4.1 · CONVERSION UNITS

METRIC ↔ METRIC	ENGLISH ↔ METRIC	ENGLISH ↔ ENGLISH
1 liter = 1,000 milliliters	1 inch = 2.54 centimeters	1 foot = 12 inches
1 kilometer = 1,000 meters	1 mile ≈ 1.6 kilometers	1 mile = 5,280 feet
1 meter = 100 centimeters	2.2 pounds ≈ 1 kilogram*	1 pint = 2 cups
1 meter = 10^9 nanometers	1.06 quarts ≈ 1 liter	1 gallon = 4 quarts
1 gram = 10^6 micrograms		1 quart = 2 pints
1 milliliter = 1 cubic centimeter		1 pound = 16 ounces
1 tonne = 1,000 kilograms		1 ton = 2,000 pounds

*Technically, the English unit of pound measures force and the metric unit of kilogram measures mass, two different quantities that cannot be directly compared. In practice, if we are working at the Earth's surface under the influence of the force of the Earth's gravity, and account for that force in the conversion factor, we can convert from pounds to kilograms and from kilograms to pounds.

TABLE 4.2 · METRIC PREFIXES*

1,000,000,000	10^9	giga
1,000,000	10^6	mega
1,000	10^3	kilo
100	10^2	hecto
10	10^1	deka
0.1	10^{-1}	deci
0.01	10^{-2}	centi
0.001	10^{-3}	milli
0.000001	10^{-6}	micro
0.000000001	10^{-9}	nano
0.000000000001	10^{-12}	pico

*Writing a prefix in front of a unit multiplies the unit by the given number. For example, a kilometer is equal to 1,000 meters. A millimeter is equal to 0.001 meters.

In unit analysis, we multiply a given measurement by a fraction that is equivalent to 1. Any fraction that has the same numerator and denominator is equal to 1. For example, $\frac{5}{5}$ is equal to 1 and $\frac{3 \text{ horses}}{3 \text{ horses}}$ is equal to 1. Similarly, $\frac{1 \text{ pound}}{16 \text{ ounces}}$ is equal to 1 since 1 pound and 16 ounces are equal measurements.

Example 1 **Write 1 foot = 12 inches as a fraction that is equal to 1.**

There are two different ways to write 1 foot = 12 inches as a fraction that is equal to 1:

$$\frac{1 \text{ foot}}{12 \text{ inches}} = 1 \quad \text{or} \quad \frac{12 \text{ inches}}{1 \text{ foot}} = 1$$

Example 2 **Write 1,000 meters = 1 kilometer as a fraction that is equal to l.**

There are two different ways to write 1,000 meters = 1 kilometer as a fraction that is equal to 1:

$$\frac{1{,}000 \text{ meters}}{1 \text{ kilometer}} = 1 \quad \text{or} \quad \frac{1 \text{ kilometer}}{1{,}000 \text{ meters}} = 1$$

In unit analysis, we choose the fraction that helps us end up with the proper arrangement of units after multiplication.

PROCEDURE

To change a measurement into other units using unit analysis:

1. Identify the needed conversion factors.
2. Write the conversion factors as fractions equal to 1.
3. Multiply the original measurement by the fractions; simplify before multiplying.
4. Multiply remaining factors.

Example 3 **Change 8.2 miles into kilometers. Round the answer to the nearest tenth of a kilometer.**

The needed conversion factor is 1 mile ≈ 1.6 kilometers (see Table 4.1).

Since we are trying to change to kilometers, we write the fraction with kilometers as the numerator and 1 mile as the denominator: $\frac{1.6 \text{ kilometers}}{1 \text{ mile}}$.

8.2 miles

$\approx \frac{8.2 \textbf{ miles}}{1} \cdot \frac{1.6 \text{ kilometers}}{1 \textbf{ mile}}$ *Multiply the original measurement by the fraction.*

$= \frac{8.2(1.6 \text{ kilometers})}{1}$ *Simplify before multiplying.*

= 13.12 kilometers *Multiply the remaining factors.*

≈ 13.1 kilometers *Round to the nearest tenth of a kilometer.*

So, 8.2 miles is about 13.1 kilometers. This is an approximation because the conversion factor is an approximation.

Unit analysis can also be used to convert measurements that are in multiple units.

Example 4 **Convert 855.9 meters per hour into centimeters per second. Round to the nearest tenth of a centimeter per second.**

The needed conversion factors are 1 meter = 100 centimeters, 1 hour = 60 minutes, and 1 minute = 60 seconds.

855.9 meters per hour

$= \frac{855.9 \text{ meters}}{1 \text{ hour}}$

$= \frac{855.9 \textbf{ m}}{1 \textbf{ h}} \cdot \frac{100 \text{ cm}}{1 \textbf{ m}} \cdot \frac{1 \textbf{ h}}{60 \textbf{ min}} \cdot \frac{1 \textbf{ min}}{60 \text{ sec}}$

$= \frac{855.9(100 \text{ cm})}{60(60 \text{ sec})}$

= 23.775 cm per sec

≈ 23.8 centimeters per second

We write the conversion factor 100 cm = 1 meter as $\frac{100 \text{ cm}}{1m}$. The meters in the numerator of the original measurement and the meters in the denominator of the conversion factor will simplify to 1.

Similarly, we write 1 hour = 60 minutes as $\frac{1 \text{ h}}{60 \text{ min}}$ and 1 minute = 60 seconds as $\frac{1 \text{ min}}{60 \text{ sec}}$.

So, 855.9 meters per hour is about 23.8 centimeters per second.

Writing the conversion factors in a fraction is a critical step in unit analysis. The units need to be positioned so that the end product will have the units in the correct place. Notice in the previous example that the fractions were written so that the unneeded units were opposite each other in a numerator and denominator.

Example 5 **A doctor orders 25 milligrams of Vistaril for a patient. The vial of Vistaril reads "Vistaril 100 mg/2 ml" (mg = milligram, ml = milliliter). Use unit analysis to determine how many milliliters should be injected into the patient.**

25 mg

$= \frac{25 \textbf{ mg}}{1} \cdot \frac{2 \text{ ml}}{100 \textbf{ mg}}$ *We do not need the unit **mg.** It is written in the numerator of one fraction and the denominator of the other fraction.*

$= \frac{25 \cdot 2 \text{ ml}}{100}$ *Simplify before multiplying.*

= 0.5 ml

Units with exponents also can be involved in measurement conversions.

Example 6 **A piece of paper has an area of 92 square inches. What is this area in square centimeters? Round the answer to the nearest square centimeter.**

The needed conversion factor is 1 in. = 2.54 cm. 2.54 cm is exactly equal to 1 inch so we do not have to use the ≈ sign. We will need to multiply by this fraction twice to eliminate the inches · inches in the original measurement.

$$92 \text{ in.}^2$$
$$= \frac{92\ \textbf{in.} \cdot \textbf{in.}}{1} \cdot \frac{2.54 \text{ cm}}{1\ \textbf{in.}} \cdot \frac{2.54 \text{ cm}}{1\ \textbf{in.}}$$
$$= \frac{92(2.54 \text{ cm})(2.54 \text{ cm})}{1}$$
$$= 593.5472 \text{ cm}^2$$
$$\approx 594 \text{ cm}^2$$

We need to eliminate inches from the numerator. We write the conversion factor 1 in. = 2.54 cm as $\frac{2.54 \text{ cm}}{1 \text{ in.}}$. The inches in the numerator of the original measurement and the centimeters in the denominator of the conversion factor will simplify to 1.

Example 7 **Convert 65 miles per hour to meters per second. Round the answer to the nearest unit.**

65 miles per hour

$$= \frac{65\ \textbf{miles}}{1\ \textbf{hour}} \cdot \frac{1\ \textbf{hour}}{60\ \textbf{min}} \cdot \frac{1\ \textbf{min}}{60 \text{ sec}} \cdot \frac{1.6\ \textbf{km}}{1\ \textbf{mile}} \cdot \frac{1{,}000 \text{ m}}{1\ \textbf{km}}$$
$$= \frac{65(1.6)1{,}000 \text{ m}}{60 \cdot 60 \text{ sec}}$$
$$= 28.\overline{8}\ \frac{\text{m}}{\text{sec}}$$

≈ 29 meters per second

CHECK YOUR UNDERSTANDING

Use unit analysis to change the measurements.

1. 5.5 kilograms into milligrams

2. 70 pounds into kilograms. Round the answer to the nearest kilogram.

3. 1.5 quarts into milliliters. Round the answer to the nearest milliliter.

4. 55 miles per hour into kilometers per second. Round to the nearest thousandth of a kilometer per second.

Answers: 1. 5,500,000 milligrams; 2. 32 kilograms; 3. 1,415 milliliters; 4. 0.024 kilometer per second.

EXERCISES SECTION 4.5

Convert the units using the process of unit analysis.

1. Convert 15 gallons to liters. Round to the nearest tenth of a liter.

2. Convert 20 pounds to kilograms. Round to the nearest kilogram.

3. Convert 22.1 meters to feet. Round to the nearest tenth of a foot.

4. Convert 6 miles to meters. Round to the nearest meter.

5. Convert 82 centimeters to inches. Round to the nearest inch.

6. Convert 110 liters to gallons. Round to the nearest tenth of a gallon.

7. Convert 32 feet per second into meters per second. Round to the nearest meter per second.

8. Convert \$1.32 per gallon into dollars per liter. Round to the nearest cent per liter.

9. Convert \$1.32 per gallon into euros per liter if the euro is equal to about \$0.9059. Round to the nearest hundredth of a euro per liter.

10. Convert 70 miles per hour to kilometers per hour. Round to the nearest tenth of a kilometer per hour.

11. Convert 40 pounds per square foot to grams per square centimeter. Round to the nearest tenth of a gram per square centimeter.

12. Convert 23.5 threads per inch to threads per meter. Round to the nearest thread per meter.

13. Convert 43 tons to tonnes (1 ton = 2,000 lb., 1 tonne = 1,000 kg). Round to the nearest tonne.

14. Convert \$1.50 per quart to dollars per liter. Round to the nearest cent per liter.

15. Convert 10 miles per hour to inches per second. Round to the nearest inch per second.

16. Convert \$8.50 per hour to dollars per 40-hour week. Round to the nearest dollar per 40-hour week.

17. The population of an endangered bird species is being monitored. The estimated population is three nesting pairs per square mile. Change this statistic to pairs per square kilometer. Round to the nearest tenth of a nesting pair. (Although birds only come in whole numbers, scientists often report this statistic to a specified place value smaller than a whole number. This helps them accurately predict populations in a large area.)

18. The density of gold is 19.29 grams per cubic centimeter. What is the density of gold in grams per cubic foot? Round to the nearest tenth of a gram per cubic foot.

19. An insurance plan presented to county employees will cost each employee $210.97 per year. Change this statistic to cost per day (365 days in a year). Round to the nearest cent per day.

20. Emily earns $38,400 per year. How much does she make per hour if she works a 40-hour week? Round to the nearest cent.

21. The population density of Utah is 23.56 people per square mile. What is the population density per square kilometer? Round to the nearest hundredth of a person.

22. It takes Mark 45 minutes to drive 45 miles to work. What is his average speed in kilometers per hour? Round to the nearest tenth of a kilometer per hour.

23. The Earth's force of gravity is 32 feet per square second. What is the force of gravity in meters per square second? Round the answer to the nearest tenth.

24. A 3-year-old child is admitted to the hospital with bacterial pneumonia. The child weighs 50 pounds and has a history of seizures. The medication orders include 200 mg of Keflex every 6 hours. Keflex is a liquid medicine that contains 125 mg of medication per 5 ml of liquid. How many milliliters of Keflex should the child receive each dose?

25. A doctor prescribes 250 milligrams of Amoxil g8h PO (every 8 hours by mouth) for 10 days. The available dosage of Amoxil is 125 milligrams per 5 milliliters. How many total milliliters will be needed for this entire course of treatment?

ERROR ANALYSIS

26. Problem: Convert \$5 per pound into dollars per kilogram.

Answer:

\$5 per pound

$$= \frac{\$5}{1 \text{ lb.}} \cdot \frac{1 \text{ kg}}{2.2 \text{ lb.}}$$

$= \$2.27$ per kilogram

Describe the mistake in the answer to this problem.

Redo the problem correctly.

27. Problem: Convert 20 feet per second into kilometers per hour.

Answer:

20 feet per second

$$= \frac{20 \text{ feet}}{1 \text{ second}} \cdot \frac{60 \text{ seconds}}{1 \text{ hour}} \cdot \frac{1 \text{ mile}}{5{,}280 \text{ feet}} \cdot \frac{1 \text{ mile}}{1.6 \text{ kilometers}}$$

$$= \frac{20 \cdot 60}{5{,}280(1.6)} \text{ kilometers per hour}$$

$= 0.142$ kilometers per hour

Describe the mistake in the answer to this problem.

Redo the problem correctly.

REVIEW

28. What is a rational number?

29. What does the perimeter of a figure measure?

30. What does the area of a figure measure?

31. What does the volume of a figure measure?

Section 4.6 Pi

LEARNING TOOLS

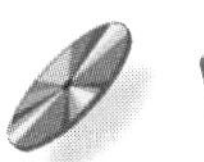

CD-ROM SSM VIDEO

Sneak Preview

Pi is an irrational number represented by the Greek symbol π. Many geometric formulas for area and volume include this number.

After completing this section, you should be able to:

1) Write a definition of π.

2) Use formulas to calculate the circumference and area of a circle, the volume and surface area of a circular cylinder, and the volume of a sphere or cone.

INVESTIGATION Circles and Pi

A **circle** is the set of all points in a plane that are at a given distance from a given point in the plane called the center. The distance around the circle, the perimeter, is called its **circumference.** A **diameter** of a circle is a line segment whose endpoints lie on the circle and that passes through the center. A **radius** is a line segment joining the center of the circle to a point on the circle.

1. Draw a circle. Label the circumference, center, diameter, and radius.

2. A group of students measured the circumference and diameter of a collection of circles of different sizes. They covered the circumference of the circle with string and then measured the string. Complete the table by calculating the quotient of the circumference and diameter for each circle.

	CIRCLE A	CIRCLE B	CIRCLE C
Circumference	25.7 cm	96 in.	100 cm
Diameter	8.1 cm	30.5 in.	32 cm
Circumference ÷ Diameter			

As these students discovered, circumference ÷ diameter for any size of circle is always close to the same number, 3.14159…. Mathematicians call this number **pi,** represent it with the Greek symbol $\boldsymbol{\pi}$**,** and have determined its exact value to many digits. However, we often round pi to the hundredths place and use 3.14 as its approximate value.

The formulas, diameter $= \dfrac{\text{circumference}}{\pi}$ and radius $= \dfrac{\text{circumference}}{2\pi}$, show the relationship of circumference, diameter or radius, and pi.

3. The distance around a circular fountain at a city park is 75 feet. Find the distance across the center of the pool. Round to the nearest whole number.

4. A lawn sprinkler sprays in a circle with the sprinkler at its center. The distance from the sprinkler to the end of the water spray is 12 feet. What is the circumference of the circle of lawn watered by the sprinkler? Round to the nearest whole number.

5. Beavers live in lodges that they construct from sticks and mud. Writing in 1894, archaeologist Lindsey Brine described a beaver lodge he observed near Lake Superior. He wrote that it was shaped like a rounded beehive, nearly 8 feet in diameter and 22 feet in circumference on the outside. Assuming that Brine correctly approximated the diameter of the beaver lodge, show whether his estimate of the circumference was too large or too small.

Answers: 3. 24 feet; 4. 75 feet; 5. His estimate was too small. $C = 8\pi \approx 25$ feet.

DISCUSSION

Circles and Pi

A **rational number** can be written as a fraction in which the numerator and the denominator are integers (and the denominator does not equal 0). For example, $\frac{3}{4}$ and $\frac{-6}{1}$ are both rational numbers. An **irrational number** cannot be written this way. Irrational numbers are nonrepeating, nonterminating decimal numbers. An important irrational number is 3.14159265359…, also known as **pi, π.** Even in ancient times, people knew that the **circumference** of the circle (the distance around the circle) divided by the **diameter** of the circle (the distance across the circle through the center) was about 3. In symbols, we write $\pi = \frac{C}{d}$ where C is the circumference of a circle and d is its diameter. To determine the value of pi to six billion digits, mathematicians use mathematical techniques of advanced algebra and calculus and the calculating speed of supercomputers. However, an approximate value of 3.14 is usually precise enough. When using the π key on a calculator, the number of decimal places shown depends on the particular calculator, and the last digit shown may be rounded.

If we multiply both sides of the equation $\pi = \frac{C}{d}$ by d, we get a formula for determining the circumference of any circle: $C = \pi d$. Since the diameter of a circle is equivalent to twice the radius of the circle, this formula can also be rewritten $C = 2\pi r$.

PROPERTY

Circumference of a circle, C

In words: Circumference = pi · diameter
In symbols: $C = \pi d$

In words: Circumference = 2 · pi · radius
In symbols: $C = 2\pi r$

Example 1 **Find the circumference of a circle with a diameter of 4.72 inches. Round the answer to the nearest hundredth.**

$C = \pi d$
$\approx (3.14)(4.72 \text{ inches})$
≈ 14.8208 inches
≈ 14.82 inches (rounded to the nearest hundredth of an inch)

We can also calculate perimeters of objects that include part of a circle using our knowledge of circumference.

Example 2 **Calculate the perimeter of a semicircular window that has a base of 36 inches. Round to the nearest tenth of an inch.**

I know . . .

P = Perimeter
$3.14 \approx \pi$
36 inches = diameter

Equations

$$\text{Perimeter} = \frac{1}{2}(\text{Circumference}) + \text{Diameter}$$

$$\text{Perimeter} = \frac{1}{2}\pi(\text{Diameter}) + \text{Diameter}$$

$$P = \frac{1}{2}\pi(36 \text{ inches}) + 36 \text{ inches}$$

Solve

$$P = \frac{1}{2}\pi(36 \text{ inches}) + 36 \text{ inches}$$
$$\approx \frac{1}{2}(3.14)(36 \text{ inches}) + 36 \text{ inches}$$
$$= \frac{113.02 \text{ inches}}{2} + 36 \text{ inches}$$
$$= 56.52 \text{ inches} + 36 \text{ inches}$$
$$= 92.52 \text{ inches}$$
≈ 92.5 inches (rounded to the nearest tenth of an inch)

Check

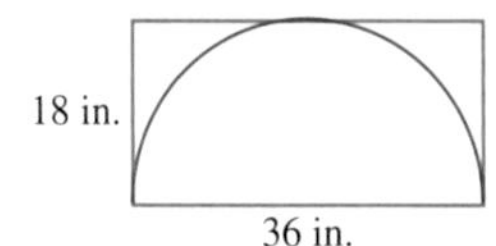

If this semicircle was a rectangle, it would be 36 inches wide and 18 inches high (half of the width). The perimeter of that rectangle would be about 2 · 40 in. + 2 · 20 in. = 80 in. + 40 in. = 120 in. and, as we expect, is a bit larger than the perimeter of the semicircle. We can also round the numbers in the original equation: $P = \frac{1}{2}(3)(40 \text{ in.}) + 40 \text{ in.}$ This equals 100 in., very close to our exact answer.

Answer

The perimeter of the window is 92.5 inches.

Pi is also included in the formula for the area of a circle: $A = \pi r^2$ where A is the area of the circle and r is a radius of the circle. The units for area are always distance^2 since the units of the radius are distance units.

PROPERTY

Area of a circle

In words: Area = pi · (radius)^2

In symbols: $A = \pi r^2$

Example 3 **A CD is 4.72 inches in diameter and 0.047 inch thick. The positioning hole in the middle is 0.59 inch in diameter. Up to 74 minutes of data can be recorded on the underside of the disk. What is the area of the underside of the CD? Round the answer to the nearest hundredth of a square inch.**

I know . . .

A = Area of underside

4.72 inches = Diameter of CD

2.36 inches = Radius of CD

0.59 inch = Diameter of center hole

0.295 inch = Radius of center hole

Formula: $A = \pi r^2 \qquad \pi \approx 3.14$

Equations

Area of underside = Area of whole CD − Area of hole

$$\text{Area} = \pi(\text{Radius of CD})^2 - \pi(\text{Radius of hole})^2$$

$$A = \pi(2.36 \text{ inches})^2 - \pi(0.295 \text{ inch})^2$$

Solve

$$\begin{aligned} A &= \pi(2.36 \text{ inches})^2 - \pi(0.295 \text{ inch})^2 \\ &\approx (3.14)(2.36 \text{ inches})(2.36 \text{ inches}) - (3.14)(0.295 \text{ inch})(0.295 \text{ inch}) \\ &= 17.488544 \text{ inches}^2 - 0.2732585 \text{ inch}^2 \\ &= 17.2152855 \text{ inches}^2 \\ &\approx 17.22 \text{ inches}^2 \end{aligned}$$

Check

To check for reasonability, we can estimate the area of the CD and ignore the area of the small hole. The CD with a diameter of 5 inches will be somewhat smaller than a square with sides of 5 inches and an area of 25 square inches. Our answer of 17.22 inches^2 is somewhat smaller than 25 square inches, so it is reasonable.

Answer

The underside of the CD has an area of about 17.22 square inches.

When using a calculator for problems like these, it is best not to round the results of the intermediate calculations. Instead, use the storage or memory features of the calculator. Also, calculators round numbers that have more digits than can be displayed by the calculator. When we see a number that fills the display of the calculator, it may be a rounded approximation of such a number.

Example 4 **Find the area of the quarter circle shown in the figure. Use $\pi \approx 3.14$ and round the answer to the nearest hundredth of a square centimeter.**

$$\begin{aligned} Q &= \frac{1}{4}\pi r^2 \\ &\approx \frac{1}{4}(3.14)(8 \text{ cm})^2 \\ &= 50.24 \text{ cm}^2 \end{aligned}$$

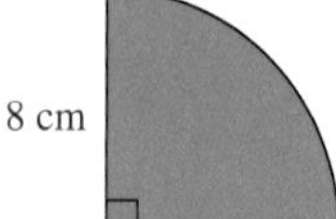

A **cylinder** is a three-dimensional object that has identical cross sections and straight sides. When we find the amount of the three-dimensional space that is occupied by the cylinder, we are finding its **volume.** The volume of any cylinder can be found by multiplying the area of its base by its height.

PROPERTY

Volume of a cylinder

In words: Volume = (Area of base) · height

In symbols: $V = Ah$

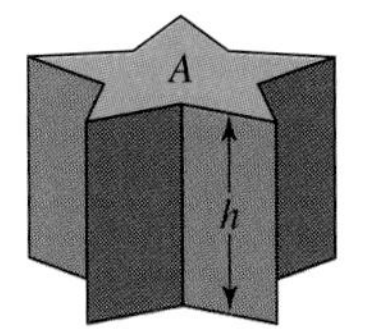

One of the more common cylinders is a **circular cylinder** where each cross section is a circle. It can be visualized as a stack of identical circles. The area (A) of the base is $A = \pi r^2$ where r = radius.

PROPERTY

Volume of a circular cylinder

In words: Volume = pi · (radius)^2 · height

In symbols: $V = \pi r^2 h$

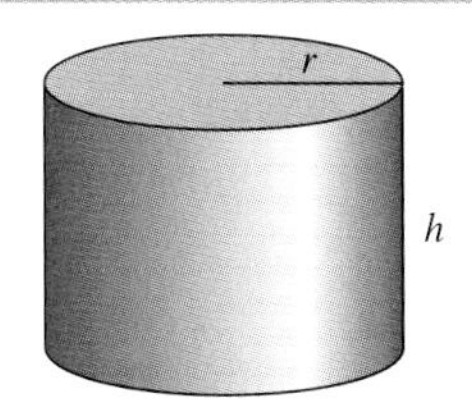

Example 5 **An architect is designing a deck to hold a circular hot tub that has a diameter of 9.75 feet and a water depth of 3.5 feet. If a cubic foot of water weighs about 62.4 pounds, how much will the water in this hot tub weigh? Round the answer to the nearest ten pounds.**

I know . . .

w = Weight of water

9.75 feet = Diameter

4.875 feet = Radius = Diameter ÷ 2

3.5 feet = Height

62.4 pounds = Weight of 1 cubic foot of water

Formula: $V = \pi r^2 h$

Equations

Weight of water = Volume of water · Weight of water per cubic foot

$$w = \pi \cdot (\text{radius})^2(\text{height})\left(\frac{\text{weight of water}}{\text{cubic feet}}\right)$$

$$w = \pi \cdot (4.875\text{ feet})^2(3.5\text{ feet})\left(\frac{62.4\text{ pounds}}{1\text{ cubic foot}}\right)$$

Solve

$$w = \pi \cdot (4.875\text{ feet})^2(3.5\text{ feet})\left(\frac{62.4\text{ pounds}}{1\text{ cubic foot}}\right)$$

$$\approx (3.14)(4.875\text{ feet})(4.875\text{ feet})(3.5\text{ feet})\left(\frac{62.4\text{ pounds}}{1\text{ cubic foot}}\right)$$

$$= (15.3075\text{ feet})(4.875\text{ feet})(3.5\text{ feet})\left(\frac{62.4\text{ pounds}}{1\text{ cubic foot}}\right)$$

$$= (74.6240625\text{ feet}^2)(3.5\text{ feet})\left(\frac{62.4\text{ pounds}}{1\text{ cubic foot}}\right)$$

$$= (261.1842188\text{ cubic feet})\left(\frac{62.4\text{ pounds}}{1\text{ cubic foot}}\right)$$

w = 16,297.89525 pounds

$w \approx$ 16,300 pounds (rounded to the nearest ten pounds)

Check

This seems huge, a little over 8 tons! But, if we round the numbers and use the formula, it checks out as reasonable:

$w = (3)(5)^2(4)(60) = 18{,}000$ pounds

Answer

The weight of the water in the hot tub is about 16,300 pounds.

Example 6 **Find the volume of a circular cylinder with a radius of 3.3 cm and a height of 10.4 cm. Use $\pi \approx 3.14$. Round to the nearest whole number.**

$$\begin{aligned} V &= \pi r^2 h \\ &\approx 3.14(3.3\text{ cm})^2(10.4\text{ cm}) \\ &= 355.62384\text{ cm}^3 \\ &\approx 356\text{ cm}^3 \end{aligned}$$

The **surface area** of a circular cylinder can be found by multiplying the circumference of the base by the height: $S = 2\pi rh$. This is the area of the tube part of the cylinder and does not include the two circular ends.

PROPERTY

Surface area of a circular cylinder

In words: Surface area = 2 · pi · (radius)² · height

In symbols: $S = 2\pi rh$

Example 7 **Lip balm is often sold in a tube that is 3.00 inches in length and 0.50 inch in diameter. To calculate the cost of various coatings for the tube, engineers need to know the surface area of the tube. Round the answer to the nearest tenth.**

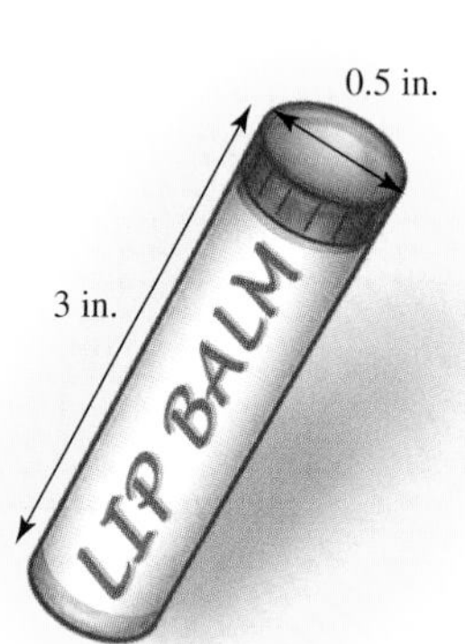

I know . . .

S = Surface area of tube

3.00 inches = Length of tube

0.50 inch = Diameter of tube

0.25 inch = Radius of tube

Formula: $S = 2\pi rh$

Equations

Surface area of tube = 2 · π · Radius of tube · Length of tube

$$S \approx 2(3.14)(0.25\text{ inch})(3\text{ inches})$$

Solve

$$\begin{aligned} S &\approx 2(3.14)(0.25\text{ inch})(3\text{ inches}) \\ &= (6.28)(0.25\text{ inch})(3\text{ inches}) \\ &= (1.57\text{ inches})(3\text{ inches}) \\ &= 4.71\text{ inches}^2 \end{aligned}$$

$S \approx 4.7\text{ inches}^2$ (rounded to the nearest tenth of a square inch)

Check

Estimating, $S \approx 2(3)(0.3\text{ inch})(3\text{ inches}) = 5.4\text{ inches}^2$, which is reasonably close to the exact answer.

Answer

The surface area of the lip balm tube is 4.7 square inches.

A **sphere** is a shape like a ball, formed from many circles of different diameter. The radius of a sphere is the distance from the center to any point on the surface of the sphere. The space inside of a sphere is called its **volume.**

PROPERTY

Volume of a sphere

In words: Volume $= \frac{4}{3} \cdot \text{pi} \cdot (\text{radius})^3$

In symbols: $V = \frac{4}{3}\pi r^3$

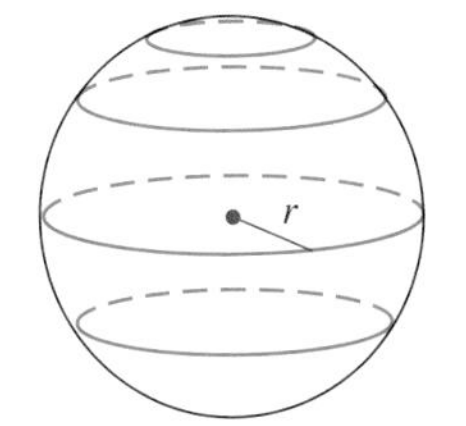

Example 8 **Find the volume of a sphere with a diameter of 3.5 inches. Round to the nearest tenth of a cubic inch.**

$$\begin{aligned} V &= \frac{4}{3}\pi r^3 \\ &\approx \left(\frac{4}{3}\right)(3.14)(1.75 \text{ in.})^3 \\ &= \frac{67.31375 \text{ in.}^3}{3} \\ &= 22.437\ldots \text{ in.}^3 \\ &\approx 22.4 \text{ in.}^3 \end{aligned}$$

When sawdust is poured out of a hopper, it naturally forms into a cone-shaped pile. Like a sphere, a **cone** is a collection of circles of different diameters. Mathematicians classify a cone as a **circular pyramid.** When we think of a pyramid, we usually visualize giant structures with a square base that rises to a point above the center of the base. However, the base of a pyramid can be any shape, including a circle.

PROPERTY

Volume of a pyramid

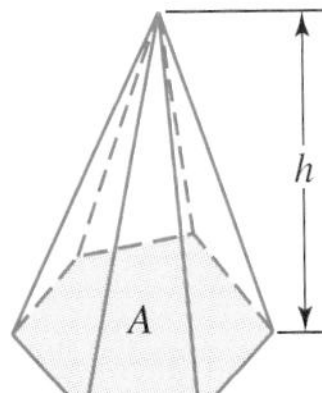

In words: Volume $= \frac{1}{3}$ (Area of base) $\cdot$ height

In symbols: $V = \frac{1}{3}Ah$

When the base of a pyramid is a circle, the pyramid is also known as a **circular cone.**

PROPERTY

Volume of a circular cone

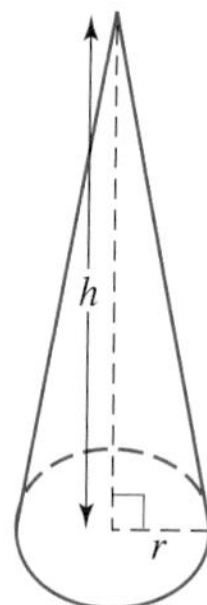

In words: Volume $= \frac{1}{3} \cdot$ pi $\cdot$ (radius)^2 $\cdot$ height

In symbols: $V = \frac{1}{3}\pi r^2 h$

Example 9 **Gail licks her vanilla ice cream cone down to the top of the cone. The diameter of the sugar cone is 1.5 inches and it is 4 inches high. A cubic inch of ice cream contains about 9.7 Calories. Calculate how many Calories of ice cream Gail will avoid if she throws away the cone, assuming that the remaining ice cream fills the cone. Round the answer to the nearest Calorie.**

I know . . .

c = Calories

1.5 inches = Diameter of cone

0.75 inches = Radius of cone

4 inches = Height of cone

$9.7 \frac{\text{calories}}{\text{inch}^3}$ = Calories per cubic inch

Formula: $V = \frac{1}{3}\pi r^2 h$

Equations

Calories $= \frac{1}{3}\pi$(Radius of cone)^2(Height of cone)(Calories per cubic inch)

$$c \approx \left(\frac{1}{3}\right)(3.14)(0.75 \text{ inch})^2(4 \text{ inches})\left(9.7\frac{\text{Calories}}{\text{inch}^3}\right)$$

Solve

$$c \approx \left(\frac{1}{3}\right)(3.14)(0.75 \text{ inch})^2(4 \text{ inches})\left(9.7\frac{\text{Calories}}{\text{inch}^3}\right)$$

$= 22.8435\ldots$ Calories

≈ 23 Calories (rounded to the nearest Calorie)

Check

An estimate of the Calories is

$$c \approx (0.3)(3)(0.8 \text{ inch})^2(4 \text{ inches})\left(10\frac{\text{Calories}}{\text{inch}^3}\right) = 23.04 \text{ Calories}$$

reasonably close to the exact answer.

Answer

Gail is throwing away about 23 Calories of ice cream.

CHECK YOUR UNDERSTANDING

1. Calculate the volume of a circular cylinder with a radius of 1.0 meter and a height of 3.0 meters. Round your answer to the nearest tenth of a cubic meter.

2. Calculate the volume of a spherical ball that has a diameter of 4.8 centimeters. Round the answer to the nearest tenth of a cubic centimeter.

3. Calculate the circumference of a circle that has a diameter of 0.5 mile. Round the answer to the nearest tenth of a mile.

4. Calculate the area of a circle that has a diameter of 0.5 mile. Round the answer to the nearest tenth of a square mile.

Answers: 1. 9.4 cubic meters; 2. 57.9 cubic centimeters; 3. 1.6 miles; 4. 0.2 square mile.

EXERCISES SECTION 4.6

Throughout this assignment use an approximate value for pi of $\pi \approx 3.14$.

Calculate the circumference of the circle with the given radius or diameter. Round to the nearest hundredth of a centimeter.

1. Radius = 6.00 centimeters

2. Radius = 6.50 centimeters

3. Radius = 60.00 centimeters

4. Radius = 65.00 centimeters

5. Diameter = 60.50 centimeters

6. Diameter = 6.50 centimeters

Calculate the radius of the circle with the given circumference. Round all radii (the plural for radius) to the hundredths place.

7. Circumference = 16.00 inches

8. Circumference = 12.00 inches

9. Circumference = 18.00 feet

10. Circumference = 5,280.00 feet

Solve the problem. Use the five-step strategy.

11. An official Major League baseball must measure between 9.00 and 9.25 inches in circumference to meet Major League standards. What are the possible diameters of a Major League baseball? Round the answer to the nearest hundredth of an inch.

12. The Caltech Millimeter Array in Owens Valley, California, consists of six dishes. Each dish has a diameter of 10.4 meters. What is the circumference of one of these dishes? Round the answer to the nearest tenth of a meter.

13. The altitude of a communications satellite in a circular orbit is 35,900 kilometers above the surface of the Earth. The distance from the Earth's surface to the center (its radius) is 6,367 kilometers. What is the circumference of the orbit of the satellite? Round the answer to the nearest hundred kilometers.

14. A can of tomatoes has a diameter of 3.0 inches and a height of 4.5 inches. What is the area of the rectangular label that covers the surface of the can? Round the answer to the nearest tenth of a square inch.

15. Write a word problem that requires the use of the circumference formula. Solve the problem.

Calculate the area of the circle with given radius or diameter. Round each area to the nearest tenth.

16. Radius = 7.0 centimeters

17. Radius = 7.5 centimeters

18. Radius = 70.0 centimeters

19. Radius = 700.0 centimeters

20. Diameter = 5,280.0 feet

21. Diameter = 3.0 feet

Solve the problem using the five-step problem-solving strategy.

22. A carpenter is cutting a circular hole through a wall for a pipe that will be wrapped with foam insulation. The diameter of the pipe is 0.50 inch and the thickness of the insulation is 0.3125 inch ($\frac{5}{16}$ inch). What is the minimum area that will need to be cut out for the pipe and insulation? Round to the nearest hundredth of a square inch.

23. A garden hose has a diameter of 0.75 inch. This large hose runs from an outdoor faucet to a Y-shaped connector from which run two smaller 0.5-inch diameter hoses. One of these hoses is used to water the grass and the other is used to water a vegetable garden. Determine whether the cross-sectional area of the two smaller hoses is enough to carry all the water coming from the larger hose.

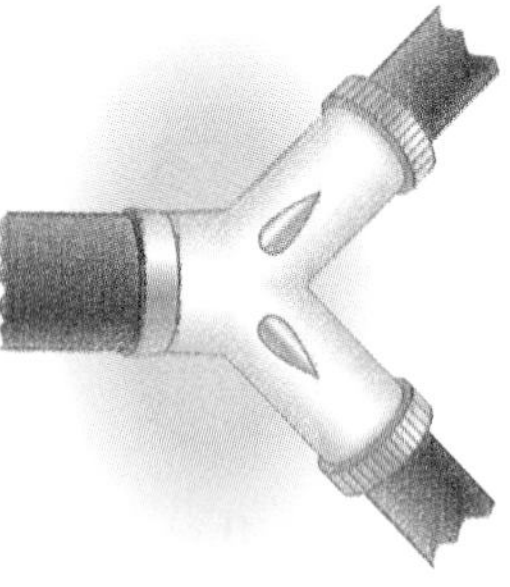

24. Dieters are often told to eat their meals from smaller plates because the portions will not appear to be as small as on a larger plate. What is the difference in area between a plate with a 10-inch diameter and a plate with a 6-inch diameter? Round to the nearest square inch.

25. The price of a large pepperoni pizza (14 inches in diameter) at Pizza Hut is \$13.79. The medium (12-inch) pizza is \$10.79. Which pizza is a better buy per square inch?

26. Papa Murphy's is a take-and-bake pizza store. A large pepperoni pizza (15-inch diameter) is \$7.99. Domino's Pizza is running a special on large pepperoni (14-inch diameter) for \$9.99. How much is saved per square inch by buying the Papa Murphy pizza? Round to the nearest cent.

27. Write a word problem that uses the area of a circle formula. Solve the problem.

Calculate the volume of the sphere with the given radius. Round the volume to the nearest tenth of a cubic centimeter.

28. Radius = 9.0 centimeters

29. Radius = 9.5 centimeters

30. Radius = 90 centimeters

31. Radius = 95.0 centimeters

Solve the problem using the five-step problem-solving strategy.

32. The diameter of a golf ball is 1.68 inches. What is the volume of a golf ball? Round to the nearest hundredth of a cubic inch.

33. In the game of paintball, players have carbon dioxide–powered guns that shoot gelatin "paintballs" that are filled with nontoxic dyes. When an opponent is hit by the paintball, the ball explodes and marks him with the paint. The most readily available size ball has an inside diameter of about 0.68 inch. About how much liquid is contained in this paintball? Round the answer to the nearest hundredth of a cubic inch.

34. Paintballs generally are sized with a diameter of 0.50 inch, 0.62 inch, or 0.68 inch. What is the difference in volume between the 0.50- and 0.62-inch paintballs? Round to the nearest hundredth of a cubic inch.

35. The first time that a standard ball was specified for a soccer game was for a game between the London Football Association and the Sheffield Association in March 1866. The required ball was a "Lillywhite's No. 5." Later, a No. 5 was specified as a ball of average circumference of not less than 27 inches and not more than 28 inches. What is the minimum volume of such a ball? Round to the nearest cubic inch.

36. The diameter of the Moon is about 2,160 miles. The diameter of the Earth at the equator is about 7,926 miles. Assuming that the Earth and Moon are essentially spheres, calculate the difference in the volume of the Earth and the Moon. Round to the nearest hundred cubic mile.

Calculate the volume of a circular cylinder with the given radius or diameter and height.

37. Radius = 4 feet, height = 2 feet. Round to the nearest cubic foot.

38. Radius = 4.5 feet, height = 2.0 feet. Round to the nearest tenth of a cubic foot.

39. Radius = 4.5 feet, height = 2.5 feet. Round to the nearest tenth of a cubic foot.

40. Diameter = 20.8 meters, height = 3.5 meters. Round to the nearest tenth of a cubic meter.

Calculate the volume of a circular cone with the given radius or diameter and height.

41. Radius = 8 inches; height = 12 inches. Round to the nearest cubic inch.

42. Diameter = 1 foot; height = 3 feet. Round to the nearest tenth of a cubic foot.

Solve the problems using the five-step problem-solving strategy.

43. A coffee can is 5.5 inches high and 4.125 inches in diameter. What is the volume of this can? Round to the nearest tenth of a cubic inch.

44. What is the volume of an engine cylinder that has a diameter of 7.5 centimeters and a height of 7.5 centimeters? Round to the nearest tenth of a cubic centimeter.

45. A hockey puck is usually made of vulcanized rubber, is 3 inches in diameter, and 1 inch thick. What volume of rubber is in a hockey puck? Round to the nearest cubic inch.

46. Tomato sauce is often sold in 15-ounce cans that have a diameter of 3.875 inches and a height of 4.375 inches. If 24 cans are packed in a carton that has a height of 9.000 inches, a width of 16.000 inches, and a length of 12.000 inches, how much empty space is there in each carton? Round the answer to the nearest tenth of a cubic inch.

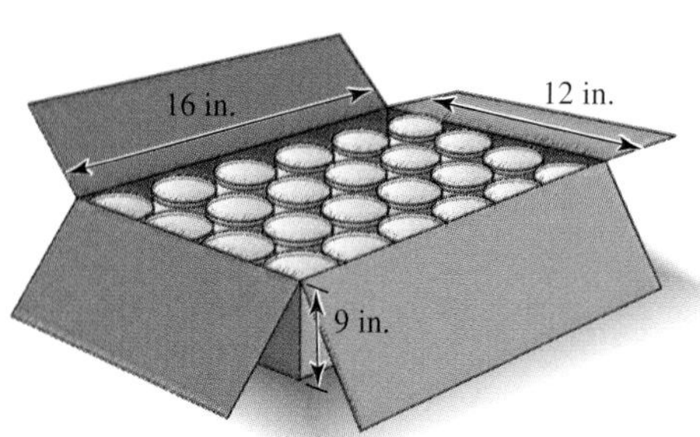

47. An ice cream cone has a half-sphere of ice cream on top. If the cone has a radius of 2.4 cm and a height of 12.2 cm, what is the combined volume of the cone and ice cream? Round to the nearest tenth of a cubic centimeter.

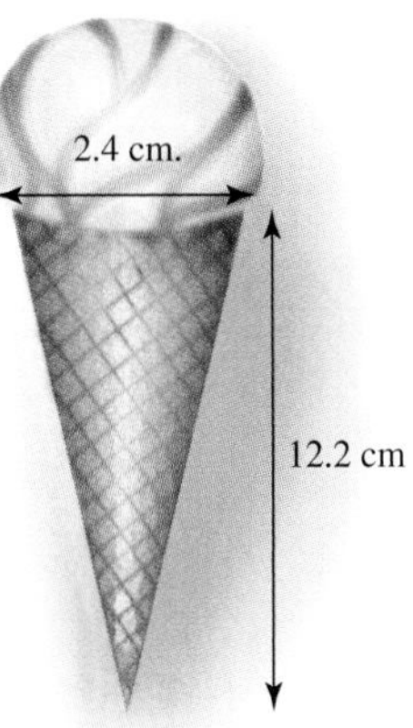

48. When sand is dumped from a dump truck it forms a cone shape. Susan had 20 m^3 of sand delivered. It was dumped into a conical pile that settled to a diameter of about 5 meters. What was the height of the pile of sand? Round to the nearest meter.

ERROR ANALYSIS

49. **Problem:** The diameter of a circle is 10 centimeters. Calculate the area of the circle. Round the answer to the nearest whole number.

Answer:

$$\begin{aligned} A &= \pi r^2 \\ &\approx 3.14(10\text{cm})(10\text{cm}) \\ &= 3.14\text{ cm}^2 \end{aligned}$$

The area of the circle is 314 square centimeters.

Describe the mistake made in solving this problem.

Redo the problem correctly.

50. **Problem:** The diameter of a circular cylinder is 11.00 inches. The height is 2.00 feet. Calculate the volume of the cylinder. Round the volume to the hundredths place.

Answer:

$$V = \pi r^2 h$$
$$\approx 3.14\,(5.50\text{ in.})(5.50\text{ in.})(2.00\text{ in.})$$
$$= 189.97\text{ in.}^3$$

The volume of the cylinder is 189.97 cubic inches.

Describe the mistake made in solving this problem.

Redo the problem correctly.

REVIEW

51. What formula is used to find the area of a square?

52. What is the sign of the product of a negative number and a negative number?

53. Rewrite 4^2 as multiplication.

54. Rewrite $(-4)^2$ as multiplication.

Section 4.7 Square Roots

LEARNING TOOLS

CD-ROM

SSM

VIDEO

Sneak Preview

When we multiply a number by itself, we are "squaring" the number. The product is called the square. When we are given a square and find the number whose square is that product, we are finding the square root. Since square roots are useful in calculating distances, finding the value of a square root was an important skill in many ancient cultures. It remains an important skill today in the building trades.

After completing this section, you should be able to:

1) Evaluate a principal square root.
2) Describe the sets of rational numbers, irrational numbers, and real numbers.
3) Evaluate formulas that include square roots.
4) Solve problems that require the use of the Pythagorean Theorem.

DISCUSSION

Square Roots

A perfect square is the product of multiplying a number by itself. For example, 16 is the square of 4 since $4 \cdot 4 = 16$. A **square root** is one of the two equal factors whose product is a perfect square. The square root of 16 is 4, since $4^2 = 16$. Notice that $(-4)^2$ also equals 16. This means that -4 is also a square root of 16. We call the positive square root the **principal square root.** The symbol for the principal square root is called a **radical sign** and has the symbol $\sqrt{}$. The number inside the radical sign is called the **radicand.** So, $\sqrt{16} = 4$. This is read "the principal square root of 16 is 4." In practice, we often just say "the square root of 16 is 4." The "principal" is implied and not said. We can also evaluate the opposite of a square root. $-\sqrt{16}$ is read "the opposite of the principal square root of 16" and is equal to -4. Numbers other than whole numbers also have square roots. For example, $\sqrt{\frac{25}{36}}$ is $\frac{5}{6}$ since $\frac{5}{6} \cdot \frac{5}{6} = \frac{25}{36}$.

Square roots of perfect squares are rational numbers. However, many square roots are irrational numbers. Before calculators were readily available, irrational square roots were determined by hand using a lengthy procedure. A calculator rounds an irrational square root to the number of digits it can hold in its display.

Example 1 **Evaluate the following square roots: $\sqrt{81}$ $\sqrt{64}$ $-\sqrt{100}$ $-\sqrt{49}$**

$$\sqrt{81} = 9 \text{ since } 9^2 = 81$$
$$\sqrt{64} = 8 \text{ since } 8^2 = 64$$
$$-\sqrt{100} = -10 \text{ since } 10^2 = 100$$
$$-\sqrt{49} = -7 \text{ since } 7^2 = 49$$

Example 2 **Use a calculator to determine $\sqrt{2}$. Round to the nearest thousandth.**

We expect $\sqrt{2}$ to be between 1 and 2 since $1^2 = 1$ and $2^2 = 4$.

$\sqrt{2} = 1.41421\ldots$ which rounds to 1.414.

If we combine the set of **rational numbers** with the set of **irrational numbers,** we have a new set, the **real numbers.** Every number on the number line is a real number.

Example 3 **Use a calculator to determine $\sqrt{-2}$.**

When $\sqrt{-2}$ is entered in a calculator, an "error" or "nonreal result" message will appear. This is because there is no irrational or rational number that multiplied by itself gives a negative product. A positive number multiplied by a positive number gives a positive product. A negative number multiplied by a negative number gives a positive product. So, $\sqrt{-2}$ has no real number solution.

Example 4 **Use a calculator to determine $-\sqrt{2}$ to 3 decimal places.**

$-\sqrt{2}$ = the opposite of the square root of 2 $= -1.414$.

CHECK YOUR UNDERSTANDING

Use the multiplication facts to find the square root.

1. $\sqrt{64}$ **2.** $-\sqrt{9}$ **3.** $\sqrt{\dfrac{81}{49}}$ **4.** $\sqrt{-25}$

Answers: 1. 8; 2. −3; 3. $\frac{9}{7}$; 4. no real number solution.

DISCUSSION

Square Roots and Formulas

Square roots are evaluated at the same level of order of operations as exponents. All simplifying inside the square root sign (also called "simplifying the radicand") is completed before evaluating the square root.

Example 5 **Solve $x = \dfrac{-5 + \sqrt{5^2 - 4 \cdot 1 \cdot 4}}{2 \cdot 1}$.**

$$x = \frac{-5 + \sqrt{5^2 - 4 \cdot 1 \cdot 4}}{2 \cdot 1}$$

$$= \frac{-5 + \sqrt{25 - 16}}{2}$$ *Evaluate inside the radical sign: evaluate exponents.*

$$= \frac{-5 + \sqrt{9}}{2}$$ *Evaluate inside the radical sign: subtract.*

$$= \frac{-5 + 3}{2}$$ *Evaluate the square root.*

$$= \frac{-2}{2}$$

$$= -1$$

Many geometric formulas include a square of the radius. When such formulas are rearranged and solved for the radius, the formula includes a square root.

Example 6 **The formula for the radius of a circle is $r = \sqrt{\frac{A}{\pi}}$ where A is the area of the circle. Calculate the radius of a circle that has an area of 0.5 square mile. Round to the nearest tenth of a mile. Use $\pi \approx 3.14$.**

$$
\begin{aligned}
r &= \sqrt{\frac{A}{\pi}} \\
&\approx \sqrt{\frac{0.5 \text{ square mile}}{3.14}} \\
&\approx \sqrt{0.1592356688 \text{ square mile}} \\
&\approx 0.3990434422 \text{ mile} \qquad \textit{$\sqrt{miles^2} = miles$} \\
&\approx 0.4 \text{ mile (rounded to the nearest tenth of a mile)}
\end{aligned}
$$

CHECK YOUR UNDERSTANDING

1. Calculate the side of a square with an area of 42.25 square centimeters using the formula $s = \sqrt{A}$ where s is the length of a side of the square and A is the area of the square.

2. Calculate the radius of a circular cylinder with a volume of 13,156 cubic inches and a height of 33.125 inches using the formula $r = \sqrt{\frac{V}{\pi h}}$, where V is the volume and h is the height of the cylinder. (This cylinder approximates a 55-gallon drum.) Round the answer to the nearest tenth of an inch.

Answers: 1. 6.5 centimeters; 2. 11.2 inches.

DISCUSSION

The Pythagorean Theorem

The theorem known as the Pythagorean theorem is an important statement about a particular kind of triangle called a **right triangle.** Right triangles have one angle that is equal to 90 degrees. The side of the triangle opposite the 90-degree angle is called the **hypotenuse.** The modern version of the Pythagorean theorem states that in any right triangle, the square of the length of the hypotenuse is equal to the sum of the squares of the lengths of the other two sides of the triangle.

Pythagoras was a Greek living in the sixth century B.C.E. Although Pythagoras gets his name on the theorem, no one knows for sure if he actually proved it. Today there are hundreds of proofs of the theorem. Chinese texts from before the first century C.E. illustrate the relationship. U.S. President-to-be James Garfield published an original proof of the theorem in 1876.

To write the Pythagorean theorem in symbols we represent the length of the hypotenuse as c and the lengths of the other two sides of the triangle as a and b.

PROPERTY

The Pythagorean Theorem

In words: For a triangle with a right angle, the square of the length of the hypotenuse is equal to the sum of the squares of the lengths of the other two sides of the triangle.

In symbols: $a^2 + b^2 = c^2$

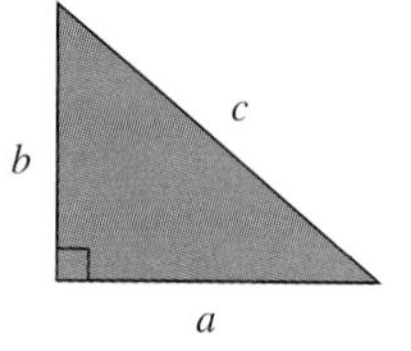

If we know any two sides of a right triangle, we can calculate the length of the third side using the Pythagorean theorem. This requires us to calculate a square root. Since square roots are often approximations, the lengths we obtain using the Pythagorean theorem will also be approximations.

Example 7 **A wooden truss for Hamad's new garage has a span of 20 feet and a height of 4 feet. What is the length of the rafter? Round to the nearest hundredth of a foot.**

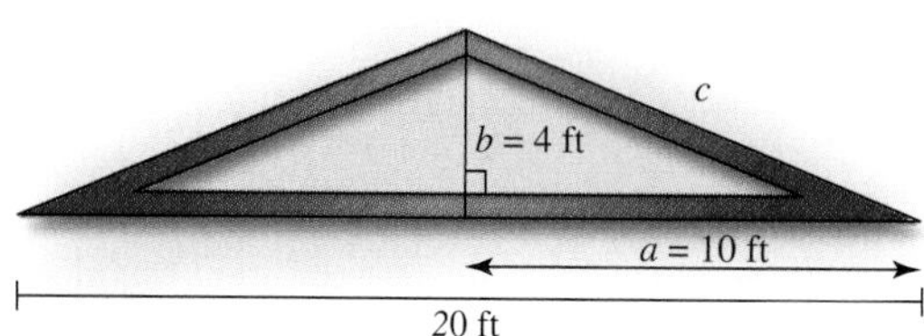

I know . . .

c = Length of rafter = Hypotenuse of right triangle

10 feet = (20 ÷ 2) feet = Half-span of garage (a)

4 feet = Height of truss (b)

Formula: $a^2 + b^2 = c^2$

Equations

$$(\text{Length of rafter})^2 = (\text{Half-span of garage})^2 + (\text{Height of truss})^2$$

$$c^2 = (10\text{ feet})^2 + (4\text{ feet})^2$$

Solve

$$\begin{aligned} c^2 &= (10\text{ ft})^2 + (4\text{ ft})^2 \\ &= 100\text{ ft}^2 + 16\text{ ft}^2 \\ &= 116\text{ ft}^2 \\ c &= \sqrt{116\text{ ft}^2} \\ &\approx 10.77\text{ ft} \end{aligned}$$

Check

Rounding the length of the base of the triangle and the height, the sum of their squares is $(10\text{ feet})^2 + (4\text{ feet})^2$ = 100 square feet + 16 square feet = 116 square feet. The square of the rounded length of the rafter is $(11\text{ feet})^2$ = 121 feet. 116 square feet is reasonably close to 121 square feet.

Answer

The length of the rafter is about 10.77 feet.

When two lines intersect at right angles, they are said to be **perpendicular** lines. Before being deeded to settlers, much of the western United States was surveyed along perpendicular section lines. Roads run on these section lines so that many roads intersect at right angles.

Example 8 **Calculate the distance by air between Hugo and Onamia to the nearest mile.**

I know . . .

c = Hugo to Onamia

150 miles = Hugo to Minden

430 miles = Minden to Onamia

Formula: $a^2 + b^2 = c^2$

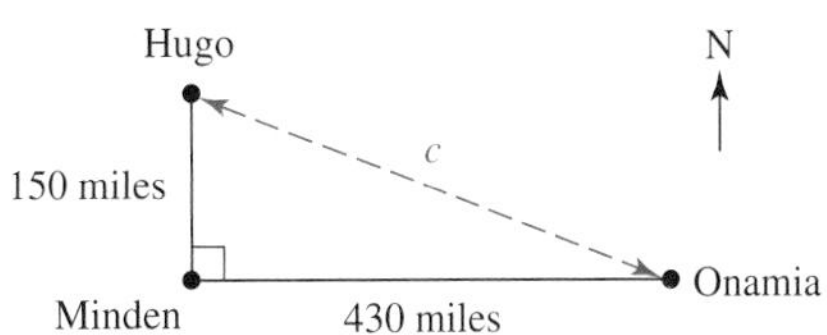

Equations

$$(\text{Hugo to Onamia})^2 = (\text{Hugo to Minden})^2 + (\text{Minden to Onamia})^2$$
$$c^2 = (150 \text{ miles})^2 + (430 \text{ miles})^2$$

Solve

$$\begin{aligned} c^2 &= (150 \text{ miles})^2 + (430 \text{ miles})^2 \\ &= 22{,}500 \text{ miles}^2 + 184{,}900 \text{ miles}^2 \\ &= 207{,}400 \text{ miles}^2 \\ c &= \sqrt{207{,}400 \text{ miles}^2} \\ &\approx 455.4119015 \text{ miles} \\ &\approx 455 \text{ miles (rounded to the nearest mile)} \end{aligned}$$

Check

We expect the direct route to be longer than either of the legs. Estimating,
$c = \sqrt{(200 \text{ miles})^2 + (400 \text{ miles})^2} = 447$ miles, very close to the exact answer.

Answer

The distance by air between Hugo and Onamia is about 455 miles.

CHECK YOUR UNDERSTANDING

1. Calculate the length of the hypotenuse of a right triangle with legs of length 4.6 meters and 6.2 meters. Round to the nearest tenth of a meter.

2. A right triangle has a hypotenuse of length 10 feet and one leg of length 6 feet. What is the length of the other leg?

Answers: 1. 7.7 meters; 2. 8 feet.

EXERCISES SECTION 4.7

Use the approximation of $\pi \approx 3.14$ in this assignment.

Find the square root of the number. If the square root is irrational, round it to the hundredths place.

1. $\sqrt{25}$
2. $\sqrt{\frac{4}{9}}$
3. $\sqrt{17}$
4. $\sqrt{0.1}$
5. $\sqrt{-28}$
6. $\sqrt{144}$
7. $\sqrt{\frac{144}{169}}$
8. $\sqrt{0.0625}$
9. $\sqrt{625}$
10. $\sqrt{62.5}$
11. $\sqrt{0.625}$
12. $\sqrt{15}$
13. $\sqrt{3}$
14. $\sqrt{1}$
15. $\sqrt{0}$
16. $\sqrt{-100}$

Evaluate the expression. Round the final answer to the hundredths place, if necessary.

17. $\sqrt{5^2 - 4 \cdot 1 \cdot 1}$
18. $\sqrt{(-4)^2 - 4 \cdot 1 \cdot 2}$
19. $\sqrt{(-5)^2 - (-2)^2}$
20. $\sqrt{3^2 + 4^2}$

21. $\sqrt{13^2 - 5^2}$

22. $\sqrt{4 \cdot 3 - 2 \cdot 5}$

23. $\sqrt{3 \cdot 7 - 5}$

24. $2 \cdot 8 + \sqrt{4 \cdot 7 - 12}$

25. $\sqrt{5^2 - 3^2} + 3 \cdot 5$

26. $\dfrac{-5 - \sqrt{5^2 - 4 \cdot 1 \cdot 1}}{2 \cdot 1}$

27. $\dfrac{-(-4) + \sqrt{(-4)^2 - 4 \cdot 1 \cdot 2}}{2 \cdot 1}$

28. $\dfrac{-3 - \sqrt{3^2 - 4 \cdot (-2) \cdot 3}}{2 \cdot (-2)}$

29. $\dfrac{2^2 + \sqrt{2^2 - 4 \cdot 3 \cdot 6}}{2 \cdot 3}$

Solve the word problems using the five-step problem-solving strategy.

30. The radius of a circle can be calculated using the formula $r = \sqrt{\dfrac{A}{\pi}}$. If a circular diving tank is to have an area of 600 square meters, calculate the diameter of the diving tank. Round the answer to the nearest tenth of a meter.

31. The radius of a circle can be calculated using the formula $r = \sqrt{\frac{A}{\pi}}$. A traveling circus is ordering portable fence for a circular arena. If the area of the arena is to be 7,850 square feet, what will the *circumference* of the arena be? Round the answer to the nearest foot.

32. When a stone is dropped from a cliff, the time it takes to hit the ground (neglecting the effect of air resistance) can be calculated using the formula $t = \sqrt{\frac{2d}{a}}$ where t is the time to hit the ground, d is the distance from the top of the cliff to the ground, and a is the acceleration of the object due to the force of gravity, $32\ \frac{\text{feet}}{\text{second}^2}$. About how long would it take a stone dropped from a cliff that is 1,350 feet tall to hit the ground? Round the answer to the nearest whole number.

33. The highest waterfall in the world is in Angel, Venezuela, and is 979 meters tall. Using the formula from Problem 32 above, calculate the approximate time it takes a drop of water to fall from the top to the bottom of the waterfall. The acceleration due to gravity in the metric system is $9.8\ \frac{\text{meters}}{\text{second}^2}$. Round the answer to the nearest whole number.

34. The horsepower of an internal combustion engine depends on both the diameter of the cylinders and the number of cylinders in the engine. With D = diameter of the cylinders, N = number of cylinders, and H = horsepower of the engine, $D = \sqrt{\frac{2.5H}{N}}$. What diameter of cylinder is required in a four-cylinder engine to produce 10 horsepower? Round to the nearest tenth of an inch.

35. Charmaine is designing a square deck with a total area of 58 square feet. Use the formula $s = \sqrt{A}$ to calculate the necessary length of the sides of the deck. Round the answer to the nearest tenth of a foot.

36. In electric circuits, the amount of current in amperes, I, depends on the power in watts, P, being drawn through the circuit and the total resistance in ohms, R, of the circuit according to this formula: $I = \sqrt{\frac{P}{R}}$. Calculate the current in a circuit if the power being drawn is 100 watts and the resistance is 144 ohms. Round the answer to the nearest tenth of an ampere (amp).

37. The radius of a circular cylinder can be calculated using the formula $r = \sqrt{\frac{V}{\pi h}}$ where V is the volume, r is the radius, and h is the height of the cylinder. What should be the diameter of a cylindrical oil tank that is to contain 25,000 cubic feet of oil and have a height of 30 feet? Round the answer to the nearest foot.

38. Write a word problem that requires using a formula with a square root. Solve the problem.

Use the Pythagorean theorem to calculate the missing side of the right triangles. Round each answer to the nearest tenth.

	HYPOTENUSE (c)	SIDE (a)	SIDE (b)
39.		6.0	8.0
40.		12.0	15.5
41.	8.0	2.7	
42.	35.0		6.0
43.		3.0	1.0
44.		200.0	500.0
45.	28.0	3.5	
46.		6.5	6.5
47.		0.6	0.9
48.		3.0	4.0

Solve using the five-step problem-solving strategy.

49. A step ladder is 10.0 feet high. If the legs are 4.0 feet apart at the base, what is the length of each side of the ladder? Round to the nearest tenth of a foot.

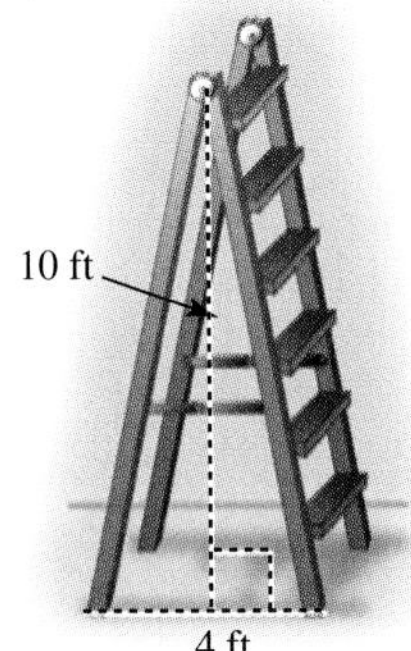

50. On a sunny day, Clarence is waiting for his bus. He looks down and notices that his shadow is stretching out in front of him just about exactly his height of 6 feet 3 inches. What is the distance from the top of Clarence's head to the top of his shadow? Round the answer to the nearest inch.

51. A rectangular window measures 2.5 meters by 1.5 meters. What is the length of the diagonal of this window? Round the answer to the nearest tenth of a meter.

52. A football field from goal line to goal line is 100 yards. The width of the field is about 53 yards. A coach requires his players to run wind sprints from one corner of the field to the opposite corner of the field on the diagonal. What is this distance? Round the answer to the nearest yard.

53. City workers are erecting a large 20-foot Christmas tree in a park and supporting it with a wire that is tied 18 inches from the top and secured to an anchored ring in the ground that is 8 feet from the base of the tree. What is the length of the wire? Round the answer to the nearest inch.

54. Calculate the length of a roof rafter if the roof has a maximum height of 10 feet, the joists of the attic floor underneath the roof are 50 feet wide, and the rafter has an overhang of 10 inches on each side of the roof. Round the answer to the nearest quarter inch.

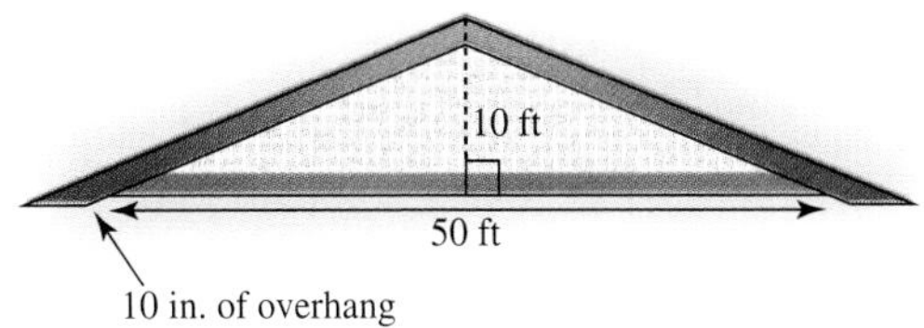

55. A folding table has a brace at each leg that unfolds to a length of 8 inches. If the brace is to extend equally on each side of the corner of the table, how far from the corner should it be attached? Round the answer to the nearest quarter inch.

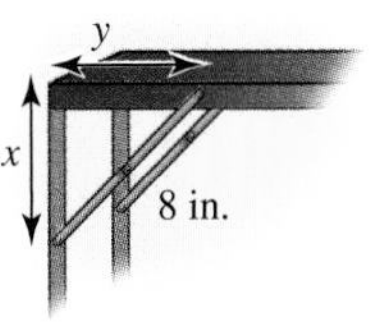

56. Write a word problem that can be solved using the Pythagorean theorem. Solve the problem.

ERROR ANALYSIS

57. Problem: The hypotenuse of a right triangle is 65 inches. The length of one side is 12 inches. Calculate the length of the remaining side. Round to the nearest inch.

Answer:
$$L = \sqrt{65^2 - 12^2}$$
$$= \sqrt{4{,}225 - 144}$$
$$= 4{,}081$$

The length of the remaining side is 4,081 inches.

Describe the mistake made in solving this problem.

Redo the problem correctly.

58. Problem: Evaluate $\sqrt{-100}$.

Answer: 10

Describe the mistake made in solving this problem.

Redo the problem correctly.

REVIEW

59. Write the prime factorization of 305.

60. Write the prime factorization of 100.

61. Simplify $\frac{305}{100}$ to lowest terms.

Read and Reflect

Measurements are an important part of many fields of study and work. Nurses administer doses of drugs in measured amounts; engineers calculate the thickness of a beam; sawmill operators adjust the angle of a blade in fractions of a degree; medical technologists mix measured solutions together. We calculate the miles per gallon of our cars, the cost of vegetables priced by the pound, and the amount of fertilizer to spread on the grass. There is a huge difference, though, between the rough estimates that we use frequently in our daily lives and the careful measurements that are required in many professions. Those careful measurements involve both precision and accuracy.

Accuracy refers to how close a measurement is to reality. If a scale isn't accurate, it cannot report the correct weight. If a person is weighed in the doctor's office and says, "That scale is off!," he is questioning the accuracy of the scale. The accuracy of a consumer Global Position System (GPS) is determined by the signals it receives from satellites and its internal calibrations. The U.S. military can limit the accuracy of a personal GPS by regulating the signals sent from the GPS satellites.

Precision refers to the detail that the measuring device is capable of measuring. If your scale is marked in pounds, you can guess your weight. You could say, "I weigh about 145.5 pounds." You know that the 0.5 pound is really a guess. It might actually be 0.7 pound or 0.3 pound. But you're sure about the 145. On the other hand, you would not go so far as to guess that you weigh 145.58 pounds. There is no way you can guess about the hundredths place when you're not sure about the tenths place. Generally speaking, a measuring device with a scale has a precision one decimal place more than the marked scale. So, if your bathroom scale is marked in pounds, you can guess to a tenth of a pound. You can't guess to a hundredth of a pound. If a ruler is marked in millimeters (0.001 meter), you can guess one more decimal place (to the 0.0001 meter). Precision is expressed in the sciences by "significant figures." If you take chemistry or physics, you will learn about significant figures very early in the course.

Notice that all of these measurements use decimal numbers. One of the advantages of using the metric system for measurement is that all measures are expressed with decimal numbers.

Pure mathematicians tend to be pretty sloppy about precision because they are used to thinking about numbers and variables rather than measurements. Scientists are anything but sloppy about it. A math book may tell you to round all the answers to the hundredths place; a chemistry book will say to round all answers following the rules of significant figures.

The next time you ask someone to be more precise, realize that you are asking for more detail. If you ask them to be more accurate, you are asking for a more correct answer.

Respond

1. A man was checking out at a grocery store and loudly complained that the automatic scan had incorrectly scanned the price of an item. Is he complaining about the accuracy or precision of the scanner? Explain.

2. A Forest Service employee went out with a measuring wheel to measure the length of a popular trail. When she came back and reported that the trail was 35.631 kilometers, her supervisor said that it wasn't possible since her measuring wheel was not that precise. Is the problem with the measurement caused by the wheel itself or by the employee? Explain.

3. The directions for a math assignment on decimal arithmetic require that all answers be rounded to the hundredths place. The problems involve measurements such as weights, lengths, volumes, and areas and involve calculations with formulas. Explain how this direction ignores the concept of precision.

4. The city of Lewiston was acquiring right-of-way from property owners to widen a road and put in sidewalks. The city code for sidewalks required a width of 8 feet. The federal government was financing a large part of the project and required that all the measurements in the project be in the metric system. When the specifications were sent to an engineering firm to be changed into drawings, a conversion table was used to change the specifications to metric. The table rounded the specifications back to the nearest meter. Hence, the 8 feet specifications were rounded to 2 meters and the drawings showed 2-meter sidewalks. As the project went to bid, the public works director discovered that the measurement of the sidewalks was actually less than the required 8 feet.

a. The drawings showed 2-meter sidewalks. What should the eight-foot measurement be in meters, rounded to the nearest tenth?

b. Was this an error in accuracy or precision? Explain.

Section	Terms and Concepts/Procedures
4.1	place value; decimal point; decimal number; rational number; terminating decimal; repeating decimal; irrational number; real number; aligning the decimal points 1. Write the word name of a decimal number. 2. Change a decimal number to an equivalent fraction. 3. Change a decimal number to an equivalent mixed number. 4. Round a decimal number to a specified place value. 5. Add or subtract decimal numbers. 6. Estimate the sum or difference of decimal numbers.
4.2	product 1. Multiply decimal numbers. 2. Estimate the product of decimal numbers.
4.3	quotient; long division 1. Divide decimal numbers. 2. Estimate the quotient of decimal numbers. 3. Simplify expressions with decimal numbers using the order of operations.
4.4	power of ten; scientific notation; place value notation 1. Write positive numbers and measurements in scientific notation. 2. Change numbers and measurements in scientific notation to place value notation.
4.5	conversion factor; unit analysis 1. Convert a measurement to an equivalent measurement in different units using unit analysis.
4.6	pi; circle; circumference; radius, diameter; area; volume; cylinder; perimeter; sphere; pyramid; cone 1. Calculate the circumference of a circle. 2. Calculate the area of a circle. 3. Calculate the volume of a circular cylinder. 4. Calculate the surface area of a circular cylinder. 5. Calculate the volume of a sphere. 6. Calculate the volume of a cone.
4.7	principal square root; rational number; irrational number; real number; Pythagorean theorem 1. Evaluate a principal square root.

SECTION 4.1 INTRODUCTION TO DECIMAL NUMBERS

1. Identify the place value of 9 in 15.559.

2. Write the word name for 17.65.

Rewrite the decimal as a fraction. Simplify to lowest terms.

3. 0.59 **4.** 10.3 **5.** 0.008 **6.** 0.065

Place a < or > between the decimals to make a true statement.

7. −0.03 0.0003 **8.** 0.1 0.09

Round the numbers to the indicated place value.

9. −1,041.925; tenths place **10.** 0.38577; thousandths place

Evaluate.

11. 0.34 + 15.9 **12.** 0.34 − 15.9 **13.** $85.16 − $4.99

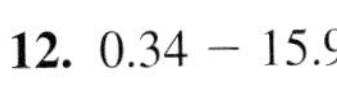

14. 0.05 meters + 0.21 [illegible] meters − 6.2 meters

Estimate the sum or difference.

15. 0.029 + 0.081 **16.** 5.06 − 2.93

SECTION 4.2 MULTIPLICATION OF DECIMAL NUMBERS

Evaluate.

17. (15)(3.9)

18. (−0.15)(0.39)

Estimate the products.

19. (15)(3.9)

20. (3.1)(0.00192)

Solve the word problem using the five-step problem-solving strategy.

21. A roll of wallpaper measures 10.05 meters long by 0.53 meter wide. Calculate the area of wallpaper in square yards. Round to the nearest hundredth.

22. The dimensions of a backboard in NBA basketball are 1.80 meters horizontally and 1.05 meters vertically, with the lower edges 2.90 meters above the floor. A rectangle is drawn behind the ring on each backboard that has an outside dimension of 0.59 meter horizontally and 0.45 meter vertically. What is the area of this rectangle? Round to the nearest hundredth.

23. According to the U.S. Department of Labor's Bureau of Labor Statistics, employer expenditures for workplace-based health care plans in 1992 were \$221.4 billion. Employee expenditures for their health care plans were \$37.2 billion. What was the total expenditure by employers and employees for the health care plans?

24. Calculate the perimeter of a rectangular swimming pool with a length of 15.1 meters and a width of 4.8 meters. Round to the nearest tenth.

SECTION 4.3 DIVISION OF DECIMAL NUMBERS

Evaluate.

25. $(-12.6) \div 0.2$

26. $108.93 \div 0.03$

27. $-0.08 \div 0.03$

28. $-0.50 \div 0.02$

Estimate the quotient.

29. $6.2 \div 0.29$

30. $0.77 \div 0.044$

Solve the word problem using the five-step problem-solving strategy.

31. Rickey Williams was named the 1998 Heisman Trophy Winner. Running for the University of Texas, Williams rushed 4,065 yards in 22 games. Calculate his rushing yardage per game. Round the answer to the tenths place.

32. The average return to an apple grower for a box of apples was about $9.50 for a 42-pound box. What was the return per pound of apples? Round to the nearest cent.

Rewrite the fraction as an equivalent decimal.

33. $\frac{5}{8}$

34. $\frac{1}{9}$

Evaluate.

35. $13.1 + 2.30 \div 0.2 - (0.4)^2$

36. $(13.1 + 2.30) \div 0.2 - (0.4)^2$

37. $1.9(1.9 - 0.38)$

38. $\frac{1.9(-0.38)}{1.9 - (-0.38)}$

SECTION 4.4 POWERS OF TEN AND SCIENTIFIC NOTATION

Evaluate.

39. $(-9.11)(-1{,}000)$

40. $91.1 \div 10$

Rewrite the measurement in scientific notation.

41. The area of the Caribbean Sea is 2,515,900 square kilometers.

42. The total number of deaths in the United States known to be caused by melanoma skin cancer in 1996 was 6,680 deaths.

43. The mass of a neutron is 0.000000000000000000000001675 kilogram.

SECTION 4.5 UNIT ANALYSIS

Convert the measurement to the given unit using unit analysis. Round to the nearest tenth.

44. $16.1 \dfrac{\text{kilograms}}{\text{meter}^2}$ to $\dfrac{\text{grams}}{\text{inch}^2}$

45. $4 \dfrac{\text{miles}}{\text{hour}}$ to $\dfrac{\text{inches}}{\text{second}}$

SECTION 4.6 PI

46. Calculate the circumference of a circle with a diameter of 4.3 feet. Round to the nearest tenth of a foot.

47. Define *pi*.

48. Calculate the area of a circle with a radius of 6.41 centimeters. Round to the nearest hundredth of a square centimeter.

49. Calculate the volume of a sphere with a diameter of 5.0 inches. Round to the nearest tenth of a cubic inch.

50. Calculate the volume of a circular cylinder with a radius of 95 millimeters and a height of 3.1 centimeters. Round the answer to the nearest cubic millimeter.

51. Calculate the volume of a circular pyramid (a cone) that has a radius of 3.0 inches and a height of 8.5 inches. Round the answer to the nearest tenth of a cubic inch.

Solve using the five-step problem-solving strategy.

52. Barb's bicycle tire has a diameter of 27 inches. How many revolutions will the tire make if Barb rides a distance of 50 *miles* on her bicycle? (1 mile = 5,280 feet) Round the answer to the nearest whole number.

53. *Sputnik* was the first satellite launched into space by the former Soviet Union in October of 1957. It was about the size of a beach ball. Its diameter was 22 inches and its weight was 183.3 pounds. Assuming that it was essentially spherical in shape, what was the volume of *Sputnik?* Round the answer to the nearest cubic inch.

54. A dome tent has a circular floor. If the floor has a diameter of 6.2 feet, what is the area of the tent floor? Round the answer to the nearest tenth of a square foot.

55. A circular compass has a diameter of 3.3 inches. What is the circumference of the compass? Round to the nearest tenth.

56. A telescope has a mirror with a diameter of 15.24 centimeters. What is the area of this mirror? Round to the nearest hundredth of a square centimeter.

57. A carton of Nestle Instant Breakfast is cylindrical in shape. What is the volume of a container that has a diameter of 5 inches and a height of $7\frac{3}{4}$ inches? Round the answer to the nearest cubic inch.

58. A mail-order catalog sells mirrored spheres called "gazing globes" which are set on top of pedestals in gardens as decoration. These spheres were very common in Victorian gardens in the late nineteenth century. Each sphere has a diameter of 8 inches. A single sphere is shipped in a packing box that measures 12 inches by 12 inches by 15 inches. What volume of packing peanuts is required to cushion the sphere in the packing box? Round the answer to the nearest cubic inch.

SECTION 4.7 SQUARE ROOTS

Evaluate the square roots. Round any irrational numbers to the thousandths place.

59. $\sqrt{25}$ **60.** $-\sqrt{25}$ **61.** $\sqrt{-25}$ **62.** $\sqrt{35}$

63. $\sqrt{\frac{16}{49}}$ **64.** $\sqrt{0.4}$ **65.** $\sqrt{15.21 \text{ square meters}}$

Use the Pythagorean theorem to calculate the missing length in a right triangle. Round all answers to the nearest hundredth of a unit.

	LENGTH OF SIDE a	LENGTH OF SIDE b	LENGTH OF HYPOTENUSE c
66.	3.00 centimeters	4.00 centimeters	
67.	99.02 feet		179.38 feet
68.		0.08 inch	0.12 inch

Solve the word problems using the five-step problem-solving system.

69. The tee on a golf hole forces golfers to hit their first shot over a pond. The distance across the pond is too long to measure directly with a tape measure. Explain how you could use the Pythagorean theorem and a tape measure to calculate the distance required to clear the pond.

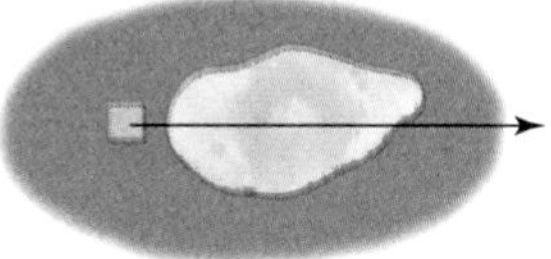

70. The distance between the bases on a Major League baseball field are 90 feet. Use the Pythagorean theorem to calculate the distance between second base and home.

1. Identify the place value of the digit 5: 24.975.

2. Solve the equation $-0.0064x = 2.065$. Round the answer to the nearest tenth.

3. Change the decimal 0.875 to a fraction in lowest terms.

4. Calculate the area of the quarter-circle shown in the diagram. Round to the nearest tenth. Use $\pi \approx 3.14$.

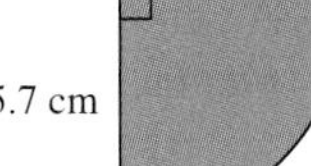

5. Evaluate (0.0067)(100,000).

6. Round 4.84578 to the nearest hundredth.

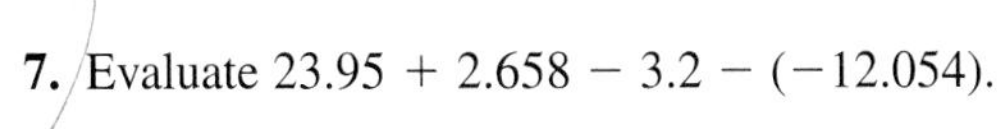

7. Evaluate $23.95 + 2.658 - 3.2 - (-12.054)$.

8. Round -3.9964 to the nearest tenth.

9. Write \$465,000,000 in scientific notation.

10. Evaluate \$53.96 + \$21.99 + \$5.63.

11. An empty beaker weighs 52.5 grams. When potassium chloride is added to the beaker, the combined weight is 58.2 grams. How much potassium chloride was added to the beaker?

12. Write 2.9×10^{-3} in place value notation.

13. Evaluate $\frac{-7.92 + 3.1(4.01)}{3.6^2 - 7.43}$. Round the answer to the nearest tenth.

14. Evaluate $-4.3 \div 2.904$. Round to the nearest hundredth.

15. A doctor prescribes 300 mg per dose of Ludiomil, 3 times a day for 5 days. The drug comes in 75-mg tablets. How many tablets should be given per dose?

Skip

16. Calculate the volume of a circular cylinder with a radius of 7 in. and a height of 5 in. Use $\pi \approx 3.14$. Round to the nearest whole number.

17. Evaluate 23.4(1.8).

18. Divide 24.8 by 10^4.

19. Evaluate 34.4 m − 23.6 m.

Skip

20. The speed of light is approximately 669,600,000 miles per hour. Write this speed in scientific notation.

21. Evaluate $12.7 + 3.2^2 - 4.08(3.1)$.

22. Use the formula $F = 1.8C + 32$ to convert 26.7°C to degrees Fahrenheit. Round to the nearest tenth of a degree.

23. A rental car company is advertising a daily rental of \$45.50 and 13¢ per mile. If Crayton rents a car for 4 days and travels 1,432.8 miles, how much will it cost for the rental car? Round to the nearest cent.

24. The speed limit on some roads in Canada is 70 kilometers per hour. What is this speed in miles per hour? Round to the nearest whole number.

25. The recommended dose of phenobarbital for an infant is 2 mg per kg of body weight per day, given in two equally divided doses. For an infant weighing 18 lb., how many mg per dose, given twice a day, is recommended? Round to the nearest tenth.

26. Estimate the quotient of $74.53 \div 0.527$.

27. Sally is building a house with a floor plan of 1,360 ft^2. She is trying to keep the cost per square foot down to \$70.50. At this price what will it cost to build the house?

28. A spherical balloon has a diameter of 6.8 in. Calculate the volume of the balloon. Use $\pi \approx 3.14$. Round to the nearest tenth.

29. Calculate the perimeter of the semicircle shown in the diagram. Use $\pi \approx 3.14$.

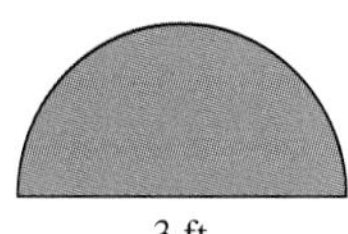

30. A helicopter landing pad is circular with a diameter of 34.8 m. Calculate the area of the landing pad. Round to the nearest tenth. Use $\pi \approx 3.14$.

31. A right triangle has a hypotenuse of length 15 cm and one leg of length 12 cm. What is the length of the other leg?

32. Evaluate $-\sqrt{36}$.

33. Calculate the volume of a cone with a diameter of 4 ft and a height of 3 ft. Round to the nearest whole number. Use $\pi \approx 3.14$.

34. Evaluate $\sqrt{4.87}$. Round to the nearest hundredth.

35. Evaluate $\sqrt{-19}$.

36. A can of tomatoes has a diameter of 3.0 inches and a height of 4.5 inches. Twelve cans are packed in a 4.6-inch high box that measures 9.1 inches by 12.2 inches. How much empty space is in a full box of cans? Round to the nearest tenth.

37. Evaluate $\sqrt{5^2 + 12^2}$.

5

Expressions and Equations

Algebra is the application of the rules of arithmetic to unknown values represented by variables. The unknown value may appear in an algebraic expression or in an equation.

In this chapter, we will learn how to solve equations. We begin by simplifying expressions that do not include variables using the order of operations. We will learn how to simplify algebraic expressions by combining like terms and how to solve equations using the properties of equality.

CHAPTER OUTLINE

Section 5.1 Evaluating Algebraic Expressions

LEARNING TOOLS

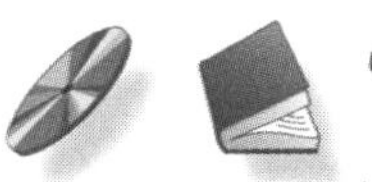

CD-ROM SSM VIDEO

Sneak Preview

A variable is a symbol that represents an unknown number. In using formulas to solve problems, we have worked with variables. In this section, we will continue to learn more about variables and evaluating expressions that include variables.

After completing this section, you should be able to:

1) Evaluate a numerical expression following the order of operations.
2) Evaluate an algebraic expression following the order of operations.

DISCUSSION

The Order of Operations

A **numerical expression** includes numbers and at least one of the arithmetic operations of addition, subtraction, multiplication, and division. For example, $3(2.4)^2 - 4.5(-1.2)$ is a numerical expression. To evaluate a numerical expression, we do the arithmetic following the order of operations.

PROCEDURE

To evaluate a numerical expression following the order of operations:

1. **Grouping:** Operations inside of grouping symbols are done first, following rules 2–4. Grouping symbols include parentheses ( ), brackets [ ], braces { }, absolute values | |, radical signs $\sqrt{}$, and fraction bars —.
2. **Exponents.** Evaluate all exponents and square roots from left to right.
3. **Multiplication or Division.** Do all multiplication and divisions from left to right.
4. **Addition or Subtraction.** Do all additions and subtractions from left to right.

Example 1 **Evaluate $3(2.4)^2 - 4.5(-1.2)$.**

$3(2.4)^2 - 4.5(-1.2)$

$= 3(5.76) - 4.5(-1.2)$ *Evaluate the exponential expression first.*

$= 17.28 + 5.4$ *Multiply and divide from left to right.*

$= 22.68$ *Add and subtract.*

In this next expression, a numerical expression is inside the radical sign. We do the operations inside the radical sign first.

Example 2 **Evaluate $\sqrt{1.3^2 - 1.2^2}$**

$\sqrt{1.3^2 - 1.2^2}$ *Work inside the radical sign first.*

$= \sqrt{1.69 - 1.44}$ *Evaluate the exponential expressions.*

$= \sqrt{0.25}$ *Subtract.*

$= 0.5$ *Evaluate the square root.*

When evaluating a fraction, first evaluate the numerator and denominator separately. The final operation will be division.

Example 3 **Evaluate $\dfrac{|3(-6) - 4(8)|}{-2^2 + 3}$.**

$\dfrac{|3(-6) - 4(8)|}{-2^2 + 3}$ *Numerator: multiply inside the absolute values. Denominator: evaluate exponential expression.*

$= \dfrac{|-18 - 32|}{-4 + 3}$ *Numerator: add and subtract inside the absolute values. Denominator: add and subtract.*

$= \dfrac{|-50|}{-1}$

$= \dfrac{50}{-1}$ *Evaluate absolute value.*

$= -50$ *Divide.*

CHECK YOUR UNDERSTANDING

Evaluate.

1. $3(-6.2) + 3.5$

2. $\sqrt{\left(\dfrac{2}{3}\right)^2 - \dfrac{1}{2}}$

3. $(6 \div 3 + 4 \cdot 2)^2$

Answers: 1. −15.1; 2. not a real number; 3. 100.

DISCUSSION

Evaluating Algebraic Expressions

An algebraic **term** is a number (7), a variable (x), or a product of a number and variable or variables ($3x$ or $3xy$). An **algebraic expression** is a combination of algebraic terms. To evaluate an algebraic expression, we replace each variable with given numbers. Place parentheses around any negative numbers that follow an operation sign such as + or −. Evaluate the resulting numerical expression, following the order of operations.

Example 4 **Evaluate $x + y + z$ when $x = -3$, $y = 8$, and $z = -5$.**

$x + y + z$

$= -3 + 8 + (-5)$ *Replace each variable with its given value.*

$= 5 - 5$ *Add or subtract from left to right.*

$= 0$ *Add or subtract from left to right.*

We often replace the variables in an algebraic expression with a measurement that includes units.

Example 5 **Evaluate $3a + 6b + c$ when $a = 1.5$ meters, $b = 2.3$ meters, and $c = -4.2$ meters.**

$3a + 6b + c$

$= 3(1.5 \text{ m}) + 6(2.3 \text{ m}) + (-4.2 \text{ m})$ *Replace each variable with its given value.*

$= 4.5 \text{ m} + 13.8 \text{ m} + (-4.2 \text{ m})$ *Multiply.*

$= 18.3 \text{ m} - 4.2 \text{ m}$ *Add or subtract from left to right.*

$= 14.1 \text{ m}$ *Add or subtract from left to right.*

In algebraic expressions, the letter n is often used to represent an unknown exponent.

Example 6 **Evaluate x^n if $x = 2$ and $n = 3$.**

x^n

$= 2^3$ *Replace x with 2 and n with 3.*

$= 2 \cdot 2 \cdot 2$

$= 8$

When replacing a variable with a fraction or a negative number, we write parentheses around the number.

Example 7 **Evaluate x^4y when $x = \frac{2}{3}$ and $y = 15$.**

x^4y

$= \left(\frac{2}{3}\right)^4 \cdot 15$ *Replace each variable with the given number.*

$= \left(\frac{2}{3}\right)^4 \cdot \frac{15}{1}$ *Rewrite the whole number as a fraction.*

$= \frac{2}{3} \cdot \frac{2}{3} \cdot \frac{2}{3} \cdot \frac{2}{\mathbf{3}} \cdot \frac{\mathbf{15}}{1}$ *Rewrite the exponential expression as multiplication.*

$= \frac{2 \cdot 2 \cdot 2 \cdot 2 \cdot \mathbf{5}}{3 \cdot 3 \cdot 3 \cdot \mathbf{1} \cdot 1}$ *Simplify before multiplying. The common factor is 5.*

$= \frac{80}{27}$

Example 8 **Evaluate $x^2 \div y$ when $x = \frac{2}{3}$ and $y = -\frac{5}{6}$.**

$x^2 \div y$

$= \left(\frac{2}{3}\right)^2 \div \left(-\frac{5}{6}\right)$ *Replace the variables with the given numbers. Separate the ÷ and − signs with parentheses.*

$= \frac{2}{3} \cdot \frac{2}{3} \div \left(-\frac{5}{6}\right)$ *Rewrite the exponential expression as multiplication.*

$= \frac{2}{3} \cdot \frac{2}{3} \cdot \left(-\frac{6}{5}\right)$ *Rewrite division as multiplication of the reciprocal.*

$= -\frac{2 \cdot 2 \cdot \mathbf{6}}{\mathbf{3} \cdot 3 \cdot 5}$ *Simplify before multiplying.*

$= -\frac{2 \cdot 2 \cdot \mathbf{2}}{\mathbf{1} \cdot 3 \cdot 5}$

$= -\frac{8}{15}$

When we multiply measurements, the resulting measurement has a new unit.

Example 9 **Evaluate $\frac{bh}{2}$ when $b = 12$ cm and $h = 4$ cm.**

$= \frac{bh}{2}$

$= \frac{12 \text{ cm} \cdot 4 \text{ cm}}{2}$ *Replace b with 12 cm and h with 4 cm.*

$= \frac{48 \text{ cm}^2}{2}$ *Multiply or divide from left to right.*

$= 24 \text{ cm}^2$ *Multiply or divide from left to right.*

If the algebraic expression is a fraction, evaluate the numerator and denominator separately. The fraction bar is acting as a grouping symbol.

Example 10 **Evaluate $\frac{2x + y}{x - y}$ when $x = 3.4$ and $y = -0.4$. Round to the nearest hundredth.**

$\frac{2x + y}{x - y}$

$= \frac{2(3.4) + (-0.4)}{3.4 - (-0.4)}$ *Replace variables with the given numbers.*

$= \frac{6.8 - 0.4}{3.4 + 0.4}$ *Numerator: multiply and divide from left to right.* *Denominator: add and subtract.*

$= \frac{6.4}{3.8}$ *Numerator: add and subtract.* *Denominator: add and subtract.*

$= 1.684\ldots$ *Divide.*

≈ 1.68 *Round to the nearest hundredth.*

CHECK YOUR UNDERSTANDING

1. Evaluate $x + 6$ when $x = 8$.

2. Evaluate $x^n - y$ when $x = 3.2$, $n = 2$, and $y = 1.5$.

3. Evaluate $2x + 3y - 5$ when $x = 7$ and $y = \frac{4}{3}$.

4. $\frac{y}{x} + z$ when $x = 2$, $y = 6$, and $z = 10$.

5. Evaluate lwh when $l = 2$ cm, $w = 4$ cm, and $h = 3$ cm.

Answers: 1. 14; 2. 8.74; 3. 13; 4. 13; 5. 24 cm³.

EXERCISES SECTION 5.1

Evaluate. Round to the hundredths place, if necessary.

1. $3 \cdot 7 - 4 \cdot 2$

2. $2.4(3.2) - 5.1^2$

3. $\sqrt{3 \cdot 6 - 5 \cdot 2}$

4. $|4 - 5 \cdot 3|$

5. $\dfrac{3^2 \cdot 4 - 5 \cdot 2}{3 \cdot 8 - 4^2}$

6. $\dfrac{4}{5} \cdot \left(-\dfrac{2}{3}\right) - \dfrac{1}{3} \cdot \left(-\dfrac{3}{5}\right)$

Evaluate the algebraic expression for the given values. Round to the hundredths place.

7. $x + y$ when $x = 420$ and $y = 10$

8. xy when $x = 420$ and $y = 10$

9. $x \div y$ when $x = 420$ and $y = 10$

10. $\dfrac{x}{y}$ when $x = 420$ and $y = 10$

11. $xy - x$ when $x = 420$ and $y = 10$

12. $x - y$ when $x = 420$ and $y = 10$

13. $x + y^2$ when $x = 420$ and $y = 10$

14. xy^2 when $x = 420$ and $y = 10$

15. $xy + y^2$ when $x = 420$ and $y = 10$

16. $x - 3y$ when $x = 420$ and $y = 10$

17. $x + y$ when $x = 2$ and $y = 8$

18. $a + b - c$ when $a = 9$, $b = 5$, and $c = 6$

19. $f - g$ when $f = 23$ and $g = 17$

20. $x - (y + z)$ when $x = 37$, $y = 12$, and $z = 8$

21. $(a + b)^2 + (c - d)^2$ when $a = 3$ meters, $b = 2.6$ meters, $c = 1.7$ meters and $d = 4$ meters

22. $(r - s + t)^2$ when $r = 2.1$, $s = -4$, and $t = 5$

23. $p + q - p - q$ when $p = 3.2$ grams and $q = 2.1$ grams

24. $(x + y) - (y - x)$ when $x = 17.6$ and $y = 22.3$

25. $(a + b + c)^2 - (a + b - c)^2$ when $a = 2$, $b = 5$, and $c = -9$

26. $\dfrac{ab - c}{a + b - c}$ when $a = 3$, $b = -8$, and $c = 5$

27. $\dfrac{|xy - z|}{x^2 - y^2}$ when $x = -4$, $y = 2$, and $z = -3$

28. $\sqrt{b^2 - 4ac}$ when $a = 2$, $b = -7$, and $c = 3$

29. $\dfrac{-b + \sqrt{b^2 - 4ac}}{2a}$ when $a = 1$, $b = -6$, and $c = -4$

30. $\dfrac{-b - \sqrt{b^2 - 4ac}}{2a}$ when $a = 1$, $b = -6$, and $c = -4$

31. $a^2 + b^2$ when $a = 12$ and $b = 2$

32. $(a + b)^2$ when $a = \dfrac{2}{3}$ and $b = -\dfrac{4}{5}$

33. $\left|\dfrac{a}{b - d}\right|$ when $a = 3.4$, $b = 0.2$, and $d = -2.5$

34. $4(c - d)^2$ when $c = -6$ and $d = -7$

35. $\frac{|bc|}{bc}$ when $b = 3.5$ and $c = -4$

36. $(abcd)^f$ when $a = 3$, $b = 1.5$, $c = -8.5$, $d = -4$, and $f = 2$

37. $a + (c^2 + b \div d) - e$ when $a = 2$, $b = 15$, $c = -4$, $d = -3$, and $e = -2$

38. $\frac{ac + ae - aa}{f}$ when $a = 3$, $c = -4$, $e = -7$, and $f = 5$

39. $\{(a + c) - 2bcd\}^2$ when $a = 9$, $b = -0.5$, $c = 1.5$, and $d = -2$

ERROR ANALYSIS

40. Problem: Evaluate $x + y$ when $x = 1$ foot and $y = 2$ feet.

Answer:

$x + y$
$= 1 + 2$
$= 3$

Describe the mistake made in solving this problem.

Redo the problem correctly.

41. Problem: Evaluate $x - yz$ when $x = 3$, $y = 4$, and $z = -2$.

Answer:

$x - yz$
$= 3 - 4(-2)$
$= -1(-2)$
$= 2$

Describe the mistake made in solving this problem.

Redo the problem correctly.

REVIEW

42. Add 4 meters + 7 meters.

43. Subtract 12 feet − 6 feet.

44. Simplify $3^2 \cdot 3^4 \cdot 3^7$. Write as an exponential expression.

Section 5.2 Simplifying Algebraic Expressions

LEARNING TOOLS

Sneak Preview

When we evaluate an algebraic expression, we replace the variables with given numbers and do the arithmetic following the order of operations. When we simplify an algebraic expression, we combine algebraic terms. Variables are not replaced with numbers.

After completing this section, you should be able to:

1) Identify like terms in an algebraic expression.

2) Simplify an algebraic expression.

INVESTIGATION Like Units and Like Terms

Measurements that have the same units are examples of **like units.** When we add or subtract such measurements, this is called **combining like units.** Only measurements that are in the same units can be combined.

1. A port district purchased 23 acres for a new business and technology park and signed an option to buy an additional 13 acres. What is the total acreage that may be purchased for the park?

In algebra, we **simplify algebraic expressions** that contain **terms.** The algebraic expression $4b + 3y$ has two terms.

4 blue cubes can represent the term $\mathbf{4b}$

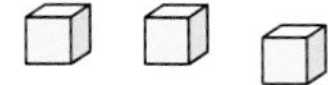

3 yellow cubes can represent the term $\mathbf{3y}$

The number in a term is called a **coefficient.** In an algebraic expression, terms are usually written in alphabetical order.

2. What is the coefficient in the term $4b$?

3. What is the coefficient in the term $3y$?

Answers: 1. 36 acres; 2. 4; 3. 3.

When we mix blue and yellow paint together, we get green paint. When we mix blue and yellow cubes together, we do not get green cubes. We still have a total of 4 blue cubes and 3 yellow cubes. Blue and yellow cubes cannot be combined. In the same way, terms that have different variables cannot be combined. $\mathbf{4b + 3y}$ cannot be **simplified** any further.

4. Write an algebraic expression that represents this collection of cubes:

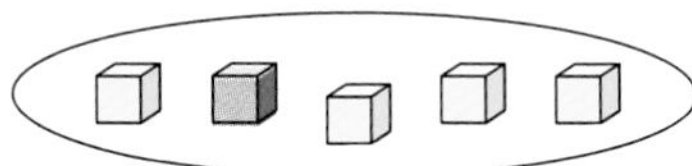

5. Write an algebraic expression that represents this collection of cubes:

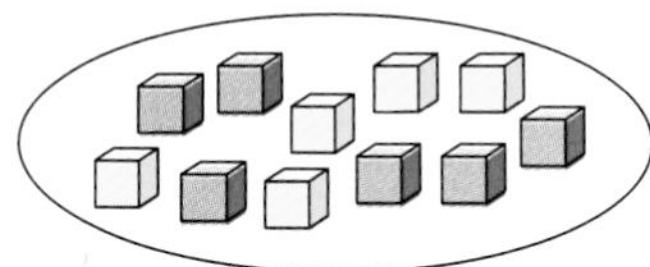

6. If we combine both groups of cubes, we get this collection. Complete this sum that represents combining these cubes.

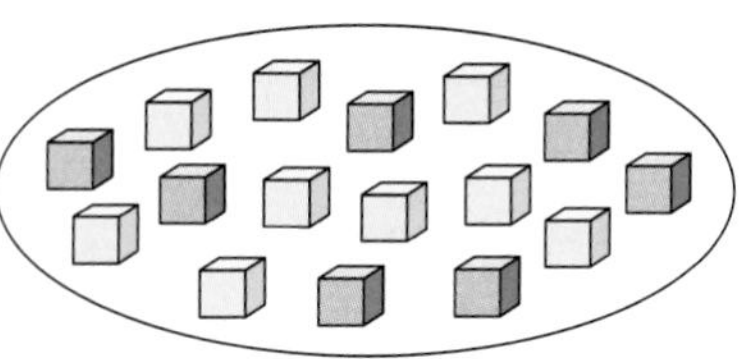

$(1b + 4y) + (6b + 5y) =$

7. Does the sum in problem 6 represent the complete collection of cubes?

8. Explain how to add two algebraic expressions.

Answers: 4. $1b + 4y$; *5.* $6b + 5y$; *6.* $7b + 9y$.

Add the algebraic expressions.

9. $(4x + 6y) + (2x + 1y)$

10. $(4w + 5x) + (3w + 2x)$

11. $(3x + 4y + 5z) + (2x + 6y + 1z)$

Answers: 9. $6x + 7y$; *10.* $7w + 7x$; *11.* $5x + 10y + 6z$

Like terms in an algebraic expression have exactly the same variable(s). We **simplify** an algebraic expression by combining any like terms in the expression.

Simplify each expression. If the terms are unlike and cannot be combined, write "already simplified."

12. $4x + 8x$

13. $12y + 15x + 3y$

14. $28q - 13q$

15. $20x + 14xy$

16. $31xyz + 2xyz$

17. $3x + 1x$

18. $3x + x$

19. $x + x + x$

20. $12w + 14y$

21. $5x + 8 + 7x + 1$

Answers: 12. $12x$; 13. $15x + 15y$; 14. $15q$; 15. already simplified; 16. $33xyz$; 17. $4x$; 18. $4x$; 19. $3x$; 20. already simplified; 21. $12x + 9$.

DISCUSSION

Simplifying Algebraic Expressions

An algebraic **term** can be a number (7), a variable (x), or a product of a number and variable or variables ($3x$ or $3xy$). Like terms can be combined in the same way as like units. To add 3 miles and 5 miles, we add the numbers and keep the unit "miles": 3 miles + 5 miles = 8 miles. Similarly, to add $4x$ and $5x$, we add the numbers and keep the variable x: $4x + 5x = 9x$. The number in a term is called a **coefficient.** In the term $9x$, 9 is the coefficient of x.

An **algebraic expression** is a combination of algebraic terms. Algebraic expressions can be simplified or evaluated. To **simplify** an algebraic expression, we combine like terms by adding or subtracting the coefficients and keeping the common variable.

Example 1 **Simplify the expression $6y + 3y$.**

$6y + 3y$
$= (6 + 3)y$ *Add the coefficients and keep the common variable.*
$= 9y$

Example 2 **Simplify the expression $45b - 72b + 6b$.**

$45b - 72b + 6b$
$= (45 - 72 + 6)b$ *Add the coefficients and keep the common variable.*
$= -21b$

When an expression contains more than one variable, only like terms can be combined.

Example 3 **Simplify the expression $6x + 5y - 9 + 7x + 3y + 12$.**

$6x + 5y - 9 + 7x + 3y + 12$
$= (6 + 7)x + (5 + 3)y + (-9 + 12)$ *Add the coefficients of like terms.*
$= 13x + 8y + 3$

CHECK YOUR UNDERSTANDING

Simplify each expression by combining like terms.

1. $2x + 3 + 6x + 7$

2. $24x + 53y + 37 + 16x + 52y + 23$

3. $35x + 25y + 72z + 24x + 53y + 29z$

4. $321a + 123b + 184a + 263b$

Answers: 1. $8x + 10$; 2. $40x + 105y + 60$; 3. $59x + 78y + 101z$; 4. $505a + 386b$.

Exponents can be used to represent repeated multiplication of a variable. For example, $x \cdot x \cdot x$ is written x^3. An expression may also contain more than one variable with an exponent. For example, $x \cdot y \cdot x \cdot y \cdot y$ can be written as x^2y^3. Generally, exponential expressions containing variables are written so that the variables are in alphabetical order. x^2 and x are not like terms, just as meters and meters^2 are not like units. Although both terms contain the same variable, they do not have the same exponent.

Example 4 **Simplify $130x^2 + 350x$.**

$130x^2 + 350x$ cannot be simplified since x and x^2 are not like terms.

Example 5 **Simplify $4.5x^2 - 6x + 7.3x^2 + 1.2x$.**

$4.5x^2 - 6x + 7.3x^2 + 1.2x$
$= (4.5 + 7.3)x^2 + (-6 + 1.2)x$ *Combine like terms.*
$= 11.8x^2 + (-4.8)x$
$= 11.8x^2 - 4.8x$

Unlike terms inside of parentheses cannot be combined. To eliminate the parentheses, we use the distributive property: $a(b + c) = ab + ac$.

Example 6 **Simplify $5(x + 3) + 8$.**

$5(x + 3) + 8$ *Unlike terms inside the parentheses cannot be combined.*
$= 5x + 5(3) + 8$ *Apply the distributive property to eliminate the parentheses.*
$= 5x + 15 + 8$ *Multiply.*
$= 5x + 23$ *Add. Since 5x and 23 are not like terms, this expression is simplified.*

Example 7 **Simplify $-6(x^2 + 5x)$.**

$-6(x^2 + 5x)$ *x^2 and x are not like terms so cannot be combined.*
$= -6 \cdot x^2 + (-6)(5x)$ *Apply the distributive property.*
$= -6x^2 - 30x$ *Simplify.*

We can also simplify expressions that include subtraction inside the parentheses.

Example 8 **Simplify $4(3x - 8)$.**

$4(3x - 8)$
$= 4(3x) - 4(8)$ *Use the distributive property.*
$= 12x - 32$

Example 9 **Simplify $-4(3x - 8)$.**

$-4(3x - 8)$
$= -4(3x) - (-4)(8)$ *Use the distributive property. Use parentheses to keep signs separated.*
$= -12x - (-32)$
$= -12x + 32$ *The opposite of a negative number is a positive number.*

If possible, simplify inside the parentheses before using the distributive property.

Example 10 **Simplify $-4(-3x - (-8))$.**

$-4(-3x - (-8))$
$= -4(-3x + 8)$ *The opposite of a negative number is a positive number.*
$= -4(-3x) + (-4)(8)$ *Use the distributive property.*
$= 12x + (-32)$ *Multiply.*
$= 12x - 32$ *Addition of an opposite can be rewritten as subtraction.*

The distributive property also allows us to divide an algebraic expression that has unlike terms inside the parentheses.

Example 11 **Simplify $(10x - 8) \div 2$.**

$(10x - 8) \div 2$
$= (10x \div 2) - (8 \div 2)$ *Use the distributive property.*
$= 5x - 4$ *Divide.*

The distributive property can be applied to any number of terms inside the parentheses.

Example 12 **Simplify $-3(7x - 5y + 8z - 9)$.**

$-3(7x - 5y + 8z - 9)$
$= -3(7x) - (-3)(5y) + (-3)(8z) - (-3)(9)$ *Use the distributive property.*
$= -21x + 15y - 24z + 27$ *Multiply.*

Example 13 **Simplify $\dfrac{4x - 8y + 24}{-4}$.**

$\dfrac{4x - 8y + 24}{-4}$
$= \dfrac{4x}{-4} - \dfrac{8y}{-4} + \dfrac{24}{-4}$ *Use the distributive property.*
$= -1x - (-2y) + (-6)$ *Divide.*
$= -x + 2y - 6$ *The opposite of a negative number is a positive number.*

CHECK YOUR UNDERSTANDING

Simplify.

1. $3(x + 1)$
2. $2(h - (-5))$
3. $7(5y - 4)$
4. $-6(x - 5y)$
5. $(4x + 8) \div 2$
6. $(6x - 3y + 12) \div (-3)$

Answers: 1. $3x + 3$; 2. $2h + 10$; 3. $35y - 28$; 4. $-6x + 30y$; 5. $2x + 4$; 6. $-2x + y - 4$.

DISCUSSION

Order of Operations and Simplifying Expressions

We simplify an algebraic expression that contains many operations by following the order of operations and combining like terms.

Example 14 **Simplify $2(x + 5) + 6x \div 3$.**

$2(x + 5) + 6x \div 3$	
$= \mathbf{2x + 10} + 6x \div 3$	*Use the distributive property to eliminate parentheses.*
$= 2x + 10 + \mathbf{2x}$	*Multiply and divide from left to right.*
$= \mathbf{4x} + 10$	*Add and subtract from left to right by combining like terms.*

Example 15 **Simplify $56x \div 2^3 + 3(6x - 4x + 5) - 4x$.**

$56x \div 2^3 + 3(6x - 4x + 5) - 4x$	
$= 56x \div 2^3 + 3(\mathbf{2x} + 5) - 4x$	*Work in parentheses; combine like terms.*
$= 56x \div \mathbf{8} + 3(2x + 5) - 4x$	*Evaluate exponents.*
$= \mathbf{7x} + 3(2x + 5) - 4x$	*Multiply or divide from left to right.*
$= 7x + \mathbf{6x + 15} - 4x$	*Multiply or divide from left to right.*
$= \mathbf{13x} + 15 - 4x$	*Add and subtract from left to right.*
$= \mathbf{9x} + 15$	*Add and subtract from left to right.*

CHECK YOUR UNDERSTANDING

Simplify the expression.

1. $3x + 8y - 2x + 9y + 7$

2. $2(4x + 5) + 3^2 - 3x$

Answers: 1. $x + 17y + 7$; 2. $5x + 19$.

EXERCISES SECTION 5.2

Identify the coefficient of x in each expression.

1. $3x + 6y$

2. $-7x + 8y - 9$

3. $-8x + 5$

4. x

Simplify the expression by combining like terms.

5. $15x + 26x$

6. $-31x - 19x + 45x + 7x + 1x$

7. $44x - 71x + 19$

8. $3x - 2y + 5x + 9y$

9. $14.6x + 5.2y + 9 - 3.1x - 18y - 15$

10. $1{,}000p + 4{,}000q + 200p + 300p$

11. $21ab - 42ab + 67ab$

12. $1{,}984h + 2{,}986h - 3{,}109h$

13. $1.2x + 1.4x$

14. $-3x + 2.4y - 11 + 2x + 1.5y + 8$

15. $33x + 40y - 105 - 24x + 15y + 81$

16. $300x + 43y - 7 + 37x - 18y$

17. $153x - 99y + 12 + 152x + 8$

18. $1{,}090x + 1{,}230y + 3{,}030$

19. $999x + 663y + 551 + 111x + 555y + 449$

20. $3{,}657x - 4{,}321x$

21. $140x + 2{,}678x$

22. $59{,}920y - 30{,}009y + 12{,}411y$

23. $852x + 804x$

24. $2.08x^2 + 6.8x$

25. $15x^2 + 11x$

Simplify using the distributive property.

26. $4(x + 3)$

27. $2(x + z)$

28. $8(8 + p)$

29. $12(1 + x)$

30. $6(a - 3)$

31. $3(2x - 4)$

32. $-3(x + 4y)$

33. $-2(x - 9)$

34. $-5(2x - 7y)$

35. $4(x - (-9))$

36. $4(2 - x)$

37. $-7(5 - y)$

38. $3(2x + 8)$

39. $-4(3x - y)$

40. $\frac{12x - 6}{2}$

41. $14(-x - 9)$

42. $\frac{-7x + 14}{-7}$

43. $3(x - 4) - 2(x + 2)$

Simplify.

44. $3(x + 6) + 5x + 10$

45. $12h \div 4 + 12 + 5(6h + 1)$

46. $12h \div 4 \cdot 2 + 5(6h + 1)$

47. $3(500a + 500b + 500c)$

48. $3.2x^2 + 8(x^2 + 4.1) - 9.6$

49. $24x \cdot 8 + 2(x + 11)$

50. $3x^2 + 8(x^2 + 6x + 4) + 9$

51. $(10x + 15x + 18x + 2x + 5x) \div 5$

52. $6.9 + 3.2(x + 4)$

53. $6 + 3(x + 4) + 7x$

54. $6(2x + 1) + 3(x + 4)$

55. $6.5(2x + 3y + 5)$

56. $2(3x - 2) + 3^2 - 4x$

57. $4^2 + 3x^2 - 7x + 5x^2$

58. $2^3(3x + 7) - 8x$

59. $4(3x^2 + 2^2 + x) + 5x^2 - 6$

ERROR ANALYSIS

60. **Problem:** Simplify $6x + 5y - 3x + 2y$.

Answer:
$$6x + 5y - 3x + 2y$$
$$= (6 + 5 - 3 + 2)xy$$
$$= 10xy$$

Describe the mistake made in solving this problem.

Redo the problem correctly.

61. **Problem:** Simplify $2(x + 9) + 5x - 3$.

Answer:
$$2(x + 9) + 5x - 3$$
$$= 2x + 9 + 5x - 3$$
$$= 7x + 6$$
$$= 13x$$

Describe the mistake made in solving this problem.

Redo the problem correctly.

REVIEW

62. Add $5 + (-5)$.

63. State the reciprocal of 7.

64. Multiply $\frac{2}{3} \cdot \frac{3}{2}$.

Section 5.3 The Properties of Equality

LEARNING TOOLS

CD-ROM SSM VIDEO

Sneak Preview

When equations have one variable that is isolated on one side of the equation, we find the solution by evaluating the other side of the equation. However, when the variable is not isolated, the properties of equality are used to find the solution.

After completing this section, you should be able to:

1) Identify whether a number is a solution of an equation.
2) Use the properties of equality to find the solution of an equation in the form $x + b = c$ or $x - b = c$.
3) Use the properties of equality to find the solution of an equation in the form $ax = c$ or $\frac{x}{a} = c$.
4) Check the solution of an equation.

DISCUSSION

Solutions of Equations

An equation is a statement that two expressions are equal. For example, "the average is equal to the sum of a group of measurements divided by the number of measurements" is an equation. Other examples of equations are $6 + 2 = 8$ and $3x + 2 = 14$.

Equations may be

1. always true, *$3 + 8 = 11$ is always true.*
2. sometimes true, *$x + 1 = 5$ is sometimes true. It is true only when $x = 4$.*
3. always false. *$6 + 2 = 9$ is an equation that is always false.*

Some equations include a **variable.** A **solution** to an equation is a number that can replace the variable in the equation and result in a true equation. Finding this number is called **solving the equation.** If there is no value that makes the equation true, the equation has **no solution.**

Example 1 **Show that $x = 5$ is a solution of the equation $3x + 2 = 17$.**

We **check** the solution by replacing the x in the equation with 5. We then evaluate each side of the equation, following the order of operations. If the final equation is true, the solution is correct.

$3x + 2 = 17$

$3 \cdot \mathbf{5} + 2 = 17$ *Replace the variable, x, with 5.*

$\mathbf{15} + 2 = 17$ *Multiply or divide from left to right.*

$\mathbf{17} = 17$ *Add or subtract from left to right.*

Since $17 = 17$ is true, $x = 5$ is a solution of the equation.

Example 2 **Show that $x = 3$ is not a solution of the equation $3x + 2 = 17$.**

$3x + 2 = 17$

$3 \cdot \mathbf{3} + 2 = 17$ *Replace the variable, x, with 3.*

$\mathbf{9} + 2 = 17$ *Multiply or divide from left to right.*

$\mathbf{11} = 17$ *Add or subtract from left to right.*

Since $11 = 17$ is false, $x = 3$ is not a solution of the equation.

CHECK YOUR UNDERSTANDING

Check the solution of each equation.

1. Equation: $4x + 5 = 13$ Solution: $x = 2$

2. Equation: $6x + 1 = 2x + 13$ Solution: $x = 3$

3. Equation: $7x + 2x = 90$ Solution: $x = 5$

4. Equation: $2(x - 4) = 192$ Solution: $x = 100$

Answers: 1. This solution is correct, $13 = 13$; 2. This solution is correct, $19 = 19$; 3. This solution is incorrect, $45 \neq 90$; 4. This solution is correct, $192 = 192$.

DISCUSSION

Solving Equations Using the Properties of Equality

Sometimes we can just look at an equation and reason out what the solution of the equation must be. For example, we could study the equation $2x + 1 = 7$ and reason that the solution must be $x = 3$. This process is called **solving an equation by inspection.** However, when equations are more complicated, we need other methods to solve them.

Equations can be thought of in terms of a teeter-totter. When both sides of a teeter-totter have exactly the same weight, the teeter-totter is balanced. When both sides of an equation have equal value, the equation is true.

2 + 3 = 5

However, if 7 is added only to the left side of this equation, the equation is no longer true.

$$2 + 3 + 7 \neq 5$$

To keep the equation true, 7 must also be added to the right side of the equation:

$$2 + 3 + 7 = 5 + 7$$

The principle of keeping both sides of an equation the same value is stated in the addition and subtraction properties of equality. **Equivalent equations** may look different but they have the same solution.

PROPERTY

The addition property of equality

In words: "Adding the same amount to both sides of an equation results in an equivalent equation."

In symbols: If $a = b$, then $a + c = b + c$.

PROPERTY

The subtraction property of equality

In words: "Subtracting the same amount from both sides of an equation results in an equivalent equation."

In symbols: If $a = b$, then $a - c = b - c$.

To find the solution of an equation, we must **isolate** the variable on one side of the equation. The solution is then the number on the opposite side of the equation. We can isolate the variable using a property of equality.

Example 3 **Use the subtraction property of equality to find the solution of the equation $d + 9 = 15$. Check the solution.**

$d + 9 = 15$ *The goal is to isolate d by itself on one side of the equation.*

$d + 9 - \mathbf{9} = 15 - \mathbf{9}$ *Subtract 9 from both sides of the equation (Subtraction Property of Equality).*

$d + 0 = 6$ *Combine like terms. Left side: 9 − 9 = 0. Right side: 15 − 9 = 6.*

$d = 6$ *The solution of the equation is d = 6.*

Check

$d + 9 = 15$

$\mathbf{6} + 9 = 15$ *Replace the variable with the solution, 6.*

$15 = 15$ *Combine like terms. Since 15 = 15 is true, the solution is correct.*

Example 4 **Solve the equation for x: $x + \frac{5}{2} = -6$. Check the solution.**

$x + \frac{5}{2} = -6$ *The goal is to isolate x by itself on one side of the equation.*

$x + \frac{5}{2} - \mathbf{\frac{5}{2}} = -6 - \mathbf{\frac{5}{2}}$ *Subtract $\frac{5}{2}$ from both sides of the equation to isolate x.*

$x + \mathbf{0} = -\frac{12}{2} - \frac{5}{2}$ *$\frac{5}{2} - \frac{5}{2} = 0$. Rewrite 6 as a fraction with a denominator of 2.*

$x = -\frac{17}{2}$ *Subtract.*

Check

$x + \frac{5}{2} = -6$

$\mathbf{-\frac{17}{2}} + \frac{5}{2} = -6$ *Replace the variable with the solution, $-\frac{17}{2}$.*

$-\frac{12}{2} = -6$

$-6 = -6$ *Since $-6 = -6$ is true, the solution is correct.*

In the previous examples, the operation in the original equation is addition. To isolate the variable, we did the opposite operation, subtraction, to both sides of the equation. In the next examples, the operation in the original equation is subtraction. To isolate the variable, we do the opposite operation, addition, to both sides of the equation.

Example 5 **Solve the equation $y - 4 = 7$ for y. Check the solution.**

$y - 4 = 7$

$y - 4 \mathbf{+ 4} = 7 \mathbf{+ 4}$ *Add 4 to both sides of the equation. (Addition Property of Equality)*

$y + 0 = 11$ *$4 - 4 = 0$ $7 + 4 = 11$*

$y = 11$

Check

$y - 4 = 7$

$\mathbf{11} - 4 = 7$ *Replace y with the solution of 11.*

$7 = 7$ *Since $7 = 7$ is true, the solution is correct.*

Example 6 **Solve the equation $-10.6 = x - 8.7$ for x. Check the solution.**

$-10.6 = x - 8.7$

$-10.6 \mathbf{+ 8.7} = x - 8.7 \mathbf{+ 8.7}$ *Add 8.7 to both sides of the equation.*

$-1.9 = x + 0$ *$-10.6 + 8.7 = -1.9$. $8.7 - 8.7 = 0$.*

$-1.9 = x$

Check

$-10.6 = x - 8.7$

$-10.6 = \mathbf{-1.9} - 8.7$ *Replace x with the solution of -1.9.*

$-10.6 = -10.6$ *True. $x = -1.9$ is the solution.*

The variable can be on either the left or the right side of the equation. $18 = x$ is the same equation as $x = 18$. This is an example of the **symmetric property.**

PROPERTY

The symmetric property

In words: "The expressions on each side of an equation can switch positions. The new equation is equivalent."

In symbols: If $x = a$, then $a = x$.

All of these examples either involved addition or subtraction of a number. Numbers in an equation are also known as **constants.** We can summarize the strategy used for solving such equations by using a and b to represent constants and x to represent the variable.

PROCEDURE

To solve an equation of the form $x + a = b$:

Subtract a from both sides of the equation.

To solve an equation of the form $x - a = b$:

Add a to both sides of the equation.

CHECK YOUR UNDERSTANDING

Solve each equation using the addition or subtraction property of equality. Do not solve by inspection. Check each solution.

1. $x + 9 = 17$

2. $y - 4.8 = 13$

3. $z - 7 = 7$

4. $6.4 + t = 23.6$

5. $12 = x - 6$

6. $\frac{2}{3} = \frac{4}{5} - x$

Answers: 1. $x = 8$; 2. $y = 17.8$; 3. $z = 14$; 4. $t = 17.2$; 5. $x = 18$; 6. $x = \frac{2}{15}$.

DISCUSSION

The Division Property of Equality

The subtraction property of equality says that any number can be subtracted from both sides of an equation without changing the solution. Since division is just a shorter way to write many subtractions of the same number, there is also a division property of equality.

PROPERTY

The division property of equality

In words: "Dividing both sides of an equation by the same nonzero amount results in an equivalent equation."

In symbols: If $a = b$, then $\frac{a}{c} = \frac{b}{c}$, $c \neq 0$.

An important exception is the number 0. Since division by zero is undefined, we cannot divide both sides of the equation by 0.

Example 7 **Demonstrate that the division property of equality can be applied to the equation $6x = 60$ without changing the solution of the equation.**

By inspection, we can determine that the solution of this equation is $x = 10$ since $6 \cdot 10 = 60$. We should be able to divide both sides of the equation by any number except zero without changing this solution Suppose we divide both sides of this equation by 3.

$6x = 60$

$6x \div \mathbf{3} = 60 \div \mathbf{3}$ *Divide both sides by 3.*

$2x = 20$

The solution of this changed equation remains $x = 10$. Dividing both sides by 3 did not change the solution. So, the division property of equality can be applied without changing the solution of the equation.

We use the division property of equality to solve equations like $6x = 60$ or $3x = 15$. These equations have a coefficient in front of the variable that is not equal to 1. By dividing both sides of the equation by the coefficient, we change that coefficient to 1. When we change the coefficient to 1, we isolate the variable and solve the equation.

Example 8 **Use the division property of equality to solve the equation $6x = -60$.**

$6x = -60$

$6x \div \mathbf{6} = -60 \div \mathbf{6}$ *Divide both sides by the coefficient, 6.*

$1x = -10$ *The coefficient in front of the variable is changed to 1.*

$x = -10$ *1x is equivalent to x.*

Check

$6x = -60$

$6(\mathbf{-10}) = -60$ *Replace the variable with our solution, −10.*

$-60 = -60$ *Multiply. Since 60 = 60 is true, the solution is correct.*

Example 9 **Use the division property of equality to solve the equation $25 = 150y$.**

$25 = 150y$

$25 \div \mathbf{150} = 150y \div \mathbf{150}$ *Divide both sides by the coefficient, 150.*

$0.1666\ldots = 1y$ *The coefficient in front of the variable is changed to 1.*

$0.1\overline{6} = y$ *1y and y are equivalent.*

$y = 0.1\overline{6}$ *The variable can be isolated on either side of the equation because of the symmetric property.*

Check

$25 = 150y$

$25 = 150(\mathbf{0.167})$ *Replace the variable with a rounded value for the solution.*

$25 \approx 25.05$ *Since 25 is very close to 25.05, the solution is correct.*

CHECK YOUR UNDERSTANDING

Use the division property of equality to solve each equation. Check each solution.

1. $4x = 40$ **2.** $64 = 200p$ **3.** $7q = 0$ **4.** $3{,}500 = 5y$

Answers: 1. x = 10; 2. p = 0.32; 3. q = 0; 4. y = 700.

DISCUSSION

The Multiplication Property of Equality

The addition property of equality says that any number can be added to both sides of an equation without changing the solution. Since multiplication is just a shorter way to write many additions of the same number, there is also a **multiplication property of equality.**

> PROPERTY
>
> **The multiplication property of equality**
>
> **In words:** "Multiplying both sides of an equation by the same amount results in an equivalent equation."
>
> **In symbols:** If $a = b$, then $ac = bc$.

When the variable in an equation is *divided* by a number, we use the multiplication property of equality to find the solution of the equation. Division of a variable can be shown in different ways. A division sign ($\div$) can be used as in $x \div 10 = 5$. Or, a horizontal bar can be used as in $\frac{x}{10} = 5$. We multiply both sides of the equation so that the coefficient in front of the variable becomes 1.

Example 10 **Use the multiplication property of equality to solve $x \div 10 = -5.5$.**

$x \div 10 = -5.5$

$\mathbf{10} \cdot x \div 10 = \mathbf{10} \cdot (-5.5)$ *Multiply both sides by the divisor, 10.*

$10x \div 10 = -55$

$1x = -55$ *10x ÷ 10 = 1x.*

$x = -55$ *1x and x are equivalent.*

Check

$x \div 10 = -5.5$

$\mathbf{-55} \div 10 = 5.5$ *Replace the variable, x, with the solution, −55.*

$-5.5 = -5.5$ *Divide. Since −5.5 = −5.5 is true, the solution is correct.*

Example 11 **Use the multiplication property of equality to solve $4 = \frac{y}{3}$.**

$4 = \frac{y}{3}$

$\mathbf{3} \cdot 4 = \mathbf{3} \cdot \frac{y}{3}$ *Multiply both sides by the divisor, 3.*

$\mathbf{3} \cdot 4 = \frac{\mathbf{3}}{\mathbf{1}} \cdot \frac{y}{3}$

$12 = \frac{3y}{3}$

$12 = 1y$ *3y ÷ 3 = 1y.*

$12 = y$ *1y and y are equivalent.*

Check

$4 = \frac{y}{3}$

$4 = \frac{\mathbf{12}}{3}$ *Replace the variable, y, with the solution, 12.*

$4 = 4$ *Divide. Since 4 = 4 is true, the solution is correct.*

PROCEDURE

To solve an equation of the form $ax = c, a \neq 0$:

Divide both sides of the equation by a.

To solve an equation of the form $\frac{x}{a} = c, a \neq 0$:

Multiply both sides of the equation by a.

CHECK YOUR UNDERSTANDING

Use the multiplication property of equality to solve each equation. Check each solution.

1. $\frac{x}{2} = 6$

2. $-5 = \frac{k}{7}$

3. $\frac{y}{40} = 60{,}000$

Use the division property of equality to solve each equation. Check each solution.

4. $4x = -12$

5. $36 = 9y$

6. $-10z = 1.2$

Answers: 1. $x = 12$; 2. $k = -35$; 3. $y = 2{,}400{,}000$; 4. $x = -3$; 5. $y = 4$; 6. $z = -0.12$.

EXERCISES SECTION 5.3

Solve each equation. Check each solution. Do not solve by inspection.

1. $x + 8 = 12$

2. $y + 7 = 12$

3. $z + 3 = 17$

4. $x - 7 = 3$

5. $y - 4.9 = 2.4$

6. $z - 9.8 = 3$

7. $a + 12 = 56$

8. $b - \frac{3}{2} = -7$

9. $c + \frac{5}{9} = -\frac{4}{3}$

10. $x + 2{,}300 = 4{,}000$

11. $45 + x = 66$

12. $22 + y = 56$

13. $a + 3{,}000 = 7{,}000$

14. $2{,}000 + b = 4{,}000$

15. $\frac{5}{6} + x = 8$

16. $34 = x + 12$

17. $33.8 = y - 45.4$

18. $450 = c - 69$

19. $240 = x - 67$

20. $21 + z = 45$

21. $3x = 90$

22. $5y = 20$

23. $z \div 4.8 = 20$

24. $65 = 5m$

25. $7c = 42$

26. $15.96 = 3b$

27. $x \div 9 = 30.64$

28. $12 = t \div 3$

29. $120x = 240$

30. $w \div 5 = 4$

31. $7x = -56$

32. $99 = 10y$

33. $\frac{x}{7} = 2$

34. $\frac{t}{5} = 9$

35. $12c = 0$

36. $6d = 4.8$

37. $6 = \frac{m}{12}$

38. $-15x = -45$

39. $1{,}200 = 60t$

40. $\frac{y}{14} = 3$

ERROR ANALYSIS

41. **Problem:** Solve $5x = 15$.

Answer:

$$5x = 15$$
$$x = 15 \cdot 5$$
$$x = 75$$

Describe the mistake in the answer to this problem.

Redo the problem correctly.

42. **Problem:** Solve the equation $y - 9 = 15$.

Answer:

$$y - 9 = 15$$
$$y = 15 - 9$$
$$y = 6$$

Describe the mistake in the answer to this problem.

Redo the problem correctly.

REVIEW

43. What are the four arithmetic operations?

44. Describe the order of operations.

45. What is a coefficient? Give an example.

Section 5.4 Equations with More Than One Operation

LEARNING TOOLS

CD-ROM SSM VIDEO

Sneak Preview

When an equation includes more than one operation, more than one property of equality is used to find its solution. When an equation includes many terms or parentheses, it may need to be simplified before properties of equality are used to find its solution.

After completing this section, you should be able to:

1) Identify the property of equality used in each step of a detailed solution of an equation.
2) Use the properties of equality to find the solution of an equation in the form $ax + b = c$ or $ax - b = c$.
3) Solve equations that must be simplified before using the properties of equality.

INVESTIGATION 1 Using the Properties of Equality

To solve an equation, we isolate the variable using the properties of equality. If more than one operation is included in the equation, we will need to use more than one property of equality. In each of the following examples, identify the properties of equality being used to isolate the variable.

EXAMPLE A: Solve $2y + 4 = 10$.

$2y + 4 = 10$

$2y + 4 \mathbf{- 4} = 10 \mathbf{- 4}$ **1.** Identify the property of equality used in this step.

$2y + 0 = 6$

$2y = 6$

$2y \mathbf{\div 2} = 6 \mathbf{\div 2}$ **2.** Identify the property of equality used in this step.

$1y = 3$

$y = 3$

3. Check the solution, $y = 3$.

EXAMPLE B: Solve $2y - 4 = 10$.

$2y - 4 = 10$

$2y - 4 \mathbf{+ 4} = 10 \mathbf{+ 4}$ **4.** Identify the property of equality used in this step.

$2y + 0 = 14$

$2y = 14$

$2y \mathbf{\div 2} = 14 \mathbf{\div 2}$ **5.** Identify the property of equality used in this step.

$1y = 7$

$y = 7$

6. Check the solution, $y = 7$.

EXAMPLE C: Solve $16 = 3x + 10$.

$16 = 3x + 10$

$16 - \mathbf{10} = 3x + 10 - \mathbf{10}$ — **7.** Identify the property of equality used in this step.

$6 = 3x + 0$

$6 = 3x$

$6 \div \mathbf{3} = 3x \div \mathbf{3}$ — **8.** Identify the property of equality used in this step.

$2 = 1x$

$2 = x$

$x = 2$

9. Check the solution, $x = 2$.

10. What property allows the solution $2 = x$ to be rewritten as $x = 2$?

Equations that require more than one property of equality to solve can be described using these **general equations:** $ax + b = c$ and $ax - b = c$.

11. Which general equation describes the equation $2y - 4 = 10$?

12. Which general equation describes the equation $2y + 4 = 10$?

13. Using the property of symmetry, which general equation describes the equation $16 = 3x + 10$?

14. Write a procedure for solving equations in the form $ax + b = c$.

15. What properties of equality are used in this procedure?

16. Write a procedure for solving equations in the form $ax - b = c$.

17. Identify the properties of equality that are used in this procedure.

Answers: 1. subtraction property of equality; 2. division property of equality; 4. addition property of equality; 5. division property of equality; 7. subtraction property of equality; 8. division property of equality; 10. symmetric property; 11. $ax - b = c$; 12. $ax + b = c$; 13. $ax + b = c$; 14. Subtract b from both sides. Divide both sides by a. Check the solution.; 15. subtraction property of equality, division property of equality; 16. Add b to both sides. Divide both sides by a; 17. addition property of equality, division property of equality.

DISCUSSION

Solving Equations Using More Than One Property of Equality

Often equations involve more than one operation. For example, $5x + 4 = 19$ involves two operations: x is *multiplied* by 5 and 4 is *added* to this product. To solve the equation, we isolate the variable using properties of equality. We first isolate the *term* that contains the variable using the addition or subtraction property of equality. We then change the coefficient in this term to 1 using the multiplication or division property of equality.

Example 1 **Solve $5x + 4 = 19$.**

$5x + 4 = 19$

$5x + 4 \mathbf{- 4} = 19 \mathbf{- 4}$ *To isolate the term 5x, use the subtraction property of equality to remove the 4.*

$5x + 0 = 15$ *$4 - 4 = 0$ $19 - 4 = 15$*

$5x = 15$

$5x \div \mathbf{5} = 15 \div \mathbf{5}$ *To change the coefficient of x to 1, divide by 5.*

$1x = 3$ *$5 \div 5 = 1$ $15 \div 5 = 3$.*

$x = 3$

Check

$5x + 4 = 19$

$5 \cdot \mathbf{3} + 4 = 19$ *Replace the variable with the solution, 3.*

$15 + 4 = 19$ *Multiply.*

$19 = 19$ *Add. Since 19 = 19 is true, the solution is correct.*

Example 2 **Solve $2x - 7 = -21$.**

$2x - 7 = -21$

$2x - 7 \mathbf{+ 7} = -21 \mathbf{+ 7}$ *Use the addition property of equality to isolate the term 2x.*

$2x + 0 = -14$

$2x = -14$

$2x \div \mathbf{2} = -14 \div \mathbf{2}$ *Use the division property of equality to isolate the x.*

$1x = -7$

$x = -7$

Check

$2x - 7 = -21$

$2(\mathbf{-7}) - 7 = -21$ *Replace the variable with the solution, −7.*

$-14 - 7 = -21$ *Multiply.*

$-21 = -21$ *Subtract. Since −21 = −21 is true, the solution is correct.*

More complex equations need to be simplified before using the properties of equality. Simplifying may involve combining like terms.

Example 3 **Solve $7x + 6 - 4x = 30$.**

$\mathbf{7x} + 6 - \mathbf{4x} = 30$	
$\mathbf{3x} + 6 = 30$	*Combine like terms: $7x - 4x = 3x$.*
$3x + 6 \mathbf{- 6} = 30 \mathbf{- 6}$	*Use the subtraction property of equality to isolate the term $3x$.*
$3x + 0 = 24$	
$3x = 24$	
$3x \div \mathbf{3} = 24 \div \mathbf{3}$	*Use the division property of equality.*
$1x = 8$	
$x = 8$	

Check

$7x + 6 - 4x = 30$	*Use the original equation to check the solution.*
$7 \cdot \mathbf{8} + 6 - 4 \cdot \mathbf{8} = 30$	*Replace the variable with the solution, 8 in both terms.*
$56 + 6 - 32 = 30$	*Multiply.*
$62 - 32 = 30$	*Add and subtract from left to right.*
$30 = 30$	*Since $30 = 30$ is true, the solution is correct.*

When there are parentheses in the equation, the distributive property can be used to remove the parentheses. Combine like terms before using the properties of equality.

Example 4 **Solve $4(y + 3) - 2 = 58$.**

$4(y + 3) - 2 = 58$	
$\mathbf{4y + 12} - 2 = 58$	*Use the distributive property to simplify; $4(y + 3) = 4y + 4 \cdot 3$.*
$4y + \mathbf{10} = 58$	*Combine like terms: $12 - 2 = 10$.*
$4y + 10 \mathbf{- 10} = 58 \mathbf{- 10}$	*Use the subtraction property of equality to isolate the term $4y$.*
$4y = 48$	
$4y \div \mathbf{4} = 48 \div \mathbf{4}$	*Use the division property of equality.*
$y = 12$	

Check

$4(y + 3) - 2 = 58$	*Use the original equation to check the solution.*
$4(\mathbf{12} + 3) - 2 = 58$	*Replace the variable with the solution, 12.*
$4(15) - 2 = 58$	*Follow order of operations. Do addition inside parentheses first.*
$60 - 2 = 58$	
$58 = 58$	*Since $58 = 58$ is true, the solution is correct.*

Once the equation has been simplified using the distributive property and/or by combining like terms, it is in the general form $ax + b = c$ or $ax - b = c$. The variable is found only on one side of the equation. An equation that has only one variable with an exponent of 1 is called a **linear equation in one variable.** Solving equations by simplifying and then using the properties of equality is summarized in the following procedure.

PROCEDURE

To solve an equation with a variable found on only one side of the equation:

1. Use the distributive property to remove any grouping symbols.
2. Simplify by combining like terms.
3. Use the addition or subtraction property of equality to isolate the term containing the variable.
4. Use the multiplication or division property of equality to isolate the variable.
5. Check the solution in the original equation.

The variable can be on either the left side or the right side of the equation. The same procedure is used.

Example 5 **Solve $68 = 2(p - 10)$.**

$68 = 2(p - 10)$	
$68 = \mathbf{2p - 20}$	*Use the distributive property to remove parentheses.*
$68 \mathbf{+ 20} = 2p - 20 \mathbf{+ 20}$	*Use the addition property of equality to isolate the term 2p.*
$88 = 2p$	
$88 \mathbf{\div 2} = 2p \mathbf{\div 2}$	*Use the division property of equality to isolate the variable p.*
$44 = p$	*The solution is 44 = p.*
$p = 44$	*Using the property of symmetry, reverse the order.*

Check

$68 = 2(p - 10)$	*Use the original equation to check the solution.*
$68 = 2(\mathbf{44} - 10)$	*Replace the variable with the solution, 44.*
$68 = 2(34)$	*Follow order of operations. Do subtraction inside parentheses first.*
$68 = 68$	*Since 68 = 68 is true, the solution is correct.*

CHECK YOUR UNDERSTANDING

Solve each equation. Check the solution.

1. $2x - 7 = 19$

2. $8y + 4 = 36$

3. $7x - 5x - 9 = 5$

4. $6 = 3(2y - 8)$

5. $0 = 3y - 9$

6. $15q = 45$

Answers: 1. $x = 13$; 2. $y = 4$; 3. $x = 7$; 4. $y = 5$; 5. $y = 3$; 6. $q = 3$.

In practical applications, equations often involve fractions and decimals. The same properties of equality are used to solve such equations.

Example 6 **Solve $3.1x + 4.5 = 2.85$. Round the solution to the nearest thousandth.**

$3.1x + 4.5 = 2.85$

$3.1x + 4.5 - \mathbf{4.5} = 2.85 - \mathbf{4.5}$ *Subtraction property of equality*

$3.1x = -1.65$

$3.1x \div \mathbf{3.1} = -1.65 \div \mathbf{3.1}$ *Division property of equality*

$1x = -0.53225\ldots$ *Divide.*

$x \approx -0.532$ *Round the quotient to the thousandths place.*

Check

$3.12(\mathbf{-0.532}) + 4.5 = 2.85$ *Replace x with the rounded solution −0.532.*

$\mathbf{-1.660} + 4.5 = 2.85$ *Multiply. Round the product to the thousandths place.*

$2.84 = 2.85$ *2.84 is almost exactly equal to 2.85. We expect that the two numbers might not be exactly equal because of rounding.*

Example 7 **Solve $0.04(120 - x) + 0.07x = 54.9$.**

$0.04(120 - x) + 0.07x = 54.9$

$\mathbf{(0.04)(120) - (0.04)(x)} + 0.07x = 54.9$ *Use the distributive property.*

$\mathbf{4.8 - 0.04x} + 0.07x = 54.9$ *Multiply or divide from left to right.*

$4.8 + \mathbf{0.03x} = 54.9$ *Combine like terms.*

$4.8 - \mathbf{4.8} + 0.03x = 54.9 - \mathbf{4.8}$ *Subtraction property of equality*

$0 + 0.03x = 50.1$

$0.03x = 50.1$

$0.03x \div \mathbf{0.03} = 50.1 \div \mathbf{0.03}$ *Division property of equality*

$1x = 1{,}670$

$x = 1{,}670$ *1x is equivalent to x.*

Check

$0.04(120 - \mathbf{1{,}670}) + 0.07(\mathbf{1{,}670}) = 54.9$ *Replace x with the solution: 1,670.*

$0.04(\mathbf{-1{,}550}) + 0.07(1{,}670) = 54.9$ *Subtract inside the parentheses.*

$\mathbf{-663.2} + 0.07(1{,}670) = 54.9$ *Multiply or divide left to right.*

$-62 + \mathbf{116.9} = 54.9$ *Multiply or divide left to right.*

$54.9 = 54.9$ *Add or subtract left to right. Since 54.9 = 54.9 is true, the solution is correct.*

When equations include fractions there are two methods that can be used to solve the equation. We can follow the same procedure we've previously used to solve equations that include whole numbers, integers, or decimals. Or, we can use the multiplication property of equality to first remove the fractions from the equation. This is called "clearing the fractions" from the equation. Both methods are shown in the next example.

Example 8 **Solve $\frac{2}{3}x - \frac{5}{6} = -\frac{5}{12}$.**

Method 1:

$$\frac{2}{3}x - \frac{5}{6} = -\frac{5}{12}$$

$$\frac{2}{3}x - \frac{5}{6} + \mathbf{\frac{5}{6}} = -\frac{5}{12} + \mathbf{\frac{5}{6}} \quad \text{Addition property of equality.}$$

$$\frac{2}{3}x = -\frac{5}{12} + \mathbf{\frac{10}{12}} \quad \text{Rewrite fractions with common denominators.}$$

$$\frac{2}{3}x = \mathbf{\frac{5}{12}}$$

$$\mathbf{\frac{3}{2}} \cdot \frac{2}{3}x = \frac{5}{12} \cdot \mathbf{\frac{3}{2}} \quad \text{Multiplication property of equality.}$$

$$1x = \frac{5 \cdot \mathbf{1}}{\mathbf{4} \cdot 2} \quad \text{Simplify before multiplying.}$$

$$x = \frac{5}{8}$$

Method 2: To eliminate the fractions from the equation we must find a number that is a multiple of all the denominators. The smallest multiple is the lowest common denominator, but any common denominator can be used to clear the fractions.

$$\frac{2}{3}x - \frac{5}{6} = -\frac{5}{12}$$

$$\mathbf{12}\left(\frac{2}{3}x - \frac{5}{6}\right) = \mathbf{12}\left(-\frac{5}{12}\right) \quad \text{Multiply both sides of the equation by the common denominator, 12.}$$

$$\frac{\mathbf{12}}{1} \cdot \frac{2}{\mathbf{3}}x - \frac{\mathbf{12}}{1} \cdot \frac{5}{\mathbf{6}} = \frac{\mathbf{12}}{1}\left(\frac{-5}{\mathbf{12}}\right) \quad \text{Apply the distibutive property.}$$

$$\frac{\mathbf{4} \cdot 2}{1 \cdot \mathbf{1}}x - \frac{\mathbf{2} \cdot 5}{1 \cdot \mathbf{1}} = \frac{\mathbf{1} \cdot (-5)}{1 \cdot \mathbf{1}} \quad \text{Simplify then multiply.}$$

$$8x - 10 = -5 \quad \text{A simplified equation.}$$

$$8x - 10 + \mathbf{10} = -5 + \mathbf{10} \quad \text{Addition property of equality.}$$

$$8x = 5$$

$$\frac{8x}{\mathbf{8}} = \frac{5}{\mathbf{8}} \quad \text{Division property of equality.}$$

$$x = \frac{5}{8}$$

Checking the solution of an equation with fractions can be very time-consuming. We can use approximate values for any fraction (round to the hundredths place) in the equation. If the solution is correct, the two sides of the evaluated equation will be approximately equal.

$$\frac{2}{3}x - \frac{5}{6} = -\frac{5}{12}$$

$$0.66x - 0.83 \approx -0.42 \quad \text{Divide each numerator by its denominator. Round.}$$

Check

$$0.66(\mathbf{0.63}) - 0.83 \approx -0.42 \quad \tfrac{5}{8} \approx 0.63.$$

$$0.42 - 0.83 \approx -0.42$$

$$-0.41 \approx -0.42$$

Since -0.41 is very close to -0.42, the solution $x = \frac{5}{8}$ is probably correct.

PROCEDURE

To solve an equation containing fractions:

Method 1

1. Simplify, if possible.
2. Apply the addition or subtraction property of equality.
3. Apply the multiplication or division property of equality.

Method 2

1. Use the distributive property to remove any parentheses.
2. Multiply both sides of the equation by a common denominator to clear the fractions.
3. Use Method 1 to solve the resulting equation.

Example 9 **Solve by clearing the fractions:** $3\left(5x - \frac{4}{5}\right) = -\frac{5}{6}$.

$$3\left(5x - \frac{4}{5}\right) = -\frac{5}{6}$$

$$3 \cdot 5x - \frac{\mathbf{3}}{\mathbf{1}} \cdot \frac{4}{5} = -\frac{5}{6} \qquad \textit{Distribute.}$$

$$15x - \frac{12}{5} = -\frac{5}{6}$$

$$\mathbf{30} \cdot \left(15x - \frac{12}{5}\right) = \mathbf{30} \cdot \left(-\frac{5}{6}\right) \qquad \textit{Multiply both sides by a common denominator: 30.}$$

$$450x - \frac{30}{1} \cdot \frac{12}{5} = \frac{\mathbf{30}}{\mathbf{1}} \cdot \left(-\frac{5}{6}\right) \qquad \textit{Distribute.}$$

$$450x - \frac{360}{5} = -\frac{150}{6}$$

$$450x - 72 = \mathbf{-25}$$

$$450x - 72 + \mathbf{72} = -25 + \mathbf{72} \qquad \textit{Addition property of equality.}$$

$$450x = 47$$

$$\frac{450x}{\mathbf{450}} = \frac{47}{\mathbf{450}} \qquad \textit{Division property of equality.}$$

$$x = \frac{47}{450}$$

Check

$$3\left(5x - \frac{4}{5}\right) = -\frac{5}{6}$$

$$3(5x - 0.8) \approx -0.83 \qquad \textit{Rewrite fractions as rounded decimals.}$$

$$3(5(\mathbf{0.10}) - 0.8) \approx -0.83 \qquad \textit{Replace } x. \ \frac{47}{450} \approx 0.10.$$

$$3(0.5 - 0.8) \approx -0.83$$

$$3(-0.3) \approx -0.83 \qquad \textit{Since } -0.9 \textit{ is very close to } -0.83, \textit{ the solution}$$

$$-0.9 \approx -0.83 \qquad x = \frac{47}{450} \textit{ is probably correct.}$$

CHECK YOUR UNDERSTANDING

Solve each equation and check the solution.

1. $-2 + 5m = -27$

2. $6x - \frac{2}{3} = -\frac{4}{7}$

3. $7.5x + 2(1.4 - 3.1x) = 9.3$

4. $\frac{1}{3}y + \frac{1}{2} = \frac{2}{5}$

Answers: 1. $m = -5$*; 2.* $x = \frac{1}{63}$*; 3.* $x = 5$*; 4.* $y = -\frac{3}{10}$*.*

EXERCISES SECTION 5.4

Solve the equation. Check the solution. Round decimal answers to the hundredths place.

1. $2x + 8 = 12$

2. $3y - 4 = 5$

3. $4w + 7 = 15$

4. $3y - 8 = 55$

5. $15 + 2x = 137$

6. $140 = 90 + 25t$

7. $6m + 6 = 42$

8. $7 + 3x = 16$

9. $17 = 7y + 3$

10. $7x + 5 = 5$

11. $12v - 8 = 28$

12. $200d - 30 = 5{,}970$

13. $2u = 15{,}750 - 430$

14. $20x + 15 = 75$

15. $240 = 90 + 25h$

16. $20x - 15 = 85$

17. $12x = 144$

18. $12(x + 2) = 144$

19. $363 = 3(12x + 1)$

20. $580 - 20 = 10x$

21. $8.4x - 6 = 90$

22. $-2(4.6x + 3) = 70$

23. $3.5 = -0.7x$

24. $3.5 = -0.7(x + 1)$

25. $3.5 = 7(x - 0.2)$

26. $2x + 83.7 = 112.1$

27. $2x - 83.7 = 112.1$

28. $79 - x = 46$

29. $79.12 - x = 46.338$

30. $79.02 + x = 46.338$

31. $3.4x = 20.74$

32. $\frac{w}{3.4} = 8.0$

33. $-2.5y + 5.4 = 23.75$

34. $0.01x - 6.9 = 44$

35. $\frac{5}{3}y = \frac{7}{4}$

36. $\frac{2x}{5} = -7$

37. $\frac{3}{4}t = 0$

38. $w + \frac{1}{3} = \frac{5}{6}$

39. $\frac{y}{4} = -\frac{6}{8}$

40. $\frac{1}{2}x = 15$

41. $x - \frac{3}{4} = 8$

42. $z + \frac{2}{7} = -4$

43. $\frac{5}{6} + x = \frac{3}{4}$

44. $\frac{1}{3}x - \frac{2}{5} = 4$

45. $\frac{5}{6}y + \frac{2}{3} = \frac{3}{4}$

46. $\frac{2}{9}d - \frac{2}{3} = \frac{5}{9}$

47. $3x + \frac{4}{5} = -\frac{2}{3}$

48. $6y - \frac{5}{6} = \frac{5}{2}$

49. $3(x - 6) = 4$

50. $4(5x + 4) = -3$

51. $\frac{2}{3}(5x - 6) = 1$

52. $\frac{6}{4}\left(\frac{2}{3} - w\right) = -\frac{5}{4}$

53. $5\left(\frac{2}{3}y - \frac{5}{6}\right) = -3$

54. $\frac{2}{3}\left(4x - \frac{1}{4}\right) = \frac{5}{12}$

55. $\frac{3w}{2} - \frac{4w}{3} = 8$

56. $3x - \frac{4}{5}x + \frac{5}{4} = \frac{1}{4}$

57. $6x - \frac{5}{7} = -\frac{3}{2}$

58. $\frac{3}{4} - \left(-\frac{2}{5}n\right) = \frac{1}{3}$

ERROR ANALYSIS

59. **Problem:** Solve $3(x + 6) = 18$.

Answer:

$$3(x + 6) = 18$$
$$3x + 6 = 18$$
$$3x + 6 - 6 = 18 - 6$$
$$3x = 12$$
$$3x \div 3 = 12 \div 3$$
$$x = 4$$

Describe the mistake in the answer to this problem.

Redo the problem correctly.

60. **Problem:** Solve $0.3x - 6.78 = -2.3$.

Answer:

$$0.3x - 6.78 = -2.3$$
$$0.3x - 6.78 + 6.78 = -2.3 + 6.78$$
$$0.3x = 9.08$$
$$0.3x \div 0.3 = 9.08 \div 0.3$$
$$x = 30.2\overline{6}$$

Describe the mistake made in solving this problem.

Redo the problem correctly.

REVIEW

61. Describe the five steps of the problem-solving method.

Section 5.5 Applications and Problem Solving

LEARNING TOOLS

CD-ROM SSM VIDEO

Sneak Preview

Throughout this text we have used a five-step problem-solving strategy to help us solve word problems. Now we can combine this strategy with our knowledge of solving equations to help us solve more advanced word problems.

After completing this section, you should be able to:

1) Solve word problems using equations that may involve more than one operation to solve.

DISCUSSION

Problem Solving

We first introduced the five-step method of problem solving in Section 1.4. The equations that were part of this process were simple equations with the unknown variable isolated on one side of the equation. Now that we are adept at solving more sophisticated equations, we can tackle more advanced word problems.

Example 1 **Karen received a classroom technology grant in the amount of \$5,000. She will buy a laptop computer for \$1,500 and software totaling \$780. The remaining money will be spent on graphing calculators for her classroom. If each calculator sells for \$89.99, how many calculators can she afford to buy with the remaining grant money?**

I know . . .

\$5,000 = Total

\$1,500 = Laptop

\$780 = Software

\$89.99 = Calculator

Number of calculators = C

Equations

Laptop + Software + (Calculator)(Number of calculators) = Total

$$\$1{,}500 + \$780 + \$89.99C = \$5{,}000$$

Solve

$$\$1{,}500 + \$780 + \$89.99C = \$5{,}000$$
$$\$2{,}280 + \$89.99C = \$5{,}000$$
$$\$2{,}280 \mathbf{- \$2{,}280} + \$89.99C = \$5{,}000 \mathbf{- \$2{,}280}$$
$$\$89.99C = \$2{,}720$$
$$\frac{\$89.99C}{\mathbf{\$89.99}} = \frac{\$2{,}720}{\mathbf{\$89.99}}$$
$$C = 30.22558\ldots$$
$$C = 30$$

Check

Each calculator costs about \$90. The total cost of the calculators is about 30(\$90) = \$2,700. Rounding the costs of the other items to the hundreds place, the sum of the laptop (\$1,500), software (\$800), and 30 calculators (\$2,700) is \$5,000. Since the total amount of money is \$5,000, our answer is reasonable.

Answer

Karen can purchase 30 graphing calculators with the remaining grant money.

When solving application problems, we need to think carefully about rounding. We cannot always follow the usual rules for rounding because it may cause us to do something that does not make sense. In the next example, we do not want to round up and spend more money than we have budgeted.

Example 2 **Kirsty's cellular phone calling plan charges $34.99 a month for the first 120 minutes. She pays 27¢ per minute for extra time. She has budgeted $60.00 a month for her cell phone. How many extra minutes can she talk each month and stay within her budget?**

I know . . .

$34.99 = Monthly charge

27¢ = $0.27 = Cost per extra minute

$60.00 = Budget

m = Extra minutes

Equations

Budget = Monthly charge + Cost of extra minutes

Budget = Monthly charge + (Cost of extra minutes)(Extra minutes)

$\$60.00 = \$34.99 + \$0.27m$

Solve

$$\$60.00 = \$34.99 + \$0.27m$$
$$\$60.00 - \mathbf{\$34.99} = \$34.99 - \mathbf{\$34.99} + \$0.27m$$
$$\$25.01 = \$0.27m$$
$$\frac{\$25.01}{\mathbf{\$0.27}} = \frac{\$0.27m}{\mathbf{\$0.27}}$$
$$92.62\ldots = m$$
$$92 \approx m \quad \textit{Round to the next lowest minute to stay within the budget.}$$

Check

If we round 92 extra minutes to 100 minutes, the cost of the extra minutes is 100($0.27) or $27. Then, the total monthly cost is about $35 + $27 or $62. This is close enough to her budget of $60 to conclude that this answer is reasonable.

Answer

Kirsty can use up to 92 extra minutes and stay within her budget.

In the previous two problems, we could have used different equations and found the same answer. To find the number of additional cell phone minutes, for example, we could have used the equation $m = \frac{\$60.00 - \$34.99}{\$0.27}$ instead of $\$60.00 = \$34.99 + \$0.27m$.

Because the two equations are equivalent, the same solution is found for each equation. Similarly, we can write a single equation that will solve a problem or we can solve the problem using a series of shorter equations. Some people think that solving a problem with a single equation is the best way; others find it easier to find one thing at a time using several equations.

Example 3 **A golf ball has a diameter of 1.680 inches. Golf balls are packaged in sleeves that consist of three balls in a rectangular box. The box has inside measurements of 1.7 inches by 1.7 inches by 5.1 inches.**

a. Use a single equation to calculate the volume of air inside the sleeve of balls to the nearest tenth of a cubic inch.

I know . . .

1.680 in. = Diameter of ball

3 = Balls in sleeve

1.7 in. = Width of box

1.7 in. = Height of box

5.1 in. = Length of box

A = Volume of air

Formulas:

Volume of ball $= \frac{4}{3}\pi r^3$

Volume of box $= lwh$

Equations

$$\text{Volume of box} = \text{Volume of balls} + \text{Volume of air}$$

$$\text{Volume of box} = 3 \cdot \text{Volume of one ball} + \text{Volume of air}$$

$$lwh = 3 \cdot \frac{4}{3}\pi r^3 + A$$

$$(1.7\text{ in.})(1.7\text{ in.})(5.1\text{ in.}) \approx 3\left(\frac{4}{3}\right)(3.14)(0.84\text{ in.})^3 + A$$

Solve

$$(1.7\text{ in.})(1.7\text{ in.})(5.1\text{ in.}) \approx 3\left(\frac{4}{3}\right)(3.14)(0.84\text{ in.})^3 + A$$

$$(1.7\text{ in.})(1.7\text{ in.})(5.1\text{ in.}) \approx 4(3.14)(0.840\text{ in.})^3 + A$$

$$14.739\text{ in.}^3 \approx 7.444\text{ in.}^3 + A$$

$$14.739\text{ in.}^3 - \mathbf{7.444\text{ in.}^3} \approx 7.444\text{ in.}^3 - \mathbf{7.444\text{ in.}^3} + A$$

$$7.295\text{ in.}^3 \approx A$$

$$7.3\text{ in.}^3 = A$$

b. Use more than one equation to calculate the volume of air inside the sleeve of balls to the nearest tenth of a cubic inch.

First, calculate the volume of the box.

Equations

Volume = (Length)(Width)(Height)

$V = (1.7\text{ in.})(1.7\text{ in.})(5.1\text{ in.})$

Solve

$V = 14.739\text{ in.}^3$

Now, calculate the volume of a ball.

Equations

Ball volume $= \frac{4}{3}\pi(\text{radius})^3$

$$b \approx \frac{4}{3}(3.14)\left(\frac{1.68\text{ in.}}{2}\right)^3$$

Solve

$b = 2.481\ldots\text{ in.}^3$

To find the volume of three balls, we just multiply the volume of one ball by 3.

$3b = 3(2.481\ldots\text{ in.}^3) = 7.444\ldots\text{ in.}^3$

Finally, we can find the volume of the air in the box.

Equations

Volume of air = Volume of box − Volume of balls

$A = 14.739\text{ in.}^3 - 7.444\ldots\text{ in.}^3$

Solve

$A \approx 7.294\ldots\text{ in.}^3$

$A \approx 7.3\text{ in.}^3$

Both methods result in the same answer. Notice that ellipses (...) follow many of the numbers in the second method. Instead of rounding the numbers, as many digits as could be carried in the calculator were stored. The calculator used to do this example allowed ten digits of each number to be used.

Check

To check, we round as many measurements as we can. The box is about 5 in. by 2 in. by 2 in. for an approximate volume of 20 in.^{3}. Each ball has a diameter of about 2 in. and π is about 3. The volume of a single ball is $\frac{4}{3}\pi r^3$.

Estimating, this is about $\frac{4}{3}(3)(1 \text{ in.})^3$ or 4 in.^{3} So, the volume of three balls is about 12 in.^{3} Subtracting the volume of the balls from the volume of the box, 20 in.^{3} − 12 in.^{3}, we get a volume of 8 in.^{3} This is reasonably close to our exact answer of 7.3 in.^{3}

Answer

A sleeve of golf balls contains about 7.3 cubic inches of air.

CHECK YOUR UNDERSTANDING

Solve using the five-step problem-solving method.

1. Hazel and Laura are writing a textbook. It currently has 1,050 pages but the publisher wants to cut it to 800 pages. If each page holds 40 lines of text, how many lines of text on average should be removed from each page to satisfy the publisher? Round the answer to the nearest whole number of lines.

2. A math instructor employed by a college receives a salary of $835 per credit hour and benefits including health insurance. The college pays $3,600 for health insurance per employee each year. Social security and retirement paid by the college equals about $\frac{11}{50}$ of the instructor's annual salary. Minh teaches 18 credits per year. He is paid in 24 equal paychecks. What is the total cost to the college for each of Minh's paychecks?

Answers: 1. Hazel and Laura should cut an average of 10 lines per page;
2. The total cost per paycheck is about $914.03.

EXERCISES SECTION 5.5

1. Violation of housing ordinances in Tigard, Oregon, can result in fines of up to $250 per violation per day. The fee for a full-time housing inspector is $70,000 per year. How many violations of $250 per day will have to be levied to pay for the salary of the housing inspector?

2. A take-over bid is being calculated for a major company that has 6,000,000 publicly traded shares. If the value of the company is $180,000,000, what should the price per share be?

3. The price for a night's lodging at a motel in Arlington, Virginia, is $135. Mario has a budget for lodging of $945. How many nights can he stay at the motel?

4. At a used car dealer, Sadie looked at a 1998 Ford Escort ZX2 priced at $9,495. When she asked the salesperson about the 1998 Ford Mustang parked in the next space, she was told it was about $5,500 higher than the Escort. What was the price of the Mustang?

5. In 1999, the hotel industry in Las Vegas added 12,500 rooms. With these new rooms, Las Vegas had a total of 121,000 hotel rooms. How many hotel rooms did the city have before the addition of these new rooms?

© *Richard Cummins/CORBIS*

6. A woman who was scarred by a poorly done liposuction procedure was awarded $33,625 for future medical costs and $150,000 for pain, suffering, and humiliation. Her original suit asked for $2,000,000. What is the difference between the amount she requested and the amount she was awarded?

7. Japan exported 17,122 cars, trucks, and buses to Asia in January 1998. This was a dramatic drop from the exports of the previous month of 22,921 vehicles. What was the decrease in exported vehicles?

8. Al has $1,420 saved in his bank account to buy a new car. His great-aunt Matilda left him 40 shares of stock in her will. He needs a total of $4,450 for the down payment on the car to keep his monthly payments manageable. How many shares of stock must he sell at the current price of $101 a share to have enough for his down payment?

9. When designing a house, sufficient venting of the attic has to be provided to allow hot air in the attic to escape outside. The formula for calculating the area of vent required for a rectangular attic with a vapor barrier between the attic and the living space is $A = \frac{lw}{300}$. George is designing a house with an attic that is 60 feet wide and 40 feet long. What area of attic vents is required for this house?

10. A newspaper article reported on claims paid by a medical center in compensation for accidental injuries or negligence, stating that the center paid out an average of \$190,113 a year or \$15,622 per claim. How many claims were paid per year?

11. Based on historical cannery records and published accounts, researchers estimate that the annual biomass of salmon returning to rivers before the arrival of settlers in Washington, Oregon, Idaho, and California was a minimum of 352 million pounds. This has declined to about 26 million pounds. What is the loss of biomass and its accompanying nutrients to the ecosystem?

© *Kennan Ward/CORBIS*

12. Wildlife managers released 38 bighorn sheep into the Hells Canyon area. They estimated the population of sheep in the area after this release to total about 820 sheep. In a previous pneumonia outbreak, about 315 sheep died. Based on this information, estimate the population of the herd before the outbreak of pneumonia.

© *Galen Rowell/CORBIS*

13. A United Way campaign raised $815,688 in pledges from its annual drive and $4,100 in cash from special events. Investments of funds during the period earned $10,925. Staff estimated that $35,000 in pledged funds would not actually be collected. Administrative staff salaries and related expenses for the year were $60,470. Actual campaign expenses totaled $45,507. What was the net gain of the campaign?

14. Wes owns an automatic sprinkler company. He estimates the cost of a job by figuring out the number of regular heads and rotating heads he needs for a job plus the fixed costs. For each regular fixed sprinkler head he charges $20. For each rotating head he charges $45. The city water connection costs $200. A back-flow limiter costs $250. A 6-station control box costs $275. A customer has $2,000 to spend on a new sprinkling system. The customer needs five rotating sprinkler heads and will purchase a city water connection, one back-flow limiter, and a 6-station control box. How many fixed heads can the customer afford to buy?

15. Barb ordered four rose bushes that cost $22.95 each and two red-twigged dogwoods that cost $43.50 each from a catalog. The total cost of the order was $202.45. How much was the shipping and handling charge?

16. Judy pays $39.95 a month for 300 minutes of cellular phone airtime. She pays 25¢ for every minute of additional airtime. In July, her total cellular phone bill was $58.95. How many additional minutes of airtime did she use?

17. An outreach student at a state college pays $94 a credit plus an additional $45 fee per semester to support the outreach center. If he is using his student loan of $1,455 to pay for the credits and the outreach center fee, how many credits can he afford to take?

18. For 30-second ads during the Super Bowl in 2000, a company paid an average price of $2,200,000 per ad. This compares to an average cost of $250,000 for a 30-second ad on a moderately successful regular prime time program. If a company purchases eight ads during the Super Bowl, how much more will this cost than the same number of ads aired during regular prime time?

19. A rectangular swimming pool is 50 meters long and 25 meters wide. A concrete deck that is 10 meters wide surrounds the pool. What is the difference in the area of the pool and the area of the concrete deck?

20. A rectangular swimming pool is 50 meters long and 25 meters wide. A concrete deck that is 10 meters wide surrounds the pool. What is the difference in the perimeter of the pool and the outer perimeter of the concrete deck?

21. Four teenage boys responsible for $300,000 in vandalism at an elementary school were ordered to pay $40,000 in restitution to the school district. One student has paid $5, one student has paid $72, one student has paid $230, and one student has paid $3,430. The district's insurance policy paid for all but a $1,000 deductible. What is the net financial gain or loss for the school district from this incident?

22. A newsletter from a health insurance company included the following table:

ACTIVITY LEVEL	CALORIES BURNED PER 15 MINUTES PER BODY WEIGHT		
	110 pounds	**150 pounds**	**190 pounds**
Intense			
Racquetball	134	188	231
Fast cycling	126	173	218
Fast swimming	117	159	203
Running	101	137	174
Moderate			
Stair climbing	91	123	157
Rowing machine	77	106	133
Low			
Golf	63	86	110
Walking, normal pace	60	83	104
Lying still	17	23	29

a. Calculate the number of Calories burned by a 150-pound woman who runs for 15 minutes, walks at a normal pace for 15 minutes, and then lies still for 30 minutes.

b. A 190-pound man uses a stair climber to keep in shape. He sets the computer on the machine to tell him when he has burned 1,000 Calories. How long does it take him to burn 1,000 Calories?

23. Adrianella poured three-fourths of a jug of distilled water into a beaker. The water had a volume of 600 milliliters. Calculate the volume of water contained in a full jug.

24. Two-thirds of Cathy's electric bill is usually due to air conditioning in the summer. In August, Cathy's electric bill was \$84. How much was probably due to air conditioning?

25. Barak was walking up a popular trail to the top of a waterfall. He paused to catch his breath at a marker pointing back in the direction he had come which read *Parking lot, $\frac{1}{4}$ mile.* He asked a family on the way down how much farther it was to the top. They replied, "You're about $\frac{2}{3}$ of the way there." About how long was the trail from the parking lot to the top of the waterfall?

26. Scott and Becky want to buy a home that will require a monthly house payment of $1,300. To qualify for a loan, the loan officer will allow them to spend a maximum of $\frac{2}{5}$ of their take-home pay. Becky is self-employed and can work more hours to increase their income if necessary. How much will their combined take-home pay per month need to be to qualify for a loan for the home?

27. Sam took a taxi from the courthouse to a law office. The taxi driver charged Sam $24. The cab's advertised rates were $2 for the first one-fifth of a mile and $4 for each additional mile. How far was it from the courthouse to the law office? Write the answer as a mixed number or integer.

28. The strength coach at a local gym has designed a lifting program for Heather. She determines the maximum amount of weight she can lift. She then does multiple repetitions lifting weights that are $\frac{7}{10}$ of this maximum lifting weight. In bench press, her personal goal is to be able to do repeated lifts of 80 pounds. What will her maximum lifting weight need to be to attain this goal? Round to the nearest pound.

29. Ken teaches 18 hours each week. This accounts for $\frac{3}{10}$ of his total work week. The rest of his time is spent preparing for class, grading, advising, meeting with students, attending committee meetings, and doing administrative duties. How many hours a week does Ken actually work?

ERROR ANALYSIS

30. **Problem:** Leah's cellular phone calling plan charges \$34.99 a month for the first 120 minutes. She pays 27¢ per minute for extra time. She has budgeted \$60.00 a month for her cell phone. How many total minutes can she talk each month and stay within her budget?

Answer: $\$60.00 = \$34.99 + \$0.27m$

$m \approx 92$

Leah can talk about 92 minutes per month.

Describe the mistake made in solving this problem.

Redo the problem correctly.

REVIEW

31. Simplify $\frac{4 \text{ meters}}{8 \text{ meters}}$.

32. Bobbie spent \$27.68 on lunches during her 5-day work week. Calculate the average amount she spent on lunch per day.

33. Build an equivalent fraction for $\frac{3}{5}$ that has a denominator of 100.

READ AND REFLECT

This text includes many "realistic" and "real-life" problems drawn from newspapers, Internet sites, magazines, professional journals, and textbooks in other subject areas. We ask you to use the five-step strategy to solve these problems. We introduce this somewhat rigid strategy to help you experience success, build confidence, and reduce any hatred and fear you have of word problems. Now it is time to broaden your thinking about problem solving.

Courtesy of Stanford University Archives, from the George Pólya Papers

George Pólya was born in Budapest, Romania, on December 13, 1887. He earned a Ph.D. in mathematics with a minor in physics from the University of Budapest in 1912. He became a professor at Stanford University in 1945 and wrote a book on problem solving called *How to Solve It* that has sold almost one million copies in 17 languages. In Pólya's view, problem solving can be broken into four processes: (1) *understanding the problem,* (2) *devising a plan* to solve the problem, (3) *carrying out the plan,* and (4) *looking back* at the problem, checking the results, and evaluating the solution. This is a broader view of problem solving than the five-step strategy used in this text. To *understand the problem* in the five-step strategy, we write down what we know using words, numbers, and variables and perhaps we draw a picture. Yet understanding a problem in real life might also require us to collect more data, do additional research, discuss the situation with other people, or think of similar problems. Similarly, in real life all problems cannot be solved using an equation. *Devising a plan* might instead involve analyzing a table, drawing and interpreting a graph, making scale diagrams, designing a computer program or using a spreadsheet to analyze data, or writing a logical argument.

Pólya also has an important strategy for problems that seem too difficult to solve. He wrote, "If you cannot solve a problem, then there is an easier problem you cannot solve: find it." In other words, break a difficult problem down into easier problems. You may not be able to solve it yet, either, but it will be easier than the original problem to figure out. This is similar to writing a series of equations rather than one big equation to solve a problem. Sometimes it is easier to do one thing at a time.

When beginning activities that involve synthesis and creativity, we often begin with a simple, even rigid, approach. For example, we learn to write sentences before paragraphs and paragraphs before essays. The five-step strategy is a simple beginning approach to solving problems. It will not be effective for all the problems you will encounter, in math books or elsewhere, but it gives you a place to start.

RESPOND

1. Write a word problem that is easy to solve with an equation. Use a newspaper, magazine, or the Internet as the source of your information.

2. Write a word problem that is not easy to solve with an equation.

3. Is the five-step problem-solving strategy a good one for you to use when solving word problems in a math class? Explain.

Section	Terms and Concepts/Procedures
5.1	numerical expression; algebraic term; algebraic expression; order of operations; grouping symbol; evaluate a numerical expression; evaluate an algebraic expression 1. Evaluate a numerical expression following the order of operations. 2. Evaluate an algebraic expression.
5.2	like terms; combining like terms; coefficient; algebraic expression; simplify an algebraic expression; distributive property 1. Simplify an algebraic expression.
5.3	equation; solution; solving an equation by inspection; isolate the variable; symmetric property; constant; addition property of equality; subtraction property of equality; multiplication property of equality; division property of equality 1. Check a possible solution of an equation. 2. Solve an equation in the form $x + b = c$. 3. Solve an equation in the form $x - b = c$. 4. Solve an equation in the form $ax = c$. 5. Solve an equation in the form $\frac{x}{a} = c$.
5.4	linear equation in one variable; multiple 1. Simplify an equation as much as possible. 2. Solve an equation in the form $ax + b = c$.
5.5	five-step problem-solving strategy 1. Solve a word problem using the five-step problem-solving strategy.

SECTION 5.1 EVALUATING ALGEBRAIC EXPRESSIONS

Evaluate. Show intermediate steps.

1. $100 + 200 \div 5 - 15$

2. $84 \div 3 \cdot 2 + 1 \cdot 2$

3. $(2 + 3)^2 - 4 + 6$

4. $24 \div (8 - 2) \cdot 2$

Evaluate.

5. $xy + 2(x - y)$ when $x = 6$ and $y = 3$

6. $x \div y + z$ when $x = 12$, $y = 4$, and $z = 1$

7. $4(x + y) + 2(y - 1)$ when $x = 7.5$ and $y = -3.2$

8. $\dfrac{xy - z}{x^2 - y^2}$ when $x = -5$, $y = 2$ and $z = -3$

SECTION 5.2 SIMPLIFYING ALGEBRAIC EXPRESSIONS

Simplify.

9. $15x + 8 + 21x + 13y + 2$

10. $70(x + 4) + 15x - 9$

11. $3(x + 2) + 7x - 5$

12. $5(2x - 9)$

13. $-4(3x - 2)$

14. $-3(5x^2 - 8x + 2)$

15. $\dfrac{5x - 10}{5}$

16. $\dfrac{12y - 12}{-12}$

SECTION 5.3 PROPERTIES OF EQUALITY

Solve. Check the solution.

17. $x + 4 = -12$

18. $y \div (-9) = -2$

19. $x - 5.9 = 3.4$

20. $\frac{x}{-2} = 6$

21. $-x = 5$

22. $-2x = -5$

23. $y + 900 = 1{,}548$

24. $70 = 25 + x$

25. $8x = 2{,}464$

26. $\frac{x}{2} = 10$

SECTION 5.4 EQUATIONS WITH MORE THAN ONE OPERATION

Solve. Check the solution. For Exercise 29, round to the hundredths place.

27. $2(3x + 1) = 26$

28. $12 = 2x - 4$

29. $4.6x - 7 = 3.5$

30. $8{,}765 = 3x + 32$

31. $\frac{5}{4}x - 6 = -\frac{2}{3}$

32. $\frac{4}{5}(5x - 8) = -\frac{5}{6}$

SECTION 5.5 APPLICATIONS AND PROBLEM SOLVING

Solve using the five-step problem-solving strategy.

33. The federal Family and Medical Leave Act allows fathers and mothers to take leave from work to provide care for a new baby. Although about 2,000,000 people a year take parental leave under the act, only 500,000 are men. About how many women take parental leave under the act per year?

34. Bianca is investigating the cost of flying six members of the Student Senate from San Diego to a convention in Philadelphia. If the budget for airfare is limited to \$2,380, what is the maximum that can be spent per ticket?

35. The cost of renting a midsize car in Minneapolis was \$285 a week plus an additional \$35 for each additional day. Calculate the cost of renting the car for 10 days.

36. Veneta is designing a square game table out of wood. The top will be a checkerboard surrounded by a 2-inch border on all sides. A checkerboard is 8 equal squares wide and eight equal squares long. If the total width of the top is to be 36 inches, how wide should each square on the checkerboard be?

37. Employees who work at the highest wage at a warehouse receive an hourly wage of $17 plus a $2,000 bonus every 6 months. Calculate the number of hours a person works per year if she makes $35,450 a year.

38. The Capri Motel has some guests that stay for extended periods of time. The rate for such stays is $75 per night for the first three nights and $35 for each additional night. What is the charge for a 45-day stay?

39. Acme Company owns a factory that requires its employees to wear uniforms and safety equipment. To discourage turnover, the company requires its employees to purchase the uniforms and safety equipment at hiring at a cost of $375. However, in each paycheck, the employee is reimbursed $25 until the cost of the equipment has been fully returned to the employee. How many paychecks does the employee need to receive to be reimbursed fully for the cost of uniforms and equipment?

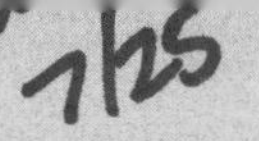

TEST CHAPTER 5

1. Evaluate $54 \div (2 + 8 \div 2) + 5^2$.

2. Evaluate $(15 - 3^2)^2$.

3. Evaluate $10 - 2(3 + 1)$.

4. Evaluate $100 \div (2 + 8) \cdot 4$.

5. Simplify $6 + 30(x + 2) + 19 - 18x$.

6. Simplify $100 + 300(x + 1) - 74 + 616x$.

7. Evaluate $6x - 7y$ when $x = 8$ and $y = 5$.

8. Evaluate $2l + 2w$ when $l = 700$ and $w = 350$.

9. **Solve. Use the five-step problem-solving strategy.**
To determine the amount of work being done by a group of faculty, a college calculates the *full time equivalents* (FTE) taught by the faculty. The formula for calculating full time equivalents is FTE $= C \div 12$, where C is the total credits taught by the faculty. The Department of Mathematics has 32 faculty who teach a total of 288 credits. How many FTE are being taught by this group of faculty?

10. Solve and check the solution: $9x = 6{,}318$.

11. Solve and check the solution: $100 = 2x + 60$.

12. Solve and check the solution: $5x - 9 = 96$.

13. Solve and check the solution: $7(x - 2) = 364$.

14. Solve and check the solution: $10(x - 1) + 12x = 188$.

15. Solve and check the solution: $2{,}000 = x - 1{,}555$.

16. Solve and check the solution: $3.05y = 64.05$.

17. Solve and check the solution: $\frac{3}{4}x - \frac{1}{8} = \frac{5}{6}$.

18. Solve and check the solution: $35.0 = x + 19.6$.

19. Solve and check the solution: $y \div 3 = 7$.

20. Solve and check the solution: $\frac{5}{7}d - 5 = -\frac{4}{9} + \frac{2}{3}$.

21. Calculate the area of a rectangle that has a length of 16 cm and a width of 5 cm.

Use the five-step problem-solving strategy to solve the word problem.

22. McDonald's restaurants in Britain apologized to millions of unhappy customers after running out of Big Macs during a weekend promotion. The promotion turned into a fiasco after massive demand over the weekend caused many of the nation's 922 outlets to run out of food and turn away long lines of customers. McDonald's bosses said they had expected to sell 2 million Big Macs during the promotion—four times the normal level—but soon found that the demand was twice that. Calculate the normal demand for Big Macs.

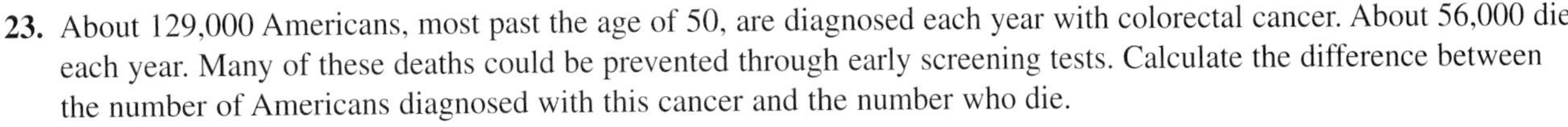

23. About 129,000 Americans, most past the age of 50, are diagnosed each year with colorectal cancer. About 56,000 die each year. Many of these deaths could be prevented through early screening tests. Calculate the difference between the number of Americans diagnosed with this cancer and the number who die.

24. John and Laurie have $2,500 for a trip to New York City. The airfare will cost $400 each. Their hotel is $135 per night and they plan to stay for five nights. They are setting aside $750 for meals and taxis. If tickets to Broadway shows cost $60 each, how many shows will they be able to attend?

6

Ratios, Rates, and Percents

Knowing how to work with decimals, fractions, and equations gives us the knowledge we need to use ratios, rates, and percents. Ratios and rates compare one quantity to another. A percent is another way to write a ratio. Since decision making and problem solving often involve comparisons, ratios, rates, and percents are essential concepts.

In this chapter, we will discover how to write information as a ratio or rate. We will learn how to use equations called proportions to solve problems that include ratios or rates. We will use this same knowledge to solve problems with the special ratio, percent. Finally, we will learn how to use circle graphs to represent information.

CHAPTER OUTLINE

Section 6.1 Introduction to Ratios and Rates

LEARNING TOOLS

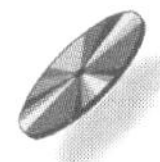

CD-ROM SSM VIDEO

Sneak Preview

"Two out of three students prefer to take notes in pen" is a ratio. $\frac{60 \text{ miles}}{1 \text{ hour}}$ is an example of a rate. The procedures for doing arithmetic with fractions can be used to solve problems involving ratios or rates.

After completing this section, you should be able to:

1) Identify ratios and rates.
2) Write a ratio as a fraction.
3) Write a rate as a fraction.
4) Calculate a unit rate.

INVESTIGATION Ratios and Rates

Read the selections, paying careful attention to the meaning of the word *ratio*.

SELECTION 1: Gear **ratios** compare the number of rotations of the drive shaft for every 100 complete rotations of the rear wheels. For example, in a car with a gear **ratio** of 305, the drive shaft rotates 305 times for every 100 complete rotations of the rear wheels.

SELECTION 2: HMOs analyze physician charges by calculating the **ratio** of the number of brief visits to the number of intermediate visits. For example, one family physician had a visit **ratio** of 29 to 10.

SELECTION 3: When mortgage lenders qualify a buyer for a home loan, two **ratios** are considered. The **front-end ratio** is the **ratio** of the monthly housing expense to the borrower's gross (pre-tax) monthly income. The **back-end ratio** is the **ratio** of the borrower's monthly debt payments (including the potential mortgage payment) to the gross monthly income.

1. What is a *ratio*?

Ratios always compare measurements that have the same units. A gear ratio compares rotations to rotations; a physician charge ratio compares visits to visits; a front-end ratio compares dollars to dollars.

A ratio can be written as a fraction. A gear ratio of 305 rotations to 100 rotations can be written as $\frac{305 \text{ rotations}}{100 \text{ rotations}}$. Because the units are the same, the ratio itself simplifies to a number without units.

$$\begin{aligned}\frac{305 \text{ rotations}}{100 \text{ rotations}} &= \frac{5 \cdot 61 \text{ rotations}}{2 \cdot 2 \cdot 5 \cdot 5 \text{ rotations}} \\ &= \frac{5}{5} \cdot \frac{61}{2 \cdot 2 \cdot 5} \cdot \frac{\text{rotations}}{\text{rotations}} \\ &= 1 \cdot \frac{61}{20} \cdot 1 \\ &= \frac{61}{20}\end{aligned}$$

Write the ratio as a fraction. Simplify the fraction to lowest terms.

2. A baitcast fishing reel has a gear ratio of one rotation of the crank to five rotations of the line spool.

3. Wade is applying for a home mortgage. His front-end ratio is \$875 monthly housing expense to \$2,300 monthly income.

4. St. Elias Alpine Guides provides guides for mountaineering expeditions. The price of the expeditions depends on the guest-to-guide ratio. A 10-day trip on Mt. Drum with a 2-to-1 guest-to-guide ratio costs \$1,995.

5. The American Prostate Society reports that two out of every three cancers are diagnosed too late to cure.

6. The aspect ratio of a car tire compares the height of the tire from the rim to the road in inches to the width of the tire in inches. The tire on Ramon's car has a height of 9 inches and a width of 12 inches.

Answers: 2. $\frac{1}{5}$; 3. $\frac{35}{92}$; 4. $\frac{2}{1}$; 5. $\frac{2}{3}$; 6. $\frac{3}{4}$.

Read the selections, paying careful attention to the meaning of the word *rate*.

SELECTION 1: Clair is paying her brother back at a **rate** of \$120 every 6 months.

SELECTION 2: The standard **rate** of speed allowed on U.S. freeways is 75 miles per hour.

SELECTION 3: The **rate** at which Internet ports should be provided for an Internet service so that subscribers do not receive busy signals is one port per ten subscribers.

SELECTION 4: By midnight each day, UPS aircraft leave the main UPS air hub in Louisville, Kentucky, at a **rate** of one aircraft every 2 minutes.

7. How is a *rate* different from a *ratio*?

Write each rate as a fraction. Simplify to lowest terms.

8. Frank took a timed typing test to determine his typing speed. Frank typed a total of 425 words in 5 minutes.

9. A Vietnamese restaurant is extremely busy at lunch but the customers tend to eat quickly. By taking orders promptly and having enough cooks in the kitchen, the restaurant averages a customer turnover rate of four tables opening for new customers every 10 minutes.

10. Everett is following a high-protein diet and is losing an average of 16 pounds every 6 weeks.

Answers: 8. $\frac{85 \text{ words}}{\text{minute}}$; 9. $\frac{2 \text{ tables}}{5 \text{ minutes}}$; 10. $\frac{8 \text{ pounds}}{3 \text{ weeks}}$.

Rates whose denominators are equal to 1 are called **unit rates.** A loan payback rate of $\frac{\$20}{1 \text{ month}}$ is a unit rate. A speed of $\frac{60 \text{ miles}}{1 \text{ hour}}$ is a unit rate. In unit rates, the number is usually written in front of the units. The units are written as a fraction. For example, $\frac{60 \text{ miles}}{1 \text{ hour}}$ is often written $60 \frac{\text{miles}}{\text{hour}}$ or as 60 miles per hour.

11. A Meals on Wheels program served 2,700 people in 30 days. Express this rate as a unit rate.

12. High Mag Gro is made from industrial waste left after production of magnesium. It is used to improve the pH balance of soil used to grow alfalfa, canola, and timothy hay. It costs $3,000 to apply High Mag Gro to 500 acres. Express the cost of applying High Mag Gro as a unit rate of dollars per acre.

Answers: **11.** $90 \frac{\text{people}}{\text{day}}$; **12.** $6 \frac{\text{dollars}}{\text{acre}}$.

DISCUSSION

Ratios and Rates

A **ratio** compares two quantities. The quantities are always in the same units. Ratios can be described in words, using colon notation, or using fraction notation. However, unlike a fraction, ratios can have a denominator of zero. For example, the number of women to men on a soccer team could be 11 to 0. This ratio shows that no men are on the team.

Example 1 **A survey of students in the college union building revealed that five out of ten students eat à la carte instead of buying a meal plan.**

In words: 5 students to 10 students

In colon notation: 5 : 10

In fraction notation: $\frac{5 \text{ students}}{10 \text{ students}}$

Ratios are usually simplified to lowest terms. Since the units are the same in the numerator and the denominator, they simplify to 1.

$$\frac{5 \text{ students}}{10 \text{ students}} = \frac{1}{2}$$

Ratio

A *ratio* is a fraction that compares two quantities with the same units of measurement.

Example 2 **A rectangular window is 60 inches long and 36 inches wide. Write the ratio of its length to its width. (Use fraction notation.)**

$$\text{Ratio} = \frac{\text{length}}{\text{width}}$$
$$= \frac{60 \text{ inches}}{36 \text{ inches}}$$
$$= \frac{5}{3}$$

36 in.

60 in.

Since both measurements in a ratio must be in the same unit, we may need to convert one of the measurements before we can write a ratio.

Example 3 **For every $1 Lucas earns, the U.S. government collects 28¢ in income taxes. Write a ratio of the taxes paid to earnings.**

We need to change one of the measurements before we can write a ratio. To change dollars into cents, we use the conversion factor 1 dollar = 100 cents.

$$\frac{\text{Taxes}}{\text{Earnings}} = \frac{28 \text{ cents}}{\$1}$$
$$= \frac{28 \text{ cents}}{100 \text{ cents}}$$
$$= \frac{7}{25}$$

CHECK YOUR UNDERSTANDING

Write each ratio in lowest terms.

1. A survey revealed that eight out of ten dogs prefer Chow Down dog food.

2. 1,800 people out of a total of 3,000 people surveyed were registered voters.

3. 15 inches out of 100 feet of ribbon were defective.

Answers: 1. $\frac{4}{5}$; 2. $\frac{3}{5}$; 3. $\frac{1}{80}$.

In contrast to ratios, **rates** compare quantities that are not measured in the same units. When simplified to lowest terms, rates still have units.

Example 4 **On a rainy day in November, Portland, Maine, recorded 4 inches of rain in a 24-hour period. Write this rate as a fraction in lowest terms.**

$$\frac{\text{Rainfall}}{\text{Time}}$$
$$= \frac{4 \text{ inches}}{24 \text{ hours}}$$
$$= \frac{1 \text{ inch}}{6 \text{ hours}}$$

Rate

A *rate* is a fraction that compares two quantities that have different units of measurement.

A **unit rate** is a rate that has a denominator of 1 unit. For example, $\frac{65 \text{ miles}}{1 \text{ hour}}$ and $\frac{3 \text{ grams}}{1 \text{ milliliter}}$ are unit rates. Unit rates are often written with the number to the left of the units. We can write $\frac{65 \text{ miles}}{1 \text{ hour}}$ as $65\frac{\text{miles}}{\text{hour}}$ and $\frac{3 \text{ grams}}{1 \text{ milliliter}}$ as $3\frac{\text{grams}}{\text{milliliter}}$.

PROCEDURE

To find a unit rate:

1. Write the rate as a fraction.
2. Divide the number in the numerator by the number in the denominator. The quotient may be a decimal number.
3. Write the measurement units as a fraction next to the quotient from Step 2.

Example 5 **In a showing of the movie *Braveheart* on network television, 48 commercials were shown in 2 hours. Write this as a unit rate.**

$$\frac{\text{Number of commercials}}{\text{Time}}$$
$$= \frac{48 \text{ commercials}}{2 \text{ hours}}$$
$$= 24 \frac{\text{commercials}}{\text{hour}}$$

Example 6 **J. C. filled the gas tank of her T-100 truck with 18.4 gallons of gas. Her trip odometer showed that she had driven 373 miles since her last fill-up. What was her gas mileage in miles per gallon on this tank of gas? Round to the nearest tenth of a mile per gallon.**

$$\frac{\text{Miles driven}}{\text{Gallons of gas used}}$$
$$= \frac{373 \text{ miles}}{18.4 \text{ gallons}}$$
$$= 20.2717\ldots \frac{\text{miles}}{\text{gallon}}$$
$$\approx 20.3 \text{ miles per gallon}$$

CHECK YOUR UNDERSTANDING

Write each rate in lowest terms.

1. The fence cost \$5 per 10 linear feet.

2. At Camelot School, there were 246 students for every 18 teachers.

Calculate the unit rate. Round to the nearest tenth.

3. Char was paid \$280 for 35 hours of work.

4. At Camelot School, there were 246 students for every 18 teachers.

Answers: 1. $\frac{\$1}{2 \text{ linear feet}}$; 2. $\frac{41 \text{ students}}{3 \text{ teachers}}$; 3. $\frac{8 \text{ dollars}}{1 \text{ hour}}$; 4. $\frac{13.7 \text{ students}}{1 \text{ teacher}}$.

EXERCISES SECTION 6.1

Express the ratio or rate in fraction notation. Simplify to lowest terms.

1. A well-designed speaker box has a depth of 9 inches, a width of 15 inches, and a height of 24 inches.

a. Write the ratio of the depth of the speaker box to the width of the speaker box.

b. Write the ratio of the height of the speaker box to the width of the speaker box.

c. Write the ratio of the depth of the speaker box to the height of the speaker box.

2. A clot that obstructs the flow of blood through an artery causes eight out of ten strokes. Write the ratio of strokes caused by a clot obstruction to the total strokes.

3. According to the Centers for Disease Control and Prevention, cases of bacterial meningitis in preschoolers declined from 12,920 cases in 1986 to 5,755 cases in 1995. Write the ratio of cases of bacterial meningitis in 1986 to the cases in 1995.

4. According to the Centers for Disease Control and Prevention, cases of *Haemophilus influenzae* type b (Hib flu) among children under age 5 dropped from about 20,000 cases in 1985 to 81 cases in 1997. The drop is attributed to the use of the Hib vaccine. Write the ratio of cases of Hib flu in 1997 to the cases in 1985.

5. In the Idaho Panhandle Forests, there are 850 abandoned mine sites. Seventy-eight of these mines have known environmental problems. Write the ratio of problem mines to the total number of mines.

6. Mattie brought 11 pounds of organic green beans to the local produce store for resale. The store owner rejected 15 ounces of the beans because they were too large. Write the ratio of rejected beans to total beans.

7. Mattie brought 11 pounds of organic green beans to the local produce store for resale. The store owner rejected 15 ounces of the beans because they were too large. Write the ratio of *accepted* beans to total beans.

8. The United States exported a total of $5,031,908,642 in telecommunications equipment in the first quarter of 1998 to its top ten markets. Canada was the leading market, importing equipment worth $729,881,456. Write the ratio of Canadian imports to total American exports.

9. A rural subdivision has 4.3 miles of roads. Only 2,500 feet of the road is paved. The rest is gravel. Write the ratio of paved road to total road in the subdivision.

10. In 1998, McDonald's Corporation announced a layoff of 525 jobs at its headquarters in suburban Chicago. Before the layoffs, 2,300 people were employed at the headquarters. Write a ratio of people who lost their jobs to the total number of jobs before layoffs.

11. According to the U.S. Census Bureau, three in ten Americans lived below the poverty line for at least 2 months of a recent 3-year period. The most likely people to be poor were families headed by single mothers. Write a ratio that compares the number of Americans living in poverty to the total number of Americans.

12. In 1 kilogram of water, 111.9 grams is from hydrogen atoms. Write the ratio of the mass of the hydrogen atoms to the total mass of the water.

13. Ninety-six rats went to space aboard the shuttle *Columbia* in 1998 to be used for research purposes. Approximately 265 million rats live in the United States. Write a ratio of the rats sent to space compared to the total rats in the United States.

14. One hundred sixty animators worked on *The Rugrats Movie.* Thirty-five animators produced *Antz*. Write a ratio that compares the animators who worked on *The Rugrats Movie* to the animators who worked on *Antz*.

15. Two out of every 100 tables served in a restaurant will not leave any tip for the server. Write a ratio that compares the number of tables who will not leave a tip to the total number of tables served.

16. Barry Noble of Newcastle, England, is the haggis-eating champion of the United Kingdom. Haggis is the national dish of Scotland and is a combination of sheep's heart, lungs, oatmeal, and spices cooked in the lining of a sheep stomach. Noble has eaten $1\frac{1}{2}$ pounds of haggis in 84 seconds. Calculate this rate as a unit rate of pounds per second. Round to the nearest hundredth of a pound per second.

17. Cucumber plants are either male or female. For every ten male cucumber plants in a given field, there are about 60 female cucumber plants. Write the ratio of male cucumber plants to female cucumber plants.

18. In 1995, the FBI reported that California had 467 reported bombings. The motives for the bombings ranged from political terrorism to personal grudges to thrill-seeking. The United States as a whole had 2,577 such bombings. Write the ratio of bombings in California in 1995 to bombings in the entire United States.

19. A river port facility charged $72 for unloading a container from a barge. Because of increased costs, it raised the fee to $80. Write the ratio of the price increase to the original cost.

20. The annual Nathan's hot dog eating championship is held on Coney Island, New York, every Fourth of July weekend. In 1997, Hirofumi Nakajima won the contest by eating $24\frac{1}{2}$ hot dogs with buns in 12 minutes. Calculate this rate as a unit rate per minute. Round to the nearest tenth of a hot dog.

21. Up to 19,200 tubes of lipstick may be produced in an 8-hour shift at a lipstick manufacturing plant. Write the unit rate of lipsticks produced per hour.

22. In 12 hours of driving, Roger drove 628 miles. Calculate Roger's average speed in miles per hour. Round to the nearest tenth of a mile per hour.

23. Jane graded 15 exams in 48 minutes. Calculate the average time it takes Jane to grade each exam. Round to the nearest tenth of a minute.

24. Thirty boaters were predicted to die as a result of accidents that involved alcohol over the 72-hour Fourth of July weekend. Calculate the unit rate of deaths per hour. Round to the nearest tenth of a death per hour.

25. The recycling program at Washington State University earned about $206,000 on 1,255 tons of recyclables in 1997. Calculate the unit rate of earning in dollars per ton. Round to the nearest hundredth of a dollar per ton.

26. Power plants in the United States generate more than 2,000 tons a year of nuclear waste. Assuming 365 days in a year, calculate the rate of waste produced per day. Round to the nearest hundredth of a ton per day.

27. According to the United Nations, nearly 6 million people were newly infected with the HIV virus in 1998. Many of these people have not yet exhibited symptoms. Calculate the unit rate of infections per day, assuming 365 days in a year. Round to the nearest whole number.

28. Peggy lost 18 pounds in 92 days. Calculate the unit rate of her weight loss per day. Round to the nearest tenth of a pound per day.

29. An intravenous setup for medication delivery drips 1,215 drops per hour. Calculate the unit rate of drops per minute. Round to the nearest drop.

30. In May of 1998, four bakers in Forli, Italy, transformed a ton of flour, 300 quarts of water, 1,540 pounds of tomatoes, and a ton of mozzarella cheese into steaming pizzas. They produced 6,940 pizzas in 12 hours. Calculate the unit rate of pizzas made per hour. Round to the nearest whole pizza.

31. A grocery store has a "bills for loose change" machine. For every \$1.18 in coins deposited in the machine, the machine returns a \$1 bill. Write a unit ratio that compares the returned money to the deposited money. Round to the nearest hundredth of a dollar.

32. Alaska musher Aliy Zirkle was the first woman to win the 1,000-mile Yukon Quest International Sled Dog Race. She won \$30,000 for her victory. The total purse was \$125,000, with prize money paid to the first 15 finishers. Write a ratio that compares the first prize money to the total prize money.

33. In March, 2000, the total tax on a carton of cigarettes in New York was \$11.10. If there are ten packs of cigarettes in a carton, calculate the unit rate of tax per pack of cigarettes. Round to the nearest cent.

34. When fuel prices are about \$1.50 a gallon, truckers who haul wheat to market charge around \$1.25 a mile. About 25¢ of that amount is used to pay for the fuel. Write a ratio of the amount of the trucking charge used for fuel to the total amount of the trucking charge.

35. In the Badlands of North Dakota, Marmarth was once a bustling railroad depot with 2,000 people. It now has a population of 134 people. Write a ratio that compares the current population to its former population.

(Courtesy of National Park Service)

36. Fifty-two high school juniors participated in a Junior Miss competition. Only six received cash prizes. Write a ratio that compares the number of cash prize winners to the total participants.

37. A family season ski pass at Brundage Mountain purchased before April 30 is \$895. The same pass purchased after November 1 is \$1,045. Write a ratio that compares the cost of the pass in April to the cost of the pass in November.

38. In 1999, the total number of electronically filed tax returns was 23,700,000. Of those returns, 2,460,000 were prepared by taxpayers rather than by professionals. Write a ratio that compares the number of returns prepared by taxpayers to the total tax returns.

39. The total cost for a prescription of 180 capsules of Prozac was $341.31. Calculate the unit rate of cost per capsule. Round to the nearest cent.

40. The co-pay paid by the patient for a prescription of 180 capsules of Prozac was $14. Calculate the unit rate of cost per capsule. Round to the nearest cent.

41. The estimated amount that will be spent on weddings in 2001 is $42,400,000,000. The estimated amount that will be spent on wedding gifts is $19,100,000,000. Write the ratio of the amount spent on weddings to the amount received in wedding gifts.

42. Between 1980 and 2000, there were 528 medicines approved by the Food and Drug Administration. Twelve of these medicines, which were recalled because of safety reasons, included Propulsid, Rezulin, Hismanal, Seldane, and Redux. Write a ratio that compares the number of medicines recalled to the number of medicines approved.

ERROR ANALYSIS

43. **Problem:** Under a proposed merit pay plan for teachers in South Dakota, school districts would provide a $1,500 bonus to selected teachers. There are about 9,000 public school teachers in the state, and 2,250 of them would get a bonus. Write a ratio comparing the number of teachers getting a bonus to the number of teachers not getting a bonus.

Answer: $\frac{2{,}250}{9{,}000}$

Describe the mistake made in solving this problem.

Redo the problem correctly.

44. Problem: Dennis Rodman returned to the NBA in February 2000. His contract required him to play 39 games for a salary of $470,000. Calculate the unit rate of his pay per game.

Answer: $\frac{39 \text{ games}}{\$470{,}000}$

$\approx \$0.00008$ per game

Describe the mistake made in solving this problem.

Redo the problem correctly.

REVIEW

45. Solve and check this equation: $\frac{x}{6} = 20$

46. Solve and check this equation: $\frac{y}{10} = \frac{3}{5}$

47. Solve and check this equation: $\frac{2}{7} = \frac{c}{100}$

48. Solve and check this equation: $\frac{3}{100} = \frac{x}{20}$

Section 6.2 Proportions

LEARNING TOOLS

CD-ROM SSM VIDEO

Sneak Preview

Two equivalent ratios or rates can be written in an equation called a proportion. Many practical problems can be solved using proportions.

After completing this section, you should be able to:

1) Determine an equivalent ratio or rate using a proportion.

2) Use proportions to solve problems involving rates or ratios.

DISCUSSION

Proportions

Two equivalent ratios or rates can be written as an equation. Such an equation is called a **proportion.** For example, 30 beats per minute is equivalent to 60 beats per 2 minutes.

As an equation, we write: $\frac{30\text{ beats}}{1\text{ minute}} = \frac{60\text{ beats}}{2\text{ minutes}}$. If we know one rate, we can use a proportion to determine an equivalent rate.

Example 1 **According to the U.S. Transportation Department, there were 41 deaths per 100 million vehicle miles in 1994 on U.S. highways. Predict how many deaths would occur for every 250 million vehicle miles traveled. Round the answer to the nearest whole number.**

I know . . .

d = Predicted number of deaths

250 = Vehicle miles traveled in millions

$$\frac{41\text{ deaths}}{100\text{ million miles}} = \text{Known death rate}$$

Equations

Set up a proportion.

$$\text{Known death rate} = \text{Predicted death rate}$$

$$\frac{41\text{ deaths}}{100\text{ million miles}} = \frac{d}{250\text{ million miles}}$$

Solve

$$\frac{41\text{ deaths}}{100\text{ million miles}} = \frac{d}{250\text{ million miles}}$$

To isolate the variable, multiply both sides by 250 million miles.

$$\left(\frac{\textbf{250 million miles}}{\textbf{1}}\right)\left(\frac{41\text{ deaths}}{100\text{ million miles}}\right) = \left(\frac{d}{250\text{ million miles}}\right)\left(\frac{\textbf{250 million miles}}{\textbf{1}}\right)$$

Simplify before multiplying.

$$\left(\frac{5}{1}\right)\left(\frac{41\text{ deaths}}{2}\right) = 1d$$

$$\frac{205\text{ deaths}}{2} = d$$

$$102.5\text{ deaths} = d$$

$$103\text{ deaths} \approx d$$

Check

250 million miles is between two and three times more than 100 million miles. We expect the deaths for 250 million miles to be between two and three times the deaths at 100 million miles. $2 \cdot 41$ deaths $= 82$ deaths and $3 \cdot 41$ deaths $= 123$ deaths. Since the exact answer (103 deaths) is between 82 deaths and 123 deaths, it seems reasonable.

Answer

The predicted death rate is 103 deaths per 250 million vehicle miles.

Proportions can be used to make predictions about the behavior of large groups of people based on polls of a random sample of people.

Example 2 **In February 2000, the Republican presidential candidates were campaigning in South Carolina. A research poll revealed that 273 of 665 likely Republican voters preferred George W. Bush as the Republican nominee. Assuming that about 1 million voters in South Carolina will vote Republican in the primary, how many votes does the poll predict will be cast for Mr. Bush? Round to the nearest voter.**

I know . . .

v = Primary voters for Bush

273 = Polled voters for Bush

665 = Total polled voters

1,000,000 = Total primary voters

Equations

Set up a proportion.

$$\frac{\text{Polled voters for Bush}}{\text{Total polled voters}} = \frac{\text{Primary voters for Bush}}{\text{Total primary voters}}$$

$$\frac{273 \text{ voters}}{665 \text{ voters}} = \frac{v}{1{,}000{,}000 \text{ voters}}$$

Solve

$$\frac{273 \text{ voters}}{665 \text{ voters}} = \frac{v}{1{,}000{,}000 \text{ voters}}$$

$$\left(\frac{\mathbf{1{,}000{,}000 \text{ voters}}}{\mathbf{1}}\right)\left(\frac{273 \text{ voters}}{665 \text{ voters}}\right) = \left(\frac{\mathbf{1{,}000{,}000 \text{ voters}}}{\mathbf{1}}\right)\left(\frac{v}{1{,}000{,}000 \text{ voters}}\right)$$

$$\frac{273{,}000{,}000 \text{ voters}}{665} = 1v$$

$$410{,}526 \text{ voters} \approx v$$

Check

The 273 polled voters for Bush account for not quite half of the total polled voters. Similarly, 410,526 primary voters is not quite half of the total primary voters. The answer seems reasonable.

Answer

The poll predicts that 410,526 out of 1,000,000 Republican primary voters will choose George W. Bush.

Government budgets are often reported using rates or ratios. We use the rate or ratio to calculate the actual money being spent.

Example 3 **In Minnesota, the state constitution requires that 18 cents of every dollar spent on lottery tickets be given to the Environmental and Natural Resources Trust Fund or the general fund. If the actual lottery ticket sales in a given year total $24 million, how much will be given to these trust funds?**

I know . . .

m = Total money for funds

\$0.18 = Money for funds

\$1.00 = Lottery sales

\$24,000,000 = Total lottery sales

Equations

Set up a proportion.

$$\frac{\text{Total money for funds}}{\text{Total lottery sales}} = \frac{\text{Money for funds}}{\text{Lottery sales}}$$

$$\frac{m}{\$24{,}000{,}000} = \frac{\$0.18}{\$1.00}$$

Solve

$$\frac{m}{\$24{,}000{,}000} = \frac{\$0.18}{\$1.00}$$

$$\left(\frac{\mathbf{\$24{,}000{,}000}}{\mathbf{1}}\right)\left(\frac{m}{\$24{,}000{,}000}\right) = \left(\frac{\mathbf{\$24{,}000{,}000}}{\mathbf{1}}\right)\left(\frac{\$0.18}{\$1.00}\right)$$

$$1m = \frac{\$4{,}320{,}000}{1}$$

$$m = \$4{,}320{,}000$$

Check

If we multiply \$0.18 by 5, we almost get \$1.00. Similarly, if we multiply \$4,320,000 by 5, we almost get \$24,000,000. The answer is reasonable.

Answer

Lottery sales totaling \$4,320,000 must be paid to the Environmental and Natural Resources Trust Fund or the general fund.

Scales on maps or drawings are often reported using colon notation. A proportion can be used to determine the actual distance represented by a drawing.

Example 4 **A landscape plan is drawn with a scale of 0.1 inch : 3 feet. A maple tree is drawn 6 inches away from an oak tree. Use a proportion to determine the actual distance this represents.**

I know . . .

d = Actual distance in feet

6 inches = Distance on plan

$\frac{3 \text{ ft}}{0.1 \text{ in.}}$ = Plan scale

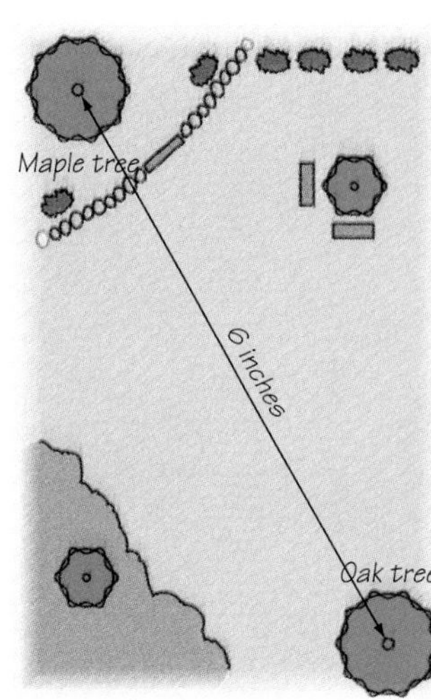

Equations

Set up a proportion.

$$\frac{\text{Actual distance in feet}}{\text{Distance on plan}} = \text{Plan scale}$$

$$\frac{d}{6 \text{ in.}} = \frac{3 \text{ ft}}{0.1 \text{ in.}}$$

Solve

$$\frac{d}{6\text{ in.}} = \frac{3\text{ ft}}{0.1\text{ in.}}$$

$$\left(\frac{\mathbf{6\text{ in.}}}{\mathbf{1}}\right)\left(\frac{d}{6\text{ in.}}\right) = \left(\frac{3\text{ ft}}{0.1\text{ in.}}\right)\left(\frac{\mathbf{6\text{ in.}}}{\mathbf{1}}\right)$$

$$1d = \frac{18\text{ ft}}{0.1}$$

$$d = 180\text{ ft}$$

Check

Since 0.1 inch on the drawing represents 3 feet, 1 inch should represent 30 feet, and 6 inches should represent 180 feet. The answer is reasonable.

Answer

The maple tree and oak tree are 180 feet apart.

CHECK YOUR UNDERSTANDING

Solve using a proportion and the five-step strategy.

1. Charlotte's car gets 24 miles per gallon of gasoline. If her gas tank holds 15 gallons of gas, how many miles can she drive on a full tank of gas?

2. Mario earns $280 for a 40-hour work week. At this rate of pay, how much money will he earn for 32 hours of work?

Answers: 1. Charlotte can drive 360 miles; 2. Mario will earn $224.

EXERCISES SECTION 6.2

Solve the problem by using a proportion and the five-step problem-solving strategy.

1. In the late 1990s, perhaps as many as 4,700,000 Americans used the prescription drugs fenfluramine and phentermine or dexfenfluramine (Redux) to help them lose weight. This combination of drugs has now been linked to heart valve damage. According to an advertisement published in a newspaper by an attorney, preliminary evidence suggests that about 30 out of 100 people who used these drugs to help them lose weight developed heart valve abnormalities. If this is true, how many Americans who used these drugs in the late 1990s will develop heart valve abnormalities from their use of these drugs?

2. According to the Florida Tourism Marketing Corporation, almost one out of ten visitors to Florida is from the New York area. About 43 million visitors went to Florida in 1996. How many of these visitors were from the New York area?

3. According to the U.S. Department of Education, in 1995–96, four out of five undergraduates reported working while they were enrolled in college. In a group of 10,000 college students, how many will be working as well as going to school?

4. According to the Centers for Disease Control and Prevention, the rate of salmonella food poisoning in 1998 caused by eating raw or undercooked eggs was 2.2 cases of salmonella per 100,000 people. How many cases of salmonella would be expected for 250,000 people? Round to the nearest whole number.

5. 44 of 100 women who suffer heart attacks die within a year; 27 out of 100 men who suffer heart attacks die within a year. This may be true because women are often treated less aggressively than men or because women get heart attacks later in life than men. If 3,500 women have a heart attack, how many are expected to die within a year?

6. At an upscale restaurant at the Mall of America, family members accompany one out of five customers. The restaurant accommodates families by offering a children's menu and a stock of coloring books, crayons, and toys. If 12,000 people eat at the restaurant in a given year, how many will be part of a family group?

7. Black truffles are a fungus that grow 3 to 12 inches underground near tree roots. They are difficult to find and must be foraged by specially trained dogs and pigs. Depending on supply, condition, and size, black truffles can sell for $400 to $750 for 16 ounces. If the current price is $575 for 16 ounces, use a proportion to calculate the cost of 3 ounces of truffle. Round to the nearest cent.

8. Large volumes of fine silt from a construction project were allowed to pollute Minnehaha Creek. About 6 tons of silt was contained in the 264,000 gallons of wastewater drained into the creek. Use a proportion to determine the amount of silt contained in 10,000 gallons of the wastewater. Round to the nearest hundredth of a ton.

9. In seven out of eight games in which Ed Belfour was the starting goalie, the Dallas Stars hockey team won the game. Assume this trend continues and Belfour will start as goalie for 32 more games. Use a proportion to predict how many of these games will be won by the Stars.

10. Three out of five snowmobile accidents in the month of January in a rural area in Wisconsin involved alcohol. Assume this pattern continues and that the total number of accidents for the winter is 60. Use a proportion to predict how many of the accidents will involve alcohol.

11. A deli clerk at a grocery store is trying to speed up her work by determining how many sandwich slices of roast beef are in a pound. She weighs six slices and finds that it weighs 0.35 of a pound. Use a proportion to determine the approximate number of slices in a pound. Round to the nearest slice.

12. Matt has a new job that requires a lot of driving. To estimate the cost of gasoline for his car, he decided to keep records for 30 days. He spent a total of $275 on gas. Use a proportion to predict how much Matt will spend on gasoline over a year (365 days). Round to the nearest dollar.

13. A survey of students in the college union building revealed that five out of ten students preferred to eat a la carte rather than enroll in a meal plan. There are about 15,750 students who eat at the college per day. Use a proportion to predict how many students eat a la carte at the college per day.

14. In the last five months of 1999, a food bank served a total of 1,133 households. Use a proportion to predict how many households would be served in a year (12 months). Round to the nearest whole number.

15. The gold content of jewelry is reported as a ratio of carats. Pure gold is 24-carat gold; 18-carat gold contains only 18 carats of pure gold to a total of 24 carats. Use a proportion to determine the weight of pure gold in an 18-carat gold bracelet that weighs 14 ounces. Round to the nearest tenth of an ounce.

16. Theobromine is the ingredient in chocolate that is toxic to dogs. A dog that eats 2 ounces of semisweet chocolate chips per kilogram of body weight has only a 50 percent chance of survival. A pug weighs 2.3 kilograms and ate half of a 6-ounce package of chocolate chips. Does he meet the criteria of being at great risk with only a 50 percent chance of survival? Explain.

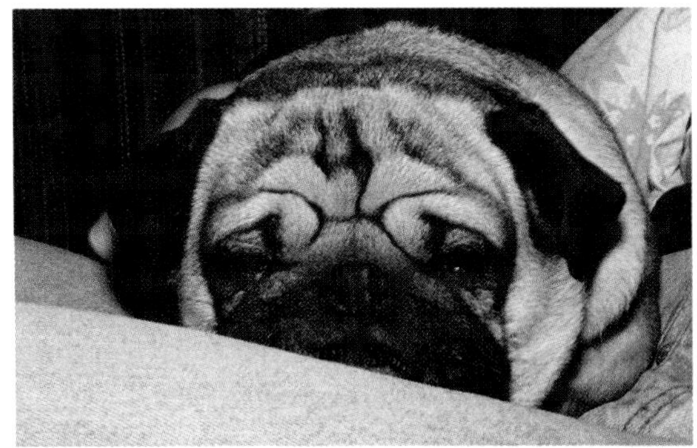

Sargie *(A. Ignoffo)*

17. Tiger Woods, professional golfer, won 10 of his last 14 tournaments in 1999. Assuming that he continues to play at this level, use a proportion to predict how many wins he will have in 100 tournaments. Round to the nearest whole number.

18. Thirty-three out of 100 hysterectomies performed in the United States each year are required because of complications of fibroids in the uterus. About 670,000 hysterectomies are performed each year. Use a proportion to determine how many of these operations were required because of fibroids.

19. According to the Centers for Disease Control and Prevention, smoking among high school students dropped in 1999. The ratio of students using tobacco products to those who did not was 284:1000. This was based on a nationwide survey of 7,529 students. Use a proportion to determine how many of the students in the survey used tobacco products. Round to the nearest student.

20. Bison are now being raised commercially in the United States. In contrast to cattle, bison do not need grain and have grazing habits that are less damaging to rangeland. Bison meat is low fat: there are 2.42 grams of fat per 100 grams of cooked lean meat compared with 9.28 grams for beef, 9.66 grams for pork, and 7.41 for chicken. Use a proportion to calculate the amount of fat in 450 grams of cooked lean bison. (450 grams is about 1 pound.) Round to the nearest hundredth of a gram.

21. Three out of four racehorses in Canada and the United States receive a diuretic drug called Lasix. The drug helps prevent small blood vessels in the lungs from rupturing during heavy exercises. A racetrack has 150 horses stabled waiting for a race. Use a proportion to predict the number of these horses who are receiving Lasix. Round to the nearest whole number of horses.

22. In 1999, the Community Action Food Bank provided 2,564 households with emergency food boxes. Of these households, 255 were homeless. In 2000, the Food Bank expects the demand for emergency food boxes to rise to about 3,000. Use a proportion to predict the number of these food boxes that will be provided to homeless households. Round to the nearest whole number of households.

23. At the Tillamook Cheese Factory, a vat holding about 25,000 quarts of milk makes about 5,500 pounds of cheese. Use a proportion to determine how many quarts of milk are needed to make 7,000 pounds of cheese. Round to the nearest quart.

24. Only one in three children enrolled at a daycare sponsored by Atlanta hotels come from families employed by the hotel industry. There are 200 children enrolled in the daycare. Use a proportion to determine the number of children who come from families employed by the hotel industry. Round to the nearest whole number.

25. For every 1,000 passengers in 1999, TWA mishandled 5.38 bags. Use a proportion to predict the number of mishandled bags for every 2,500 passengers. Round to the nearest hundredth of a bag.

26. In 1999, Delta Airlines reported on-time arrival for 78 out of 100 flights. Use a proportion to predict the number of on-time arrivals for 650 flights. Round to the nearest whole number.

27. In designing a new 750-room hotel, planners want three out of five rooms to be nonsmoking. Use a proportion to calculate the number of rooms that should be nonsmoking. Round to the nearest whole number.

28. The Infectious Diseases Society of America polled hospital-based infectious-disease doctors regarding difficulty in obtaining supplies of antibiotics. For every 25 members, 19 reported experiencing shortages in Penicillin G. Use a proportion to determine how many of the 700 doctors responding to the survey experienced shortages in Penicillin G. Round to the nearest whole number.

29. Tracy made 1 out of 3 free throws in the first 4 games of the basketball season. If she has 45 more free throws in the next 12 games and continues to shoot with the same accuracy, use a proportion to predict how many free throws she will make.

30. To pass a school bond levy in Troy, 65 out of 100 votes must be in favor of the levy. There are 7,561 votes cast in the election. Use a proportion to calculate how many must be in favor of the levy for it to pass. Round to the next highest whole number.

31. Two out of three first-time offenders sent to a "boot camp" prison are approved for parole after 180 days. A judge sentences 185 offenders to the boot camp. How many can be predicted to be approved for parole after 180 days? Round to the nearest whole number.

32. A state health department reported that there was an annual average of 104 pregnancies per 1,000 girls between ages 15 and 19 in a rural county. Use a proportion to predict the number of pregnancies in a population of 350 girls in this age group living in this county. Round to the nearest whole number.

33. The Women, Infants, and Children health program reported that 4 out of 25 children between 3 and 5 years of age had the disorder known as baby bottle tooth decay. Use a proportion to predict how many of the 580 children enrolled in the program will have this disorder. Round to the nearest child.

34. A land swap between a lumber company and the U.S. Forest Service trades 2 acres of Forest Service property for every 5 acres of private land. However, the Forest Service property has valuable mature timber. If the Forest Service is willing to trade 1,235 acres, use a proportion to determine how many acres of private land it will receive in the swap. Round to the nearest acre.

35. Unemployment in a city is 4.9 unemployed people per 100 people in the workforce. If there are 65,000 people in the workforce, use a proportion to predict how many are unemployed.

36. A steelhead fishing report based on interviews of some of the people fishing on the Upper Clearwater River stated that eight fish were caught in a total of 180 hours of fishing. Use a proportion to predict how many fish would be caught in a total of 500 hours of fishing. Round to the nearest fish.

37. At a public hearing on breaching of the four lower Snake River dams, the wait to testify was extremely long. 172 people chose to record their testimony on tape. Of these people, 141 were opposed to breaching the dams. If this testimony was representative of the local population of 55,000 people, use a proportion to predict how many people would be opposed to breaching. Round to the nearest whole number.

38. The Centers for Disease Control and Prevention reported in March 2000 that nearly one out of four Texans have no health insurance. If the population of a Texas county is 480,642 people, use a proportion to predict how many of these people do not have health insurance. Round to the nearest whole number.

39. About three out of ten car and light truck crashes are side-impact crashes. For a total of 575 car and light truck crashes, use a proportion to calculate the number that are side-impact crashes. Round to the nearest whole number.

40. A trail mix recipe calls for 2 pounds of toasted oats, 8 ounces of raisins, 8 ounces of almonds, and 4 ounces of chocolate chips. Use a proportion to determine how many ounces of chocolate chips are needed to make 5 pounds of trail mix. Round to the nearest whole number.

ERROR ANALYSIS

41. Problem: The scale on a map is 1 inch : 5 miles. The distance between two cities on the map is 3.5 inches. Determine the distance between the two cities in miles.

Answer:

$$\frac{5 \text{ miles}}{1 \text{ in.}} = \frac{d}{3.5 \text{ in.}}$$

$$\left(\frac{1}{5 \text{ miles}}\right)\left(\frac{5 \text{ miles}}{1 \text{ in.}}\right) = \left(\frac{d}{3.5 \text{ in.}}\right)\left(\frac{1}{5 \text{ miles}}\right)$$

$$d = \frac{1 \text{ mile}}{17.5}$$

$$d \approx 0.06 \text{ miles}$$

Describe the mistake made in solving this problem.

Redo the problem correctly.

42. Problem: One out of ten middle-school students surveyed used tobacco products. Use a proportion to predict how many students in a middle school of 635 students use tobacco products. Round to the nearest whole number.

Answer:

$$\frac{10 \text{ students}}{1 \text{ student}} = \frac{x}{635 \text{ students}}$$

$$\left(\frac{635 \text{ students}}{1}\right)\left(\frac{10 \text{ students}}{1 \text{ student}}\right) = \left(\frac{x}{635 \text{ students}}\right)\left(\frac{635 \text{ students}}{1}\right)$$

$$6{,}350 \text{ students} = x$$

Describe the mistake made in solving this problem.

Redo the problem correctly.

REVIEW

43. Rewrite $\frac{7}{100}$ as a decimal.

44. Rewrite $\frac{53}{1000}$ as a decimal.

45. Rewrite $\frac{211}{100}$ as a decimal.

46. Rewrite 0.23 as a fraction.

47. Rewrite 0.07 as a fraction.

Section 6.3 Introduction to Percents

LEARNING TOOLS

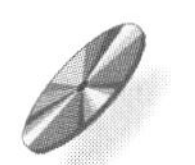

CD-ROM SSM VIDEO

Sneak Preview

Percents are frequently used in daily life. From news reports to bank statements to advertisements, information is reported using percents.

After completing this section, you should be able to:

1) Write a ratio as a percent.

2) Write a percent as a ratio.

3) Write a percent as a decimal.

4) Calculate percent increases and percent decreases.

INVESTIGATION Percents

1. Write a fraction that compares the number of sad faces to the total number of faces.

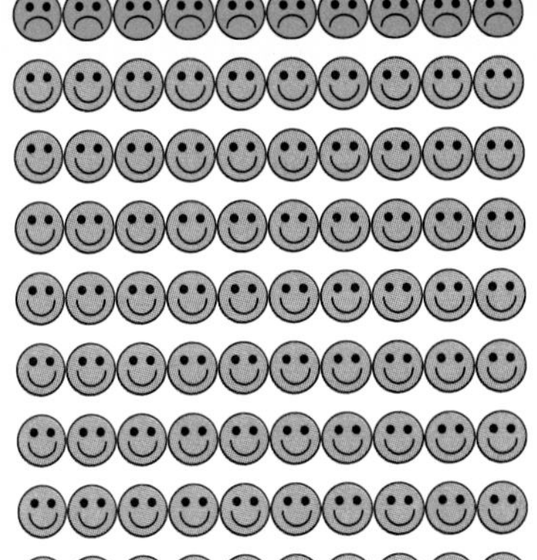

2. Change this fraction into an equivalent decimal.

The fraction $\frac{10}{100}$ and the decimal number 0.10 are equivalent. Both numbers tell us that we have 10 out of a possible 100 parts. Another equivalent way to express this value is *percent notation* (%): $10\% = \frac{10}{100} = 0.10$. A **percent** is a part of a whole, expressed in hundredths.

3. Write a fraction that compares the number of happy faces to the total number of faces.

4. Represent the fraction of happy faces to the total number of faces as a percent.

A percent is a ratio with a denominator of 100. For example, according to a survey done for the Lutheran Brotherhood, 17 percent of couples disagree about the use of credit. Seventeen percent is another way to write $\frac{17 \text{ couples}}{100 \text{ couples}}$. This ratio simplifies to $\frac{17}{100}$. To write this ratio as a percent, we write the numerator followed by a percent sign, 17%.

Rewrite each percent as a ratio. Do not simplify the ratio to lowest terms.

5. Sixty percent of today's Major League baseball players are not playing for their original teams.

6. The United States exports about $11 billion in weapons to other countries. Egypt is the leading importer of U.S. weapons, accounting for about 9 percent of exported weapons.

7. Forty-two percent of adult baseball fans think that the 1998 New York Yankees are the best baseball team in the history of baseball.

Answers: 5. $\frac{60}{100}$; 6. $\frac{9}{100}$; 7. $\frac{42}{100}$.

8. A rate such as 60 $\frac{\text{miles}}{\text{hour}}$ cannot be rewritten as a percent. Explain why.

The ratio $\frac{10}{100}$ and 10% are equivalent. We can also write a decimal number that is equivalent to a percent. Since $\frac{10}{100} = 0.1$ and $\frac{10}{100} = 10\%$, we know that $0.1 = 10\%$. Similarly, $24\% = 0.24$ and $125\% = 1.25$.

9. We can quickly rewrite a percent as a decimal by moving the decimal point. How many places and in what direction should the decimal point be moved?

Complete the table.

	Ratio	Percent	Equivalent Decimal
10.	$\frac{30}{100}$		
11.	$\frac{23}{100}$		
12.		63%	
13.		5%	
14.		132%	
15.			0.45
16.			0.01

Answers: 10. 30%, 0.30; 11. 23%, 0.23; 12. $\frac{63}{100}$, 0.63; 13. $\frac{5}{100}$, 0.05; 14. $\frac{132}{100}$, 1.32; 15. $\frac{45}{100}$, 45%; 16. $\frac{1}{100}$, 1%.

A percent with a decimal point can also be rewritten as an equivalent decimal or ratio. However, the denominator of the ratio may not be 100.

Complete the table.

	Percent	Equivalent Decimal	Ratio
17.	99.9%		
18.	0.1%		
19.	10.5%		
20.	0.07%		
21.	5%		

Answers: 17. 0.999, $\frac{999}{1{,}000}$; 18. 0.001, $\frac{1}{1{,}000}$; 19. 0.105, $\frac{105}{1{,}000} = \frac{21}{200}$; 20. 0.0007, $\frac{7}{10{,}000}$; 21. 0.05, $\frac{5}{100}$.

DISCUSSION

Percent Notation

A percent is equivalent to a ratio with a denominator of 100. The symbol for percent is %. Percents can be written either as fractions or as decimals. For example, $7\% = \frac{7}{100} = 0.07$ and $100\% = \frac{100}{100} = 1$.

Example 1 **According to the Environmental Protection Agency, 54.4% of aluminum packaging material was recycled in 1997. What percent of aluminum packaging material was not recycled in 1997?**

The total amount of aluminum packaging used is 100%.

$$\begin{aligned}\text{Amount not recycled} &= \text{Total} - \text{Amount recycled}\\ &= 100\% - 54.4\%\\ &= 45.6\%\end{aligned}$$

A percent is a fraction. Any fraction can be rewritten as a decimal by dividing the numerator by the denominator. Since a percent always has a denominator of 100, a percent can be changed into a decimal by dividing by 100. This is equivalent to moving the decimal point two places to the left.

PROCEDURE

To change a percent to a decimal:

Move the decimal point two places to the left and omit the percent sign (%).

Example 2 **A community college reports that enrollment in the fall semester is 124% of the previous fall enrollment. Write this percent as a decimal.**

124% = 1.24

Example 3 **The unemployment rate dropped 0.4% from June to July. Write this percent as a decimal.**

0.4% = 0.004 *If necessary, write extra zeros to move the decimal point.*

To change a decimal number to a percent, we multiply by 100. This is equivalent to moving the decimal point two places to the right.

PROCEDURE

To change a decimal to a percent:

Move the decimal point two places to the right and write a percent sign (%).

Example 4 **A credit union pays interest at a rate of 0.055. Write this rate as a percent.**

0.055 = 5.5% *Move the decimal point two places to the right.*

CHECK YOUR UNDERSTANDING

Change each percent to a decimal.

1. 5% **2.** 79% **3.** 128% **4.** 0.008%

Change each decimal to a percent.

5. 0.07 **6.** 0.83 **7.** 0.9 **8.** 2.76

Answers: 1. 0.05; 2. 0.79; 3. 1.28; 4. 0.00008; 5. 7%; 6. 83%; 7. 90%; 8. 276%.

Any ratio can be written as a percent by finding an equivalent ratio with a denominator of 100.

Example 5 **Change $\frac{3}{4}$ to a percent using a proportion.**

$$\frac{3}{4} = \frac{x}{100}$$

$$\frac{100}{1} \cdot \frac{3}{4} = \frac{x}{100} \cdot \frac{100}{1}$$

$$\frac{300}{4} = x$$

$$75 = x$$

$\frac{3}{4}$ is equivalent to 75%.

PROCEDURE

To change a ratio to a percent using a proportion:

1. Write an equivalent ratio with a denominator of 100 and a numerator of x.
2. Write a proportion with the two ratios and solve for the percent, x.

Example 6 **In 1990, it was estimated that 13,000 of the 130,000 people diagnosed with diabetes developed kidney disease. Use a proportion to find the percent of people with diabetes who developed kidney disease.**

$$\frac{13{,}000 \text{ diabetics with kidney disease}}{130{,}000 \text{ diabetics}} = \frac{x}{100 \text{ diabetics}}$$

$$\frac{\mathbf{100\ diabetics}}{\mathbf{1}} \cdot \frac{13{,}000 \text{ diabetics with kidney disease}}{130{,}000 \text{ diabetics}} = \frac{x}{100 \text{ diabetics}} \cdot \frac{\mathbf{100\ diabetics}}{\mathbf{1}}$$

$$10 \text{ diabetics with kidney disease} = x$$

For every 100 diabetics, 10 of them will develop kidney disease. So, 10 percent of people with diabetes developed kidney disease.

A ratio can also be changed to a percent by first dividing to write it as a decimal, and then converting the decimal to a percent.

PROCEDURE

To change a ratio to a percent by division:

1. Divide the numerator by the denominator.
2. Move the decimal point two places to the right and write a percent sign (%).

Example 7 **The American Prostate Society reports that two out of every three prostate cancers are diagnosed too late to cure. Write this as a ratio. Change the ratio to a percent. Round to the nearest tenth of a percent.**

$\frac{2}{3}$

$= 2 \div 3$ *Divide the numerator by the denominator.*

$= 0.6666\ldots$

$= 66.66\ldots\%$ *Move the decimal point two places to the right.*

$\approx 66.7\%$ *Round to the nearest tenth of a percent.*

Still another way to change a ratio to a percent is to multiply the ratio by 100%. We can multiply any number by 1 without changing its value. Since 100% is equal to 1, multiplying by 100% will not change the value of the ratio.

Example 8 **Change $\frac{2}{3}$ to a percent. Write the percent as a mixed number.**

$\frac{2}{3}$

$= \frac{2}{3} \cdot \frac{100\%}{1}$ *Multiply the ratio by 100%.*

$= \frac{200}{3}\%$

$= 66\frac{2}{3}\%$ *Rewrite the improper fraction as a mixed number.*

PROCEDURE

To change a ratio to a percent by multiplying by 100%:

1. Multiply the ratio by 100%.
2. Simplify to an integer or a mixed number.

CHECK YOUR UNDERSTANDING

Change each ratio to a percent.

1. $\frac{7}{5}$ **2.** $\frac{1}{2}$ **3.** $1\frac{5}{8}$ **4.** $\frac{1}{3}$

Answers: 1. 140%; 2. 50%; 3. 162.5%; 4. $33\frac{1}{3}$% or $33.\overline{3}$.

Percents are often used to describe change. Using a ratio, an increase or decrease is compared to the original amount. This ratio is then changed to a percent.

PROCEDURE

To describe a change using a percent:

1. Calculate the increase or decrease.
2. Write a ratio: $\frac{\text{increase}}{\text{original amount}}$ or $\frac{\text{decrease}}{\text{original amount}}$.
3. Rewrite the ratio as a percent.

Example 9 **The crime of murder has declined dramatically in large U.S. cities. In 1997, there were 767 murders in New York City. In 1998, there were 616 murders in New York City, a decrease of 151 murders. Calculate the percent decrease in the murders from 1997 to 1998. Round to the nearest percent.**

$$\frac{\text{Decrease in murders}}{\text{Murders in 1997}}$$
$$= \frac{151 \text{ murders}}{767 \text{ murders}}$$
$$\approx 0.19687092568$$
$$= 19.687092568\%$$
$$\approx 20\% \text{ (round to the nearest percent)}$$

The previous example found a percent decrease. The next example illustrates how to find a percent increase.

Example 10 **Construction of a new basketball arena and ice rink planned for the University of Rhode Island needed \$10.4 million more than the original estimate of \$55.6 million to be completed. Calculate the percent increase in the cost of the arena. Round to the nearest percent.**

$$\frac{\text{Increase in cost}}{\text{Original estimate}}$$
$$= \frac{\$10.4 \text{ million}}{\$55.6 \text{ million}}$$
$$\approx 0.18705035971$$
$$= 18.705035971\%$$
$$\approx 19\%$$

If the change is more than the original amount, the percent change will be greater than 100%. The ratio will be an improper fraction.

Example 11 **In 1963, just 417 breeding pairs of bald eagles remained in the lower 48 states. By 1999, 5,800 pairs of bald eagles flourished and the species was removed from the endangered list. This was an increase of 5,383 pairs. Calculate the percent increase of bald eagles from 1963 to 1999. Round to the nearest percent.**

$$\frac{\text{Increase in eagles}}{\text{Original number}}$$
$$= \frac{5{,}383}{417}$$
$$\approx 12.908872$$
$$= 1{,}290.8872\%$$
$$\approx 1{,}291\%$$

CHECK YOUR UNDERSTANDING

Express each increase or decrease as a percent of the original amount. Round to the nearest percent.

1. In 1900, the average U.S. life span was 47 years. In 1999, the average U.S. life span had increased by 30 years to 77 years. What was the percent increase in life span?

2. In Dayton, Ohio, public school enrollment in 1981 was 44,165 students. By 1999, enrollment had dropped by 20,982 students. What was the percent decrease in enrollment?

Answers: 1. 64% increase; 2. 48% decrease.

EXERCISES SECTION 6.3

Complete the table (do not simplify the fractions to lowest terms).

	RATIO	DECIMAL	PERCENT
1.	$\frac{3}{100}$		
3.			50%
5.		0.71	
7.			4.3%
9.			130%
11.			0.5%
13.		1.00	
15.	$\frac{0}{100}$		
17.			$55\frac{5}{9}\%$
19.	$\frac{3}{7}$		

	RATIO	DECIMAL	PERCENT
2.		0.32	
4.	$\frac{99}{100}$		
6.			8%
8.	$\frac{81}{100}$		
10.	$\frac{1}{100}$		
12.		0.64	
14.	$\frac{110}{100}$		
16.	$\frac{2}{3}$		
18.		2.5	
20.			$122\frac{1}{3}\%$

Complete the mathematical statement by writing an = , < , or > between the two numbers.

21. 0.45 50%

22. 21% $\frac{22}{100}$

23. $\frac{22}{100}$ 22.0

24. $\frac{22}{100}$ 0.22

25. 132% 0.132

26. 1% 1.0

27. Explain how to change a percent into an equivalent decimal.

28. Explain how to change a percent into an equivalent ratio.

29. Public schools nationwide are making steady progress in their efforts to get connected to the Internet. The table shows the percentage of schools wired to the Internet in 1995, 1996, and 1997. (Source: U.S. Department of Education.)

	1995	1996	1997
ALL SCHOOLS	50%	65%	78%
Elementary	46%	61%	75%
Secondary	65%	77%	89%
City	47%	64%	74%
Rural	48%	60%	79%
Less than 6% minority	52%	65%	84%
50% or more minority	40%	56%	63%

a. What was the overall percent increase in schools wired to the Internet from 1995 to 1997?

b. What was the difference in the percent of secondary schools wired to the Internet and the percent of elementary schools wired to the Internet in 1997?

c. Which type of school has made the most progress getting wired to the Internet?

d. Which type of school has made the least progress getting wired to the Internet?

Rewrite the percentage as a ratio. Simplify the ratio to lowest terms. (Some ratios may already be in lowest terms.)

30. Clearcutting is a way of harvesting lumber that removes all the marketable timber from a forested area. It is the quickest and cheapest way to get logs to mills. Clearcutting accounts for 89% of Oregon's timber harvest.

31. About 14% of U.S. workers belonged to labor unions in 1998.

32. Thirty-seven percent of movie theater tickets are sold to Americans who are younger than 25 years old.

33. The national debt increased an average of 23.6% per year while Ronald Reagan was President of the United States (1981–1989).

34. Processed cheese is subject to legal standards. Products labeled as "cheese food" or "cheese spread" must be 51 percent cheese. The remainder of the product is water or gums that make the product spreadable.

35. In 1995, actor Christopher Reeve severely injured his spinal cord. Confined to a wheelchair, he has publicized the need for more research on spinal injuries. The increase in spending by the National Institutes of Health on such research increased by 26% between 1995 and 1998.

36. In a telephone survey, 93% of sixth-graders could explain what a modem does while only 23% of executives could do so. Write the percent of sixth-graders as a fraction.

37. A cemetery overlooking Omaha Beach on the coast of France is the burial place of many soldiers who died during the Allied invasion of 1944. Of a total of 1,300,000 visitors to the cemetery in 1997, about 195,000 were American.

a. Write a ratio of American visitors to total visitors.

b. What percent of the visitors to the cemetery were Americans?

c. What percent of the visitors to the cemetery were not Americans?

38. In Fayetteville, North Carolina, criminals who are on probation may be assigned to a 90-day program supervised by a former Army drill sergeant. In the past 6 years, 3,100 men and women entered the program. Of these, 2,480 men and women completed the requirements.

a. Write a ratio of people who completed the program to all people who entered the program.

b. What percent of participants completed the program?

c. What percent of participants did not complete the program?

39. Four in ten adult baseball fans said that the 1998 World Series champion New York Yankees were the best baseball team in history.

a. Write the ratio of baseball fans who believed that the Yankees were the best team in history to the total number of baseball fans.

b. What percent of the fans believed that the 1998 Yankees were the best team?

c. What percent of the fans did not believe that the 1998 Yankees were the best team?

40. In November 1998, Premera Blue Cross announced it would no longer offer health insurance in the state of Washington to customers who buy their insurance individually rather than through an employer. At the time of the announcement, Premera insured about 125,000 such customers. Within the state of Washington, about 208,000 people purchased their health insurance individually.

a. Write a ratio of people who bought individual insurance through Premera Blue Cross to all the people who buy insurance individually in the state.

b. What percent of people with individual health policies purchased them from Premera Blue Cross? Round to the nearest percent.

41. The number of families receiving welfare in Clarkston, Washington, fell from 621 families in October 1997 to 468 families in December 1998. Calculate the percent decrease in families receiving welfare. Round to the nearest percent.

42. To preserve the quality of Yosemite National Park, Interior Secretary Bruce Babbitt recommended that the number of parking spaces for day use should be reduced from about 1,600 to 550. Calculate the percent reduction in parking spaces. Round to the nearest percent.

(Courtesy of National Park Service)

43. A national rule went into effect in April 2000 that limited the use of jet skis to 21 of 379 national parks. Calculate the percent of national parks that allow the use of jet skis. Round to the nearest percent.

44. In 1998, the average annual salary for a New York City schoolteacher was $44,500. In 1999, the average annual salary increased to $47,000. Calculate the percent increase in salary. Round to the nearest percent.

45. The rise of the value of stocks brought wealth to Americans who owned mutual funds in the late 1990s. However, only 29% of households with incomes less than $50,000 own mutual funds. In contrast, 70% of households with incomes more than $50,000 own mutual funds. The total household assets in mutual funds rose from $2,501 billion in 1998 to $3,104 billion in 1999. Calculate the percent increase in the total household assets in mutual funds from 1998 to 1999. Round to the nearest percent.

46. Microstrategy is a Virginia software and Internet company that experienced a huge growth in stock value over a 12-month period, rising from $7.34 per share to $333 per share. Calculate the percent increase in the value of a share of stock. Round to the nearest percent. (Source: *TIME,* April 3, 2000.)

47. The software company Microstrategy suffered a huge decrease in its stock value after challenges were made to its accounting practices including its statement of revenues for 1998 and 1999. A share of the company's stock fell from $313 prior to the announcement of changes in earnings to $129 per share. Calculate the percent decrease in the value of a share of stock. Round to the nearest percent. (Source: *TIME,* April 3, 2000.)

For problems 48–50 use the following information and table.

When a consumer buys a gallon of gasoline, the cost per gallon can be divided into money for taxes, money for refining and marketing, and money for the purchase of the crude oil itself. The table shows the breakdown of this division with adjustments for inflation.

YEAR	TAXES	REFINING/MARKETING	CRUDE OIL
1970	\$0.49	\$0.72	\$0.35
1980	\$0.28	\$0.85	\$1.38
1990	\$0.35	\$0.54	\$0.69
March 2000	\$0.42	\$0.41	\$0.74

48. Calculate the percent increase in the price of crude oil used to make a gallon of gasoline from 1970 to 1980. Round to the nearest tenth of a percent.

49. Calculate the percent decrease in the taxes collected from a gallon of gas from 1970 to March 2000. Round to the nearest tenth of a percent.

50. The amount of money paid for refining and marketing is where American oil companies and service station owners find their profit. Calculate the percent decrease in the amount of money for a gallon of gas that goes to refining and marketing from 1970 to March 2000. Round to the nearest tenth of a percent.

51. The price for soybeans in May 1997 was \$8.40 per bushel. The price in December 1999 was \$4.29 per bushel. Calculate the percent decrease in the price per bushel. Round to the nearest percent.

ERROR ANALYSIS

52. **Problem:** Write 9 out of 10 as a percent.

Answer: 9%

Describe the mistake made in solving this problem.

Redo the problem correctly.

53. **Problem:** Write 25% as a fraction in lowest terms.

Answer: $\frac{1}{25}$

Describe the mistake made in solving this problem.

Redo the problem correctly.

REVIEW

54. Rewrite this fraction as an equivalent decimal: $\frac{120}{6{,}000}$.

55. Rewrite this fraction as a decimal, rounded to the nearest tenth: $\frac{36}{52}$.

56. Rewrite this fraction as a decimal, rounded to the nearest thousandth: $\frac{28{,}836}{164{,}500}$.

Section 6.4 Applications of Percents

LEARNING TOOLS

CD-ROM SSM VIDEO

Sneak Preview

Percents are used frequently in everyday life, particularly in finances. They are also often used instead of actual numbers in the media.

After completing this section, you should be able to:

1) Solve problems involving a percent.
2) Calculate simple interest.
3) Calculate markup or markdown.
4) Calculate sales tax on a purchase.
5) Solve other word problems using a percent.

DISCUSSION

Solving Problems Involving Percents

A fraction of a number can be found by multiplying the number by the fraction.

Example 1 **Find $\frac{2}{3}$ of 18.**

$\frac{2}{3}$ of 18

$= \frac{2}{3} \cdot \frac{18}{1}$ *To find a fraction of a number, multiply.*

$= \frac{2}{1} \cdot \frac{6}{1}$ *Simplify before multiplying.*

$= 12$

To find a percent of a number, we can change the percent to either a decimal or a fraction. We then multiply the number by the decimal or fraction.

Example 2 **What is 45% of 200?**

$(45\%)(200)$ *Rewrite as multiplication.*

$= (0.45)(200)$ *Rewrite the percent as a decimal.*

$= 90$

PROCEDURE

To find a percent of a number:

1. Change the percent to a fraction or a decimal.
2. Multiply by the number.

We know that two equivalent ratios can be written as an equation. Such an equation is called a **proportion.** Since a percent is a ratio, we can use proportions to solve problems involving percents. We could use a proportion to solve the problem in Example 2.

Example 3 **What is 45% of 200?**

$$x = 45\% \text{ of } 200 \quad \textit{Assign a variable.}$$

$$\frac{x}{200} = \frac{45}{100} \quad \textit{Set up a proportion.}$$

$$\left(\frac{\mathbf{200}}{\mathbf{1}}\right)\left(\frac{x}{200}\right) = \left(\frac{\mathbf{200}}{\mathbf{1}}\right)\left(\frac{45}{100}\right) \quad \textit{Solve the proportion.}$$

$$1x = \frac{200 \cdot 45}{100}$$

$$x = 2 \cdot 45 \quad \textit{Simplify before multiplying.}$$

$$x = 90$$

So, 45% of 200 is 90.

PROCEDURE

To solve a problem involving percent using a proportion:

1. Write a proportion that sets the percent ratio equal to another ratio.
2. Solve the proportion.

Using a proportion to solve the problem "What is 45% of 200?" is not the quickest method. However, it helps us see how to set the percent ratio equal to another ratio. Solving a proportion is an excellent strategy to use in the following problem.

Example 4 **20 is 40% of what number?**

The unknown number is equivalent to 100%. The number that is equivalent to 40% is 20. So, 20 is written in the numerator of its ratio and 40 is written in the numerator of the percent ratio. Similarly, x is written in the denominator of its ratio, and 100 is written in the denominator of the percent ratio.

$$x = \text{the number} \quad \textit{Assign a variable.}$$

$$\frac{20}{x} = \frac{40}{100} \quad \textit{Set up a proportion.}$$

$$\left(\frac{\mathbf{100x}}{\mathbf{1}}\right)\left(\frac{20}{x}\right) = \left(\frac{40}{100}\right)\left(\frac{\mathbf{100x}}{\mathbf{1}}\right) \quad \textit{Solve the proportion by clearing the fraction.}$$

$$2{,}000 = 40x$$

$$2000 \div \mathbf{40} = 40x \div \mathbf{40}$$

$$50 = 1x$$

$$50 = x$$

So, 20 is 40% of 50.

Example 5 **15% of some number is 12. What is this number?**

$$x = \text{the number} \quad \textit{Assign a variable.}$$

$$\frac{12}{x} = \frac{15}{100} \quad \textit{Set up a proportion.}$$

$$\left(\frac{\mathbf{100x}}{\mathbf{1}}\right)\left(\frac{12}{x}\right) = \left(\frac{15}{100}\right)\left(\frac{\mathbf{100x}}{\mathbf{1}}\right) \quad \textit{Solve the proportion by clearing the fraction.}$$

$$1{,}200 = 15x$$

$$1{,}200 \div \mathbf{15} = 15x \div \mathbf{15}$$

$$80 = 1x$$

$$80 = x$$

The number is 80.

CHECK YOUR UNDERSTANDING

Solve using a proportion.

1. What is 28% of 1,500?

2. 30 is 10% of what number?

3. 75 is 5% of what number?

4. What is 75% of 200?

Answers: 1. 420; 2. 300; 3. 1,500; 4. 150.

Percents may be used to describe the quality of a production process.

Example 6 **A quality control test on a production line of motherboards for computers found that 2% of the motherboards were defective. If 6,000 motherboards are produced weekly, predict how many will be defective. Do not use a proportion to solve.**

© *Mark Richards/PhotoEdit*

I know . . .

M = Defective motherboards

2% = 0.02 = Percent defective

6,000 = Total motherboards

Equations

Defective motherboards = (Percent defective)(Total motherboards)

$M = (0.02)(6{,}000)$

Solve

$M = (0.02)(6{,}000)$

$= 120$

Check

To check, we can calculate the percent represented by 120 motherboards of a total of 6,000 motherboards. If this percent is about 2%, then the answer is reasonable.

$\frac{120 \text{ motherboards}}{6{,}000 \text{ motherboards}}$

$= 120 \div 6{,}000$

$= 0.02$

$= 2\%$

Answer

The company should expect 120 of the motherboards to be defective.

Sometimes the variable will appear in the denominator of a ratio. To solve a proportion with such a ratio, clear the denominators from the equation before solving.

Example 7 **In the first basic reading-and-writing test given to teacher candidates in Massachusetts in 1998, 1,120 of the candidates failed the test. This was 56% of the total candidates who took the test. How many candidates took the test? Use a proportion to solve.**

I know . . .

c = Total candidates

$$56\% = \frac{56}{100} = \text{Percent who failed}$$

1,120 = Candidates who failed

Equations

$$\frac{\text{Candidates who failed}}{\text{Total candidates}} = \text{Percent who failed}$$

$$\frac{1{,}120}{c} = \frac{56}{100}$$

Solve

$$\frac{1{,}120}{c} = \frac{56}{100}$$

$$\left(\frac{\mathbf{100c}}{\mathbf{1}}\right)\left(\frac{1120}{c}\right) = \left(\frac{56}{100}\right)\left(\frac{\mathbf{100c}}{\mathbf{1}}\right)$$ *To clear the denominators from the equation, multiply both sides by 100c.*

$$112{,}000 = 56c$$

$$112{,}000 \div \mathbf{56} = 56c \div \mathbf{56}$$ *Use the division property of equality to isolate the variable.*

$$2{,}000 = 1c$$

$$2{,}000 = c$$

Check

To check, we can calculate the percent represented by 1,120 failures of a total of 2,000 tests. This should be close to 56%.

$$\frac{1120}{2000}$$

$= 1120 \div 2000$

$= 0.56$

$= 56\%$

Answer

A total of 2,000 candidates took the test.

CHECK YOUR UNDERSTANDING

Solve.

1. The sales tax in Minnesota is 6.5%. Calculate the sales tax charged on a $2,599 computer. Round to the nearest cent.

2. The package for a 50-foot modular extension cord for a phone line states that the wire length is only approximate. The length can vary by plus or minus two percent (±2%). What is the shortest possible length the phone cord should be?

Answers: 1. The sales tax is $168.94; 2. Two percent of 50 feet is 1 foot. The shortest possible length the phone cord should be is 49 feet.

Percents are often used to calculate the **markdown** on sale merchandise. The markdown is equal to the original price multiplied by the percent markdown. In the following example, we see two different approaches for solving a problem involving markdown.

Example 8 **The regular price of a four-piece queen-size comforter set including an oversized comforter, coordinating dust ruffle, and two pillow shams was $310. A newspaper ad announced a sale of 77% off the regular price. Calculate the sale price.**

I know . . .

s = Sale price

$310 = Original price

77% = 0.77 = Percent markdown

Approach 1

Equations

Sale price = Original price − Markdown

$s = \$310 - (0.77)(\$310)$

Solve

$$\begin{aligned} s &= \$310 - (0.77)(\$310) \\ &= \$310 - \$238.70 \\ &= \$71.30 \end{aligned}$$

We can also calculate the sale price without calculating the markdown. We instead use the percent that the sale price is of the original price.

Approach 2

Equations

Sale price = (Percent sale price is of original price)(Original price)

$s = (0.23)(\$310)$ *100% − 77% = 23% or 0.23*

Solve

$s = (0.23)(\$310)$

$= \$71.30$

Both approaches result in the correct answer.

Check

Estimating using rounding, the difference becomes \$300 − \$240 = \$60. This is close to the exact answer of \$71.30. The answer is reasonable.

Answer

The sale price of the comforter set is \$71.30.

CHECK YOUR UNDERSTANDING

1. A pair of Rollerblade in-line skates is marked down 33%. The original price was \$119.99.

a. Calculate the markdown on the skates. Round the answer to nearest cent.

b. Calculate the sale price of the skates. Round the answer to the nearest cent.

2. A travel agent receives a 45% discount on a hotel room at a national chain if she provides the reservation agent with her travel agent identification number. If the original price of the room per night is \$115, calculate the price of the room to the travel agent. Round to the nearest dollar.

Answers: 1a. The markdown is \$39.60; 1b. The sale price is \$80.39; 2. The room will cost \$63.

In the next example, we know the percent markdown and the sale price of an item. We do not know its original price. In the first approach, we think of the sale price as the difference of the original price and the markdown price. The markdown is found by multiplying the percent markdown by the original price.

Example 9 **An Internet music site is having a 35%-off sale. The advertised price on the Internet for a double CD set of the soundtrack of *Les Miserables* is $20. What was the original price of the CD set?**

I know . . .

p = Original price

35% = Markdown percent

$20 = Sale price

Approach 1

Equations

Sale price = Original price − Markdown

$$\$20 = p - (0.35)p$$

Solve

$$\$20 = p - (0.35)p$$

$$\$20 = 1p - (0.35)p \quad \textit{The coefficient in front of p is 1.}$$

$$\$20 = 0.65p$$

$$\$20 \div \mathbf{0.65} = 0.65p \div \mathbf{0.65}$$

$$\$30.7692307692 \approx 1p$$

$$\$30.77 \approx p \quad \textit{Rounded to the nearest cent.}$$

Another way to solve this problem is to determine the percent that the sale price is of the original price.

Approach 2

Equations

Sale price = (Percent sale price is of the original price)(Original price)

$$\$20 = 0.65p$$

Solve

$$\$20 = 0.65p$$

$$\$20 \div \mathbf{0.65} = 0.65p \div \mathbf{0.65}$$

$$\$30.7692307692 \approx 1p$$

$$\$30.77 \approx p \quad \textit{Rounded to the nearest cent.}$$

Check

If we find 35% of this original price, we should get the markdown. (0.35)($30.77) = $10.77. Adding $10.77 + $20, we get the same original price as we just calculated, $30.77. This answer is reasonable.

Answer

The original price of the CDs was $30.77.

Example 10 **A sporting goods store was holding an end-of-season sale on ski equipment and snowboards. What was the original price of the snowboard package? Round to the nearest cent.**

SALE!!! SNOWBOARD PACKAGE
Board, Boots, Bindings
SALE PRICE: $269.97
60% off!

I know . . .

p = Original price

60% = Markdown percent

\$269.97 = Sale price

Approach 1

Equations

Sale price = Original price − Markdown

$\$269.97 = p - 0.60p$

Solve

$\$269.97 = p - 0.60p$

$\$269.97 = 0.40p$ *1p − 0.60p = 0.40p*

$\$269.97 \div \mathbf{0.40} = 0.40p \div \mathbf{0.40}$

$\$674.925 \approx 1p$

$\$674.93 \approx p$ *Round to the nearest cent.*

Approach 2

Equations

Sale price = (Percent sale price is of the original price)(Original price)

$\$269.97 = 0.40p$

Solve

$\$269.97 = 0.40p$

$\$269.97 \div \mathbf{0.40} = 0.40p \div \mathbf{0.40}$

$\$674.925 \approx 1p$

$\$674.93 \approx p$ *Round to the nearest cent.*

Check

If we find 60% of this original price, we should get the markdown. (0.60)(\$674.93) = \$404.96. Adding the markdown to the sale price, \$404.96 + \$269.97, we get the same original price as we just calculated, \$674.93. This answer is reasonable.

Answer

The original price of the snowboard package was \$674.93.

It may take longer to work through and understand an example that is solved using two different approaches. However, this helps us learn new ways to think about a problem. Problems in mathematics often can be solved in a variety of equally correct ways.

CHECK YOUR UNDERSTANDING

1. A watch was advertised as 35% off. The sales price was $42.21. Calculate the original price of the watch. Round to the nearest cent.

2. A five-piece dining set was advertised at a price of $499, a markdown of 17%. What was the original price of the dining room set? Round to the nearest dollar.

Answers: 1. The original price of the watch was $64.94; 2. The original price of the dining room set was $601.20.

Percents are used to calculate interest earned on an investment. A formula to calculate interest earned depends on how often the interest is calculated and added to the investment. Banks and other lenders usually calculate interest using formulas that include exponents. These formulas allow the interest earned to be continuously added to the invested amount, the **principal.** This type of interest is called compound interest. When the interest is calculated only once during the time of the investment, the interest is referred to as **simple interest.** The interest rate in simple interest is usually written in percent earned per year. The time of investment is usually written in years or a fraction of years.

Formula for Simple Interest

Interest = (Principal)(Interest rate per year)(Time of investment in years)

$I = PRT$

It is usually easiest to calculate simple interest with the interest rate written as a decimal.

Example 11 **An investment of $500 was deposited in a savings account for a period of 2 years at an interest rate of $5\frac{3}{4}\%$. Calculate the simple interest earned on this investment. Round the interest to the nearest cent.**

I know . . .

I = Simple interest

$500 = Principal (amount invested) = P

$5\frac{3}{4}\% = 5.75\% = 0.0575$ = Interest rate = R

2 = Years invested

Equations

Interest = (Principal)(Interest rate per year)(Time of investment in years)

$I = PRT$

$I = (\$500)(0.0575)(2)$

Solve

$I = (\$500)(0.0575)(2)$

$= \$57.50$

Check

At an interest rate of 10% for 1 year, we would earn ($500)(0.1)(1) or $50.

The problem asks for about 5% interest for 2 years. This is a total gain of 10% so we expect this answer to be close to $50. Since $57.50 is very close to $50, the answer is reasonable.

Answer

The amount of simple interest earned is $57.50.

The time used in this formula must be in years. If the time is given in weeks, it must be converted into years. To convert from weeks to years, divide the given number of weeks by the number of weeks in a year, 52.

Example 12 **An investment of $500 was deposited in a savings account for a period of 36 weeks at an interest rate of 3%. Calculate the simple interest earned on this investment. Round the time in years to the nearest hundredth.**

I know . . .

I = Simple interest

$500 = Principal (amount invested) = P

3% = 0.03 = Interest rate = R

$\frac{36}{52} \approx 0.69$ = Years invested = T *Time in weeks ÷ 52.*

Equations

Interest = (Principal)(Interest rate per year)(Time of investment in years)

$I = PRT$

$I \approx (\$500)(0.03)(0.69)$

Solve

$I \approx (\$500)(0.03)(0.69)$

$\approx \$10.35$

Check

At an interest rate of 3% for 1 year, we would expect interest equal to (\$500)(0.03)(1) = \$15.00. Since we are investing 0.69 of a year, we expect our total to be less than \$15.00. We expect more than \$7.50, the interest earned in $\frac{1}{2}$ of a year. Since \$7.50 < \$10.35 < \$15.00, our answer is reasonable.

Answer

The amount of simple interest earned is \$10.35.

When something increases, we are often interested in comparing this increase to the original amount. We can calculate a percent increase using a percent in decimal form.

Example 13 **In 1995, 164,500 adverse reactions to prescription drugs were voluntarily reported to the Food and Drug Administration. In 1996, 193,336 such reactions were reported. To the nearest tenth of a percent, calculate the percent increase in reported adverse reactions to prescription drugs.**

I know . . .

p = Percent increase in reactions

164,500 = Reactions in 1995

193,336 = Reactions in 1996

28,836 = Increase in reactions *193,336 − 164,500 = 28,836*

Equations

$$\text{Percent increase} = \frac{\text{Increase in reactions}}{\text{Reactions in 1995}}$$

$$p = \frac{28{,}836}{164{,}500}$$

Solve

$$\begin{aligned} p &= \frac{28{,}836}{164{,}500} \\ &= 28{,}836 \div 164{,}500 \\ &\approx 0.1752948328 \\ &\approx 17.5\% \quad \text{Rounded to the nearest tenth of a percent.} \end{aligned}$$

Check

We can estimate the percent by rounding the increase in reactions and the reactions in 1995.

$$\begin{aligned} p &= \frac{28{,}836}{164{,}500} \\ &\approx \frac{30{,}000}{200{,}000} \\ &\approx 0.15 \\ &\approx 15\% \end{aligned}$$

This is reasonably close to our exact answer of 17.5%.

Answer

The number of adverse prescription drug reactions increased about 17.5% from 1995 to 1996.

Example 14 **From 1990 to 1997, the annual production of municipal solid waste in the United States increased by 6%. Solid waste production in 1990 was 205 million tons. What was the increase in annual production of waste?**

Lewiston, Idaho, solid waste transfer station

I know . . .

p = Increase in production

6% = 0.06 = Percent increase

205 million tons = Solid waste production in 1990

Equations

Increase in production = (Percent increase)(Production in 1990)

$$p = (0.06)(205 \text{ million tons})$$

Solve

$p = (0.06)(205 \text{ million tons})$

$= 12.3 \text{ million tons}$

Check

10% of 205 million tons is about 20.5 million tons. 10% is a little less than double 6% so we expect 20.5 million tons to be a little less than double 12.3 million tons (24.6 million tons). The answer is reasonable.

Answer

The increase in annual production of municipal waste between 1990 and 1997 was 12.3 million tons.

In some situations, we may know the percent and the amount it represents, but not the total amount.

Example 15 **In a newspaper article about health insurance, it was reported that almost 600,000 people in Washington State, 11% of the population, did not have health insurance. What population of Washington State was used to calculate this percent? Round the answer to the nearest hundred thousand.**

I know . . .

p = Total population

600,000 = Population without insurance

11% = 0.11 = Percent without insurance

Equations

Population without insurance = (Percent without insurance)(Total population)

$$600{,}000 = 0.11p$$

Solve

$$600{,}000 = 0.11p$$
$$600{,}000 \div \mathbf{0.11} = 0.11p \div \mathbf{0.11}$$
$$5{,}454{,}545.45 \approx 1p$$
$$5{,}500{,}000 \approx p \quad \textit{Round to the nearest hundred thousand.}$$

Check

Rounding 11% to 10%, we know that the entire population should be about 10 times greater than the population without health insurance.

$10 \cdot 600{,}000 = 6{,}000{,}000$. This is reasonably close to our exact answer.

Answer

The population of Washington State that was used to calculate the percent is about 5,500,000 people.

The media often reports information that is a combination of actual numbers and percents. We can use the percent to calculate the actual numbers that are not reported.

Example 16 **A newspaper article reported that the number of dog bites that required medical attention rose 37% between 1986 and 1996 to 800,000 bites. To the nearest ten thousand, how many dog bites required medical attention in 1986?**

I know . . .

b = Bites in 1986

37% = 0.37 = Percent increase in bites

800,000 = Bites in 1996

Equations

Bites in 1986 + Increase in bites = Bites in 1996

Bites in 1986 + (Bites in 1986)(Percent increase in bites) = Bites in 1996

$$b + 0.37b = 800{,}000$$

Solve

$$b + 0.37b = 800{,}000$$
$$1.37b = 800{,}000$$
$$1.37b \div \mathbf{1.37} = 800{,}000 \div \mathbf{1.37}$$
$$1b \approx 583{,}941.6058$$
$$b \approx 580{,}000 \quad \textit{Round to the nearest ten thousand.}$$

Check

Rounding, a 40% increase in 600,000 bites is 240,000 bites. Adding this increase to the 1986 dog bites results in an estimated 840,000 dog bites in 1996, reasonably close to the actual number of 800,000 bites.

Answer

According to the information in the newspaper article, there were about 580,000 dog bites that required medical attention in 1986.

CHECK YOUR UNDERSTANDING

1. Ben bought a new truck for $28,400. He expects the value of the truck to depreciate 12% in the first year. What is the value of the truck after the first year?

2. Kim and Kathy had dinner at Mariposa. The bill was $136.64. They left a tip of $30. What percent of the bill was the tip? Round to the nearest percent.

Answers: 1. The truck will lose $3,408 in value. The value of the truck after the first year will be $24,992; 2. The tip was 22% of the total bill.

EXERCISES SECTION 6.4

Solve using the five-step problem-solving strategy.

1. Carbohydrate loading for optimal athletic performance is done by reducing exercise for about 3 days before competition and increasing carbohydrate intake to between 65% and 80% of Calories eaten. Scott is going to run in a marathon. He usually eats 2,800 Calories a day and decides to carbo-load at the 80% level. How many Calories of carbohydrate should he eat during each of the three days?

2. Victor and Sally Ganz spent a total of $50 million on art over a period of 50 years. When auctioned off at Christie's in November 1997, their collection fetched $206.5 million. What was the percent increase in value of this collection over the 50-year period? (Source: *TIME*, November 24, 1997.) Round the percent to the nearest whole number.

3. Shopko advertised a 70%-off sale on 14-carat gold and all Black Hills gold jewelry for Mother's Day. The regular price of a 14-carat gold Circles of Life necklace is $169.99. What should the sale price be?

4. Health care providers watch for warning signs of malnutrition in their older patients. One of every five patients has trouble walking, shopping, buying, and cooking food. What percent of these patients experience these difficulties?

5. The economy of Mexico depends heavily on sales of its oil by state-owned Petroleos Mexicanos SA (Pemex). The price of oil in 1998 averaged $11 per barrel. However, the price fell as low as $7.50 a barrel in the last part of November. What percent is this price below the average price? Round the percent to the nearest whole number.

6. In 1997, the Secret Service arrested 9,455 people for crimes involving identity fraud, up from 8,806 in 1995. In identity fraud, a criminal uses personal information such as a social security number and a mother's maiden name to access bank accounts or make fraudulent credit card charges. The cost of such crimes in 1997 was $745 million, up from $442 million in 1995. What was the percent increase in the cost of such crimes between 1995 and 1997? Round the percent to the nearest whole number.

7. Starbucks Coffee gives CARE, the international humanitarian organization, $4 of the $25.95 Starbucks charges for a CARE-earmarked package of four coffees. Calculate the percent of the total price that is donated to CARE. Round the percent to the nearest whole number.

8. Between January 1973 and January 1998, Pink Floyd's album *Dark Side of the Moon* was on *Billboard's* Top 200 Albums chart for 741 weeks. What percent of time during these 25 years was the album on the chart? Round the percent to the nearest whole number.

9. *The Atlantic Monthly,* January 1999, reported: "The incidence of tuberculosis in Russia has skyrocketed. The number of *deaths* ascribed to tuberculosis in Russia in 1996 (24,877) was almost 15 percent greater than the number of new *infections* (usually nonfatal) that year in the United States." Calculate the number of new infections that year in the United States. Round to the nearest whole number.

10. In its annual report, a manufacturing company reported a net decrease in employees of 14% caused by layoffs and retirements. Its current number of employees is 49,020. How many employees did it have before the layoffs and retirements? Round to the nearest whole number.

11. To preserve the quality of Yosemite National Park, Interior Secretary Bruce Babbitt recommended that the number of lodging rooms in the valley should be reduced from 1,260 to 981 by removing cabins at Curry Village and Housekeeping Camp. Calculate the percent decrease in lodging rooms. Round the percent to the nearest whole number.

12. According to the United Network for Organ Sharing, in mid-June 1998 about 56,000 people were waiting for transplants. There were 10,750 registrations for livers, 390 for pancreases, and 88 for intestines. Every day, ten people die waiting for a transplant to become available. What percent of the patients were waiting for liver transplants? Round the percent to the nearest whole number.

13. According to a Federal Trade Commission study, check-out scanners record the wrong price on 1 out of every 28 items on sale at supermarkets and department stores. What percent of the prices are scanned incorrectly? Round to the nearest tenth of a percent.

14. On December 15, 1998, financial services company Citigroup, Inc. announced it would cut 10,600 jobs, a total of 6 percent of its global work force. How many people did the company employ globally before the job cuts?

15. Ricky Williams, tailback from the University of Texas football team and Heisman trophy winner, concluded his career by breaking or sharing 20 NCAA records. His career rushing yardage total was 6,279 yards. Tony Dorsett was the former record holder, rushing 6,082 yards. Calculate the percent increase in the new record set by Williams. Round to the nearest tenth of a percent.

16. A survey of 1,059 children ages 8 to 14 by *Zillions,* the Consumer Report magazine for children, found that only 43% receive a weekly allowance. The children surveyed receive an average of $5.82 a week in allowance, with older kids getting more money than younger ones. How many of the surveyed children received an allowance? Round to the nearest whole number.

17. In order to save salmon runs in the Northwest, some scientists advocate the breaching of four hydroelectric dams on the Snake River. The Bonneville Power Administration reported that electricity bills could rise a minimum of 4% per month if the dams could no longer produce electricity. The new average electric bill would be a minimum of $57.20. What was the original average electric bill used in this calculation?

18. Write a word problem that requires the use of a percent. Solve the problem.

19. In the late 1980s, about 710 million board feet of timber was harvested each year on Forest Service lands in Idaho. In 1998, the timber harvest on these lands shrank to 220 million board feet. By what percent did the timber harvest decrease? Round the percent to the nearest tenth.

20. Nine states have no income tax: Alaska, Florida, Nevada, New Hampshire, South Dakota, Tennessee, Texas, Washington, and Wyoming. What percent of states have no income tax? Round the percent to the nearest whole number.

21. In 1997–98, the Miami-Dade County public schools enrolled 345,861 students in 302 schools. This was reported as a 1.45% increase over the previous year. What was the enrollment the previous year? Round to the nearest whole number.

22. The sales tax rate in a city in Washington State was 8.85%. The price of a television is $299.99.

a. How much sales tax will be paid on the television? Round to the nearest cent.

b. Calculate the final price of the television. Round to the nearest cent.

23. The U.S. Census Bureau counted about 2,000,000 American Indians and Alaska Natives. However, it also believes that it missed about 4.5% of this population. Calculate what the population count of American Indians and Alaska Natives should have been based on this information. Round to the nearest whole number.

24. When physicians operate on the wrong organ or limb, this is called *wrong-site surgery.* Now many surgeons autograph patients' bodies at the surgical site in a pre-operation visit to avoid such mistakes. From 1984 to 1995, the Physicians Insurers Association of America counted 225 claims for wrong-site surgery by its 110,000 doctors. Calculating a percent of claims per doctor is probably not a valuable statistic. Explain why.

25. A public health department initiated a study of the most dangerous highways in a state. Highway 99 was chosen because it has a high traffic volume and the highest number of alcohol-related accidents. Between 1992 and 1996, 954 accidents were recorded resulting in 965 injuries and 21 deaths. One of the department employees calculated that 2% of accidents resulted in a death using this proportion: $\frac{21}{954} = \frac{p}{100}$. What is wrong with the employee's thinking?

26. A full-size mattress set is advertised at a price of $558, marked down 15%. What was the original price of the mattress set? Round to the nearest dollar.

27. A bathroom sink faucet was marked down $5 from its regular price to a sale price of $49.99. What was the percent markdown? Round to the nearest percent.

28. A 12-gallon wet/dry vacuum was advertised as 15% off. If the regular price was $89.99, what was the sale price? Round to the nearest cent.

29. A department store had a special sale for credit card customers, offering an additional 10% off sale prices in the store. A sweater originally sold for $49.99. It was on sale at 30% off the regular price. What was the final price of the sweater to a credit card customer? Round to the nearest cent.

30. Rusty Wallace was the 1989 NASCAR Winston Cup champion. The bar graph shows his wins by racetrack as of April 2000.

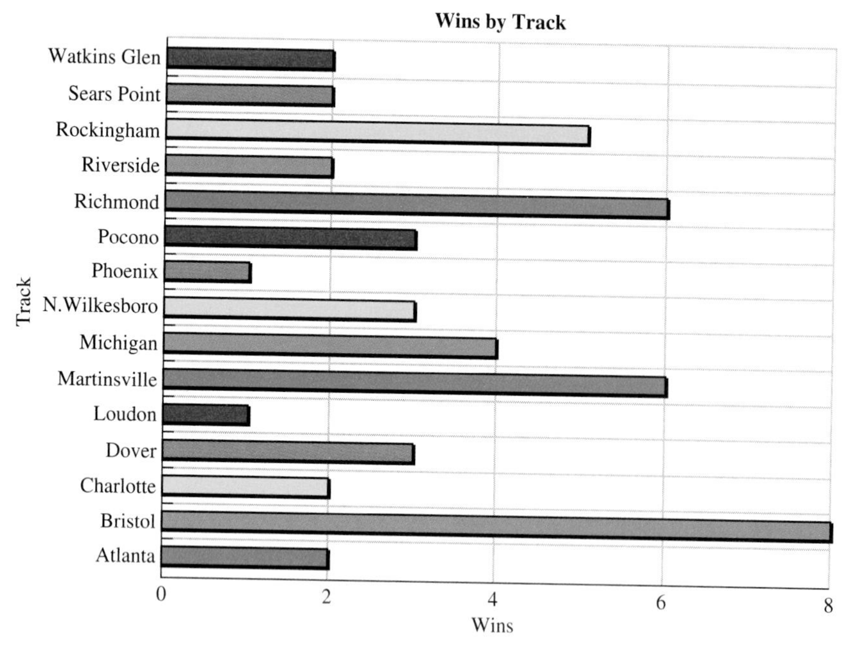

a. Calculate the percent of his wins that were raced at the Bristol track.

b. Calculate the percent of wins that were raced at the Pocono track.

31. Calculate the simple interest earned on $6,000 invested for 3 years at 5% interest.

32. Calculate the simple interest earned on $6,000 invested for 3 years at $5\frac{3}{4}\%$ interest.

33. Calculate the simple interest earned on $8,500 invested for 6 months at $4\frac{1}{2}\%$.

34. The wholesale price of a beginning set of golf clubs is \$200. A golf pro plans to mark up the price by 25%. What will the final price of the clubs be? Round to the nearest cent.

35. A farmer grew 26,000 bushels of barley and 52,000 bushels of wheat. What percentage of his total yield in bushels was the barley? Round to the nearest percent.

36. Natural river flows for salmon migration are supplemented each year with water released from Idaho reservoirs. About 1,200,000 acre-feet is released from Dworshak Reservoir and 427,000 acre-feet comes from southern Idaho irrigation reservoirs. Calculate the percent of the supplemental water that comes from southern Idaho irrigation reservoirs. Round to the nearest percent.

(Courtesy of U.S. Army Corps of Engineers)

37. The Community Action Food Bank served a total of 6,349 individuals with emergency food boxes in 1999. Of these individuals, 2,351 were under the age of 18. Calculate the percent of individuals served who were under age 18. Round to the nearest percent.

38. The total cost of Bob Dole's failed 1996 campaign for the U.S. presidency was \$42,000,000. Through January 31, 2000, George W. Bush had spent \$50,000,000 on his presidential campaign. Calculate the percent increase to this point in the Bush campaign expenditures compared to the expense of Dole's entire campaign. Round to the nearest percent.

39. Alvin is doing a rough estimate of his state income taxes. He knows he will pay about 8% of his adjusted federal income of \$36,000. Estimate his state income taxes.

40. An ad for a credit card includes the following table:

OUTSTANDING BALANCE	ANNUAL INTEREST PAID WITH OTHER CARDS AT 15.9% APR	ANNUAL INTEREST PAID WITH YOUR AT&T UNIVERSAL CARD AT 4.9% APR	ANNUAL INTEREST SAVINGS BY TRANSFERRING BALANCES TO YOUR AT&T UNIVERSAL CARD
$1,000	$159	$49	$110
$2,000	$318	$98	$220
$5,000	$795	$245	$550

a. Calculate the percent savings by transferring a $1,000 balance from the "other cards" to an AT&T card that charges a lower interest rate.

b. Calculate the percent savings by transferring a $5,000 balance from the "other cards" to an AT&T card that charges a lower interest rate.

41. Hoover Dam is on the Arizona/Nevada border and was placed in service in 1936. It contains 3.25 million cubic yards of concrete. Grand Coulee Dam is in Washington State and was placed in service in 1941. It contains 11.975 million cubic yards of concrete. Calculate the percent increase in cubic yards of concrete used to construct Grand Coulee Dam from the amount that was used to construct Hoover Dam. Round to the nearest percent.

42. A charter company in Depoe Bay, Oregon, advertises the following fishing rates in its advertising flyer. The flyer includes a coupon for $5 off any 5-hour fishing trip.

FISHING TRIP	GROUPS (8+) (PER PERSON)	ADULT (PER PERSON)
5-hour bottom fishing	$45.00	$55.00
6-hour fishing with crabbing	$62.00	$72.00
5-hour Coho Salmon	$50.00	$60.00
9-hour Chinook Salmon	$125.00	$125.00
12-hour Tuna	$140.00	$150.00

a. Calculate the percent savings the coupon offers on a 5-hour Coho Salmon trip at the group rate. Round to the nearest percent.

b. Calculate the percent savings the coupon offers on a 5-hour Coho Salmon trip at the per person rate. Round to the nearest percent.

43. Grand Coulee Dam is the largest hydropower producer in the United States with a capacity of 6,809 megawatts. The Guri Dam in Venezuela has a capacity of 10,000 megawatts and the Itaipu Dam on the Paraguay/Brazil border has a capacity of 12,500 megawatts. Calculate the percent increase in power of the Guri Dam compared to the Grand Coulee Dam. Round to the nearest percent.

44. The wholesale price of a textbook is $45. The bookstore will mark the text up 15% to pay for its overhead including salaries. What will the price of the book be to a student? Round to the nearest cent.

45. In January 2000 the Fox network broadcast *Who Wants to Marry a Multimillionaire?* Ten million viewers saw the first half hour of the broadcast. The last half hour was watched by 22.8 million viewers. Calculate the percent increase in viewers. Round to the nearest percent.

46. The following table was included in a special advertising feature on women's health sponsored by U.S. pharmaceutical companies. (Source: *TIME,* December 6, 1999.)

DISEASE	NUMBER OF DRUGS IN DEVELOPMENT	NUMBER OF WOMEN AFFECTED
Alzheimer's disease	23	4 million patients, mostly women
Breast cancer	60	175,000 new cases each year
Depression	17	12.5 million
Diabetes	19	8.1 million
Lupus	4	117,000
Migraine headache	10	12.6 million
Multiple sclerosis	14	200,000+
Osteoporosis	24	8 million
Ovarian cancer	38	25,200 new cases each year
Rheumatoid arthritis	24	1.5 million

a. Calculate the total number of drugs in development listed in the table.

b. Calculate the percent of this total number that is intended for the treatment of breast cancer. Round to the nearest hundredth of a percent.

c. Calculate the percent of this total that is intended for the treatment of osteoporosis. Round to the nearest hundredth of a percent.

47. Cynthia Moss has been doing research on elephants in Kenya for nearly 28 years. When she arrived in 1972, there were 700 elephants in Amboseli National Park. By 2000, there were 1,100 elephants. Calculate the percent increase in elephants. Round to the nearest percent.

48. The *Industry Standard* is an Internet magazine with a circulation of 150,000. Nearly 60% of this circulation is free to subscribers who must only give personal financial information that is sold to marketers. Determine the number of subscribers receiving a free subscription.

ERROR ANALYSIS

49. **Problem:** Calculate 16% of \$1,392.

Answer: (16)(1,392) = \$22,272

Describe the mistake made in solving this problem.

Redo the problem correctly.

50. **Problem:** The sale price of a pair of jeans is 40% off the original price of \$39.99. Calculate the sale price.

Answer: (\$39.99)(0.40) = \$15.99. The sale price is \$15.99.

Describe the mistake made in solving this problem.

Redo the problem correctly.

REVIEW

51. Draw a circle. Divide it into four equal parts. Shade in one of the parts.

a. Write a fraction that represents one of these parts.

b. Change the fraction in Part a to a percent.

52. Draw a circle. Divide it into six equal parts. Shade in one of the parts.

a. Write a fraction that represents one of these parts.

b. Change the fraction in Part a to a percent.

53. Draw a circle. Divide it into eight equal parts. Shade in one of the parts.

a. Write a fraction that represents one of these parts.

b. Change the fraction in Part a to a percent.

Section 6.5 Circle Graphs

LEARNING TOOLS

CD-ROM SSM VIDEO

Sneak Preview

Circle graphs, also known as pie charts, are an effective visual way to present data expressed in percents.

After completing this section, you should be able to:

1) Use information presented in a circle graph to solve word problems.

2) Construct a circle graph.

DISCUSSION

Circle Graphs and Percents

A circle graph (also called a pie chart) illustrates a whole set divided into parts or percents. The entire circle represents 100% of a set.

Example 1 **According to the American Medical Association, in 1996, 64% of doctors in the United States under age 35 were men and 36% were women. Represent this information as a circle graph.**

Thirty-six percent of the circle area represents the percentage of doctors who are women. Sixty-four percent of the circle area represents the percentage of doctors who are men.

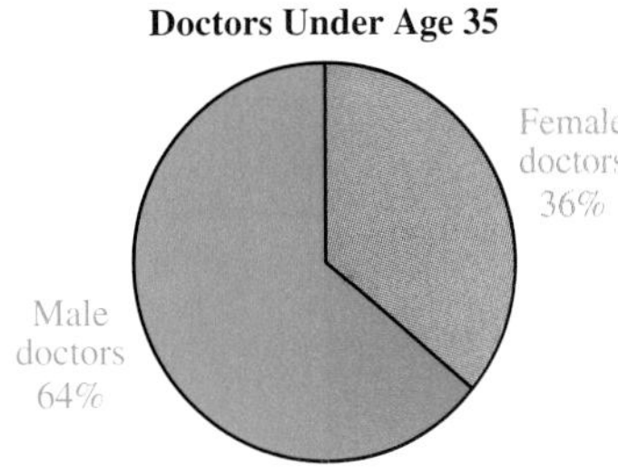

Sometimes the information we need to solve a problem is found in a circle graph. The percents in the circle graph can be used with the other information included in the problem to find a solution. In the next example, the problem is solved using a proportion.

Example 2 **In 2001, 67 million households were connected to the Internet. To the nearest million, how many of these households were in North America?**

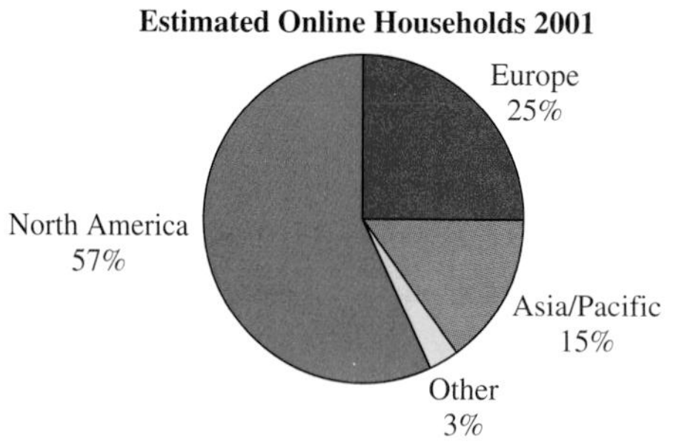

I know . . .

N = Connected North American households, in millions

67 = Total connected households, in millions

$57\% = \frac{57}{100}$ = Percent of connected households that are North American. *From graph.*

Equations

$$\frac{\text{Connected North American households}}{\text{Total connected households}} = \text{Percent connected that are North American}$$

$$\frac{N}{67} = \frac{57}{100}$$

Solve

$$\frac{N}{67} = \frac{57}{100}$$

$$\frac{\mathbf{67}}{\mathbf{1}} \cdot \frac{N}{\mathbf{67}} = \frac{67}{1} \cdot \frac{57}{100}$$

$$N = \frac{3{,}819}{100}$$

$$= 38.19$$

$$\approx 38$$

Check

Since 57% of connected households are North American, we expect our answer to be somewhat more than one half of the total. 67 million divided by 2 is about 34 million, which is reasonably close to the exact answer.

Answer

About 38 million households connected to the Internet are North American.

Instead of using a proportion, we may be able to solve a problem by multiplying by the percent in decimal form.

Example 3 **The Metropolitan Achievement Test is a test used by urban school districts to measure how students perform in math, reading, and other areas compared to a national sample of students. The test is designed so that a typical school should expect a certain percentage of its students to score in the "below average," the "average," and the "above average" categories. A school has 801 students. Calculate the number of students who should score at the below average level. Round to the nearest student.**

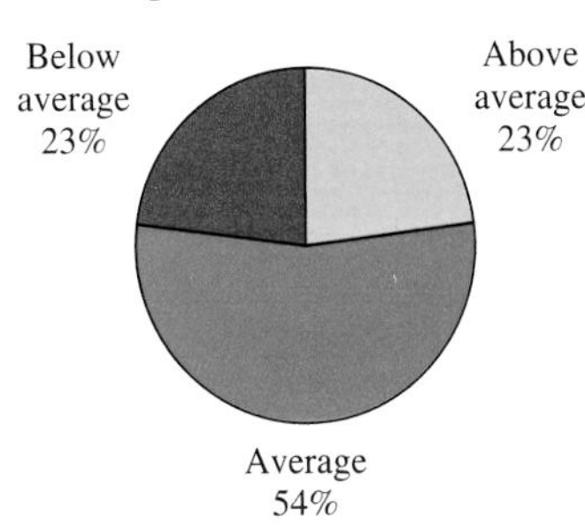

I know . . .

A = Students with below average scores

801 = Total students

23% = 0.23 = Percent of students with below average scores *From the graph.*

Equations

Students with below average scores = (Percent with below average scores)(Total students)

$$A = (0.23)(801)$$

Solve

$$A = (0.23)(801)$$
$$= 184.23$$
$$\approx 184$$

Check

23% is about 25% or one-fourth of the students. The total students, 801, is about 800. One-fourth of 800 students is 200 students, close to the exact answer of 184 students. The answer is reasonable.

Answer

About 184 students should be expected to have below average scores on the MAT.

CHECK YOUR UNDERSTANDING

1. A total of 5,800,000 new HIV infections were reported worldwide in 1998. How many of these infections were of children?

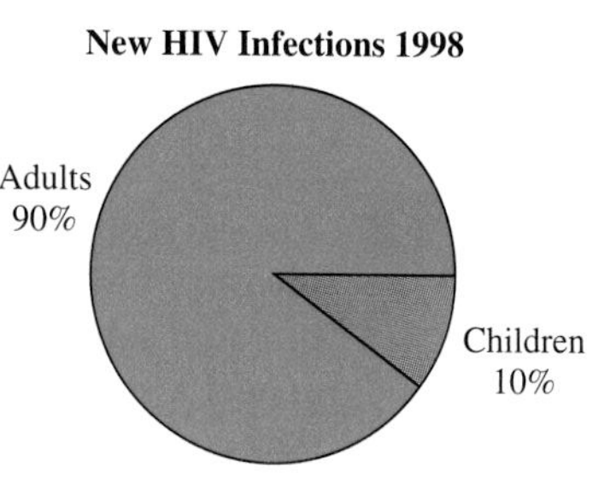

Answer: 1. 580,000 infections were of children.

To show percent information using a circle graph, we need to draw the correct size sections of the circle to correspond to the different percents. We know that the angle measure of any circle is equal to a measure of 360° (Section 1.1). We can find the number of degrees that represent a certain percent of the circumference.

Example 4 **Calculate the number of degrees that are equal to 50% of the circumference of a circle. Round to the nearest degree.**

x = Number of degrees equal to 50% of the circumference

$= 50\%$ of $360°$

$= (0.50)(360°)$ *Rewrite the percent in decimal form.*

$= 180°$

Example 5 **Calculate the number of degrees that are equal to 11% of the circumference of a circle. Round to the nearest degree.**

x = Number of degrees equal to 11% of the circumference

$= 11\%$ of $360°$

$= (0.11)(360°)$

$= 39.6°$

$\approx 40°$

To draw a circle, we can use a compass. The stiff metal leg of the compass is placed where the center of the circle is to be. Holding this metal leg planted, trace a circle using the pencil attached to the other leg.

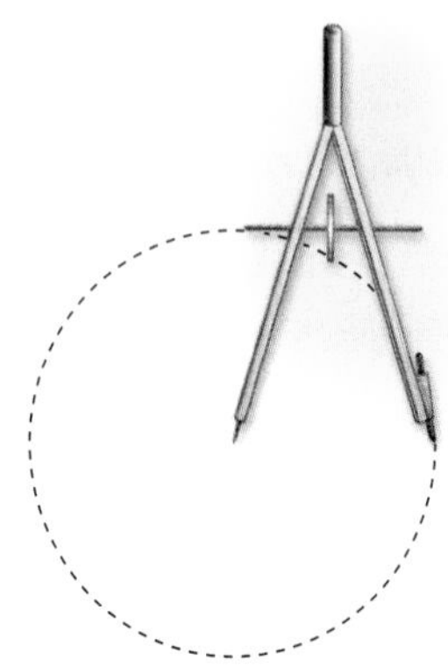

To measure a certain distance on the circumference in degrees, we can use a protractor. A protractor can measure 180° or one half of the circumference of a circle. Place the small open circle of the protractor on the center of the circle. Mark 0° on the circumference. Make a mark at a distance equal to the needed number of degrees. Now draw a line from the center of the circle to each mark. This "piece of pie" is equivalent to the percent represented by this number of degrees.

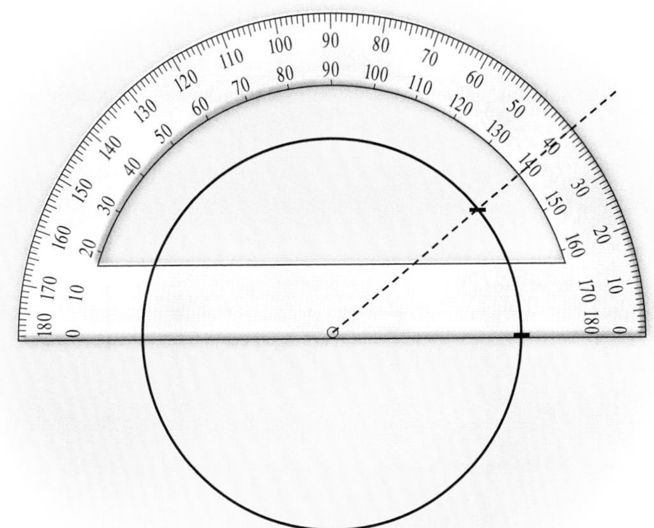

To mark off 40°, put the open circle of the protractor on the center of the circle. Using the top scale, mark 0°. Go counterclockwise to 40° on the protractor. A line connecting 40° on the protractor and the center of the circle will intersect the circle on the circumference at a distance of 40° from the first mark of 0°.

PROCEDURE

To draw a circle graph:

1. Calculate the number of degrees of the circumference of the circle that corresponds to each percent.
2. Draw a circle with a compass.
3. Use a protractor to determine the size of each piece of the circle graph.
4. Include a title and labels for each section of the graph. A different color or pattern is recommended for each section.

Example 6 **According to the *Princeton Review, The Best 331 Colleges 2000 Edition,* the student body of Temple University is 28% African American, 10% Asian, 4% Hispanic, 3% International, and 55% Caucasian. Present this information in a circle graph.**

Calculate the number of degrees of the circumference of the circle that correspond to each percent.

28% African American	$x = (0.28)(360°)$ $= 100.8°$ $\approx$ **101°**
10% Asian	$x = (0.10)(360°)$ $=$ **36°**
4% Hispanic	$x = (0.04)(360°)$ $= 14.4°$ $\approx$ **14°**
3% International	$x = (0.03)(360°)$ $= 10.8°$ $\approx$ **11°**
55% Caucasian	$x = (0.55)(360°)$ $=$ **198°**

Use a protractor to determine the size of each piece of the circle graph.

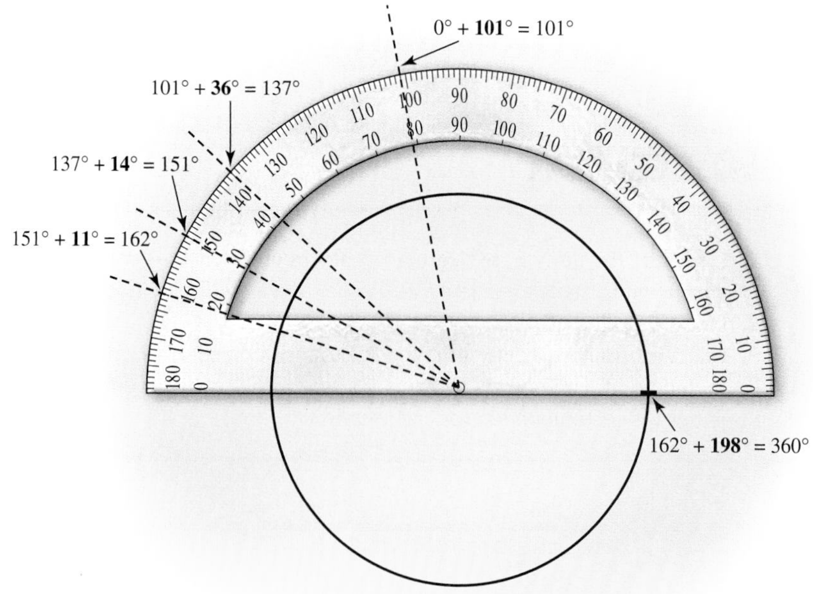

Include a title and labels for each section of the graph. A different color or pattern is recommended for each section.

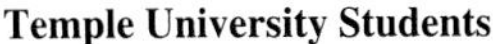

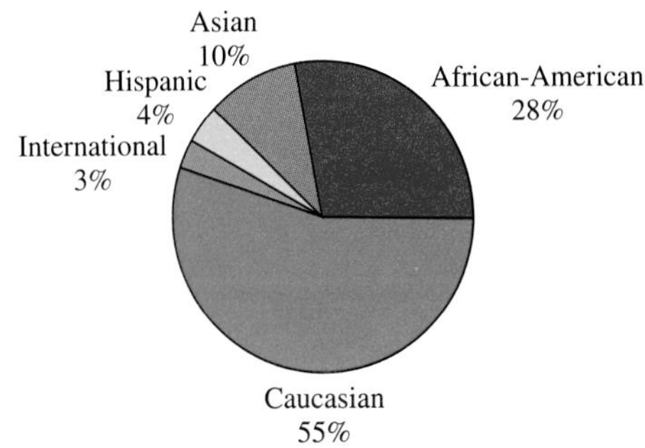

CHECK YOUR UNDERSTANDING

Calculate the number of degrees of a circle that are equal to the given percents. Round to the nearest degree.

1. 20% **2.** 33% **3.** 98% **4.** 78%

Answers: 1. 72°; 2. 119°; 3. 353°; 4. 281°.

EXERCISES SECTION 6.5

Throughout this assignment, round all percents to the nearest whole number.

1.

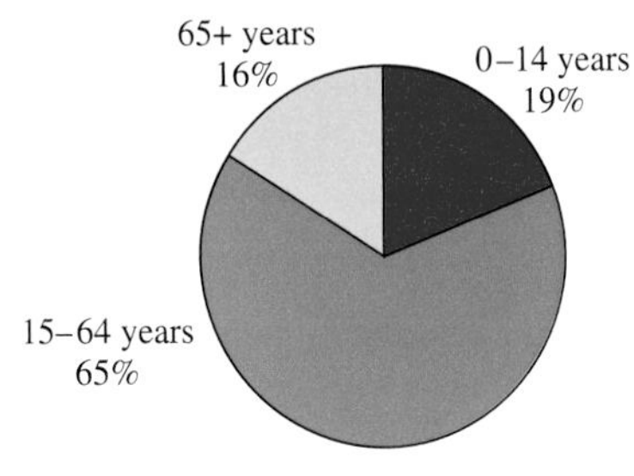

a. How many people in France in 1998 were 65 years or over?

b. How many people in France in 1998 were 15–64 years old?

c. How many people in France in 1998 were 0–14 years old?

2.

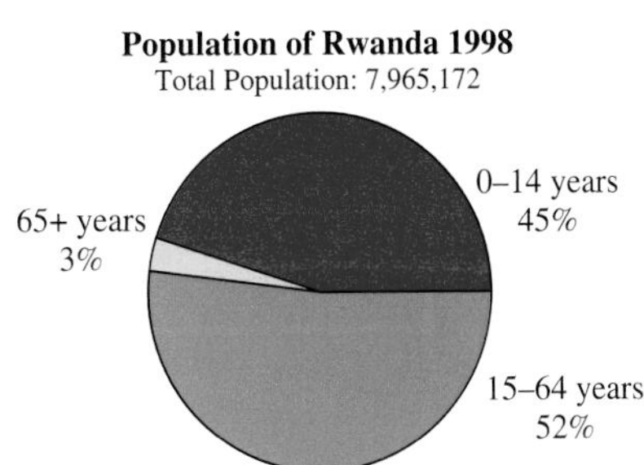

a. How many people in Rwanda in 1998 were 65 years or over?

b. How many people in Rwanda in 1998 were 15–64 years old?

c. How many people in Rwanda in 1998 were 0–14 years old?

3.

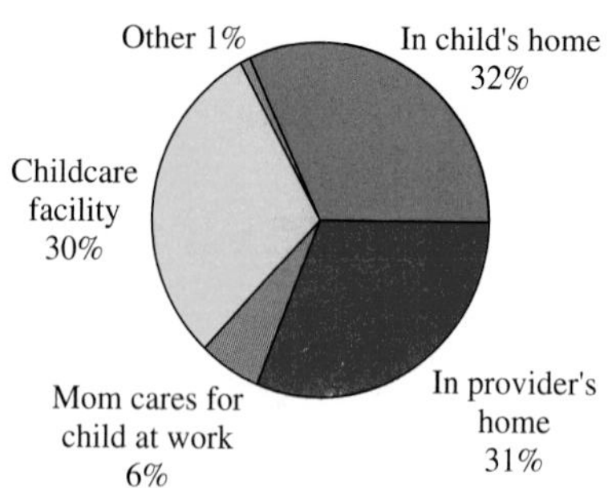

a. Calculate the number of children who were cared for in their homes.

b. Calculate the number of children who were cared for in the home of a provider.

c. Calculate the number of children who were cared for in a childcare facility such as a daycare center or preschool.

4. Complete the following circle graph using the supplied data.

MOST LIKELY CAUSES FOR A GOOD EMPLOYEE TO QUIT A JOB	
Limited advancement potential	41%
Lack of recognition	25%
Low salary/benefits	15%
Unhappy with management	10%
Bored with job	5%
Don't know	4%

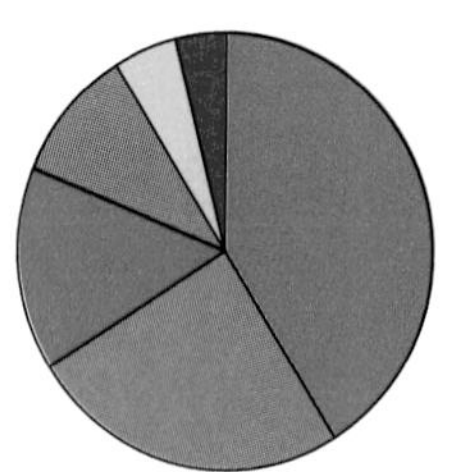

5. Complete the circle graph.

PLAYER, SCHOOL	FIRST PLACE VOTES 1998 HEISMAN TROPHY	PERCENTAGE OF FIRST PLACE VOTES
Ricky Williams, Texas	714 votes	85
Michael Bishop, Kansas State	41 vote	5
Cade McNown, UCLA	28 votes	3
Tim Couch, Kentucky	26 votes	3
Donovan McNabb, Syracuse	13 votes	2
Others	16 votes	2

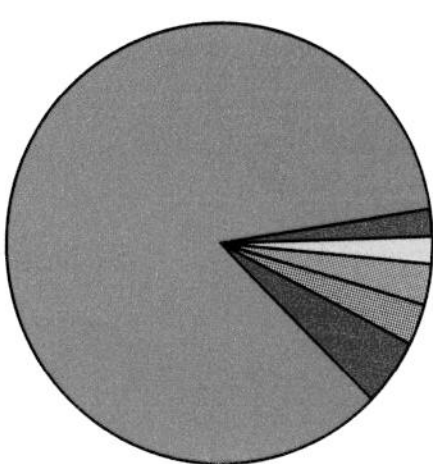

6. Create a circle graph using this information.

SHARE OF THE BREAKFAST CEREAL MARKET	
Malt-O-Meal, Inc.	4%
Kellogg Co.	31%
General Mills/Chex	27%
Quaker Co.	9%
Philip Morris Co./Post	17%
Store brands	11%
All others	1%

7. According to the *Princeton Review The Best 331 Colleges 2000 Edition,* the student body of the University of Texas at Austin is 4% African American, 14% Asian, 14% Hispanic, 3% International, and 65% Caucasian.

   a. Calculate the number of degrees in the circumference of a circle that are equal to the given percents. Round to the nearest degree.

b. Construct a circle graph to present this information.

8. A solution that is 25% calcium chloride, 3% sodium silicate, 3% ground silica, and 69% water can be added to concrete during mixing. This helps to waterproof the concrete. Construct a circle graph to present the composition of the solution.

9. The daily diet recommended by a dietitian for an obese 52-year-old woman with type II diabetes includes 199 grams of carbohydrates, 61 grams of protein, and 20 grams of fat.

a. Calculate the percent of the daily diet that is carbohydrates.

b. Calculate the percent of the daily diet that is protein.

c. Calculate the percent of the daily diet that is fat.

d. Calculate the number of degrees in the circumference of a circle that are equal to the given percents.

e. Present the information about the diet with a circle graph.

10. The table shows the percent of women and men who received a certain average number of e-mails per day in July 1999.

AVERAGE NUMBER OF E-MAILS RECEIVED PER DAY	WOMEN	MEN
1 to 5	15%	23%
6 to 10	27%	29%
11 to 20	28%	25%
21 to 50	19%	17%
Over 50	10%	7%
None	1%	0.2%

Source: *USA Today*, February 3, 2000.

a. Construct a circle graph that shows the information in this table for women.

b. Construct a circle graph that shows the information in this table for men.

c. Predict how many of a group of 500 women would receive 1 to 5 e-mails per day.

11. 82% of the student body at an elementary school qualifies for free or reduced lunches. Present this information using a circle graph.

12. The table shows the number of productions of frequently performed operas in North America during the 1999–2000 season.

OPERA	COMPOSER	NUMBER OF PRODUCTIONS
Madame Butterfly	Puccini	23
The Barber of Seville	Rossini	22
Tosca	Puccini	19
La Boheme	Puccini	18
Don Giovanni	Mozart	15

Source: *USA Today*, February 3, 2000.

a. Calculate the percent of the total number of productions that were of operas composed by Puccini. Round to the nearest percent.

b. Calculate the percent of the total number of productions that were of operas composed by Rossini. Round to the nearest percent.

c. Calculate the percent of the total number of productions that were of operas composed by Mozart. Round to the nearest percent.

d. Draw a circle graph that shows the percent of the total number of productions by each composer.

13. The average American was expected to spend a total of $1,558 on holiday expenses in 1999. These expenses included gifts ($1,088), entertaining ($188), travel ($151), cards, gift wrap, and decorations ($77), and other ($54).

a. Calculate the percent of the total holiday expenses to be spent on gifts.

b. Calculate the percent of the total holiday expenses to be spent on entertaining.

c. Calculate the percent of the total holiday expenses to be spent on travel.

d. Calculate the percent of the total holiday expenses to be spent on cards, gift wrap, and decorations.

e. Calculate the percent of the total holiday expenses to be spent on "other."

f. Present the information from Problem 13a–e in a circle graph.

14. According to the Bureau of Labor Statistics, the average household spent $35,535 in 1998. (The average of the number of people per household is 2.5 people.) The bar graph shows the expenditures by category.

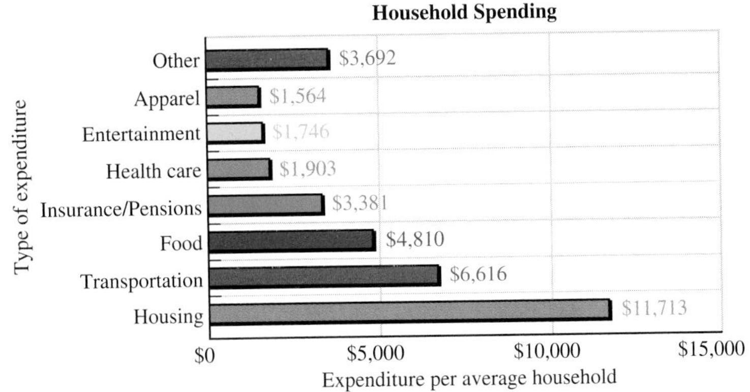

a. Calculate the percent of total spending for each type of expenditure. Round to the nearest percent.

b. Present the information in Problem 14a as a circle graph.

15. In November 1999, a telephone poll of 1,021 adults by Yankelovich Partners Inc. for Time/CNN revealed 19% had bigger than usual plans for the arrival of the New Year in 2000, 8% had smaller than usual plans, 72% had about the same plans, and 1% were not sure about their plans.

a. Present this information in a circle graph.

b. About how many of the polled adults had bigger than usual plans for the 2000 New Year?

16. Prior to January 1, 2000, a great deal of time and money was spent by companies and individuals to prepare for any computer problems caused when the date changed from 1999 to 2000. In late November of 1999, 10% of surveyed Americans were very concerned about the "Y2K-bug," 31% were somewhat concerned, 30% were not very concerned, 28% were not concerned, and 1% were not sure. Present this information in a circle graph.

17. Global AIDS figures released in 1998 showed the following information:

New HIV infections in 1998	5,800,000
New HIV infections in children (<15) in 1998	590,000
People living with AIDS/HIV	33,300,000
Deaths due to AIDS/HIV in 1998	2,500,000
Cumulative AIDS/HIV deaths	14,500,000

a. Write the ratio of new infections of HIV/AIDS in children in 1998 to the total number of new HIV/AIDS infections in 1998. Reduce to lowest terms.

b. About what percentage of new infections in 1998 occurred in children?

c.. What percentage of all HIV/AIDS related deaths occurred in 1998? Round the answer to the nearest percent.

d. Construct a circle graph showing the number of new HIV infections in children versus adults in 1998.

18. A small convention center has 264 total guest rooms. 80 of the rooms are smoking standard rooms, 80 of the rooms are nonsmoking standard rooms, 52 are smoking deluxe rooms, and 52 are nonsmoking deluxe rooms.

a. Calculate the percent of the total rooms that were smoking.

b. Calculate the percent of the total rooms that were nonsmoking.

c. Present the information about the guest rooms in a circle graph.

ERROR ANALYSIS

19. **Problem:** A travel agent is making a presentation on extra expenses for cruises. She estimates that her clients spend \$50 on beverages, \$40 for tips, and \$60 on extra entertainment on a 3-day cruise. She decides to make a circle graph of these estimates.

Answer: As she draws the graph, she marks off 50° for the beverages, 40° for tips, and 60° for the extra entertainment. She labels the rest of the circle graph as "other."

Describe the mistake made in solving this problem.

Redo the problem correctly.

20. Problem: A superintendent is making a presentation to the school board on the needs of students at an elementary school. 52% of the children qualify for reduced or free lunches. 36% have a native language other than English. 17% of the students are identified as learning disabled. He decides to use a circle graph to show the information.

Answer: Since the percents add up to more than 100%, the superintendent rounds them enough so that they will fit in the circle. He then makes the following circle graph.

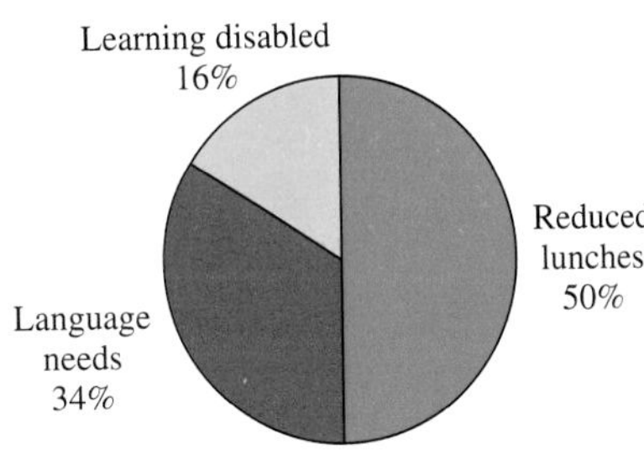

Describe the mistake made in solving this problem.

Redo the problem correctly.

REVIEW

21. Solve the equation $3p + 4 = 19$. Check the solution.

22. Solve the equation $3p - 4 = -19$. Check the solution.

23. Solve the equation $\frac{2}{3}y = 50$. Check the solution.

READ AND REFLECT

The United States has a representative democracy. Citizens are supposed to learn about important issues and work to elect people to office who best represent their views. However, these important issues often involve numbers and mathematical interpretations of numbers. In the following selections from newspapers and magazines, information is expressed as percents, ratios, fractions, averages, or probabilities. An understanding of these mathematical concepts is essential to use this information to make political decisions.

SELECTION 1

"Mubarak said an average of 6,452 children under the age of 5 were now dying each month. The death rate was 92 per 1,000 births in 1998 compared to 24 in 1989 . . . deaths among people over 5 had risen from 1,600 to 7,600 a month over the same period." (Source: *USA TODAY,* August 11, 1998.)

SELECTION 2

"Before we started burning oil and coal and gas, the atmosphere contained about 280 parts CO_2 per million. Now the figure is about 360." (Source: McKibben, B., *The Atlantic Monthly,* May 1998.)

SELECTION 3

"Roughly seven of 10 Republicans want President Clinton out of office, the sooner the better, and seven of 10 Democrats won't stand for it—period, end of story." (Source: Montgomery, R., Knight Ridder Newspapers, October 18, 1998.)

SELECTION 4

"Lawmakers wrapped up work Thursday on a farm bailout that will give producers a 50 percent bonus in their direct subsidy payments and provide another $2.6 billion in relief to growers who lost crops to drought and disease." (Source: Associated Press, October 16, 1998.)

SELECTION 5

"The players, already the highest-paid pro athletes, with an average annual salary of $2.6 million, want to retain the ability to earn unlimited pay as free agents. . . . The owners insist on a firm salary cap for each team, to reduce the share of revenue that goes to players. (Last year it was 57% of about $1.7 billion.)" (Source: *TIME,* October 19, 1998.)

SELECTION 6

"The *increase* in human population in the 1990s has exceeded the *total* population in 1600. The population has grown more since 1950 than it did during the previous four million years." (Source: McKibben, B., *The Atlantic Monthly,* May 1998.)

These selections contain information important to citizens deciding positions on international sanctions, national politics, population control, air pollution, or an NBA player lockout. However, unless citizens understand the implications and limitations of percents, ratios, averages, and probabilities, the information can be misinterpreted. Further, many times such information is presented in a way that is designed to sway people to take a particular position. People who do not understand mathematics are vulnerable to being swayed by incomplete, distorted, or biased statistics.

Some knowledge of mathematics is essential for responsible citizenship. The question facing mathematics teachers, school boards, curriculum committees, and state boards of education is "How much mathematics is enough?" If the task of such people is to set the minimum standards for the citizens of our nation, what should they require that students learn? Currently many states are involved in the process of writing "exiting standards" for high school graduation. The minimum standards of mathematical knowledge needed for citizenship are part of that process.

RESPOND

1. Find a current newspaper or magazine article that expresses information using *only* raw numbers that are not a percentage, ratio, or other mathematical expression. Underline the numerical information.

2. Find a current newspaper or magazine article that expresses information that has been expressed mathematically as a percentage, ratio, probability, and so on. Underline the relevant information.

3. Some states are now requiring exit exams for students who are seeking a high school diploma. Such exams require that students successfully calculate percentages, averages, and ratios, construct and read graphs, and interpret probabilities. Do you think that this type of exam should be required for high school graduation? Explain.

4. Generally, a college degree signifies that the person who has completed the degree has a different type of knowledge and/or skills than a person with a high school diploma. How do you think the mathematics knowledge of a person with a college degree should be different from the knowledge of a person with a high school diploma? Be specific.

5. Discuss the pros and cons of a required college mathematics exit exam.

Section	Terms and Concepts/Procedures
6.1	ratio; rate; unit rate **1.** Write a ratio as a fraction. **2.** Write a rate as a fraction. **3.** Calculate a unit rate.
6.2	proportion; colon notation **1.** Determine an equivalent ratio using a proportion. **2.** Determine an equivalent rate using a proportion.
6.3	percent **1.** Write a ratio as a percent. **2.** Write a decimal as a percent. **3.** Write a percent as a ratio. **4.** Write a percent as a decimal. **5.** Calculate a percent increase or decrease.
6.4	markdown; simple interest; principal **1.** Calculate simple interest. **2.** Calculate markup or markdown. **3.** Calculate sales tax on a purchase.
6.5	circle graph **1.** Obtain data from a circle graph. **2.** Construct a circle graph.

SECTION 6.1 INTRODUCTION TO RATIOS AND RATES

1. Explain the difference between a ratio and a rate.

2. Explain why a percent is a ratio, not a rate.

3. What is necessary for a rate to be a unit rate?

4. Of twelve pairs of jeans advertised in a sale, only three pairs cost less than $25. Write a ratio that compares the number of jeans that cost less than $25 to the total number of jeans. Simplify to lowest terms.

5. A total of about 500,000 people visit Crater Lake National Park in Oregon each year. Only 50,000 people come to the park during the 6-month winter season. Write a ratio that compares the winter visitors to the total visitors. Simplify to lowest terms.

6. A currency exchange offered 9.33 Mexican pesos for 1 U.S. dollar. Write this information as a rate using colon notation.

7. Karen drove 420 miles in 4 hours. Write this information as a unit rate of speed. Simplify to lowest terms.

8. Roger earned $448 for 40 hours of work. What is his unit rate of pay?

SECTION 6.2 PROPORTIONS

9. One out of ten visitors to Crater Lake in Oregon visit in the 6-month winter season. If the total visitors per year is about 500,000, use a proportion to calculate the number of winter visitors.

10. Decades of mining in the Silver Valley pushed lead and other heavy metals like arsenic and cadmium downstream into lakes and rivers. The lead is then deposited on beaches where people gather to fish or picnic. Lead contamination on the shores of the upper Spokane River peaked at 2,360 parts per million. Use a proportion to calculate the amount of lead in 2 pounds of sand.

11. Infant mortality in a state declined from 8.2 deaths per 1,000 live births in 1991–92 to 7.2 in 1996–98. Use a proportion to predict the number of deaths per 2,500 live births. Round to the nearest tenth.

12. A U.S. standard brick weighs about 4.5 pounds. Use a proportion to calculate the number of bricks in a ton. (A ton is 2,000 pounds.)

13. The North Central District Health Department reported that 1 out of 20 food establishments inspected required an on-site follow up inspection. If 740 establishments were inspected, use a proportion to calculate the number that required a follow up inspection.

SECTION 6.3 INTRODUCTION TO PERCENTS

14. Explain how to change a percent into an equivalent decimal.

15. Explain how to change a percent into an equivalent ratio.

16. Change 54% to a decimal.

17. Write 7% as a fraction.

18. Change $\frac{13}{100}$ to a percent.

19. Change 0.3% to a decimal.

20. MBTE (methyl butyl tertiary ether) is a widely used gasoline additive that makes cars pollute less but may cause cancer in people. The additive was found in 66 of about 1,000 public water systems in Maryland.

a. Write a ratio of water systems contaminated with the additive to the total water systems tested.

b. Calculate the percent of the water systems that were contaminated.

21. The weather condition called El Niño coincided with increased lightning storms for those Americans who live near the Gulf of Mexico. There were 33 days with lightning out of a total of 90 days studied during the winter of 1997–98.

a. Write a ratio of days with lightning to total days.

b. What percent of days had lightning? Round to the nearest percent.

22. The total fund balance for the LCSC Foundation was $662,777 on January 31, 1996. On January 1, 2000, the total fund balance was $2,980,458. Calculate the percentage increase in the fund balance. Round to the nearest percent.

23. The Planning and Zoning Commission voted to allow campaign signs on city rights of way 30 days prior to an election. The existing code allowed campaign signs to be posted for 45 days prior to an election. Calculate the percent decrease in the number of days campaign signs may be posted. Round to the nearest percent.

24. After the crash of an Alaska Airlines plane into the Pacific Ocean because of possible failure in the horizontal stabilizer, the FAA ordered the inspections of planes in the MD-80, MD-90, DC-9, and Boeing 717 series. Of 960 inspections completed, 27 revealed problems that required additional evaluation or repair. Gritty material was found in the jackscrew grease of some planes. Jackscrews were replaced on 15 airplanes and stabilizer assemblies were lubricated in others.

a. Write a ratio that compares the airplanes that required additional evaluation or repair to the total airplanes inspected.

b. What percent of the inspected airplanes required additional evaluation or repair? Round to the nearest tenth of a percent.

c. Write a ratio that compares the number of airplanes that required replacement of jackscrews to the total airplanes inspected.

d. What percent of the inspected airplanes required replacement of jackscrews? Round to the nearest tenth of a percent.

25. The number of minority students enrolled at a college was 867 in the fall of 1999. In the spring of 1999, the enrollment was 788 students. In the spring of 1998, the enrollment of minority students was 734.

a. Calculate the percent increase in enrollment of minority students from the spring of 1999 to the fall of 1999.

b. Calculate the percent increase in enrollment of minority students from the spring of 1998 to the fall of 1999.

SECTION 6.4 APPLICATIONS OF PERCENTS

26. Calculate 35% of 600.

27. Use a proportion to answer the question: 82 is 4% of what number?

28. Jodi Oftelie, Director of College Advancement at Lewis-Clark State College, reported that "650 alumni donors contributed in a phonathon—10% of those who were asked." Use a proportion to determine the total number of alumni who were contacted in the phonathon.

29. Community advocacy groups asked a state legislature to extend the minimum wage to farmworkers. Of 33,000 such workers, the groups estimated at least 17% are not making minimum wage. Use a proportion to estimate the number of workers not making minimum wage. Round to the nearest ten.

30. The sale price of a fleece vest was $13.97 at a department store. The store claimed that this was a 75% savings of the original price. Calculate the original price.

31. Calculate the simple interest earned on $750 invested at 4% for 3 years.

32. The United Way reported that the total amount raised in a local county in 1999 was a 5% increase over the 1998 total of $680,000. What was the total amount raised in 1999?

33. Bill Gates paid $1,076,231 in property taxes on his home and five acres of land on the Lake Washington waterfront. In 1999, Gates paid $615,000 in property taxes that were based on an assessment before construction of the 48,000-square-foot house and grounds was completed. Calculate the percent increase in the property taxes paid. Round to the nearest percent.

34. The number of daycares inspected by a county health department rose from 87 in 1997 to 149 in 1999. Calculate the percent increase in daycares inspected. Round to the nearest tenth of a percent.

35. A pair of SilverTab® cargo pants was marked down 35%. The original price was $52. Calculate the sales price. Round to the nearest cent.

36. Renee is traveling out of state and stops at a fast-food restaurant. The big hamburger on the menu is listed at $0.99. She buys three hamburgers and her total is $3.15. What is the tax rate charged on restaurant food in this state? Round to the nearest tenth of a percent.

37. The Department of Commerce announced that estimated retail sales for November 1998 were $229.4 billion, a 0.6 percent increase from the previous month. What were the sales in the previous month? Round to the nearest tenth of a billion dollars.

38. Violent crime in Baltimore County decreased during the first 6 months of 1998 with 3,047 cases compared to 3,561 cases in the first six months of 1997. What was the percent decrease in violent crime? Round to the nearest tenth of a percent.

39. The season before Larry Bird began playing for the Boston Celtics, 1978–79, the Celtics won 29 games and lost 53 games. In 1979, Bird came to Boston. The Celtics' record improved to 61 wins and 21 losses.

a. What percent of games did the Celtics win in the 1978–79 season?

b. What percent of games did the Celtics win in the 1979–80 season?

40. The sales tax in Michigan is 6%. Calculate the sales tax on a four-wheeler ATV that costs $3,250.

41. In 1998, there were 838 players on major league baseball opening day rosters. Only 111 of the players played for the same team in 1994. What percent of players did not change teams? Round to the nearest percent.

42. Abandoned mines in the West are sources of toxic heavy metal pollution in streams. Of the 480 abandoned mines in the Coeur d'Alene Ranger District, 78 have known environmental problems. What percent do *not* have known environmental problems? Round to the nearest percent.

SECTION 6.5 CIRCLE GRAPHS

43. The circle graph shows the projected racial and ethnic origins of the U.S. civilian workforce in 2006. The total number of workers will be about 148,847,000 people. How many of these workers will be of Hispanic background?

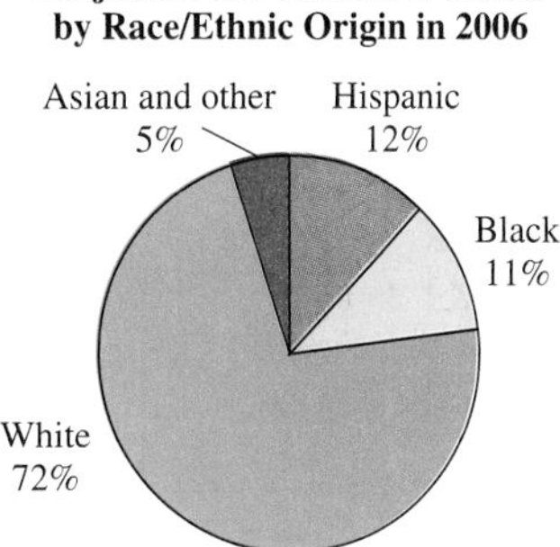

44. The expenditures of a health department in 1999 are shown in the table.

ENVIRONMENTAL HEALTH	HEALTH EDUCATION	ADMINISTRATION	PHYSICAL HEALTH
$515,746	$125,057	$882,394	$1,492,794

a. Calculate the percent of total expenditures that are spent on environmental health. Round to the nearest percent.

b. Calculate the percent of total expenditures that are spent on health education. Round to the nearest percent.

c. Calculate the percent of total expenditures that are spent on administration. Round to the nearest percent.

d. Calculate the percent of total expenditures that are spent on physical health. Round to the nearest percent.

e. Complete the circle graph below. Include a title and data labels. The sum of the values may not equal 100%, due to rounding.

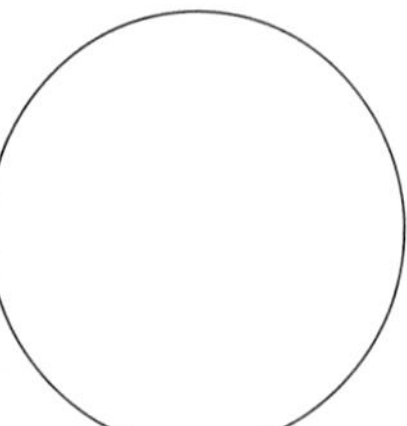

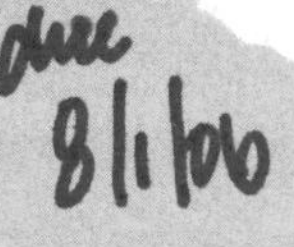

1. Explain the difference between a ratio and a rate.

2. Write 64% as a decimal.

3. Write 71% as a ratio in fraction form.

4. Write $\frac{45}{100}$ as a percent.

5. Up to 25% of the U.S. population carries the bacterium that causes bacterial meningitis. These people have no symptoms of the disease; they are known as "carriers." When an outbreak of bacterial meningitis occurs, a local health district may distribute antibiotics to lower the normal number of carriers in the population. A county has about 456,000 people. How many could be carriers of bacterial meningitis?

6. India has 152,792 post offices, more than any other country in the world. The United States has about 28,000 post offices. Write the ratio of U.S. post offices to Indian post offices in lowest terms.

7. In February 2000, the U.S. House of Representatives voted to send to the Senate a bill that would cut income taxes $182 billion over 10 years for all married taxpayers. 268 Representatives voted yes; 158 Representatives voted no.

   a. Write a ratio of Representatives voting No to the total Representatives voting.

   b. Use a proportion to determine the percent of the total voting Representatives who voted No. Round to the nearest percent.

8. A community action agency keeps data on the amount of vacant housing available in a small city. In 1997 there were 13,047 homes in the city. Of these, 391 were vacant, awaiting sale. Use a proportion to determine the percent of vacant homes. Round to the nearest percent.

9. Caleb drove 137 miles in 3 hours. Write this rate in fraction form.

10. The year 2000 was declared a Holy Year by the Pope of the Roman Catholic Church. As many as 30 million pilgrims were expected to visit the city of Rome during the year. Write a unit rate that describes the average number of pilgrims expected per month.

11. Calculate the simple interest earned on \$5,500 invested at $6\frac{3}{4}\%$ for 3 years.

12. Voters in Washington State approved tax-cut Initiative 695 in Fall 1999. This tax cut caused a \$750 million loss in tax revenues, primarily from reductions in taxes paid for vehicle license plates. About 75% of the loss in tax revenues directly affected budgets for state highways, ferries, rail, local road projects, and transit. What was the dollar amount of this loss of revenues?

13. Karen bought ten acres of land to build a vacation home. The price of the land was \$73,000. After digging multiple wells, she realized that there was no water available. She sold the land for \$12,000. What was the percent loss in the value of the land? Round to the nearest percent.

14. The table shows the population and land area in square miles of Utah from 1880 to 1980.

YEAR	POPULATION	LAND AREA GROWTH
1880	20,768	—
1890	44,843	—
1900	53,531	—
1910	92,777	47.5
1920	118,110	51.1
1930	140,267	52.0
1940	149,934	52.5
1950	182,121	53.9
1960	189,454	56.1
1970	175,885	59.3
1980	163,033	75.2

a. What was the percent increase in population from 1880 to 1980?

b. What was the percent increase in land area from 1910 to 1980?

15. A correction department director reported that 1,730 inmates participated in the department's adult basic education programs in the last budget year. 296 inmates earned high school diplomas or GEDs. Use a proportion to determine what percent of the participating inmates earned a high school diploma or GED. Round to the nearest percent.

16. A craft store advertised skeins of yarn for \$0.99 each. The regular price was \$1.49 per skein. Calculate the percent markdown per skein. Round to the nearest percent.

17. The circle graph shows the results of a survey in the student union to determine the favorite soft drink of students.

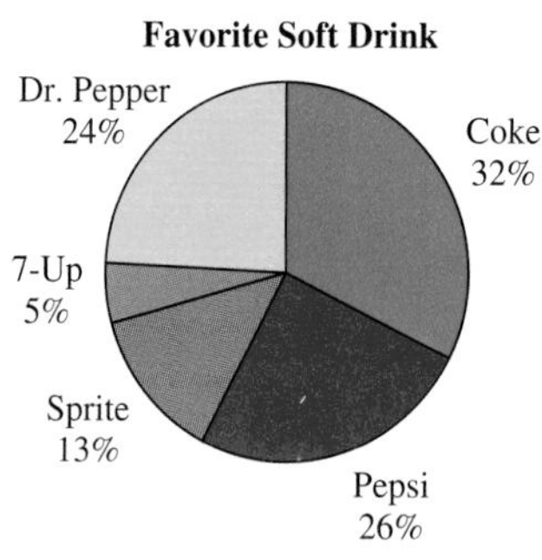

a. What is the favorite soda in the survey?

b. The survey was completed by 250 students. How many of the students preferred 7-Up?

c. If there are a total of 10,000 students on campus, predict how many of the students would prefer Pepsi.

18. The sales tax in California is 6%. Calculate the sales tax on a portable phone that costs $69.98. Round to the nearest cent.

19. One out of 25 people who own stock is a homemaker. Predict how many homemakers will be found in a group of 5,400 stock owners. Round to the nearest person.

20. In 1999, the population of the United States was about 268,922,000. Thirteen percent of the population was 65 years and over. Use a proportion to determine how many people were 65 years and over. Round to the nearest hundred.

21. Calculate the simple interest earned on $85,300 invested at 4% for 2 years.

22. The wholesale cost of a portable CD player to a store owner is \$58.00. The store sells the CD player for \$89.99. Calculate the percent markup on the CD player. Round to the nearest tenth of a percent.

23. The sales tax in Arkansas is 4.625%. Calculate the sales tax on a car that costs \$13,599. Round to the nearest cent.

24. During the off season, the occupancy ratio in hotels in a coastal city is 1 : 5. In other words, 1 out of 5 hotel rooms is occupied. The town has a total of 26,900 hotel rooms. Calculate the number of occupied hotel rooms.

25. In February 2000, 444 Democrats and likely Democratic leaners were asked, "If you were asked to vote for a Democratic nominee for President today, for whom would you vote?" About 271 people chose Al Gore, 93 chose Bill Bradley, and the remaining either were not sure or chose other candidates. (Source: *TIME,* February 14, 2000.)

a. Determine the percent of people polled who preferred Al Gore. Round to the nearest percent.

b. Determine the percent of people polled who preferred Bill Bradley. Round to the nearest percent.

c. Determine the percent of people polled who were not sure or chose other candidates. Round to the nearest percent.

d. Present this information in a circle graph. Include a title and data labels.

Appendix 1

Geometry

Prior to 500 B.C.E., mathematics involved only the practical use of numbers and arithmetic, in areas such as trade, farming, and taxes. From 500 to 300 B.C.E., the ancient Greeks looked at mathematics from the viewpoint of geometry, and mathematics involved the study of both numbers and shapes. Mathematics was studied and valued not only because it was useful, but also because of its beauty and logical structure. The formal ideas of axioms, definitions, theorems, and proof were introduced. The work of Euclid of Alexandria (circa 300 B.C.E.) is the oldest complete manual of ancient Greek geometry that survives. This geometry is, therefore, often called **Euclidean geometry.**

In this appendix, we will learn some of the basic terms and concepts used in Euclidean geometry. These concepts will prepare us for any future work in logic and proof. They are also useful in solving many application problems that involve shapes and measurement. This material is a summary of the geometry found throughout the text.

APPENDIX OUTLINE

Section A1.1 Essential Terms and Definitions

LEARNING TOOLS

CD-ROM SSM VIDEO

Sneak Preview

As in other disciplines, a **definition** in mathematics explains what a word means using other previously defined words. Of course, in order to define one word, we use other words, and to define these words we must use still other words. In the end, some words have to simply be accepted as "undefined." In this section, we will learn the words that are usually left undefined in Euclidean geometry. We will also learn other important definitions.

After completing this section, you should be able to:

1) List three words that are usually left undefined in geometry.

2) Define: parallel lines, line segment, ray, angle, vertex, perpendicular lines, circle, center of a circle, radius, chord, diameter, degree, right angle, acute angle, obtuse angle, straight angle, supplementary angles, complementary angles, polygon, triangle, equilateral triangle, isosceles triangle, right triangle, hypotenuse, quadrilateral, square, rectangle, trapezoid, rectangular solid, cube, pyramid, circular cylinder, cone, and sphere.

DISCUSSION

Undefined Terms

In plane Euclidean geometry, three words are usually left undefined: *point, plane,* and *line.* Although we will not precisely define these words, we can describe them so that we share an understanding of their meaning.

A point is drawn on paper with a dot. A dot on paper can be large or small. A **point** in mathematics has no size; it just represents a position. A **line** is straight and extends forever in both directions. It is made up of an infinitely large set of points. Since points have no size, a line has no width. (Of course, it looks like it does but this is just a picture of an idea.) Two lines that never cross each other are called **parallel lines.**

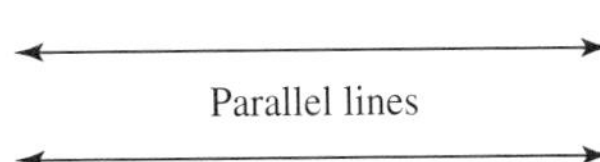

A **plane** is a flat surface that extends forever in all directions. The floor of a classroom is part of a plane. A plane has no thickness.

DISCUSSION

Line Segments, Circles, and Angles

We can use the undefined words *point, line,* and *plane* to write other important definitions. Draw a straight line. Pick two points on this line and call them A and B.

The **line segment** $\overline{AB}$ is the set of points containing A, B, and all the points lying between A and B. A and B are called **endpoints.** Unlike a line that travels on forever in both directions, a line segment stops at each endpoint.

A B

The length of a line segment is the distance between its endpoints. Unless a unit of measurement is specified, the length is simply a number.

Again, draw a straight line. Pick a point on that line. Now erase all the points on the line to one side of that point. What remains is called a **ray.**

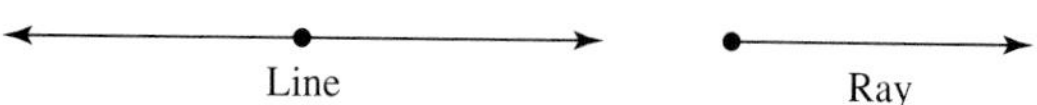

Rays are often named by stating the endpoint and another point on the line. For example, a ray might have an endpoint named A and another point named B. This ray can be named $\overrightarrow{AB}$.

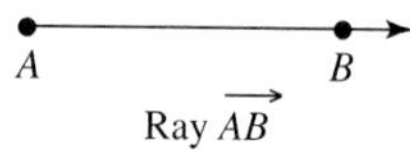

Ray $\overrightarrow{AB}$

The notation $\overrightarrow{AB}$ is used to show that the ray starts at the endpoint A and travels through the point B.

If we draw two different rays that have the same endpoint, we create an **angle.** Angles can be named using the letters of three points, with the endpoint of each ray in the middle. Angles can also be named using just the shared endpoint of the rays. The symbol for angle is $\angle$.

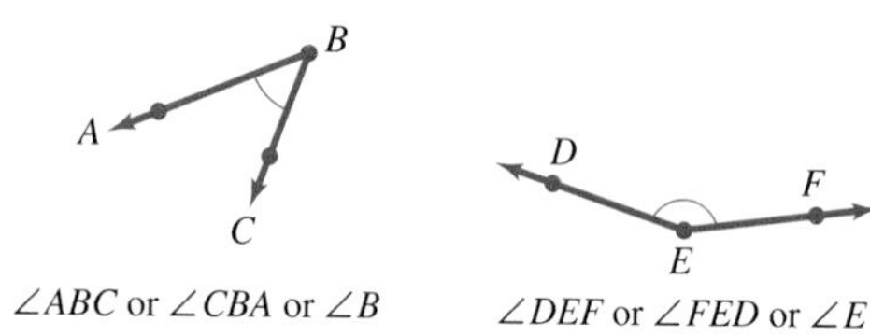

$\angle ABC$ or $\angle CBA$ or $\angle B$ $\angle DEF$ or $\angle FED$ or $\angle E$

A **circle** is the set of all points in a plane that are at the same distance from a point called the **center.** The distance around any circle can be divided into 360 equal parts. The angle at the center of the circle for each of these parts is called a **degree.** If we start at 0 on the right-hand side of the circle, the point directly opposite from 0 degrees will be 180 degrees away. When we arrive back at the starting point, we will have traveled 360 degrees. The symbol for degree is °. The degree symbol is written to the right of the measure of the angle. So, 360 degrees is written 360°.

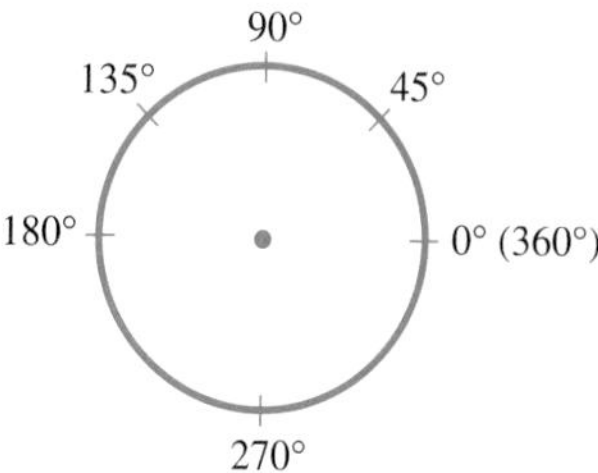

If the endpoint of an angle is the center of a circle, we can measure the angle in degrees.

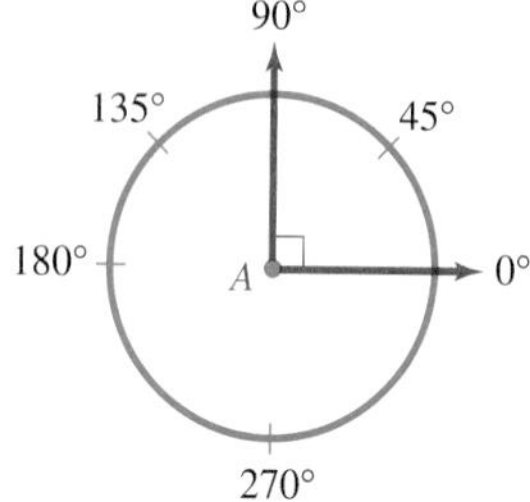

The measure of $\angle A$ is 90°. Any angle that measures 90° is called a **right angle.** Any angle that has a measure less than 90° is called an **acute angle.** Any angle that has a measure greater than 90° but less than 180° is called an **obtuse angle.** An angle that is exactly 180° is called a **straight angle.**

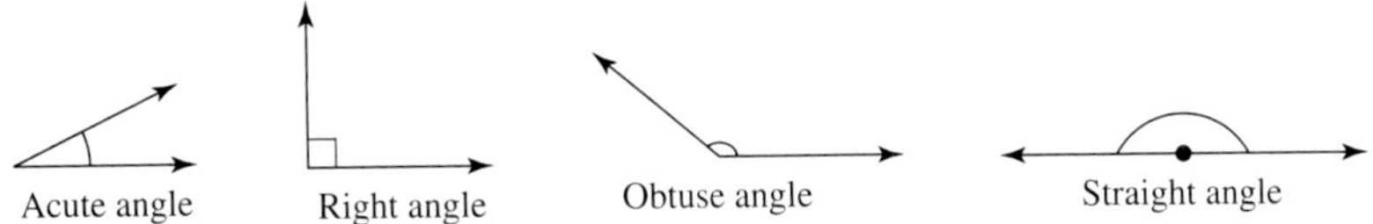

When drawing a right angle, we show that it measures exactly 90° by drawing a ⌐ connecting the two rays of the angle. When two lines intersect and create a 90° angle, we say that the lines are **perpendicular.**

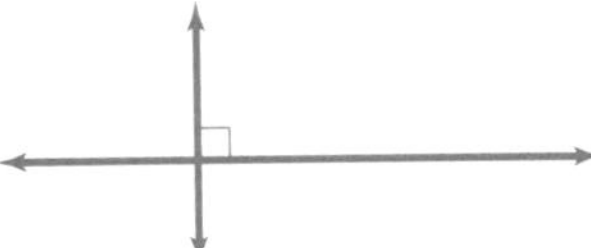

If the sum of the measure of two angles is 180°, the angles are **supplementary angles.**

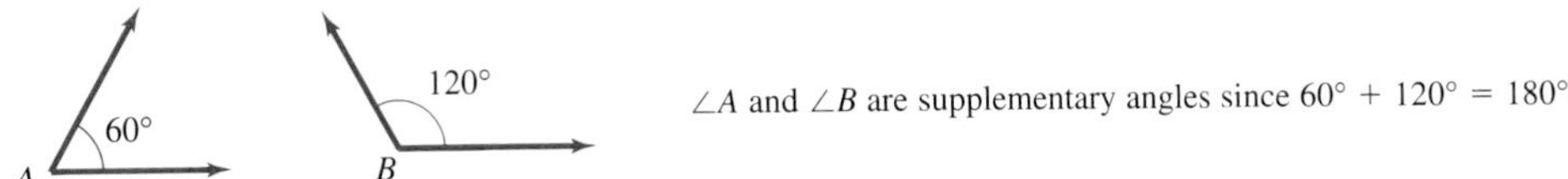

If the sum of the measure of two angles is 90°, the angles are **complementary angles.**

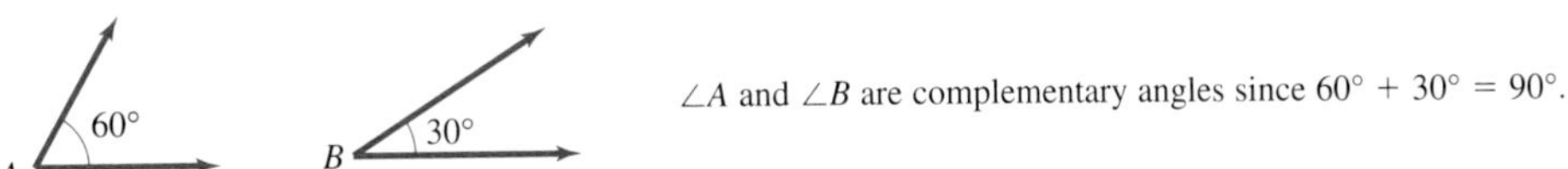

A **radius** is a line segment that joins the center of a circle to a point on the circle.

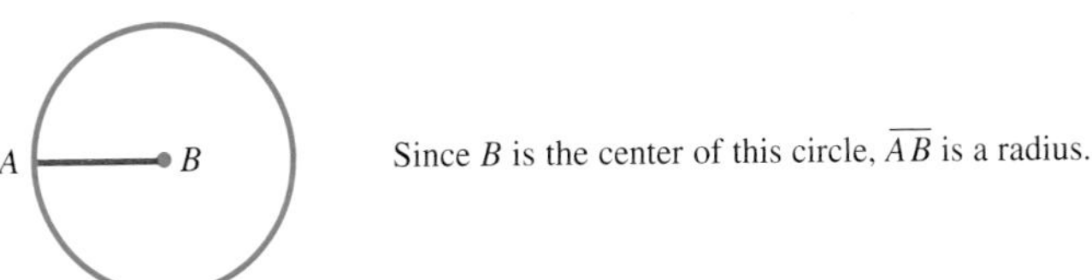

A **chord** is a segment whose endpoints both lie on a circle.

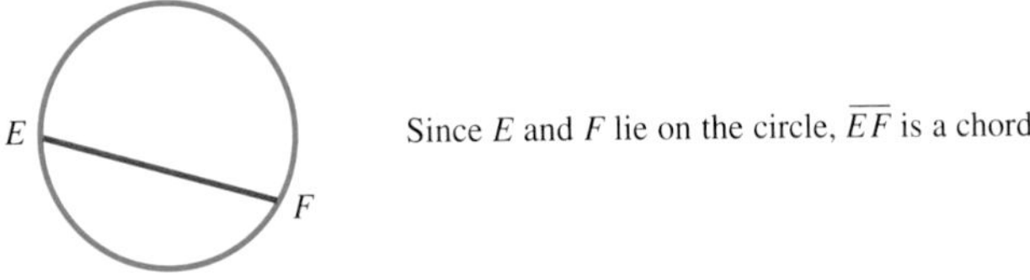

A **diameter** is a chord that passes through the center of a circle.

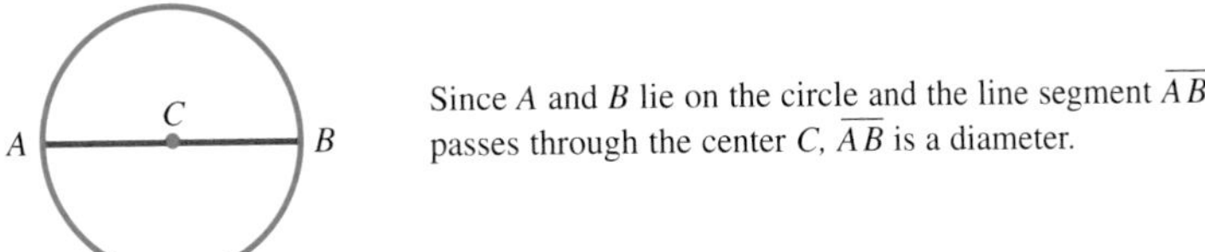

CHECK YOUR UNDERSTANDING

1. Name this angle:

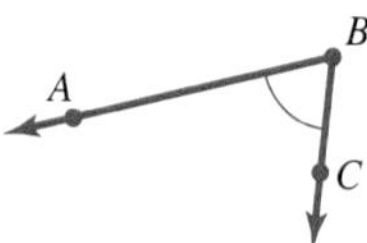

2. What is the difference between a ray and a line?

3. Classify an angle with a measure of 35° as acute, obtuse, right, or straight.

4. Classify an angle with a measure of 120° as acute, obtuse, right, or straight.

5. What is the difference between a radius and a diameter?

Answers: 1. ∠ABC or ∠CBA or ∠B; 2. A line extends forever in both directions; a ray has an endpoint and only extends forever in one direction; 3. This is an acute angle; 4. This is an obtuse angle; 5. A radius is a line segment with an endpoint on the circle and on the center of the circle; a diameter is a segment whose endpoints are both on the circle and which passes directly through the center of the circle.

DISCUSSION

Polygons

When we connect the endpoints of line segments to build a closed flat shape, we are constructing a **polygon.** (The prefix *poly-* means many. A polygon has many sides.) For example, when we connect the endpoints of three line segments, we create a polygon called a **triangle.**

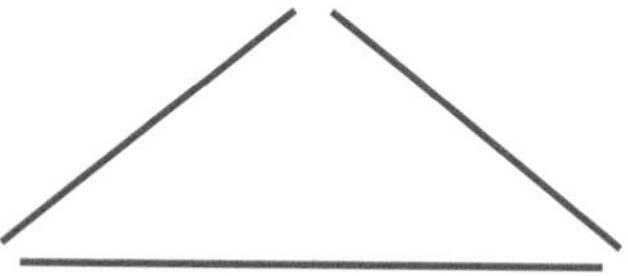

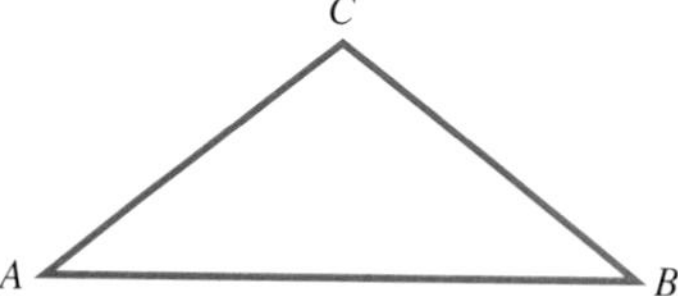

Triangles always have three sides and three angles. (The prefix *tri-* means three.) The sum of the measure of the angles always equals 180°. If all three sides of the triangle are equal in length, it is an **equilateral triangle.** If two sides of the triangle are equal in length, it is an **isosceles triangle.** If one of the angles in the triangle measures 90°, the triangle is a **right triangle.** The side opposite the right angle is called the **hypotenuse.**

Equilateral triangle

Isosceles triangle

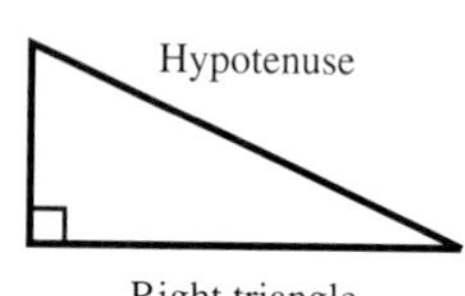

Right triangle

A polygon with four sides is called a **quadrilateral.** (The prefix *quad-* means four.) The sum of the measure of the angles of a quadrilateral is always 360°. A **trapezoid** is a quadrilateral with exactly two parallel sides. The parallel sides are called **bases.** The other two sides are called legs. A **parallelogram** is a quadrilateral in which both pairs of opposite sides are parallel. A **rectangle** is a parallelogram whose angles are right angles. A **square** is a rectangle in which all the sides have the same length.

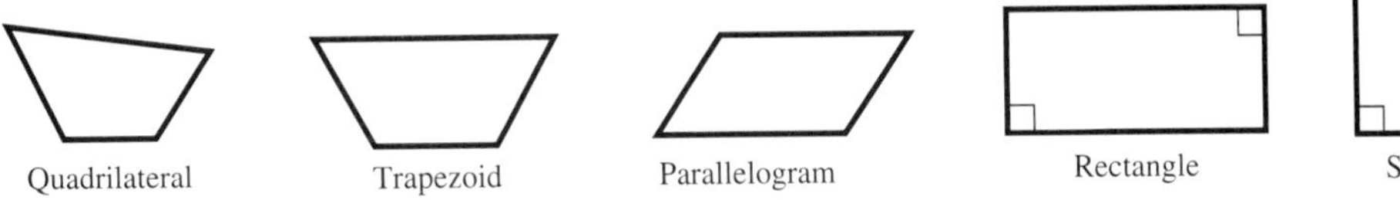

CHECK YOUR UNDERSTANDING

1. What is the difference between a square and a rectangle?
2. What is the difference between an equilateral triangle and an isosceles triangle?
3. Is a rectangle also a trapezoid? Explain.
4. Can an isosceles triangle be a right triangle? Explain.

Answers: 1. A square is a special kind of rectangle, one in which all four sides are the same length; 2. An equilateral triangle has three sides that have the same length; an isosceles triangle only has two sides that have the same length; 3. No, a trapezoid has exactly two sides that are parallel. Because a rectangle is a parallelogram, both pairs of opposite sides are parallel; 4. Yes, a triangle with two sides that are the same length can include a right angle.

DISCUSSION

Solids

Polygons are flat figures that are contained in one plane. We say that polygons are **two-dimensional** figures. When a shape has three dimensions, it is often referred to in geometry as a **solid.** A **sphere, rectangular solid (cuboid), pyramid,** and **cylinder** are all examples of solids.

A **sphere** is a solid that looks like a ball, formed from many circles of different diameter. The radius of a sphere is the distance from the center to any point on the surface of the sphere.

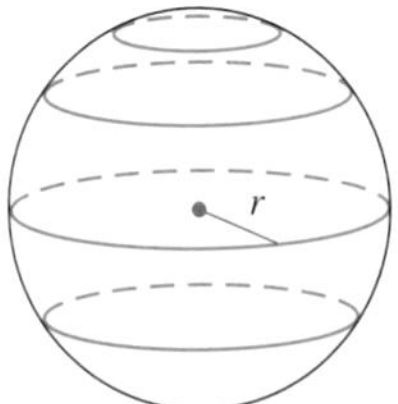

A **cylinder** has identical cross sections and straight sides. One of the more common cylinders is a **circular cylinder** where each cross section is a circle. It can be visualized as a stack of identical circles.

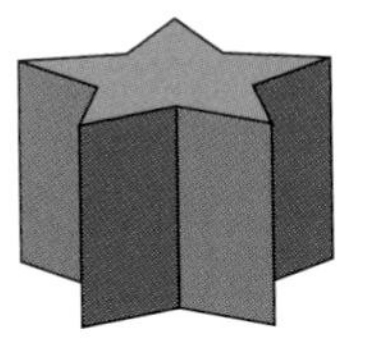

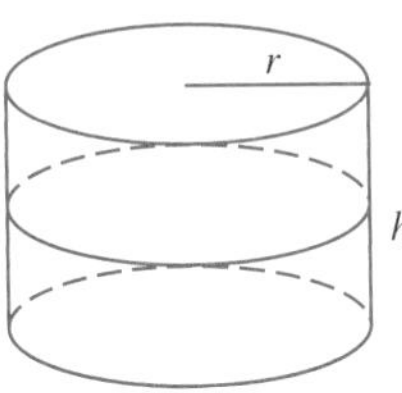

A **rectangular solid** or **cuboid** is a shape like a box, formed from many rectangles of the same size. A rectangular solid has length, width, and height. If the length, width, and height are equal lengths, the solid is called a **cube.**

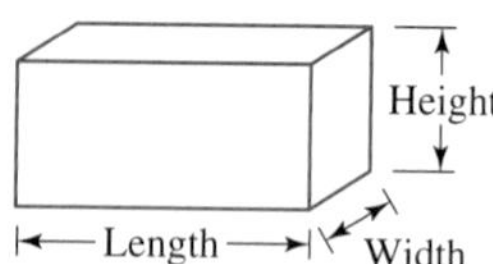

A **pyramid** can have a base of any shape. All points on the base are connected by straight lines that meet at a point in a different plane. This pyramid has a base that is a square.

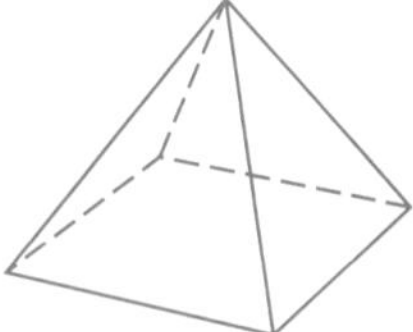

This pyramid has a base that is a triangle.

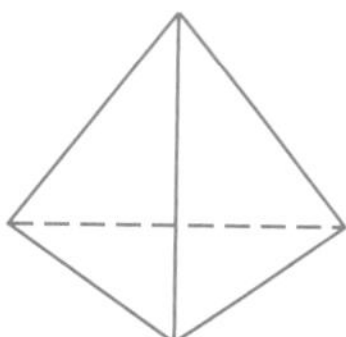

This pyramid has a base that is a circle. We commonly call this type of pyramid a **cone.**

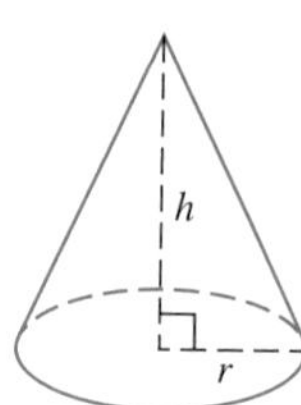

CHECK YOUR UNDERSTANDING

1. What is the difference between a polygon and a solid?

2. What solid is represented by a shipping box for a computer?

3. What solid is represented by a can of soup?

4. What solid is represented by a basketball?

Answers: 1. A polygon is a two-dimensional shape; a solid is three dimensional; 2. a rectangular solid (cuboid); 3. a circular cylinder; 4. a sphere.

EXERCISES SECTION A1.1

1. What is the difference between a line segment and a line?

2. What is the difference between a ray and a line?

3. Explain why all chords of a circle are not diameters of the circle.

4. The radius of a circle is 8 cm. What is the diameter of the circle?

5. Explain the difference between an isosceles triangle and an equilateral triangle.

6. What is the measure of an angle that is complementary to an angle that measures 63°?

7. What is the measure of an angle that is supplementary to an angle that measures 63°?

8. Two lines are drawn across a busy street to mark a crosswalk. Are these lines parallel lines or are they perpendicular lines?

9. Explain how a trapezoid is different from a rectangle.

10. Explain how a cone is different from a square pyramid.

Additional Exercises:

Section 1.1 #62, 63; Section 1.3 #66–73; Section 1.4 #34, 35.

Section A1.2 Perimeter and Circumference

LEARNING TOOLS

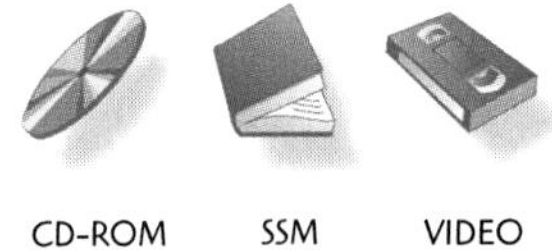

CD-ROM SSM VIDEO

Sneak Preview

In many problem situations, we may need to know the distance around a shape. This distance is called the perimeter.

After completing this section, you should be able to:

1) Define perimeter and circumference.
2) Calculate the perimeter of a square, rectangle, or triangle.
3) Define pi.
4) Calculate the circumference of a circle.

DISCUSSION

Perimeter

The distance around a shape or figure is called its **perimeter.** In a rectangle, the opposite sides are parallel and have the same length. We can call two sides the length and represent each with the variable, L. We can call the other two sides the width and represent each with the variable, W. The expression $L + L + W + W$ is equal to the distance around the rectangle, the perimeter P. Since $L + L = 2L$ and $W + W = 2W$, the formula becomes $P = 2L + 2W$.

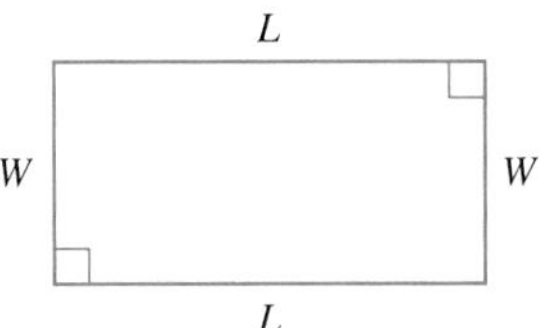

A square is a special kind of rectangle with four sides of equal length. If we call the length of any side s, the perimeter is equal to the sum of $s + s + s + s$. Since $s + s + s + s$ can be rewritten as $4s$, the formula becomes $P = 4s$.

Example 1 **Find the perimeter of a rectangle that has a length of 7 meters and a width of 5 meters.**

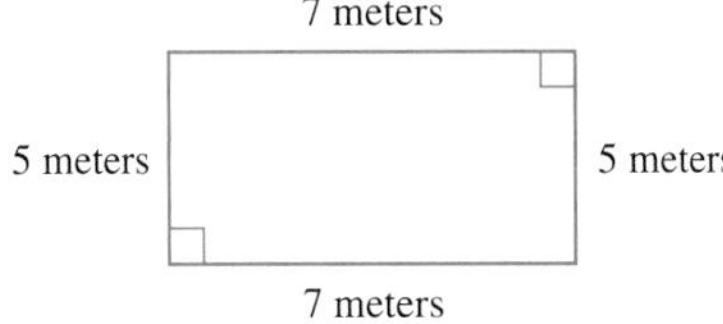

$$\begin{aligned} P &= L + L + W + W \\ &= 7 \text{ meters} + 7 \text{ meters} + 5 \text{ meters} + 5 \text{ meters} \\ &= 24 \text{ meters} \end{aligned}$$

The perimeter of a triangle is equal to the sum of the length of each side of the triangle. If these sides are called A, B, and C, the perimeter is equal to $A + B + C$. An **isosceles triangle** is a triangle in which at least two of the three sides are equal in length.

Example 2 **Find the perimeter of an isosceles triangle with these sides: $A = 3$ inches, $B = 3$ inches, and $C = 2$ inches.**

$$\begin{aligned} P &= A + B + C \\ &= 3 \text{ inches} + 3 \text{ inches} + 2 \text{ inches} \\ &= (3 + 3 + 2) \text{ inches} \\ &= 8 \text{ inches} \end{aligned}$$

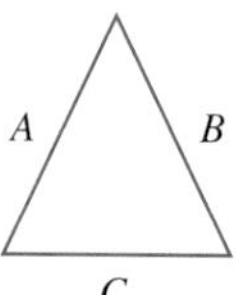

The perimeter of any figure can be found by determining the distance around the figure.

Example 3 **To estimate the amount of concrete to be used in footings for a house, a contractor needs to determine the perimeter of the foundation. Use the drawing to determine the perimeter of the foundation of this house plan.**

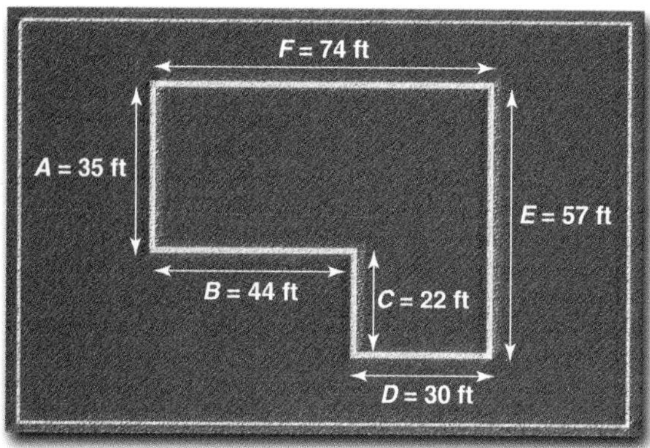

$$\begin{aligned} \text{Perimeter} &= \text{Sum of all sides of the figure} \\ P &= A + B + C + D + E + F \\ &= 35 \text{ ft} + 44 \text{ ft} + 22 \text{ ft} + 30 \text{ ft} + 57 \text{ ft} + 74 \text{ ft} \\ &= 262 \text{ ft} \end{aligned}$$

The perimeter of the foundation is 262 feet.

CHECK YOUR UNDERSTANDING

1. Find the perimeter of a square that is $\frac{3}{4}$ inches long.

2. Find the perimeter of a rectangle with height 6.2 meters and width 3.9 meters.

3. Find the perimeter of an equilateral triangle with a side that is 5 cm long.

Answers: 1. 3 inches; 2. 20.2 meters; 3. 15 cm.

DISCUSSION

Circles, Pi, and Circumference

The distance around a circle can be called its perimeter but usually is referred to as its **circumference.** Since ancient times people have known that the **circumference** of a circle divided by the **diameter** of the circle was about 3. No matter how large or small the circle, this number remains the same. It is called **pi.** The definition of pi is $\pi = \frac{C}{d}$ where C is the circumference and d is the diameter. Pi is a nonrepeating, nonterminating decimal. Calculating the value of pi directly by measuring the circumference and diameter is limited by the precision of the measuring device. To determine the value of pi to many decimal places, mathematicians have used other mathematical techniques of advanced algebra and calculus. Using these techniques and the calculating speed of supercomputers, pi is now known to 6 billion digits. However, for most purposes, an approximate value of 3.14 is sufficient. If pi is entered using a pi key on a calculator, typically the approximate value of 3.14159265359 is used. The number of decimal places after the point will depend on the calculator and the last digit shown may be rounded.

If we multiply both sides of the equation $\pi = \frac{C}{d}$ by d, we get a formula for determining the circumference of any circle: $C = \pi d$. Since the diameter of a circle is equivalent to twice the radius of the circle, this formula can also be rewritten $C = 2\pi r$.

PROPERTY

Circumference of a circle

In words: Circumference of a circle = pi · diameter

In symbols: $C = \pi d$

Example 4 **Find the circumference of a circle that has a diameter of 6 inches. Round to the nearest tenth of an inch.**

$$\begin{aligned} C &= \pi d \\ &= (3.14)(6 \text{ inches}) \\ &= 18.84 \text{ inches} \\ &\approx 18.8 \text{ inches} \end{aligned}$$

CHECK YOUR UNDERSTANDING

1. Find the circumference of a circle that has a diameter of 8.2 inches. Round the circumference to the nearest tenth.

2. Find the circumference of a circle that has a radius of 0.75 kilometer. Round to the nearest hundredth of a kilometer.

Answers: 1. 25.7 inches; 2. 4.71 kilometers.

EXERCISES SECTION A1.2

1. Define and give an approximate value for pi.

2. Calculate the circumference of a circle with a radius of 4 cm. Round to the nearest tenth of a centimeter.

3. Calculate the circumference of a circle with a diameter of 4 cm. Round to the nearest tenth of a centimeter.

4. A semicircle is one-half of a circle. What is the circumference of a semicircle with diameter of 3 feet? Round to the nearest foot.

5. The length of one side of a square is 15 millimeters. What is the perimeter of this square?

6. A rectangle has a length of 4.5 inches and a width of 3.2 inches. What is the perimeter of this rectangle?

7. An isosceles triangle has two equal sides that are 4 cm long and another side that is 1.6 cm long. What is the perimeter of this triangle?

8. Calculate the perimeter of the outside edge of this figure:

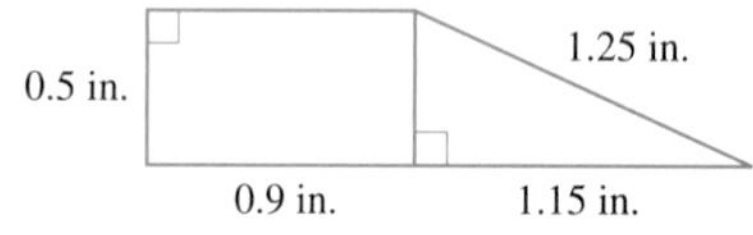

9. The side of an equilateral triangle is $1\frac{3}{4}$ inches. What is the perimeter of this triangle?

10. Calculate the perimeter of the outside edge of this figure. Each side of the square and the diameter of the circle measure 0.7 inch. Round to the nearest tenth of an inch.

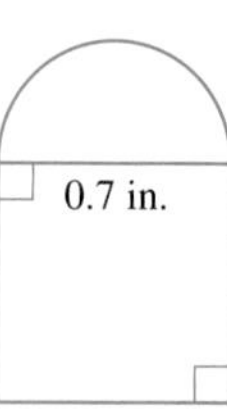

Additional Exercises:

Section 1.3 #28–31, 74–82; Section 1.4 #34, 36; Section 4.6 #1–13; Section 5.5 #20.

Section A1.3 Area

LEARNING TOOLS

CD-ROM SSM VIDEO

Sneak Preview

We often use **area** to describe the characteristics of a surface. We might say that a parking lot has an area of 100 square meters. We can use formulas to calculate the area of squares, rectangles, triangles, trapezoids, and circles.

After completing this section, you should be able to:

1) Calculate the area of a square and a rectangle.
2) Calculate the area of a triangle and a trapezoid.
3) Calculate the area of a circle.

DISCUSSION

Area of Rectangles and Squares

Area is a measure of the amount of space on a two-dimensional, flat surface and is measured in square units, written *unit*^{2}.

This square has an area of 1 cm^2 (read "one square centimeter"). It is an example of a **unit square:** a square whose sides each have a measure of 1 unit.

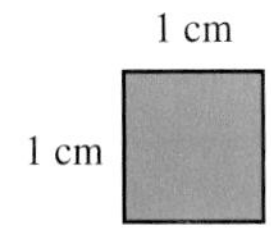

This rectangle is 3 cm long and 2 cm wide. It is made up of 6 squares that are each 1 cm long and 1 cm wide and have an area of 1 cm^2. Its area is 6 cm^2.

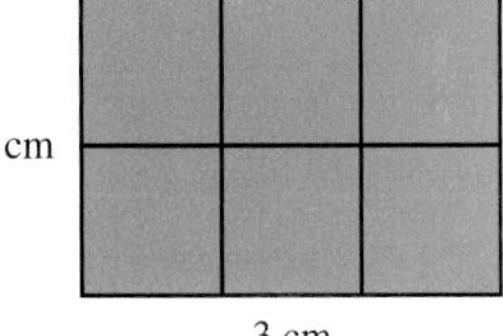

The area of the rectangle can be calculated by multiplying the length by the width of the rectangle.

PROPERTY

Area of a rectangle

In words: Area of a rectangle = Length · Width

In symbols: $A = lw$

Example 1 **Calculate the area of a rectangle with length 8 meters and width 3 meters.**

Area = Length · Width
= lw
= 8 meters · 3 meters
= 24 meters^2 or 24 m^2

The area of the rectangle is 24 square meters.

Since a square is just a special rectangle, we can use the same formula to calculate the area of a square. Since both the length and the width are the same, the formula for calculating area of a rectangle can be rewritten as $A = \text{side} \cdot \text{side}$ or $A = s^2$.

PROPERTY

Area of a square

In words: Area of a square = Side · Side

In symbols: $A = s^2$

Example 2 **Calculate the area of a square with length 3.1 inches. Round to the nearest tenth of a square inch.**

$$\begin{aligned} \text{Area} &= \text{Side} \cdot \text{Side} \\ A &= s^2 \\ &= (3.1 \text{ inches})^2 \\ &= 9.61 \text{ inches}^2 \\ &\approx 9.6 \text{ inches}^2 \end{aligned}$$

The area of the square is 9.6 square inches.

In order to calculate a useful area of any polygon, the length and width must be in the same units of measurement.

Example 3 **Calculate the area of a rectangle with length 8 inches and length 2 feet.**

$$\text{Width} = 8 \text{ inches} \qquad \begin{aligned} \text{Length} &= 2 \text{ feet} \\ &= 2(12 \text{ inches}) \\ &= 24 \text{ inches} \end{aligned}$$

$$\begin{aligned} \text{Area} &= \text{Length} \cdot \text{Width} \\ &= 24 \text{ inches} \cdot 8 \text{ inches} \\ &= 192 \text{ inches}^2 \end{aligned}$$

The rectangle has an area of 192 square inches.

CHECK YOUR UNDERSTANDING

1. Calculate the area of a rectangle with a length of 5 miles and a width of 3 miles.

2. Calculate the area of a rectangle with a length of 6 inches and a width of 3 feet.

3. Calculate the area of a square with a length of 4.9 meters. Round to the nearest square meter.

Answers: 1. 15 miles^2; 2. 1.5 feet^2 or 216 inches^2; 3. 24 meters^2.

DISCUSSION

Area of Triangles and Trapezoids

The area of a rectangle is equal to its length multiplied by its width, $A = lw$. If we draw a diagonal from opposite corners of a rectangle, we divide the rectangle into two right triangles. The area of each triangle is equal to one-half of the area of the rectangle, $A = \frac{1}{2}lw$. If we change the variables to b for the base of the triangle and h for the height of the triangle, the formula for area of this right triangle becomes $A = \frac{1}{2}bh$.

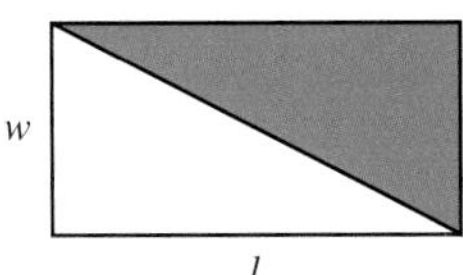

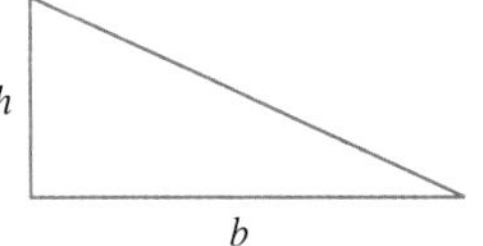

We can extend this formula to find the area of all triangles, not just right triangles. The base is one side of the triangle. The height is the distance of a perpendicular segment whose endpoints are the opposite vertex and a point on the base.

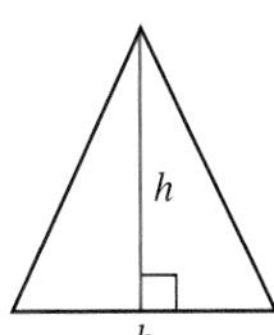

PROPERTY

Area of a triangle

In words: Area of a triangle $= \dfrac{1}{2} \cdot$ Base $\cdot$ Height

In symbols: $A = \dfrac{1}{2}bh$

Example 4 **Calculate the area of a triangle whose height is 4 centimeters and whose base is 8 centimeters.**

$$
\begin{aligned}
A &= \frac{1}{2}bh \\
&= \frac{1}{2}(4\text{ cm})(8\text{ cm}) \\
&= \frac{32}{2}\text{cm}^2 \\
&= 16\text{ cm}^2
\end{aligned}
$$

The area of the triangle is 16 square centimeters.

A trapezoid is a quadrilateral with exactly two parallel sides called bases. The height of the trapezoid is the shortest distance between the two parallel sides. The area of a trapezoid can be calculated using the formula $A = \frac{1}{2}h(b_1 + b_2)$ where A is the area, h is the height, b_1 is the length of one of the bases, and b_2 is the length of the other base.

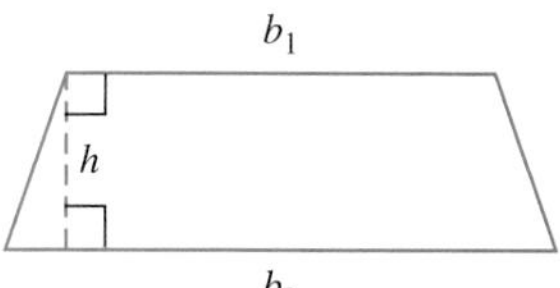

PROPERTY

Area of a trapezoid

In words: Area $= \frac{1}{2} \cdot$ Height $\cdot$ Sum of the bases

In symbols: $A = \frac{1}{2}h(b_1 + b_2)$

Example 5 **Calculate the area of a trapezoid with a height of 2 inches, a base of 8 inches, and another base that is 5 inches.**

$$
\begin{aligned}
A &= \frac{1}{2}h(b_1 + b_2) \\
&= \frac{1}{2}(2\text{ in.})(8\text{ in.} + 5\text{ in.}) \\
&= \frac{1}{2}(2\text{ in.})(13\text{ in.}) \\
&= \frac{26}{2}\text{in.}^2 \\
&= 13\text{ in.}^2
\end{aligned}
$$

The area of the trapezoid is 13 square inches.

CHECK YOUR UNDERSTANDING

1. Calculate the area of a triangle with a base of 20 inches and a height of 14 inches.

2. Calculate the area of this triangle:

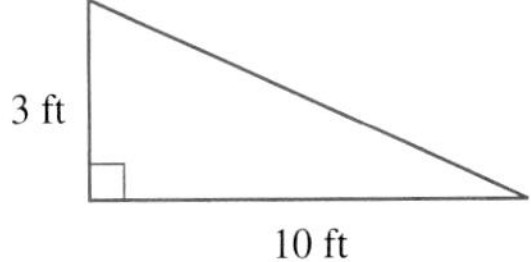

3. Calculate the area of a trapezoid that has a height of 6 cm, one base that is 10 cm, and another base that is 12 cm.

Answers: 1. 140 in^2; 2. 15 ft^2; 3. 66 cm^2.

DISCUSSION

Area of a Circle

Pi is part of the formula for calculating the area of a circle: $A = \pi r^2$ where A is the area of the circle and r is a radius of the circle. The units for area are always distance^2 since the units of the radius are distance units.

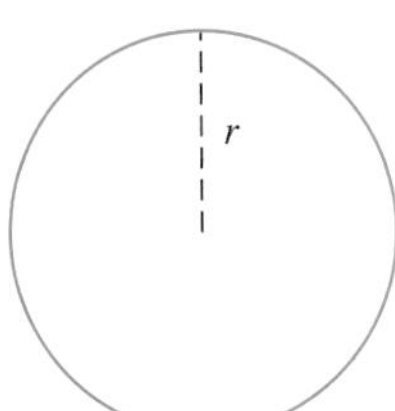

PROPERTY

Area of a circle

In words: Area = pi · (radius)^2

In symbols: $A = \pi r^2$

Example 6 **Determine the area of a circle with a diameter of 6 inches. Round to the nearest tenth of a square inch.**

The radius is equal to the diameter divided by 2. So, the radius of this circle is 6 inches ÷ 2 = 3 inches.

$$\begin{aligned} A &= \pi r^2 \\ &\approx (3.14)(3 \text{ inches})^2 \\ &= (3.14)(9 \text{ inches}^2) \\ &= 28.26 \text{ inches}^2 \\ &\approx 28.3 \text{ inches}^2 \end{aligned}$$

The area of the circle is 28.3 square inches.

CHECK YOUR UNDERSTANDING

1. Calculate the area of a circle that has a radius of 4.2 cm. Round to the nearest square centimeter.

2. A circle has a diameter of 8 meters. Calculate the area of the circle, rounding to the nearest tenth of a square meter.

Answers: 1. 55 cm^2; 2. 50.2 m^2.

EXERCISES SECTION A1.3

1. Calculate the area of a triangle with a base of 6 inches and a height of 8 inches.

2. Calculate the area of a square with a side of 3.4 inches.

3. Calculate the area of a rectangle with a length of 7.1 cm and a height of 6.2 cm.

4. Calculate the area of a trapezoid with a height of 8.1 cm, a base of 15 cm, and another base of 13 cm.

5. Calculate the area of a circle with a radius of 5 cm. Round to the nearest tenth of a square centimeter.

6. Calculate the area of a circle with a diameter of $\frac{3}{4}$ inch. Round to the nearest tenth of a square inch.

7. Calculate the area of a circle whose circumference is 10.5 meters. Round to the nearest tenth of a square meter.

8. One side of a right triangle measures 6 inches; a second side measures 8 inches; the third side measures 10 inches. Calculate the area of this triangle.

9. A rectangle has a height of 4 cm and a length of 6 cm. A trapezoid fits inside this rectangle with two of its vertices matching the vertices of the rectangle. One base of the trapezoid is 6 cm. The other base of the trapezoid is 3 cm. Calculate the difference in area between the rectangle and the trapezoid.

10. A rectangle has a length of 16 inches and a width of 8 inches. It is divided into four equal smaller rectangles. What is the area of one of these smaller rectangles?

Additional Exercises:

Section 1.7 #15–27, 40–44; Section 3.3 #50–54; Section 3.6 #57, 58, 59; Section 3.7 #38, 41, 42; Section 4.6 #16–26, 49; Section 5.5 #9, 19.

Section A1.4 Volume

LEARNING TOOLS

CD-ROM SSM VIDEO

Sneak Preview

Volume is often used to describe the amount of space occupied by a three-dimensional shape. We might describe the volume of a packing box as 0.3 cubic meter. We can use formulas to calculate the volume of rectangular solids, spheres, circular cylinders, and pyramids.

After completing this section, you should be able to:

1) Calculate the volume of a rectangular solid (cuboid).
2) Calculate the volume of a sphere and a circular cylinder.
3) Calculate the volume of a pyramid, including a cone.

DISCUSSION

Volume of a Rectangular Solid

Volume is a measure of the amount of space inside a three-dimensional object and is measured in cubic units, written $unit^3$.

This cube has a volume of 1 cm³ (read "one cubic centimeter"). It is an example of a **unit cube:** a cube whose sides each have a measure of 1 unit.

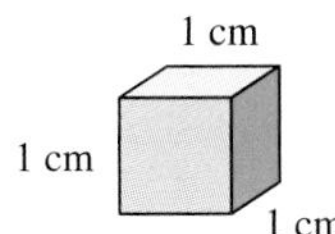

If we place six cubes with length 1 centimeter next to each other in two rows of three, we form a box. The box is 3 centimeters long, 2 centimeters wide, and 1 centimeter high. Such a box is an example of a **cuboid** or **rectangular solid.** (A cube is a cuboid with equal length, width, and height.)

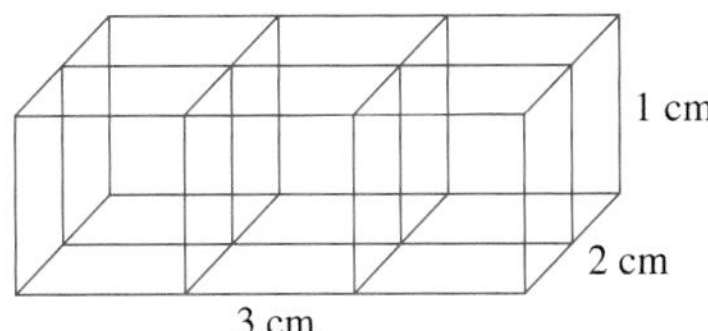

To find the volume of this cuboid, we can add the volume of all the unit cubes.

$$\begin{aligned}\text{Volume of cuboid} &= \text{Sum of volume of unit cubes}\\ &= 1\text{ cm}^3 + 1\text{ cm}^3 + 1\text{ cm}^3 + 1\text{ cm}^3 + 1\text{ cm}^3 + 1\text{ cm}^3\\ &= 6\text{ cm}^3 \quad \textit{Combine like terms.}\end{aligned}$$

The volume of the cuboid is also equal to the product of the length, width, and height of the cuboid.

$$\begin{aligned}\text{Volume of cuboid} &= \text{Length of cuboid} \cdot \text{Width of cuboid} \cdot \text{Height of cuboid}\\ &= 3\text{ cm} \cdot 2\text{ cm} \cdot 1\text{ cm}\\ &= (3 \cdot 2 \cdot 1)(\text{cm} \cdot \text{cm} \cdot \text{cm})\\ &= 6\text{ cm}^3 \quad \textit{This volume is equal to the sum of the volume of the individual cubes.}\end{aligned}$$

PROPERTY

Volume of a cuboid

In words: Volume = Length · Width · Height

In symbols: $V = lwh$

In the SI system, volume is measured in cubic units such as cm^3 and m^3. It also is measured in units based on liter. For example, 1,000 liters is equivalent to 1 cubic meter; a milliliter is equivalent to 1 cubic centimeter (cm^3). Another abbreviation used for cubic centimeter is *cc*. The auto industry measures engine volume in either *cc* (a 350-cc motorcycle) or in liters (a 4-liter engine). Health care professionals may refer to liquid medicine volume in cubic centimeters using *cc* as the abbreviation. For example, an order for liquid Tylenol for a child might read "5 *cc* every 4 hours." This is equivalent to "5 ml every 4 hours."

In the English system, liquid volume is measured in gallons, quarts, pints, cups, tablespoons, and teaspoons. Solid volume is measured in cubic units like cubic feet or cubic inches.

PROPERTY

Equivalent measures

$$1{,}000 \text{ liters} = 1 \text{ meter}^3$$
$$1 \text{ cm}^3 = 1 \text{ ml}$$
$$1 \text{ cm}^3 = 1 \text{ cc}$$
$$5 \text{ cm}^3 = 1 \text{ teaspoon}$$

Example 1 **Find the volume of a cube with a side of 3 inches.**

$$\begin{aligned}\text{Volume} &= \text{Length} \cdot \text{Width} \cdot \text{Height}\\ &= 3 \text{ in.} \cdot 3 \text{ in.} \cdot 3 \text{ in.}\\ &= 27 \text{ in.}^3\end{aligned}$$

The volume of the cube is 27 cubic inches.

Example 2 **Find the volume of a cuboid that is 2 feet by 18 inches by 14 inches.**

To calculate volume, all measurements must be in the same units. Convert 2 feet into inches using the conversion factor 1 foot = 12 inches.

$$\begin{aligned}&2 \text{ feet}\\ &= 2(12 \text{ inches})\\ &= 24 \text{ inches}\end{aligned}$$

$$\begin{aligned}\text{Volume} &= \text{Length} \cdot \text{Width} \cdot \text{Height}\\ &= 24 \text{ in.} \cdot 18 \text{ in.} \cdot 14 \text{ in.}\\ &= 6{,}048 \text{ in.}^3\end{aligned}$$

The volume of the cuboid is 6,048 cubic inches.

CHECK YOUR UNDERSTANDING

1. Find the volume of a cuboid with length 2 inches, width 3 inches, and height 4 inches.

2. Find the volume of a cube with side length of 6 millimeters.

3. Find the volume of a rectangular briefcase that measures 24 inches by 12 inches by 5 inches.

Answers: 1. 24 in^3; 2. 216 mm^3; 3. 1,440 in^3.

DISCUSSION

Volume of Spheres and Circular Cylinders

A **sphere** is formed from many circles of different diameter. The radius of a sphere is the distance from the center to any point on the surface of the sphere. The volume of a sphere can be calculated using the formula $V = \frac{4}{3}\pi r^3$ where r is the radius of the sphere.

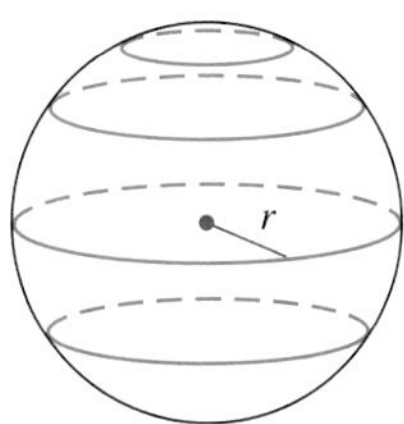

PROPERTY

Volume of a sphere

In words: $\text{Volume} = \frac{4}{3} \cdot \text{pi} \cdot (\text{radius})^3$

In symbols: $V = \frac{4}{3}\pi r^3$

Example 3 **The diameter of a spherical candle is 8.2 inches. Calculate the volume of wax used to make the candle. Round to the nearest tenth of a cubic inch.**

The radius is equal to the diameter divided by 2. So, the radius of this sphere is 8.2 inches ÷ 2 = 4.1 inches.

$$\begin{aligned} V &= \frac{4}{3}\pi r^3 \\ &\approx \frac{4}{3}(3.14)(4.1 \text{ in.})^3 \\ &= \frac{4}{3}(3.14)(68.921 \text{ in.}^3) \\ &= \frac{865.64776}{3}\text{in.}^3 \\ &\approx 288.5 \text{ in.}^3 \end{aligned}$$

The volume of wax used to make the candle is about 288.5 cubic inches.

A cylinder is a three-dimensional object that has identical cross sections and straight sides. For example, a cuboid is a cylinder with a rectangular base. The volume of any cylinder can be found by multiplying the area of its base by its height: $V = Ah$. If each cross section of a cylinder is a circle, the area (A) of the base is $A = \pi r^2$ where r is the radius of the circle. The formula for the volume (V) of a circular cylinder with a height of h then becomes $V = \pi r^2 h$.

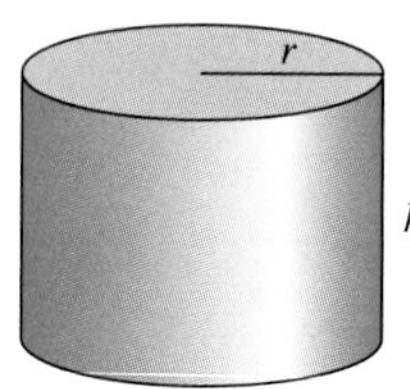

PROPERTY

Volume of a circular cylinder

In words: Volume = pi · radius² · height

In symbols: $V = \pi r^2 h$

Example 4 **A cylinder has a diameter of 10 meters and a height of 2 meters. What is the volume of this cylinder? Round to the nearest cubic meter.**

The radius is equal to the diameter divided by 2. So, the radius of this cylinder is 10 meters ÷ 2 = 5 meters.

$$\begin{aligned} V &= \pi r^2 h \\ &\approx (3.14)(5 \text{ meters})^2(2 \text{ meters}) \\ &= 157 \text{ meters}^3 \end{aligned}$$

The volume of this cylinder is about 157 cubic meters.

CHECK YOUR UNDERSTANDING

1. Calculate the volume of a ball that has a radius of 12 cm. Round to the nearest tenth of a cubic centimeter.

2. Calculate the volume of a ball that has a diameter of 12 cm. Round to the nearest cubic centimeter.

3. Calculate the volume of a cylinder that has a diameter of 2 inches and a height of 6 inches. Round to the nearest cubic inch.

Answers: 1. 7,234.6 cm^3; 2. 904 cm^3; 3. 19 in.^{3}

DISCUSSION

Volume of a Pyramid or Cone

A pyramid is a three-dimensional shape in which every point on the perimeter of the base is joined to the same point in a different plane. The volume of the pyramid is equal to $\frac{1}{3}Ah$ where A is the area of the base and h is the height of the pyramid. (The height is the distance of a line that passes through the vertex and is perpendicular to the base.)

PROPERTY

Volume of a pyramid

In words: $\text{Volume} = \frac{1}{3}(\text{Area of base}) \cdot \text{height}$

In symbols: $V = \frac{1}{3}Ah$

If a pyramid has a square base, the area of the base is equal to the product of the two sides.

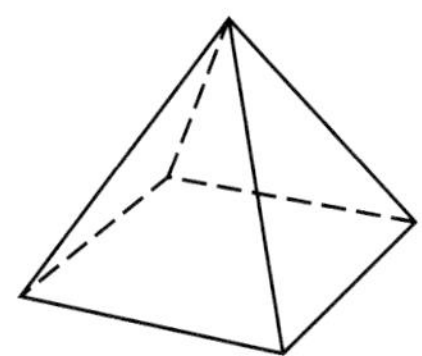

PROPERTY

Volume of a square pyramid

In words: Volume $= \frac{1}{3}$(side)^2(height)

In symbols: $V = \frac{1}{3}s^2h$

Example 5 **The side of the base of a square pyramid is 8 inches. The height is 12 inches. Find its volume.**

$$V = \frac{1}{3}s^2h$$
$$= \frac{1}{3}(8 \text{ in.})^2(12 \text{ in.})$$
$$= \frac{768 \text{ in.}^3}{3}$$
$$= 256 \text{ in.}^3$$

The volume of the pyramid is 256 cubic inches.

If a pyramid has a circular base, it is often called a cone. The area of the base is equal to πr^2.

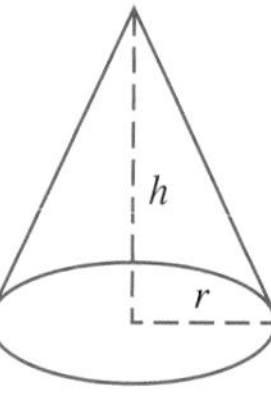

PROPERTY

Volume of a cone

In words: Volume $= \frac{1}{3}$pi $\cdot$ (radius)^2(height)

In symbols: $V = \frac{1}{3}\pi r^2h$

Example 6 **Find the volume of a cone with a diameter of 5 inches and a height of 12 inches. Round to the nearest tenth of a cubic inch.**

The radius is equal to the diameter divided by 2. So, the radius of this cone is 5 meters ÷ 2 = 2.5 meters.

$$V = \frac{1}{3}\pi r^2h$$
$$\approx \frac{1}{3}(3.14)(2.5 \text{ in.})^2(12 \text{ in.})$$
$$= \frac{235.5 \text{ in.}^3}{3}$$
$$= 78.5 \text{ in.}^3$$

The volume of the cone is about 78.5 cubic inches.

CHECK YOUR UNDERSTANDING

1. Find the volume of a square pyramid with a side length of 3 feet and a height of 4 feet. Round to the nearest cubic foot.

2. Find the volume of a circular pyramid with a radius of 6 meters and a height of 5 meters. Round to the nearest tenth of a cubic meter.

3. Find the volume of a cone with a diameter of 7 cm and a height of 10 cm. Round to the nearest cubic centimeter.

Answers: 1. 12 ft^3; 2. 188.4 m^3; 3. 128 cm^3.

EXERCISES SECTION A1.4

Use $\pi \approx 3.14$.

1. Calculate the volume of a rectangular solid that is 3 inches long, 2 inches wide, and 1 inch high.

2. Calculate the volume of a cube with a side of 8.25 inches. Round to the nearest tenth of a cubic inch.

3. Calculate the volume of a sphere with a radius of 5 cm. Round to the nearest cubic centimeter.

4. Calculate the volume of a sphere with a diameter of 5 cm. Round to the nearest cubic centimeter.

5. Calculate the volume of a circular cylinder with a radius of 8.2 inches and a height of 12 inches. Round to the nearest tenth of a cubic inch.

6. Calculate the volume of a circular cylinder with a height of 2 feet and a diameter of 14 inches. Round to the nearest tenth of a cubic inch.

7. The base of a square pyramid has a side of length 12 meters. It has a height of 10 meters. Calculate the volume of the pyramid, rounding to the nearest cubic meter.

8. A pyramid has a base that is an equilateral triangle. Each side of the base measures 4 inches. The height of the base triangle is 4.5 inches. The height of the pyramid is 6.1 inches. Calculate the volume of the pyramid, rounding to the nearest tenth of a cubic inch.

9. A cone has a radius of 8 cm and a height of 12 cm. Calculate the volume of the cone to the nearest cubic centimeter.

10. A cone has a diameter of 4 cm and a height of 12 cm. Calculate the volume of the cone to the nearest cubic centimeter.

Additional Exercises:

Section 1.7 #28–39, 45; Section 3.3 #46, 55, 56; Section 3.7 #40; Section 4.6 #28–48, 50.

Section A1.5 Similar Polygons

LEARNING TOOLS

Sneak Preview

Two polygons that are the same shape but are different sizes are said to be **similar.** If we know the length of the sides of a polygon, we can determine the lengths of the sides of a similar polygon. This can be very helpful in solving application problems.

After completing this section, you should be able to:

1) Determine if two polygons are similar.

2) Find an unknown length in two similar polygons.

DISCUSSION

Similar Polygons

The corresponding angles of **similar polygons** are the same measure and the lengths of corresponding sides are proportional. If two polygons are similar, we should be able to imagine enlarging one of the polygons until it exactly matches the shape of the other polygon. In enlarging the polygon, we keep the measurement of the angles the same. Only the lengths of the sides change.

For example, these two squares are similar. Each of the angles in both squares is 90°. The lengths of the sides in the larger square are all two times longer than the lengths of the sides in the smaller square. We can imagine enlarging any square until it exactly matches the shape of any larger square. All squares are similar to each other.

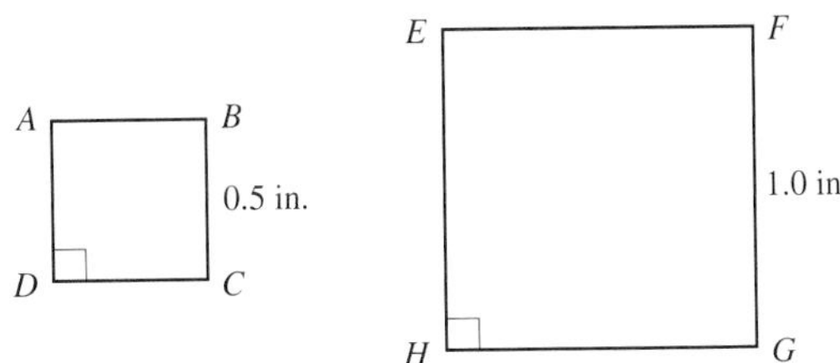

However, all triangles are not similar. If two triangles are similar, we should be able to imagine enlarging one triangle until it exactly matches the shape of the other triangle. For example, triangles *JKL* and *MNO* are not similar.

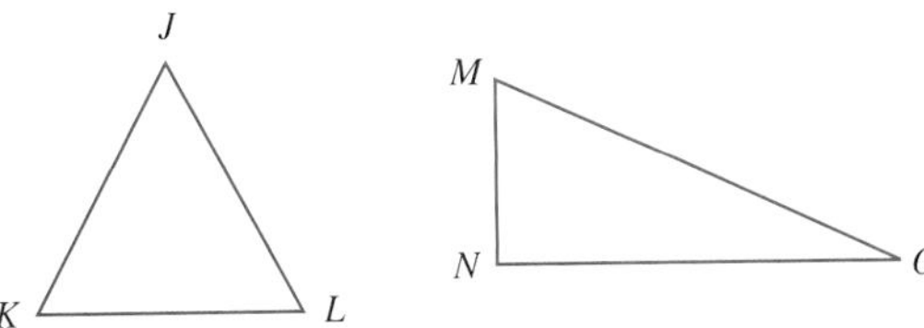

However, triangle *ABC* and triangle *DEF* are similar. Although triangle *ABC* is smaller than *DEF,* we can imagine enlarging it without changing its shape until it matches *DEF.*

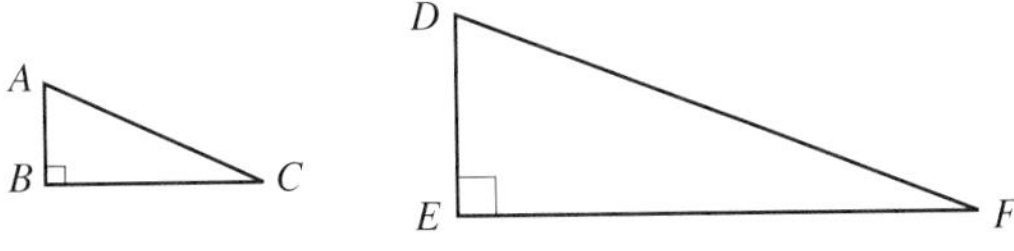

Angles in two triangles that have the same relative position are called **corresponding angles.** In the triangles *ABC* and *DEF,* there are three pairs of corresponding angles: $\angle A$ and $\angle D$, $\angle B$ and $\angle E$, and $\angle C$ and $\angle F$. If triangles are similar, the measures of the corresponding angles are equal.

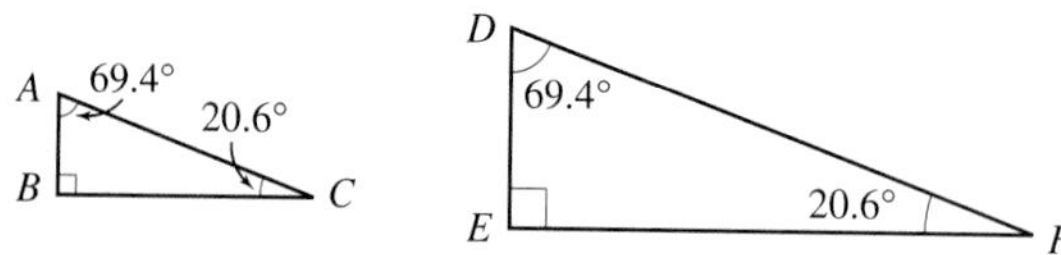

The **corresponding sides** of similar triangles are proportional. In other words, the ratio of each pair of corresponding sides is equal.

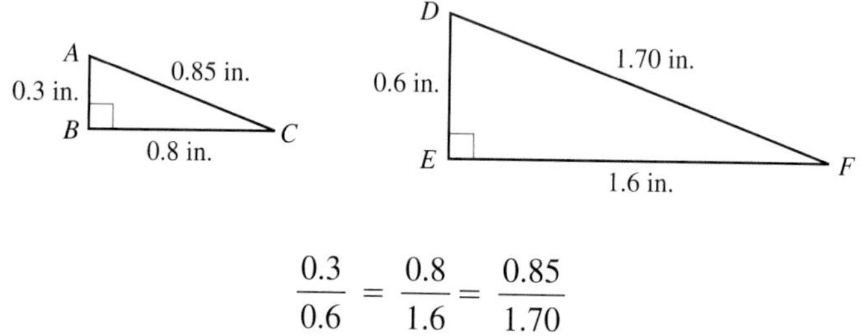

$$\frac{0.3}{0.6} = \frac{0.8}{1.6} = \frac{0.85}{1.70}$$

Each of these ratios simplifies to $\frac{1}{2}$. So, the corresponding sides of these similar triangles are proportional.

CHECK YOUR UNDERSTANDING

Answer as true or false.

1. If two triangles are similar, all the angles in the triangles have the same measurement.

2. If two triangles are similar, the corresponding angles in the triangles have the same measurement.

3. If two triangles are similar, the lengths of corresponding sides are equal.

4. If two triangles are similar, the lengths of corresponding sides of the triangles are all equal to the same ratio.

Answers: 1. false; 2. true; 3. false; 4. true.

DISCUSSION

Proportions and Similar Triangles

If two triangles are similar, their corresponding sides are proportional. This allows us to calculate the unknown length of one triangle using a proportion.

Example 1 **Determine the unknown length in the two similar triangles.**

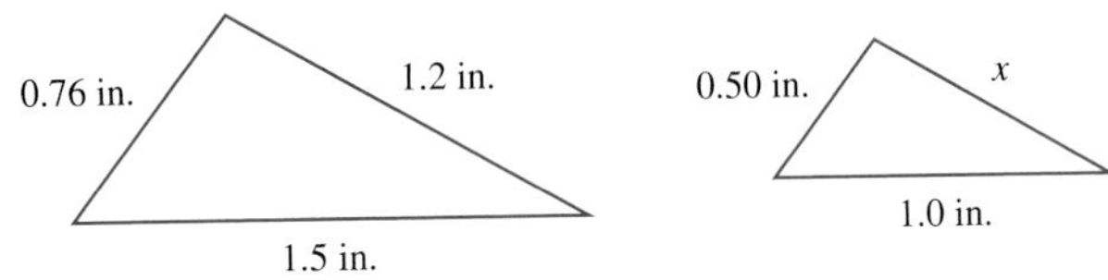

The side that corresponds to the unknown length is 1.2 inches. There are two possible proportions we can use: $\frac{x}{1.2} = \frac{0.5}{0.76}$ or $\frac{x}{1.2} = \frac{1.0}{1.5}$. It does not matter which proportion we select.

We can solve $\frac{x}{1.2} = \frac{1.0}{1.5}$ using the multiplication property of equality.

$$\frac{x}{1.2} = \frac{1.0}{1.5}$$

$$\mathbf{1.2} \cdot \frac{x}{1.2} = \frac{1.0}{1.5} \cdot \mathbf{1.2} \quad \textit{Multiply both sides of the equation by 1.2.}$$

$$1x = \frac{1.2}{1.5}$$

$$x = 0.8$$

The unknown length is 0.8 inch.

We can use proportions to calculate unknown lengths for any pair of similar figures. This is especially useful when considering scale models or photographs. For example, a house plan is similar to the actual house it represents. If we enlarged the house plan enough, it would exactly match the actual house.

Example 2 **Many grocery stores and bookstores sell house plan magazines. In this house plan, the width of the house in the drawing is 3.25 inches, and the length of the house is 4.5 inches. The actual house is 28 feet wide and 38 feet long. Determine the length of the master bedroom in the actual house. Round to the nearest tenth.**

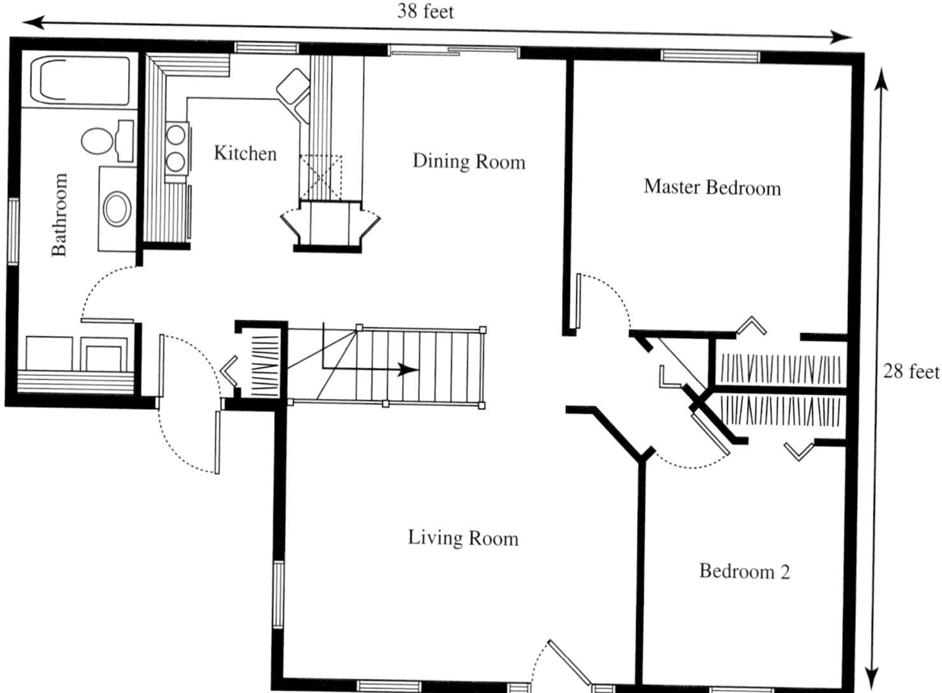

Measuring with a ruler, the length of the master bedroom on the plan is 1.5 inches. To find the length of the actual bedroom, we can use either of these proportions:

$$\frac{38\text{ feet}}{4.5\text{ inches}} = \frac{L}{1.5\text{ inches}} \quad \text{or} \quad \frac{28\text{ feet}}{3.25\text{ inches}} = \frac{L}{1.5\text{ inches}}$$

If we choose the proportion to the left, we solve using the multiplication property of equality.

$$\frac{38\text{ feet}}{4.5\text{ inches}} = \frac{L}{1.5\text{ inches}}$$

$$\mathbf{1.5\text{ inches}} \cdot \frac{38\text{ feet}}{4.5\text{ inches}} = \frac{L}{1.5\text{ inches}} \cdot \mathbf{1.5\text{ inches}}$$

$$\frac{38\text{ feet}}{3} = 1L$$

$$12.7\text{ feet} = L$$

The length of the master bedroom in the house is about 12.7 feet.

CHECK YOUR UNDERSTANDING

1. These triangles are similar. Find the missing length. Round to the nearest hundredth of an inch.

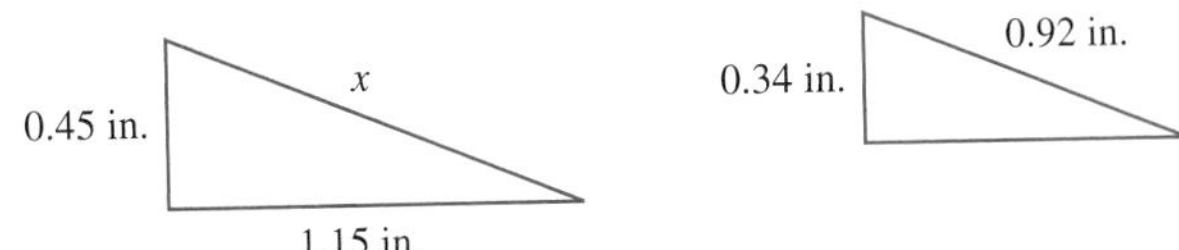

2. Determine the length and width of the bathroom in the house in Example 2.

Answers: 1. 1.22 in.; 2. about 15.8 feet long and 6.3 feet wide.

EXERCISES SECTION A1.5

1. The two triangles are similar. Determine the unknown length.

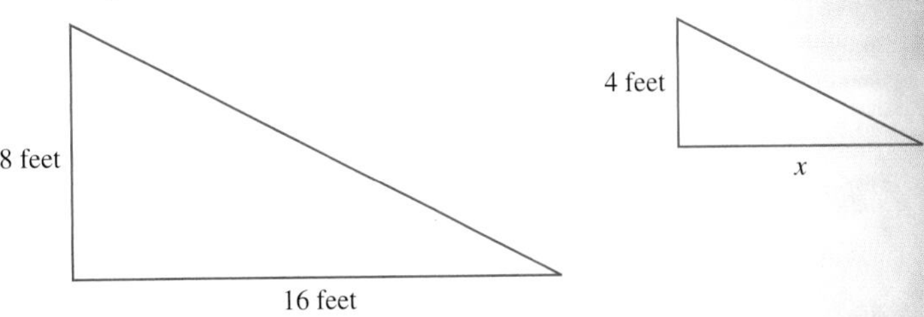

2. The two triangles are similar. Determine the unknown length.

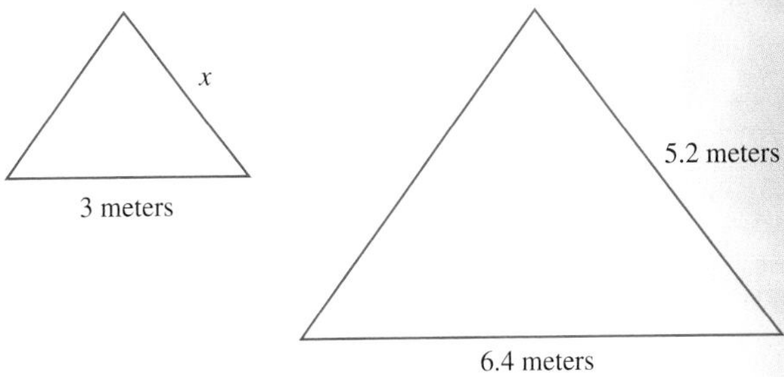

3. The two rectangles are similar. Determine the unknown length.

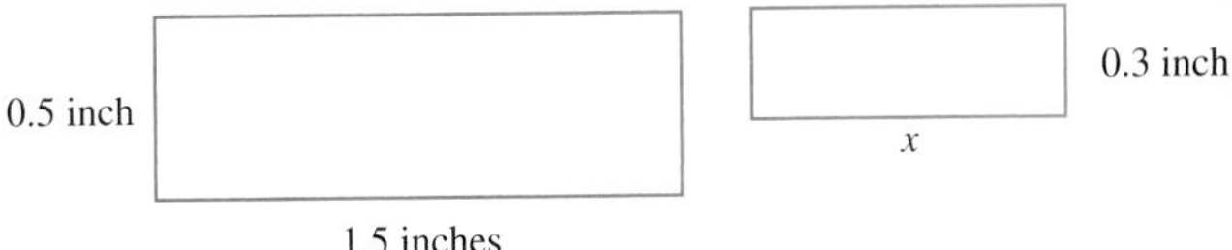

4. A quilt pattern has hexagon (a polygon with six equal sides) pieces. Each side of the hexagon is 4 centimeters long. Carolyn is changing the pattern and making the hexagons 35% larger.

a. How long will each side of the new hexagon be?

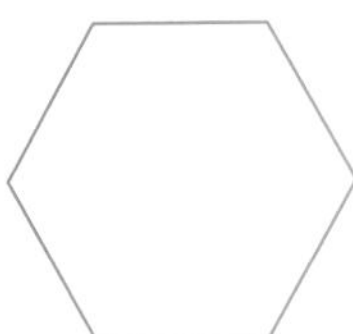

b. Are the original hexagons and the new hexagons similar? Explain.

5. Photographs are often printed as 4-inch by 6-inch prints, and are sometimes enlarged to 8-inch by 10-inch prints. Are these two sizes of prints similar rectangles?

6. Standard letter paper in the United States is 8.5 inches by 11 inches. In the United Kingdom standard letter size is A4, which is 210 mm by 294 mm. Are these two letter types similar rectangles? Explain.

7. A house plan shows a rectangular living room with dimensions of 4.5 inches by 5.75 inches. If the width of the actual living room is 18 feet, find the length of the actual living room.

8. The aspect ratio on a laptop computer screen is 800:600. The aspect ratio on a 17-inch flat screen monitor is 1,280:960. Do these two aspect ratios represent screens that are similar rectangles? Explain.

9. In a landscape artist's rendering, a rectangular garden plot measures 3.5 inches by 1.5 inches. If the width of the plot is actually 14 feet, what is the length of the plot? Round to the nearest tenth.

10. A surveyor is drawing a to-scale plan of a vacant lot. The sides of the lot have lengths of 84 feet, 68 feet, 48 feet, and 72 feet. He starts by drawing the longest side as an 8-inch line. What length of line should he use for the other three sides? Round to the nearest hundredth.

Appendix 2

Using Technology

In the United States, the basic skills that most people believe should be mastered in school include reading, writing, speaking, and arithmetic. Arithmetic is a broad subject but often is described as addition, subtraction, multiplication, and division of whole numbers, integers, fractions, and decimals. Even though inexpensive calculators are available to most Americans, basic skills are still essential knowledge. A calculator is not always available or convenient to use. We often need to check calculations done by others with a quick mental estimate.

However, in many situations, a calculator is very helpful. It allows us to concentrate on the process of solving a problem rather than on the arithmetic involved in finding the solution. When doing repeated calculations with large numbers, our work is often more accurate if we use a calculator.

In this appendix, we will learn some of the basic key strokes to use on a scientific calculator by completing example problems from *Investigating Basic Mathematics.*

APPENDIX OUTLINE

Section A2.1 Adding, Subtracting, Multiplying, and Dividing Whole Numbers with a Calculator

LEARNING TOOLS

CD-ROM

SSM

VIDEO [3.3]

Sneak Preview

All calculators can add, subtract, multiply, and divide whole numbers.

After completing this section, you should be able to:

1) Turn a calculator on and off.

2) Clear an entry in a calculator.

3) Complete addition, subtraction, multiplication, and division of whole numbers on a calculator.

4) Evaluate an exponential expression on a calculator.

5) Evaluate a numerical expression that includes parentheses on a calculator.

DISCUSSION

Introduction to a Scientific or Statistics Calculator

A scientific or statistics calculator allows you to perform many mathematical computations. There are many keys and symbols, some familiar and some you probably have never seen before and may never even use. The calculator has keys with commands or numbers printed directly on them. If you press a key, the calculator performs that command or enters that number. There are also commands printed above the keys directly on the face of the calculator. These commands are usually in a different color, often yellow. You tell the calculator to perform the yellow commands by first pressing the [2nd] button and then pressing the key directly below the command. The [2nd] button is usually the same color as the commands.

To turn on the calculator, press the [ON] key. It is often in the uppermost left corner or in the lowermost left corner of the calculator. Some calculators have the word [ON] printed on the key; other calculators print the word ON directly above the key on the face of the calculator. Some have a [C] printed on the key. Although many calculators have a separate [OFF] key, some solar powered calculators do not have an [OFF] key. They simply shut down after a period of nonuse.

The keys used for adding [+], subtracting [−], multiplying [×], and dividing [÷] are usually found in a vertical column on the right-hand side of a calculator. The equals sign [=] is found at the bottom of the column.

When completing an addition problem, the entered numbers appear in the **entry line** of the calculator. Some calculators have a screen that is large enough to display many entry lines; others can display only one entry at a time.

Example 1 **Solve g = \$1,774 + \$256 + \$38 + \$0 + \$281.**
(Section 1.4, Example 4)

We enter the numbers and operations in order from left to right.

Press [1][7][7][4] *The number 1774 appears in the entry line. No comma appears.*
Press [+] *The entry line remains 1774.*
Press [2][5][6] *The entry line is now 256.*
Press [+] *The entry line now shows the sum of the first two numbers, 2030.*
Press [3][8] *The entry line is now 38.*
Press [+] *The entry line now shows the sum of the first three numbers, 2068.*
Press [0] *The entry line is now 0. We could omit this step because the sum of any number and 0 is always the number.*
Press [+] *The entry line now shows the sum of the first four numbers, 2068.*
Press [2][8][1] *The entry line is now 281.*
Press [=] *The entry line now shows the sum of all the numbers, 2349.*

The sum of \$1,774 + \$256 + \$38 + \$0 + \$281 is \$2,349. The calculator does not include units like dollars or insert commas in the appropriate place. We need to remember to include these details in the final answer.

When we are ready to start another problem, it is best to clear the entry line. Often, two keys are available for clearing. The [CLEAR] or [AC] button clears the entry line and any entered operations. The [CE/C] key clears only the entry line. If we have pressed [+] and then a number, the [CE/C] key clears only the number. The [+] command remains.

Example 2 **Solve $a = 180° - 70° - 70°$.**
(Section 1.4, Example 5)

We enter the numbers and operations in order from left to right.

Press [1][8][0] *The number 180 appears in the entry line.*
Press [−] *The entry line remains 180.*
Press [7][0] *The entry line is now 70.*
Press [−] *The entry line shows the difference of the first two numbers, 110.*
Press [7][0] *The entry line is now 70.*
Press [=] *The entry line shows the final difference of the numbers, 40.*

The solution to this equation is $a = 40°$.

If we make a mistake during entry of a number, we can clear only the entry line using the [CE/C] key. On some calculators, we can also delete a single digit using the "back arrow" key [←]. To multiply numbers, we use the [×] key. The · notation for multiplication is not found on a scientific calculator.

Example 3 **Solve $L = 6 \cdot 5{,}280$ feet.**
(Section 1.5, Example 28)

We enter the numbers and operations in order from left to right.

Press [6] *The number 6 appears in the entry line.*
Press [×] *The entry line remains 6.*
Press [5][2][8][0] *The entry line is now 5280.*
Press [=] *The entry line shows the product of the numbers, 31680.*

The length of this hiking trail is 31,680 feet.

CHECK YOUR UNDERSTANDING

Evaluate. Use a calculator.

1. \$41,200 − \$23,800

2. 4(750 milliliters)

3. 7,934 inches + 34 inches + 6,886 inches + 658 inches

Answers: 1. \$17,400; 2. 3,000 ml; 3. 15,512 in.

DISCUSSION

Exponents and Division

There may be more than one key on a scientific calculator that can be used to evaluate exponential expressions. When the exponent is two, a key labeled [x^2] (said "x squared") can be used.

Example 4 **Find the area of a square with length 21 centimeters.**
(Section 1.7, Discussion: Area and Volume, Check Your Understanding, Problem 2)

$$A = s \cdot s$$
$$= s^2$$
$$= (21 \text{ centimeters})^2$$

Press [2] [1] *The number 21 appears in the entry line.*
Press [x^2] *The square of 21 appears, 441.*

$(21 \text{ centimeters})^2 = 441 \text{ centimeters}^2$

To evaluate expressions with exponents other than 2, a calculator may have a [y^x] key, described as the "y to the x" key. (On some calculators, this key may alternatively be [x^y].)

Example 5 **Evaluate 2^3.**
(Section 1.6, Example 1)

Press [2] *The number 2 appears in the entry line.*
Press [y^x] *The entry line remains 2.*
Press [3] *The number 3 appears in the entry line.*
Press [=] *The entry line is now 8, the value of 2^3.*

On other calculators, exponents are entered using a [^] key, often described as a "caret" key.

Example 6 **Evaluate 2^3.**
(Section 1.6, Example 1)

Press [2] *The number 2 appears in the entry line.*
Press [^] *The entry line remains 2.*
Press [3] *The number 3 appears in the entry line.*
Press [=] *The entry line is now 8, the value of 2^3.*

To divide numbers, we use the ÷ key.

Example 7 **Solve $s = 679 \div 7$.**
(Section 1.8, Example 21)

We enter the numbers and operations in order from left to right.

Press 6 7 9 *The number 679 appears in the entry line.*
Press ÷ *The entry line remains 679.*
Press 7 *The entry line is now 7.*
Press = *The entry line shows the quotient of the numbers, 97.*

There are 97 spaces in the parking lot.

In long division of whole numbers, the quotient may include a whole number remainder. For example, the quotient of 328 ÷ 34 is 9 R(22). A calculator does not show a whole number remainder in a quotient. Instead, it calculates the quotient as a decimal number. Until we are ready to work with decimal numbers, we can use a calculator to determine the whole number part of the quotient. We can use the memory feature of the calculator to find the remainder part of the quotient.

Example 8 **Use long division to find the quotient of 328 ÷ 34.**
(Section 1.8, Example 19)

We enter the numbers and operations in order from left to right.

Press 3 2 8 *The number 328 appears in the entry line.*
Press ÷ *The entry line remains 328.*
Press 3 4 *The entry line is now 34.*
Press = *The entry line shows the quotient of the numbers, 9.6470588.*

We know that the whole number part of the quotient is 9. To find the remainder multiply the whole number part of the quotient by the divisor. Subtract this product from the original dividend.

Press 9 *The whole number part of the quotient, 9, appears in the entry line.*
Press × *The entry line remains 9.*
Press 3 4 *The entry line is now 34.*
Press = *The entry line shows the product of the numbers, 306.*

If the calculator has a "store to memory" key such as STO, pressing this key stores the product in its memory. Of course, the product can also be "stored" by writing it down on a piece of paper.

Press STO *The product, 306, is stored in memory.*
Press CE/C *The entry line is cleared.*
Press 3 2 8 *The entry line is 328, the original dividend.*
Press − *The entry line remains 328.*
Press RCL *The calculator retrieves 306 from memory. The entry line is now 306.*
Press = *The entry line shows the difference of the numbers, 22. This is the remainder.*

The quotient of 328 ÷ 34 is 9 R(22).

Division by zero is undefined. If we try to divide by zero on a calculator, the entry line shows some kind of error message. Some calculators display ERROR in the entry line; others display a small E to the left of the entry line.

CHECK YOUR UNDERSTANDING

Evaluate. Use a calculator.

1. 2^4

2. 3^2

3. 5^0

4. 72 feet ÷ 12

5. 600,000 ÷ 120

6. 7,823 ÷ 27

Answers: 1. 16; 2. 9; 3. 1; 4. 6 ft; 5. 5,000; 6. 289 R(20).

DISCUSSION

Order of Operations

Most scientific calculators allow multiple entries of numbers and operations. The calculator is programmed to follow order of operations in such calculations.

Example 9 **Evaluate $20 - 6 \div 2 + 3 \cdot 2$.**
(Section 1.9, Example 2)

We enter the numbers and operations in order from left to right.

Press [2] [0]	*The number 20 appears in the entry line.*
Press [−]	*The entry line remains 20.*
Press [6]	*The entry line is now 6.*
Press [÷]	*The entry line remains 6.*
Press [2]	*The entry line is now 2.*
Press [+]	*The entry line may show an intermediate sum of 17.*
Press [3]	*The entry line is now 3.*
Press [×]	*The entry line remains 3.*
Press [2]	*The entry line is now 2.*
Press [=]	*The entry line shows the result of all the operations, 23.*

Following order of operations, $20 - 6 \div 2 + 3 \cdot 2 = 23$.

When parentheses are found in an expression, they are entered by pressing the appropriate [(] or [)] key.

Example 10 **Evaluate $2^2 + 3^2 + 13 \div (5 \cdot 3 - 2)$.**
(Section 1.9, Example 4)

We enter the parentheses, numbers, and operations in order from left to right.

Press [2]	*The entry line is 2.*
Press [x^2]	*The entry line is now 4, the result of 2^2.*
Press [+]	*The entry line is now 4, the result of 2^2.*
Press [3]	*The entry line is now 3.*
Press [x^2]	*The entry line is now 9, the result of 3^2.*
Press [+]	*The entry line may show an intermediate sum of 13.*
Press [1] [3]	*The entry line is now 13.*
Press [÷]	*The entry line remains 13.*
Press [(]	*The entry line remains 13.*
Press [5]	*The entry line is now 5.*
Press [×]	*The entry line remains 5.*
Press [3]	*The entry line is now 3.*
Press [−]	*The entry line shows the intermediate product, 15.*
Press [2]	*The entry line is now 2.*
Press [)]	*The entry line shows 13.*
Press [=]	*The entry line shows the result of all the operations, 14.*

$2^2 + 3^2 + 13 \div (5 \cdot 3 - 2) = 14$

A scientific calculator may not recognize the implied multiplication in the expression $2(5 - 1)$. When entering this expression, the implied multiplication must be directly entered as $2 \cdot (5 - 1)$.

Example 11 **Evaluate $2(5 - 1)^2 \div (24 \div 3)$.**
(Section 1.9, Check Your Understanding, Problem 4)

We enter the parentheses, numbers, and operations in order from left to right. We include the implied operation of multiplication.

Press [2]	*The entry line is 2.*
Press [×]	*We include implied multiplication. The entry line remains 2.*
Press [(]	*The entry line remains 2.*
Press [5]	*The entry line is now 5.*
Press [−]	*The entry line remains 5.*
Press [1]	*The entry line is now 1.*
Press [)]	*The entry line shows the intermediate result of 4.*
Press [x^2]	*The entry line shows 16, the result of squaring the value inside the parentheses.*
Press [÷]	*The entry line shows the intermediate result of 32.*
Press [(]	*The entry line remains 32.*
Press [2] [4]	*The entry line is now 24.*
Press [÷]	*The entry line remains 24.*
Press [3]	*The entry line is now 3.*
Press [)]	*The entry line shows the intermediate result of 8.*
Press [=]	*The entry line shows the final result of 4.*

$2(5 - 1)^2 \div (24 \div 3) = 4$

CHECK YOUR UNDERSTANDING

Evaluate. Use a calculator.

1. $6 + 4 \div 2 - 3 \cdot 2$

2. $(5 - 3)^2$

3. $(4 \cdot 2 + 12 \div 4)^2 - 7 \cdot 3$

4. $24 \div 8 \cdot 2 + 8 \cdot 7 \div 4$

Answers: 1. 2; 2. 4; 3. 100; 4. 20.

Section A2.2 Adding, Subtracting, Multiplying, and Dividing Integers with a Calculator

LEARNING TOOLS

CD-ROM SSM VIDEO

Sneak Preview

Many calculators can add, subtract, multiply, and divide integers.

After completing this section, you should be able to:

1) Complete addition, subtraction, multiplication, and division of integers on a calculator.

2) Evaluate a numerical expression that includes integers on a calculator.

DISCUSSION

Negative Numbers

The [−] key is used for subtraction, not to enter a negative number. If we need to enter a negative number, we first enter it without the negative sign. We then press the [+/−] key. This makes the number a negative number.

Example 1 **Add −1,943 + 212.**
(Section 2.2, Example 4)

We enter the numbers and operations in order from left to right.

Press [1] [9] [4] [3] *The number 1943 appears in the entry line. No comma appears.*
Press [+/−] *The entry line changes to −1943.*
Press [+] *The entry line remains −1943.*
Press [2] [1] [2] *The entry line is now 212.*
Press [=] *The entry line now shows the sum, −1731.*

$-1{,}943 + 212 = -1{,}731$

Parentheses are often included in expressions with negative numbers. If the purpose of the parentheses is only to separate + and − signs, then they do not have to be entered in the calculator.

Example 2 **Add 7 + (−5) + (−8) + 9 + 2 + (−7).**
(Section 2.2, Example 5)

We enter the numbers and operations in order from left to right. All of the parentheses in this expression are used to separate + and − signs. They are not used to indicate order of operations and do not have to be entered. However, you will get the same answer if you do enter them.

Press [7] *The number 7 appears in the entry line.*
Press [+] *The entry line remains 7.*
Press [5] *The number 5 appears in the entry line.*
Press [+/−] *The entry line changes to −5.*
Press [+] *The entry line shows the intermediate result, 2.*
Press [8] *The entry line is now 8.*
Press [+/−] *The entry line changes to −8.*
Press [+] *The entry line shows the intermediate result, −6.*
Press [9] *The entry line is now 9.*
Press [+] *The entry line shows the intermediate result, 3.*
Press [2] *The entry line is now 2.*
Press [+] *The entry line shows the intermediate result, 5.*
Press [7] *The number 7 appears in the entry line.*
Press [+/−] *The entry line changes to −7.*
Press [=] *The entry line now shows the final sum, −2.*

$7 + (-5) + (-8) + 9 + 2 + (-7) = -2$

When subtracting integers, we may need to use both the [−] key and the [+/−] key.

Example 3 **Subtract −30 − 80.**
(Section 2.3, Example 3)

We enter the numbers and operations in order from left to right.

Press [3] [0] *The number 30 appears in the entry line.*
Press [+/−] *The entry line changes to −30.*
Press [−] *The entry line remains −30.*
Press [8] [0] *The entry line is now 80.*
Press [=] *The entry line now shows the final result, −110.*

$-30 - 80 = -110$

In expressions that include subtraction of a negative number, parentheses are used to separate the signs. On most scientific calculators, entering these parentheses is optional. They are included, however, in the following example.

Example 4 **Subtract $-30 - (-80)$.**
(Section 2.3, Example 4)

We enter the numbers and operations in order from left to right.

Press [3] [0] *The number 30 appears in the entry line.*
Press [+/−] *The entry line changes to −30.*
Press [−] *The entry line remains −30.*
Press [(] *The entry line becomes 0.*
Press [8] [0] *The entry line is now 80.*
Press [+/−] *The entry line changes to −80.*
Press [)] *The entry line remains −80.*
Press [=] *The entry line now shows the final result, 50.*

$-30 - (-80) = 50$

We also use the [+/−] key when multiplying or dividing integers. When multiplication is implied by parentheses, we must use the [×] key to enter multiplication. Entering the parentheses is optional even when the numbers are negative.

Example 5 **Evaluate $(-9)(-5)$.**
(Section 2.4, Example 5)

We enter the numbers and operations in order from left to right.

Press [9] *The number 9 appears in the entry line.*
Press [+/−] *The entry line changes to −9.*
Press [×] *The entry line remains −9.*
Press [5] *The entry line is now 5.*
Press [+/−] *The entry line changes to −5.*
Press [=] *The entry line now shows the final result, 45.*

$(-9)(-5) = 45$

In the next example, the purpose of the parentheses is only to separate the ÷ and − signs. Entering the parentheses is optional.

Example 6 **Evaluate $-12 \div (-6)$.**
(Section 2.4, Example 2)

We enter the numbers and operations in order from left to right.

Press [1] [2] *The number 12 appears in the entry line.*
Press [+/−] *The entry line changes to −12.*
Press [÷] *The entry line remains −12.*
Press [6] *The entry line is now 6.*
Press [+/−] *The entry line changes to −6.*
Press [=] *The entry line now shows the final result, 2.*

$-12 \div (-6) = 2$

CHECK YOUR UNDERSTANDING

Evaluate. Use a calculator.

1. $15 - 3$

2. $-15 - 3$

3. $-15 - (-3)$

4. $-15 + 3$

5. $-13 \div 0$

6. $-8(-8)$

7. $-56 \div 7$

8. $(-3)(-5)(-2)$

Answers: 1. 12; 2. −18; 3. −12; 4. −12; 5. undefined; 6. 64; 7. −8; 8. −30.

DISCUSSION

Exponential Expressions

Many scientific calculators are not programmed to evaluate exponential expressions with bases that are negative numbers. We cannot always enter the numbers from left to right and get the correct result. For example, -3^2 is read "the opposite of 3 to the second power" or "the opposite of 3 squared." We must evaluate the exponent first and then make the expression negative.

Example 7 **Evaluate -3^2.**
(Section 2.5, Example 1)

Press [3] *The number 3 appears in the entry line.*
Press [x^2] *The entry line is now 9.*
Press [+/–] *The entry line changes to −9. This is the opposite of 3^2.*

If the base is a negative number, then parentheses must be entered. Most calculators will give a result of "error."

Example 8 **Evaluate $(-3)^2$.**
(Section 2.5, Example 2)

Press [(]
Press [3] *The number 3 appears in the entry line.*
Press [+/–] *The entry line is now −3.*
Press [)] *The entry line remains −3.*
Press [x^2] *The entry line shows the final result 9 or shows ERROR.*

This limitation of some calculators for negative numbers and exponents extends to evaluation of more complicated expressions. With these calculators, we cannot directly enter $(-2)^3 \div (3 \cdot 6 - 2 \cdot 11)$ into the calculator. However, we can use a calculator to evaluate parts of the expression.

Example 9 **Evaluate $(-2)^3 \div (3 \cdot 6 - 2 \cdot 11)$.**
(Section 2.5, Example 8)

If we cannot directly evaluate the exponential expression, we can evaluate the part of the expression in the parentheses: $(3 \cdot 6 - 2 \cdot 11)$.

Press [(]	
Press [3]	*The number 3 appears in the entry line.*
Press [×]	*The entry line remains 3.*
Press [6]	*The entry line changes to 6.*
Press [−]	*The entry line shows the intermediate result, 18.*
Press [2]	*The number 2 appears in the entry line.*
Press [×]	*The entry line remains 2.*
Press [1] [1]	*The entry line changes to 11.*
Press [)]	*The entry line shows the final result, −4.*
Press [=]	*The entry line shows the final result, −4. The = command does not change the answer.*

The original expression can be rewritten as $(-2)^3 \div (-4)$.

Now, use multiplication to evaluate the exponential expression.

Press [2]	*The number 2 appears in the entry line.*
Press [+/−]	*The entry line changes to −2.*
Press [×]	*The entry line remains −2.*
Press [2]	*The entry line changes to 2.*
Press [+/−]	*The entry line changes to −2.*
Press [×]	*The entry line shows the intermediate result, 4.*
Press [2]	*The entry line is now 2.*
Press [+/−]	*The entry line changes to −2.*
Press [=]	*The entry line shows the final result, −8.*

The original expression can be rewritten as $-8 \div (-4)$.

Finish evaluating the expression by completing the final division.

Press [8]	*The number 8 appears in the entry line.*
Press [+/−]	*The entry line changes to −8.*
Press [÷]	*The entry line remains −8.*
Press [4]	*The entry line is now 4.*
Press [+/−]	*The entry line changes to −4.*
Press [=]	*The entry line shows the final answer, 2.*

$(-2)^3 \div (3 \cdot 6 - 2 \cdot 11) = 2$

We can use a calculator to evaluate exponential expressions with whole number bases. However, when the base is a negative number or the expression is included in a more complicated expression, we may not get the correct result using some scientific calculators.

CHECK YOUR UNDERSTANDING

Evaluate. Use a calculator.

1. $3 + (-20) \div 5 + 9$

2. $10 + 20 \div (-5) - 2^3$

3. $3(-6^2 - 4) \div 2(4)$

4. $3^3 - (-3)^2$

Answers: 1. 8; 2. −2; 3. −240; 4. 18.

Section A2.3 Fraction Arithmetic with a Calculator

LEARNING TOOLS

CD-ROM SSM VIDEO

Sneak Preview

A calculator with a fraction key(s) can add, subtract, multiply, and divide fractions. The result of the calculations is always simplified by the calculator to lowest terms.

After completing this section, you should be able to:

1) Add, subtract, multiply, and divide fractions using a calculator.

DISCUSSION

Arithmetic with Proper and Improper Fractions

The [a b/c] key is used to enter a fraction or a mixed number into a calculator. The calculator automatically simplifies any entered fraction into lowest terms.

Example 1 **Simplify $\frac{40}{72}$.**

(Section 3.2, Example 16)

Press [4] [0] *The number 40 appears in the entry line.*
Press [a b/c] *A symbol that represents a fraction bar appears behind the 40 ⌋.*
Press [7] [2] *The entry line becomes 40 ⌋ 72.*
Press [=] *The entry line shows the final simplified result, 5 ⌋ 9.*

$\frac{40}{72}$ simplifies to $\frac{5}{9}$.

To multiply two fractions, we enter each fraction using the [a b/c] key.

Example 2 **Evaluate $\frac{3}{5} \cdot \frac{2}{3}$.**

(Section 3.3, Example 1)

Press [3] *The number 3 appears in the entry line.*
Press [a b/c] *A symbol that represents a fraction bar appears behind the 3 ⌋.*
Press [5] *The entry line becomes 3 ⌋ 5.*
Press [×] *The entry line remains 3 ⌋ 5.*
Press [2] *The number 2 appears in the entry line.*
Press [a b/c] *A symbol that represents a fraction bar appears behind the 2 ⌋.*
Press [3] *The entry line becomes 2 ⌋ 3.*
Press [=] *The entry line shows the final simplified result, 2 ⌋ 5.*

$$\frac{3}{5} \cdot \frac{2}{3} = \frac{2}{5}$$

As we did with integers, negative fractions can be entered using the [+/-] key.

Example 3 **Evaluate** $-\frac{5}{16} \div \left(-\frac{7}{4}\right)$.

(Section 3.4, Example 7)

Press [5]	*The number 5 appears in the entry line.*
Press [a b/c]	*A symbol that represents a fraction bar appears behind the 5 ⌟.*
Press [1] [6]	*The entry line becomes 5 ⌟ 16.*
Press [+/−]	*The entry line becomes −5 ⌟ 16.*
Press [÷]	*The entry line remains −5 ⌟ 16.*
Press [7]	*The number 7 appears in the entry line.*
Press [a b/c]	*A symbol that represents a fraction bar appears behind the 7 ⌟.*
Press [4]	*The entry line becomes 7 ⌟ 4.*
Press [+/−]	*The entry line becomes −7 ⌟ 4.*
Press [=]	*The entry line shows the final simplified result, 5 ⌟ 28.*

$$-\frac{5}{16} \div \left(-\frac{7}{4}\right) = \frac{5}{28}$$

We follow the same procedures when both integers and fractions are present.

Example 4 **Solve** $B = 6 \text{ ft} \div \frac{3}{4} \text{ ft}$.

(Section 3.4, Example 9)

Press [6]	*The number 6 appears in the entry line.*
Press [÷]	*The entry line remains 6.*
Press [3]	*The number 3 appears in the entry line.*
Press [a b/c]	*A symbol that represents a fraction bar appears behind the 3 ⌟.*
Press [4]	*The entry line becomes 3 ⌟ 4.*
Press [=]	*The entry line shows the final simplified result, 8.*

$$B = 6 \text{ ft} \div \frac{3}{4} \text{ ft}$$
$$= 8$$

Addition and subtraction with fractions can also be completed using the [a b/c] key.

Example 5 **Evaluate** $\frac{3}{14} + \frac{1}{10} - \frac{2}{35}$.

(Section 3.5, Example 10)

Press [3]	*The number 3 appears in the entry line.*
Press [a b/c]	*A symbol that represents a fraction bar appears behind the 3 ⌟.*
Press [1] [4]	*The entry line becomes 3 ⌟ 14.*
Press [+]	*The entry line remains 3 ⌟ 14.*
Press [1]	*The number 1 appears in the entry line.*
Press [a b/c]	*A symbol that represents a fraction bar appears behind the 1 ⌟.*
Press [1] [0]	*The entry line becomes 1 ⌟ 10.*
Press [−]	*The entry line shows the intermediate result of 11 ⌟ 35.*
Press [2]	*The number 2 appears in the entry line.*
Press [a b/c]	*A symbol that represents a fraction bar appears behind the 2 ⌟.*
Press [3] [5]	*The entry line becomes 2 ⌟ 35.*
Press [=]	*The entry line shows the final simplified result, 9 ⌟ 35.*

$$\frac{3}{14} + \frac{1}{10} - \frac{2}{35} = \frac{9}{35}$$

CHECK YOUR UNDERSTANDING

Evaluate. Use a calculator.

1. $\dfrac{4}{5} \cdot \dfrac{3}{7}$

2. $\left(-\dfrac{6}{5}\right)(20)$

3. $\dfrac{3}{7} \div \dfrac{2}{5}$

4. $\dfrac{2}{3} + \dfrac{3}{16} - \dfrac{1}{36}$

Answers: 1. $\dfrac{12}{35}$; 2. -24; 3. $\dfrac{15}{14}$; 4. $\dfrac{119}{144}$.

DISCUSSION

Arithmetic with Mixed Numbers

To enter a mixed number, the integer is entered first. The [a b/c] key is then used to enter the fraction portion of the mixed number. The [d/c] key is used to change a mixed number into an improper fraction or an improper fraction into a mixed number.

Example 6 **Rewrite $3\dfrac{5}{8}$ as an improper fraction.**

(Section 3.6, Example 4)

Press [3]	*The number 3 appears in the entry line.*
Press [a b/c]	*A symbol that represents a fraction bar appears behind the 3 ⌟.*
Press [5]	*The entry line becomes 3 ⌟ 5.*
Press [a b/c]	*The entry line becomes 3 _ 5 ⌟.*
Press [8]	*The entry line becomes 3 _ 5 ⌟ 8.*
Press [2nd]	*The entry line remains 3 _ 5 ⌟ 8.*
Press [d/c]	*The entry line shows the equivalent improper fraction, 29 ⌟ 8.*

$$3\frac{5}{8} = \frac{29}{8}$$

An improper fraction can also be rewritten as a mixed number using a calculator.

Example 7 **Rewrite $\dfrac{23}{7}$ as a mixed number.**

(Section 3.6, Example 2)

Press [2] [3]	*The number 23 appears in the entry line.*
Press [a b/c]	*A symbol that represents a fraction bar appears behind the 23 ⌟.*
Press [7]	*The entry line becomes 23 ⌟ 7.*
Press [=]	*The entry line becomes 3 _ 2 ⌟ 7.*

$$\frac{23}{7} = 3\frac{2}{7}$$

When multiplying or dividing mixed numbers without a calculator, we must first rewrite the mixed numbers as improper fractions. If a calculator with an [a b/c] key is used, this is not necessary.

Example 8 **Find the product of $3\frac{3}{4}$ and $4\frac{4}{9}$.**

(Section 3.6, Example 10)

Press [3]	*The number 3 appears in the entry line.*
Press [a b/c]	*A symbol that represents a fraction bar appears behind the 3 ⌋.*
Press [3]	*The entry line becomes 3 ⌋ 3.*
Press [a b/c]	*The entry line becomes 3 ⌋ 3 ⌋.*
Press [4]	*The entry line becomes 3 _ 3 ⌋ 4.*
Press [×]	*The entry line remains 3 _ 3 ⌋ 4.*
Press [4]	*The number 4 appears in the entry line.*
Press [a b/c]	*A symbol that represents a fraction bar appears behind the 4 ⌋.*
Press [4]	*The entry line becomes 4 ⌋ 4.*
Press [a b/c]	*The entry line becomes 4 _ 4 ⌋.*
Press [9]	*The entry line becomes 4 _ 4 ⌋ 9.*
Press [=]	*The entry line shows the final simplified result, 16 _ 2 ⌋ 3.*

$$\left(3\frac{3}{4}\right)\left(4\frac{4}{9}\right) = 16\frac{2}{3}$$

Expressions that include fractions and mixed numbers can also be evaluated.

Example 9 **Solve $L = 20$ inches $- 12\frac{5}{8}$ inches $- \frac{1}{8}$ inch.**

(Section 3.7, Example 7)

Press [2] [0]	*The number 20 appears in the entry line.*
Press [−]	*The entry line remains 20.*
Press [1] [2]	*The number 12 appears in the entry line.*
Press [a b/c]	*A symbol that represents a fraction bar appears behind the 12 ⌋.*
Press [5]	*The entry line becomes 12 ⌋ 5.*
Press [a b/c]	*The entry line becomes 12 _ 5 ⌋.*
Press [8]	*The entry line becomes 12 _ 5 ⌋ 8.*
Press [−]	*The entry line shows the intermediate result of 7 _ 3 ⌋ 8.*
Press [1]	*The number 1 appears in the entry line.*
Press [a b/c]	*A symbol that represents a fraction bar appears behind the 1 ⌋.*
Press [8]	*The entry line becomes 1 ⌋ 8.*
Press [=]	*The entry line shows the final result of 7 _ 1 ⌋ 4.*

$$L = 20 \text{ inches} - 12\frac{5}{8} \text{ inches} - \frac{1}{8} \text{ inch} = 7\frac{1}{4} \text{ inches}$$

CHECK YOUR UNDERSTANDING

Evaluate. Use a calculator.

1. Rewrite $\frac{17}{3}$ as a mixed number.

2. Rewrite $-9\frac{2}{5}$ as an improper fraction.

Evaluate.

3. $5\frac{1}{2} \div 1\frac{3}{4}$

4. $2\frac{1}{5} + 4\frac{3}{8}$

Answers: 1. $5\frac{2}{3}$; 2. $-\frac{47}{5}$; 3. $3\frac{1}{7}$; 4. $6\frac{23}{40}$.

Section A2.4 Decimal Arithmetic with a Calculator

LEARNING TOOLS

CD-ROM

SSM

VIDEO

Sneak Preview

Many calculators can add, subtract, multiply, and divide decimal numbers. A scientific calculator can also evaluate square roots.

After completing this section, you should be able to:

1) Add, subtract, multiply, and divide decimal numbers using a calculator.
2) Use the [π] key.
3) Evaluate a square root with a calculator.

DISCUSSION

Arithmetic With Decimal Numbers

We add, subtract, multiply, and divide decimal numbers using most of the same procedures used with integers. The decimal point is usually found in the bottom row of keys.

Example 1 **Solve d = 67.5°F − (−3.8°F).**
(Section 4.1, Example 11)

Press [6] [7] [.] [5] *The number 67.5 appears in the entry line.*
Press [−] *The entry line remains 67.5.*
Press [3] [.] [8] *The number 3.8 appears in the entry line.*
Press [+/−] *The entry line changes to −3.8.*
Press [=] *The entry line shows the final result, 71.3.*

d = 67.5°F − (−3.8°F) = 71.3°F

Example 2 **Solve r = (40,000)($0.22).**
(Section 4.2, Example 4)

Press [4] [0] [0] [0] [0] *The number 40000 appears in the entry line.*
Press [×] *The entry line remains 40000.*
Press [0] [.] [2] [2] *The number .22 appears in the entry line.*
Press [=] *The entry line shows the final result, 8800.*

r = (40,000)($0.22) = $8,800

Example 3 **Solve m = 12.011 grams + 2(15.994 grams).**
(Section 4.3, Example 11)

Press [1] [2] [.] [0] [1] [1] *The number 12.011 appears in the entry line.*
Press [+] *The entry line remains 12.011.*
Press [2] *The number 2 appears in the entry line.*
Press [×] *The entry line remains the same.*
Press [1] [5] [.] [9] [9] [4] *The number 15.994 appears in the entry line.*
Press [=] *The entry line shows the final result, 43.999.*

m = 12.011 grams + 2(15.994 grams) = 43.999 grams

To evaluate some quotients without writing down intermediate results, we need to add parentheses to the expression before evaluating.

Example 4 **Evaluate** $\dfrac{\mathbf{3(1.2) - (0.3)^2}}{\mathbf{1.2(0.3) + (-2.1)}}$**. Round to the nearest tenth.**

(Section 4.3, Example 13)

We need to tell the calculator to evaluate the numerator completely and then evaluate the denominator completely before dividing. We do this by rewriting the expression as $[3(1.2) - (0.3)^2] \div [1.2(0.3) + (-2.1)]$. We enter the brackets [ ] using the (and) keys.

Press (

Press 3 — *The number 3 appears in the entry line.*

Press × — *The entry line remains 3.*

Press 1 . 2 — *The number 1.2 appears in the entry line.*

Press − — *The entry line remains 1.2.*

Press (— *The entry line remains 1.2.*

Press 0 . 3 — *The number 0.3 appears in the entry line.*

Press) — *The entry line remains 0.3.*

Press x^2 — *0.09 appears in the entry line; the result of 0.3^2.*

Press) — *The intermediate result, 3.51, appears in the entry line.*

Press ÷ — *The entry line remains 3.51.*

Press (— *The entry line remains 3.51.*

Press 1 . 2 — *The number 1.2 appears in the entry line.*

Press × — *The entry line remains 1.2.*

Press 0 . 3 — *The number 0.3 appears in the entry line.*

Press + — *The entry line remains 0.3.*

Press 2 . 1 — *The number 2.1 appears in the entry line.*

Press +/− — *The entry line changes to −2.1.*

Press) — *The entry line shows the intermediate result, −1.74.*

Press = — *The number −2.0172414... appears.*

Rounding to the nearest tenth, $\dfrac{3(1.2) - (0.3)^2}{1.2(0.3) + (-2.1)} \approx -2.0$.

A calculator works with a finite (limited) number of digits. If there are more digits in the result than the calculator can use, it rounds the result to the maximum number of digits. Every time the calculator rounds, error is introduced into the result. When we evaluate an expression, error is also introduced if we round intermediate results. Unless we use the same number of places as a calculator, we actually introduce more error than the calculator. Reduce error by completing as many steps as possible without additional rounding.

CHECK YOUR UNDERSTANDING

Evaluate.

1. $15.5 \div 3.1 + 6(4.1 - 0.9)$

2. (55.847 grams)(2) + (15.994 grams)(3)

3. $\dfrac{(0.240)(0.144)}{0.240 + 0.144}$

4. (\$0.45)(80) + \$22.50

Answers: 1. 24.2; 2. 159.676 grams; 3. 0.09; 4. \$58.50.

DISCUSSION

Pi

Many scientific calculators include a [π] key. Pi is an irrational number that is equal to 3.14159265359 . . . . When evaluating formulas, we often use the approximate value of 3.14 for pi. However, we can enter a more precise value of pi using the [π] key. The number of digits in pi that appear in the entry line depends on the precision of the calculator.

Example 5 **Solve $C = \pi(4.72 \text{ inches})$. Round to the nearest hundredth of an inch.**
(Section 4.6, Example 1)

Press [π]	*The number 3.1415927 appears in the entry line.*
Press [×]	*The entry line remains 3.1415927.*
Press [4] [.] [7] [2]	*The entry line is now 4.72.*
Press [=]	*The entry line shows the final result, 14.828317.*

Rounding to the hundredths place, $C = \pi(4.72 \text{ inches}) = 14.83$ inches.

Example 6 **Solve $V = \pi(3.3 \text{ cm})^2 (10.4 \text{ cm})$. Round to the nearest cubic centimeter.**
(Section 4.6, Example 6)

Press [π]	*The number 3.1415927 appears in the entry line.*
Press [×]	*The entry line remains 3.1415927.*
Press [3] [.] [3]	*The entry line is now 3.3.*
Press [y^x]	*The entry line remains 3.3.*
Press [2]	*The entry line is now 2.*
Press [×]	*The entry line shows the intermediate result, 34.211944.*
Press [1] [0] [.] [4]	*The entry line is now 10.4.*
Press [=]	*The entry line shows the final result, 355.80422.*

Rounding to the nearest cubic centimeter, $V = \pi(3.3 \text{ cm})^2 (10.4 \text{ cm}) \approx 356 \text{ cm}^3$.

In many practical situations, our measurements are far less precise than the value of pi available on a calculator. This is why we often use the approximate value of 3.14 for pi.

CHECK YOUR UNDERSTANDING

Evaluate. Use the pi key on a calculator.

1. $\pi(8.5 \text{ inches})$ Round to the nearest tenth of an inch.

2. $\pi(7.61 \text{ meters})^2$ Round to the nearest hundredth of a square meter.

3. $2\pi(6 \text{ feet})(12 \text{ feet})$ Round to the nearest square foot.

4. $\frac{4}{3}\pi(2.18 \text{ cm})^2$ Round to the nearest hundredth of a cubic centimeter.

Answers. 1. 26.7 in.; 2. 181.94 m²; 3. 452 ft²; 4. 19.91 cm³.

DISCUSSION

Square Roots

We can use a scientific calculator to find the principal square root of any real number that is greater than or equal to zero. On many calculators, there is a principal square root key, [√].

Example 7 **Use a calculator to determine $\sqrt{81}$.**
(Section 4.7, Example 1)

Press [8] [1] *The number 81 appears in the entry line.*
Press [√] *The entry line shows the final result, 9.*

$\sqrt{81} = 9$

The square root command, [√], is sometimes found printed above another key, often the [x^2] key. To use this command, we must first press the [2nd] key and then press the key with [√] printed above it.

Example 8 **Use a calculator to evaluate $\sqrt{2}$. Round to the nearest thousandth.**
(Section 4.7, Example 2)

Press [2] *The number 2 appears in the entry line.*
Press [2nd] *The entry line remains 2.*
Press [√] *The entry line shows the final result, 1.4142136. . . .*

Rounding to the nearest thousandth, $\sqrt{2} = 1.414$.

If a fraction has a rational square root, we can use a calculator to find the square root as a fraction or as its decimal number equivalent. With many calculators, the denominator of the fraction must be less than 1,000. We first determine the square root in decimal form, using parentheses for the operations inside the square root. We can then use the fraction-to-decimal exchange key, [F↔D], to change the result into fraction form. The decimal exchange key is often accessed by using the [2nd] key. If the result is a mixed number, we can then change the result into an improper fraction using the [d/c] key.

Example 9 **Use a calculator to evaluate $\sqrt{\frac{81}{49}}$. Express the answer as an improper fraction.**
(Section 4.7, Check Your Understanding, Problem 3)

Press [(] *The number 0 remains in the entry line.*
Press [8] [1] *The number 81 appears in the entry line.*
Press [÷] *The number 81 remains in the entry line.*
Press [4] [9] *The number 49 appears in the entry line.*
Press [)] *The intermediate result 1.653061224. . . appears in the entry line.*
Press [√] *The entry line shows the final result in decimal form, 1.285714286. . . .*
Press [2nd] *The entry line is unchanged.*
Press [F↔D] *The entry line shows the final result as a mixed number, $1\frac{2}{7}$.*
Press [2nd] *The entry line is unchanged.*
Press [d/c] *The entry line shows the final result as an improper fraction, $\frac{9}{7}$.*

$$\sqrt{\frac{81}{49}} = \frac{9}{7}$$

Parentheses are also needed to use the Pythagorean theorem, $a^2 + b^2 = c^2$.

Example 10 **Solve $c^2 = (150 \text{ miles})^2 + (430 \text{ miles})^2$. Round to the nearest mile.**
(Section 4.7, Example 8)

$c^2 = (150 \text{ miles})^2 + (430 \text{ miles})^2$
$c = \sqrt{(150^2 + 430^2)}$

Press [(] *The entry line remains 0.*
Press [1] [5] [0] *The number 150 appears in the entry line.*
Press [x^2] *The entry line becomes 22500.*
Press [+] *The entry line remains 22500.*
Press [4] [3] [0] *The entry line is now 430.*
Press [x^2] *The entry line is now 184,900.*
Press [)] *The entry line shows the intermediate result, 207,400.*
Press [$\sqrt{\ }$] *The entry line contains the final result, 455.14119015.*

Rounding to the nearest mile, the distance is 455 miles.

CHECK YOUR UNDERSTANDING

Evaluate or solve. Use a calculator. Round to the nearest tenth.

1. $\sqrt{35}$ **2.** $\sqrt{-14}$ **3.** $\sqrt{144 \text{ meters}^2}$

4. Solve for c: $c = \sqrt{(4.6 \text{ meters})^2 + (6.2 \text{ meters})^2}$.

Answers: 1. 5.9; 2. no real number; 3. 12 m; 4. $c \approx 7.7$ m.

Answers to Odd-Numbered Exercises

CHAPTER 1

Section 1.1

1. Answers will vary. 1,234 is a whole number with 4 digits; **3.** This set is finite because the set of numbers does not go on forever; **5.** hundreds; **7.** ones; **9.** hundred thousands; **11.** ones; **13.** millions; **15.** 1,089; **17.** five thousand, nine hundred eighty-one; **19.** five hundred two; **21.** three million, six hundred eighty-seven thousand, five hundred sixty-four; **23.** 8,000,000 + 900,000 + 10,000 + 3,000 + 700 + 70 + 5;
25. 900,000 + 40,000 + 3,000 + 900 + 30 + 3;

27. 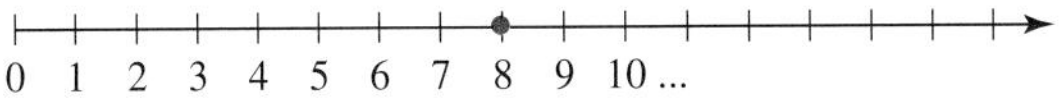 0 1 2 3 4 5 6 7 8 9 10 ...

29. false; **31.** false; **33.** true; **35.** $21 < 72$; **37.** $45 > 34$; **39.** $14 > 8$;
41. $6{,}999 < 7{,}003 < 7{,}284 < 7{,}874 < 7{,}964$; **43.** The height is about 8 feet; **45.** *Estimates:* 22 hours, 1.5 million words. *Exact:* 3 hours, 1,260 changes, 2 years, 11 new forms, 177 others;
47. 9,000; **49.** 88,000; **51.** 0; **53.** 4,000; **55.** 100; **57.** 5,700; **59.** $1,121,000,000,000;

61. **BILLS IN CIRCULATION**

TYPE OF BILL	NUMBER OF BILLS	NUMBER OF BILLS ROUNDED TO NEAREST MILLION
$1	3,571,913,726	3,572,000,000
$5	987,814,668	988,000,000
$10	1,136,337,194	1,136,000,000
$20	2,579,310,289	2,579,000,000
$50	1,530,339,488	1,530,000,000
$100	765,169,744	765,000,000
$1,000	1,544,720	2,000,000
$10,000	347	0

63. 90°; **65.** greater than signs were used instead of less than signs; $99 < 990 < 999 < 9{,}900$.

Section 1.2

1a.

NAME	WEIGHT (POUNDS)	ROUNDED WEIGHT
Aerni	45	50
Barnes	71	70
Beastie	86	90
Champ	97	100
Escamilla	71	70
Flo	67	70
Harley	45	50
Pirate	61	60
Sugar	71	70
Chuckie	92	90
Duke	48	50
Paddy	12	10
Sarge	52	50
Mitzi	49	50
Sam	38	40
Ruby	47	50
Rudy	70	70
Luka	74	70

1b.

ROUNDED WEIGHT	NUMBER OF OCCURRENCES
10	1
20	0
30	0
40	1
50	6
60	1
70	6
80	0
90	2
100	1

1c.

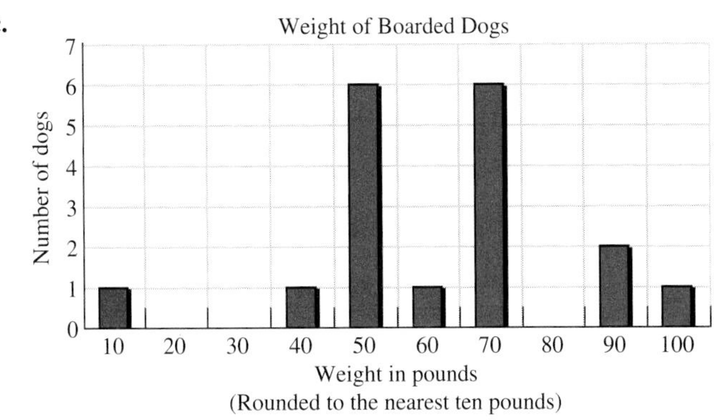

1d. The data has two modes at 50 pounds and 70 pounds. We say that this data is "bi-modal";

3a.

VALUE OF BOOK ($)	92	45	76	17	27	45	31	98	89	65	17	27	45	98	62	17	70	46
ROUNDED VALUE ($)	90	50	80	20	30	50	30	100	90	70	20	30	50	100	60	20	70	50

3b.

ROUNDED VALUE OF BOOK ($)	10	20	30	40	50	60	70	80	90	100
NUMBER OF OCCURRENCES	0	3	3	0	4	1	2	1	2	2

3c.

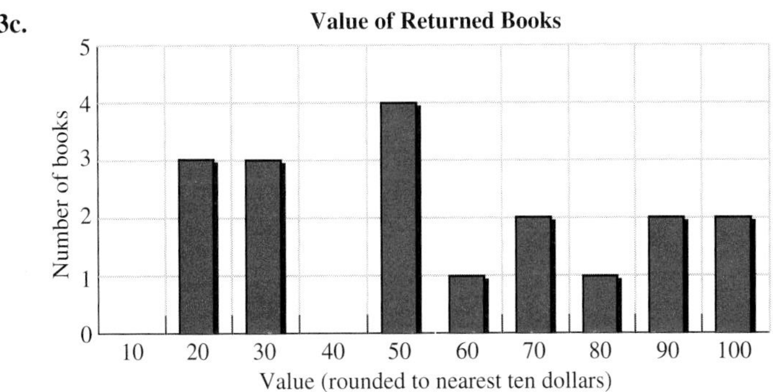

3d. This data has two modes at \$17 and \$45; **3e.** The mode of the rounded value of returned books is \$50;
5a. Africa; **5b.** 50 violations; **5c.** 80 violations;

7a.

ETHNICITY OF U.S. PHYSICIANS	PERCENT
White	79
African-American	3
Hispanic	5
Asian	10
Other	3

7b.

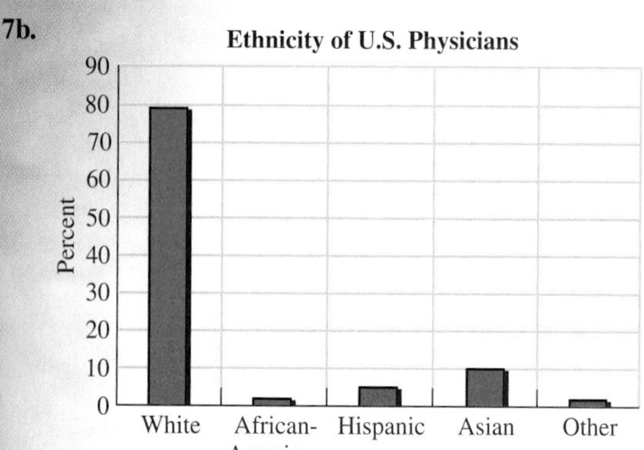

9a.

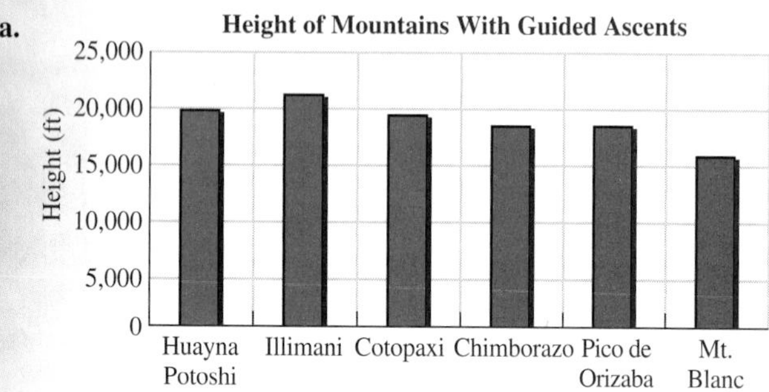

9b.

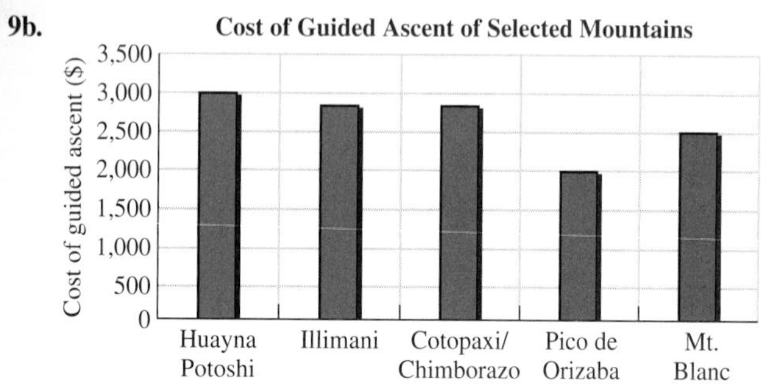

11a.

YEAR	LIVING KIDNEY DONORS
1994	3,008
1995	3,360
1996	3,606
1997	3,856
1998	4,154

11b.

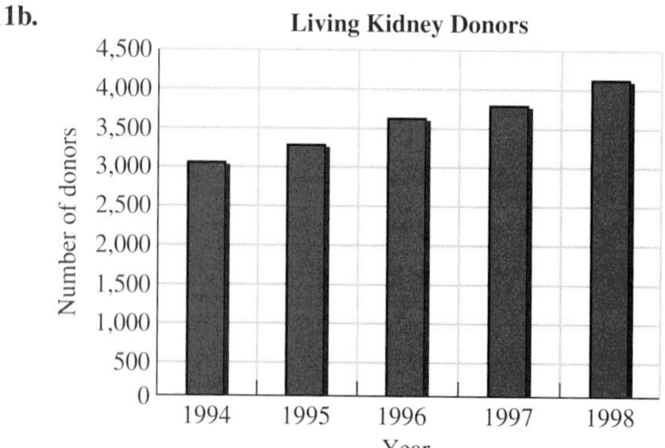

13.

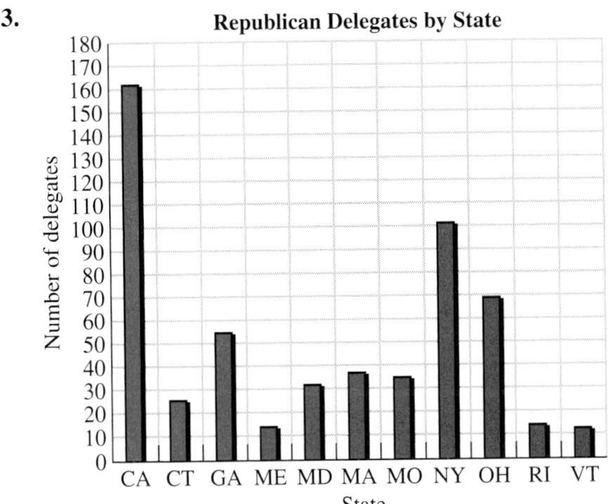

15a.

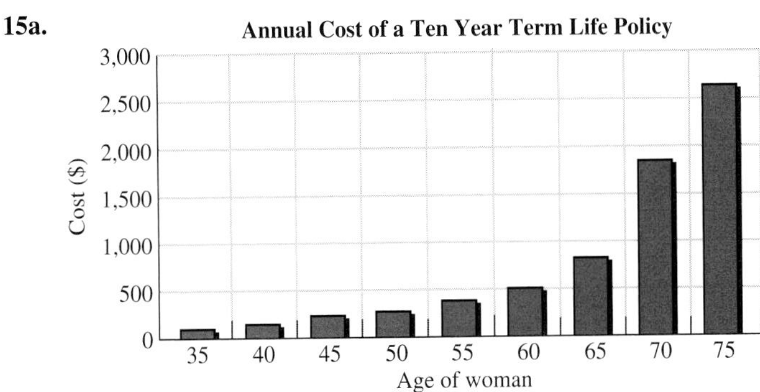

15b.

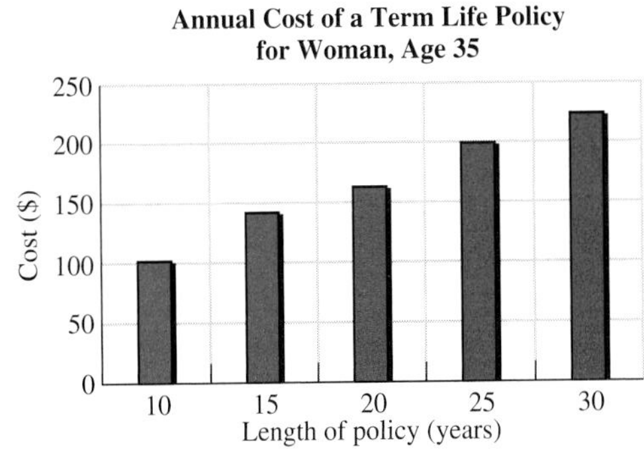

17. Changing the scale of the graph to intervals of \$5 instead of \$2 and/or changing the Return (\$) so that it starts at \$0 instead of \$564 would show that Trust Bank's CD is almost as good a value as the CD from Security Bank; **19.** 40,000 + 0 + 0 + 0 + 2; **21.** 7,920.

Section 1.3

1.

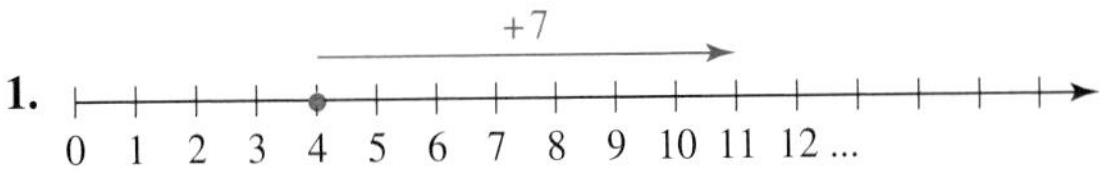

3.

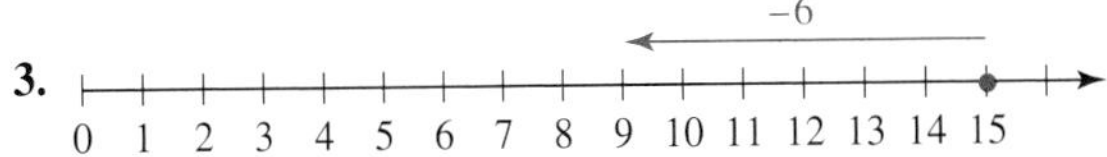

5. 6,300; **7.** 11; **9.** 21; **11.** Changing the order of addition does not change the sum. For example, 3 + 5 = 5 + 3; **13a.** 0; **13b.** 6; **13c.** Yes, because (10 − 7) − 3 = 0 and 10 − (7 − 3) = 6 . Since 0 does not equal 6, this is a counterexample. Subtraction is not associative; **15.** 9; **17.** 8; **19.** 800; **21.** 17 light-years; **23.** \$6,000,000; **25.** 21 feet; **27.** 15 miles; **29.** 16 meters; **31.** 28 feet; **33.** 1 inch; **35.** 90; **37.** 10,000,000; **39.** 88,000; **41.** \$3,000,000; **43.** \$2,000,000,000; **45a.** thousands; **45b.** \$2,000; **45c.** \$3,000; **45d.** \$4,000; **45e.** \$3,808;

47a. ten-thousands; **47b.** \$30,000; **47c.** \$10,000; **47d.** \$40,000; **47e.** \$38,568; **49a.** hundreds; **49b.** 400; **49c.** 300; **49d.** 100; **49e.** 96; **51a.** ten thousands; **51b.** 80,000; **51c.** 30,000; **51d.** 50,000; **51e.** 51,641; **53.** Their combined income was about \$50,000; **55.** 1 foot 9 inches; **57.** 21 hours 33 minutes; **59.** 3 gallons 2 quarts 1 cup; **61.** 1 pound 15 ounces; **63.** 1 mile 4,880 feet; **65.** 6 gallons; **67.** The sum of two right angles is 180°. Complementary angles are angles whose sum is 90°; **69.** Yes. An acute angle has a measure less than 90°. The sum of two acute angles can be exactly 90° so that they are complementary angles. For example, 60° + 30° = 90°; **71.** The sum of the measure of two complementary angles is 90°. The measure of an obtuse angle is greater than 90°. The sum of two obtuse angles must be greater than 90° and so these angles cannot be complementary; **73.** 104°; **75.** 10 feet 5 inches; **77.** 141 feet 4 inches;

79a.

3 ft
1 in.
2 ft

79b. 124 inches or 10 feet 4 inches;

81a.

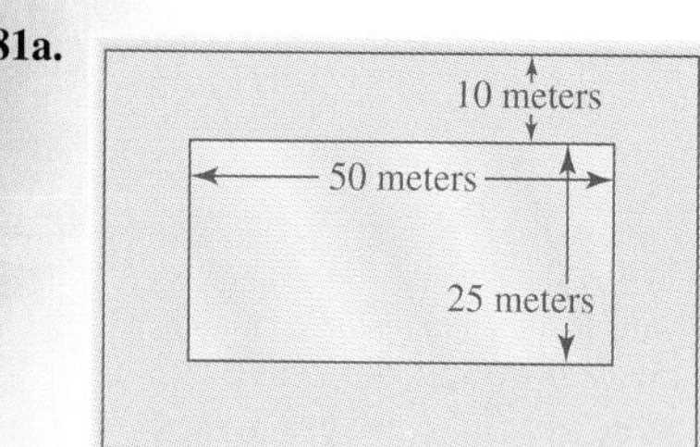

81b. 100 meters; **81c.** 230 meters; **83.** *Mistake:* The sum was rounded instead of the numbers being added. *Correction:* 80; **85.** *Mistake:* The numbers were not rounded to the highest place value found in both of the numbers, the thousands place. *Correction:* \$1,000; **87.** 56,100 < 65,001 < 65,010 < 65,100.

Section 1.4

1. Juanita needs 46 credits to graduate; **3.** The total cost of the set and the box is \$6,998; **5.** Mitch will drive 621 miles on his trip; **7.** Buying a scratch and scuff cover will save \$76; **9.** The total amount of these donations is \$4,014,000,000; **11.** The increase in population was 142,265 people; **13.** The difference in annual hours worked per person was 83 hours; **15.** A frequent binge drinker would consume 36 drinks in two weeks; **17.** The Navy should have had 7,649 pilots; **19.** The difference in weight is 12 ounces; **21.** 216 years passed between the two balloon flights; **23.** The price of a dozen long-stemmed roses is \$85; **25.** \$1,638 of the operating expenses was for fuel; **27.** Of the county's expenses, \$391,483 will not be reimbursed; **29.** The discount on the mattress set is \$450; **31.** The total net income was \$13,618; **33.** The total ticket revenues were \$272,600,000; **35.** The total measure of the angles is 180°; **37.** *Mistake:* The table only contains information about metro areas in the West. *Correction:* To calculate the number of permits issued outside the west, we need to know the total number of permits issued in the West. We cannot find out the correct answer without more information; **39.** Determine the highest place value of the largest number. Round both numbers to this place value. Add the rounded numbers; **41.** $a + b = b + a$.

Section 1.5

1. $3 \cdot 4 = 12$; **3.** The order of multiplication does not affect the product; **5.** a fundamental principle of our number system; **7.** associative property of multiplication; **9.** 24; **11.** 10; **13.** 16; **15.** 20; **17.** 60; **19.** 18; **21.** 5,000; **23.** 2,800; **25.** 54,000; **27.** 100,000; **29.** The revenue from the specials is \$120,000; **31.** 24,000; **33.** 45,000; **35.** 200; **37.** 150,000; **39.** 180,000; **41.** When estimating sums, we round all the numbers to the highest place value in the largest number. When estimating products, we round each number to its highest place value; **43.** The cost of buying milk for a year is about \$5,000; **45.** The fees collected will be about \$800,000; **47.** 332; **49.** 1,287; **51.** 54,510; **53.** 9,801; **55.** The cost of this purchase is \$7,375; **57.** The annual cost of disposal is \$794,311; **59.** The total amount of water flushed would be 3,222 gallons; **61.** He will not meet the requirements because he will only have written about 9,048 words; **63.** 14 hours 35 minutes 42 seconds; **65.** 13 yards 9 inches; **67.** 3,520 ounces; **69.** 120 minutes; **71.** 12 feet; **73.** 3,200 ounces; **75.** *Mistake:* The wrong number of zeros was attached the product of 3 and 6. *Correction:* 18,000 feet; **77.** 10,000; **79.** Changing the grouping in multiplication does not change the product.

Section 1.6

1. 3^4; **3.** 5^2; **5.** 1^{11}; **7.** 1; **9.** 1,000; **11.** 1; **13.** 25; **15.** 512; **17.** undefined; **19.** 2^4; **21.** 3^3; **23.** 10^{11}; **25.** 5^{10}; **27.** 0^{21}; **29.** 60; **31.** 16; **33.** 30,000; **35.** 21,000; **37.** 7,400,000 pounds; **39.** 70,000,000,000,000,000,000,000 kilograms; **41.** 3,000,000,000 possible ways; **43.** 1,910,000,000,000,000,000,000 descendants; **45.** 25,000 quills; **47.** 300,000,000 meters per second; **49.** 186,000 miles per second; **51.** *Mistake:* The exponents were multiplied. They should have been added. *Correction:* 4^5; **53.** 48 inches; **55.** 12 quarts.

Section 1.7

1. 40,000 meters; **3.** 40,000 milliliters; **5.** 2,000 milliliters; **7.** 25,000 milliliters; **9.** 531,000 meters; **11.** 80 millimeters; **13.** 36,000 nanometers; **15.** 128 cm^2; **17.** 12 $inches^2$; **19.** 17,000 cm^2; **21.** 2,294 $inches^2$; **23.** 1,016,064 $feet^2$; **25.** 17,000 $feet^2$; **27.** 432 $inches^2$; **29.** 2,688 $inches^3$; **31.** 147,730 cm^3; **33.** 1,701 cm^3; **35.** 4,560 $feet^3$; **37.** 1,080 $feet^3$; **39a.** Each family will collect 2,304 cubic inches of newspaper; **39b.** The total volume of newspaper collected is 3,110,400 cubic inches; **41a.** The area is 950 square inches; **41b.** The area is 475,000 square inches; **41c.** The total area is 1,900,000 square inches; **41d.** The weight is 240 pounds; **43a.** The area of the floor is 192 square feet; **43b.** The area of one side of a panel is 32 square feet; **43c.** The total area of one side of this plywood is 224 square feet; **43d.** Roberto needs 192 square feet. If he does not waste more than 32 square feet in cutting, he has enough; **45.** *Mistake:* All of the measurements are not in the same units. *Correction:* 6,000,000 mm^3; **47.** 0; **49.** The product of any number and zero is always zero. For example, $4 \cdot 0 = 0$.

Section 1.8

1. $40 \div 10 = 4$; **3.** 2 R(3); **5.** 6; **7.** 1 R(8); **9.** 3 R(2); **11.** 4 R(1); **13.** 0; **15.** 420; **17.** 350; **19.** undefined; **21.** Each animal will be allocated $1,200; **23.** Each person will have to pay $56,450,020; **25.** 2 feet; **27.** 9 hours; **29.** 16; **31.** 16 feet 2 inches; **33.** 10; **35.** 6; **37.** 20; **39.** Glenn will need 3 pounds 2 ounces of ground pork for each part of the recipe; **41.** Each person had to distribute 500 tokens; **43.** The average speed of the truck was 73 miles per hour; **45.** 138 R(1); **47.** 811 R(10); **49.** 4 R(14); **51.** 152; **53.** 118 R(1); **55.** 37 R(12); **57.** 532 R(10); **59.** 283 R(16); **61a.** Each steamer would have to travel 153 miles; **61b.** A distance of 22 miles would be left over; **63.** *Mistake:* After subtracting 57 − 56, the 1 was not written down. *Correction:* 82 R(3); **65.** A variable is a letter used to represent a number; **67.** $a \cdot (b \cdot c) = (a \cdot b) \cdot c$.

Section 1.9

1. Working from left to right, evaluate inside grouping symbols, evaluate exponents, multiply or divide, add or subtract; **3.** 15; **5.** 26; **7.** 50; **9.** 32; **11.** 3; **13.** 2; **15.** 3; **17.** 20; **19.** undefined; **21.** 34 feet; **23.** 25; **25.** 2; **27.** The average height is 27,240 feet; **29.** The average acreage burned per fire was 225 acres; **31.** The average amount of gold bullion produced per month was $13,333; **33.** The average length of these home runs was 514 feet; **35.** The average physician charge is $39; **37.** *Mistake:* The exponential expression 4^2 was evaluated incorrectly as 8. *Correction:* 11; **39.** *Mistake:* Some of the measurements are in hours, the others are in minutes. The units must be the same in order to find the average. *Correction:* 130 minutes; **41.** 400.

Review Exercises

1. tens; **3.** 3,275; **5.** $4 < 15$; **7.** $35 < 41 < 57$ or $57 > 41 > 35$; **9.** 2,060; **11.** 2,000; **13.** 100,000 students;

15.

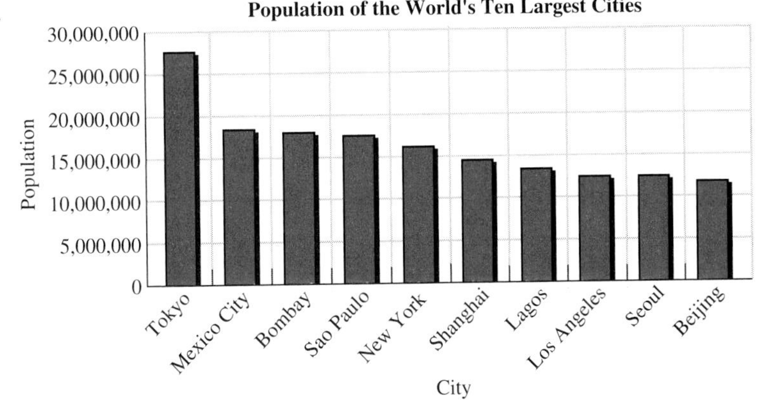

17. The mode is 12 students.

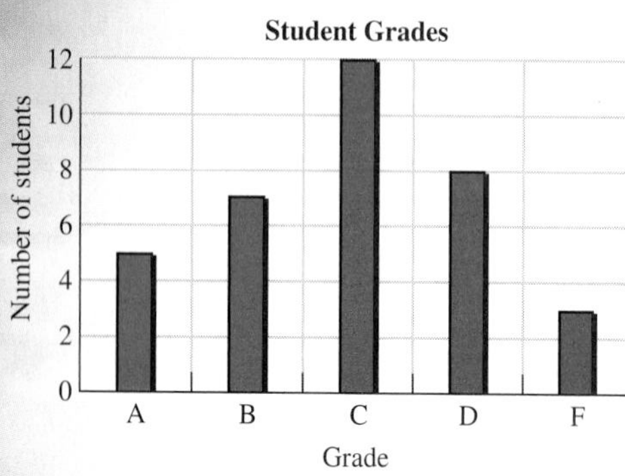

19. Changing the grouping of numbers being added does not change the sum; **21.** 19,200 pounds; **23.** 0; **25.** 11,000; **27.** $600,000,000; **29.** 751; **31.** 205; **33.** 289; **35.** 14 feet 3 inches; **37.** 87°; **39.** 47 minutes; **41.** 121°; **43.** 22 cm; **45.** The difference between the total revenues and profits was $16,740,000; **47.** The combined annual sales of these two drugs is $2,938,000,000; **49.** Changing the order of multiplication does not change the product; **51.** 1,000,000; **53.** These fleas can lay up to 400,000 eggs; **55.** 15,000; **57.** The cost is about $1,000; **59.** 1,045; **61.** The value of these tickets was $12,656; **63.** The total length is 49 feet 4 inches; **65.** 3,600 seconds; **67.** 6 pounds 8 ounces; **69.** 16; **71.** 1; **73.** 5^{11}; **75.** 18,000 miles; **77.** 16,000 liters; **79.** 840 inches^2; **81.** The area of the slide was 7,500 square feet; **83.** $28 - 7 - 7 - 7 - 7 = 0$. The quotient is 4; **85.** 5,000; **87.** 990; **89.** 77; **91.** 6 feet 3 inches; **93.** There were 6,240,000 sterile male flies released per square mile; **95.** 125; **97.** 27; **99.** 10.

Chapter Test

1. A number is written with symbols called digits; **3.** hundreds; **5.** five thousand, six hundred ninety-five; **7.** Inequalities cannot be used to compare measurements that are in different units; **9.** 9,520; **11.** 1,000;

13. 0 1 2 3 4 5 6 7 8 9 10 ...

15a.

TREE	WEIGHT
Ash	45
Birch	32
Cherry	44
Hickory	48
White pine	28
Hard maple	42
White maple	33

15b.

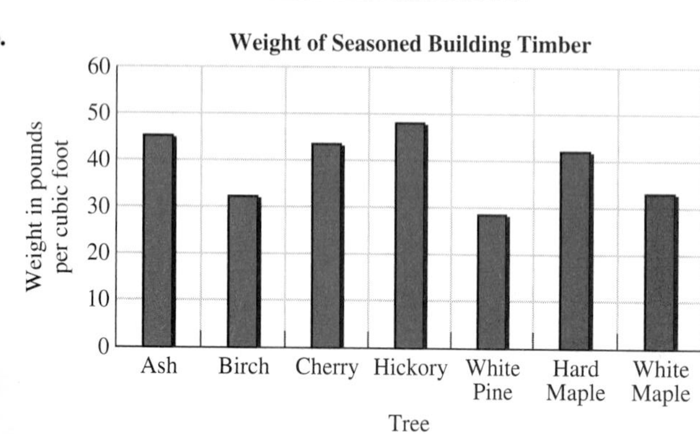

17.

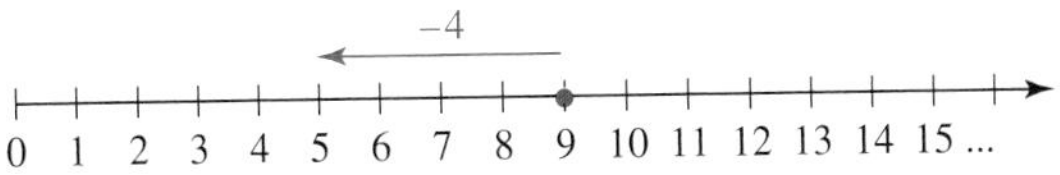

19. Luis spends a minimum of 2,200 minutes on the phone each month; **21.** The measurements must be in the same units; **23.** 6 feet 1 inch; **25.** 24; **27.** 353; **29.** 54; **31.** 28,000,000,000; **33.** 31,072; **35.** 84 inches; **37.** 125; **39.** 3^{12}; **41.** 185,000 meters; **43.** The area of the screen was 1,200 square feet; **45.** $25 - 8 - 8 - 8 = 1$; **47.** 4 R(3). Check: $4 \cdot 9 + 3 = 36 + 3 = 39$; **49.** 4,500,000; **51.** 761 R(35); **53.** 48 feet.

CHAPTER 2

Section 2.1

1. +\$2; **3.** −\$500; **5.** +14,494 feet; **7.** +40 yards; **9.** +3 hours;

11.

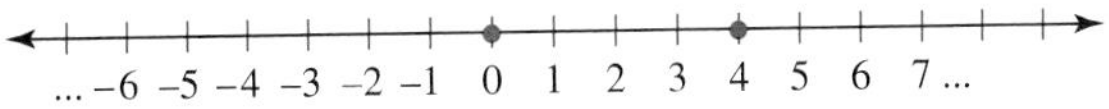

13.

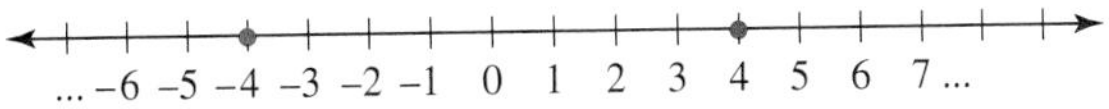

15.

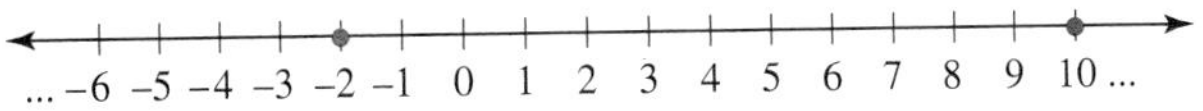

17. >; **19.** <; **21.** >; **23.** >; **25.** >; **27.** >; **29.** >; **31.** <; **33.** 30 miles south; **35.** A height of 20 meters; **37.** negative 1,005; −1,005; **39.** the opposite of negative 45; 45; **41.** the opposite of negative 8,927; 8,927; **43.** the opposite of the opposite of negative 3; −3; **45.** −13; **47.** −791; **49.** 11; **51.** $t = 12$ or $t = -12$; **53.** $s = 0$; **55.** tens: −440; hundreds: −400; thousands: 0; **57.** tens: 8,050; hundreds: 8,100; thousands, 8,000; **59.** *Mistake:* −567 was rounded to the hundreds place instead of to the thousands place. *Correction:* −567 rounds to −1,000;

61.

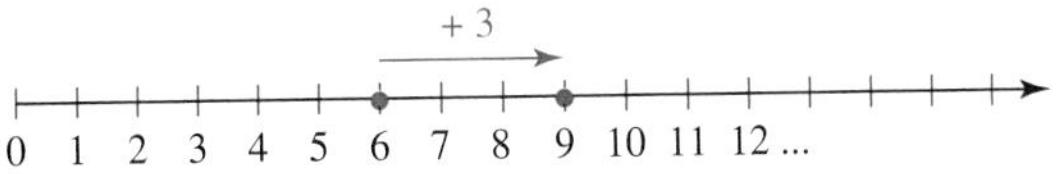

63. In words: Changing the grouping in addition does not affect the sum. In symbols: $a + (b + c) = (a + b) + c$.

Section 2.2

1. 8

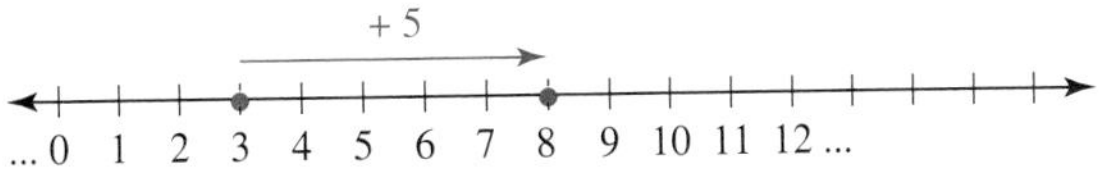

3. −2

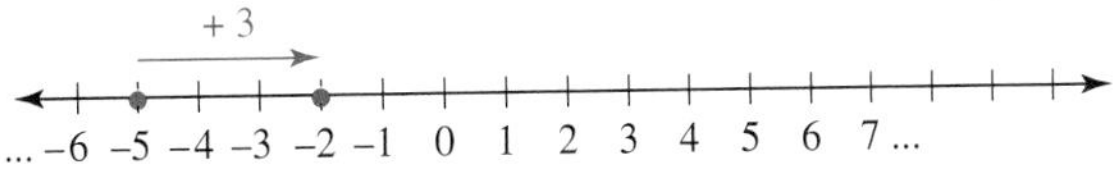

5. −5

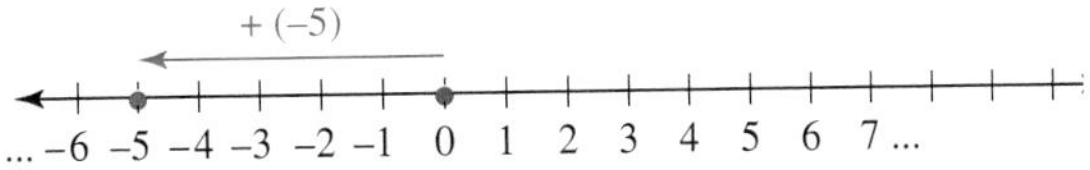

7. 8

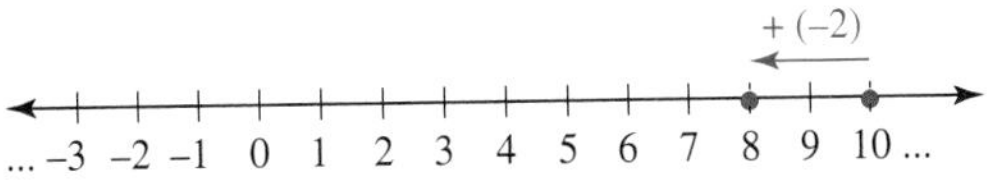

9. −12

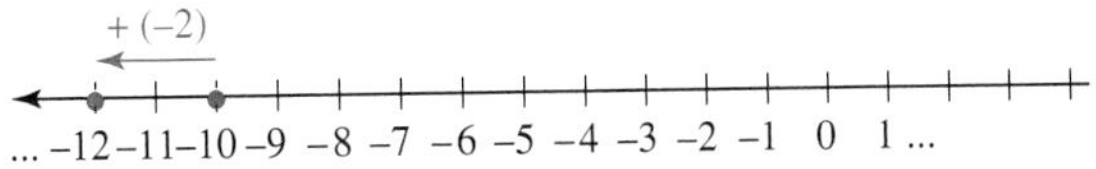

11. 15 meters; **13.** −15°C; **15.** −9; **17.** 2°F; **19.** −2; **21.** −16; **23.** 3; **25.** −3; **27.** −4; **29.** Phil has $528 left in his account; **31.** Curtis still has $1,199; **33.** −$4; **35.** −4; **37.** 20; **39.** −180; **41.** The elevator finally stopped on the 14th floor, assuming that the hotel has a 13th floor; **43.** $x = 370$; **45.** The new balance in her account is $312; **47.** Answers will vary; **49.** The decrease in value of the stock was −$73 a share; **51.** −6; **53.** 7.

Section 2.3

1. 4; **3.** −$4; **5.** −4; **7.** −4; **9.** 10; **11.** −20; **13.** 19; **15.** −19; **17.** 5; **19.** −17; **21.** −50; **23.** −158; **25.** 484; **27.** −6,329; **29.** −12; **31.** The difference in temperature was 17°C;

33.

CHECK NO.	PAYEE	DEBIT AMOUNT	CREDIT AMOUNT	BALANCE
				$532
454	Safeway	$38		$494
455	Bookstore	$162		$332
ATM	Cash Withdrawal	$45		$287
456	Ray J. White Properties	$475		−$188
457	TR Video	$6		−$194
458	Domino's Pizza	$16		−$210
459	GM Finance	$384		−$594
DEP	Deposit		$831	$237
460	Lamont's	$72		$165
461	TR Video	$6		$159
462	Valley Medical Center	$125		$34
463	Kmart Pharmacy	$15		$19
464	Albertson's	$82		−$63
465	Bob's Pet Store	$20		−$83
466	Ray J. White Properties	$475		−$558

35. $x = -22$; **37.** $m = 38$; **39.** The new balance in her account is −$75; **41.** The difference in temperatures was −4°F; **43.** answers will vary; **45.** Emily had to hike uphill 1,755 feet; **47.** *Mistake:* The −8 was incorrectly written as +8. *Correction:* $-7 - 8 = -15$; **49.** 36; **51.** 5.

Section 2.4

1. −6; **3.** 24; **5.** −120; **7.** −20,000; **9.** −2,400,000; **11.** −2,790; **13.** 0; **15.** −44; **17.** −289; **19.** 0; **21.** −8; **23.** 50; **25.** −5; **27.** −50,000; **29.** undefined; **31.** 0; **33.** 45; **35.** −45; **37.** 12; **39.** 12; **41.** 54; **43.** −30; **45.** Kendra will have $1,320 less take home pay per year; **47.** It will take 30 days to switch totally over to decaffeinated coffee; **49.** The total savings is $68,590; **51.** *Mistake:* −(−4) was incorrectly written as multiplication. *Correction:* $-2 - (-4) = -2 + 4 = 2$; **53.** 9; **55.** 24.

Section 2.5

1. 16; **3.** 25; **5.** −4; **7.** −64; **9.** 0; **11.** −1; **13.** −64; **15.** −49; **17.** 4^8; **19.** $(-5)^4$; **21.** $(-4)^{10}$; **23.** 4^{15}; **25.** 9^{11}; **27.** 1,024; **29.** 27; **31.** 3; **33.** −10; **35.** −9; **37.** 60; **39.** −28; **41.** 13; **43.** 1; **45.** −10; **47.** There are now 49 cups of punch in the punch bowl; **49.** After 5 days, Grant will have spent $111,110. The sixth day, he plans to spend $1,000,000 and will not have enough money;

51a.

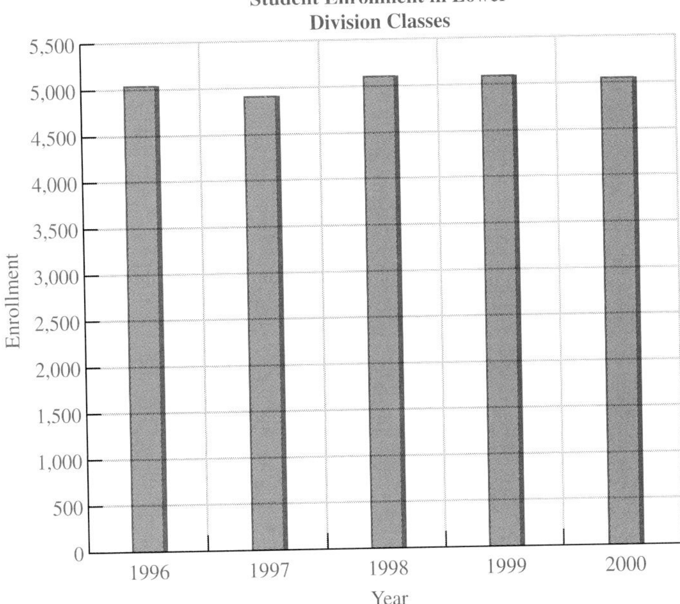

51b.

TIME	1996–1997	1997–1998	1998–1999	1999–2000
INCREASE OR DECREASE	−130	197	−18	60

51c.

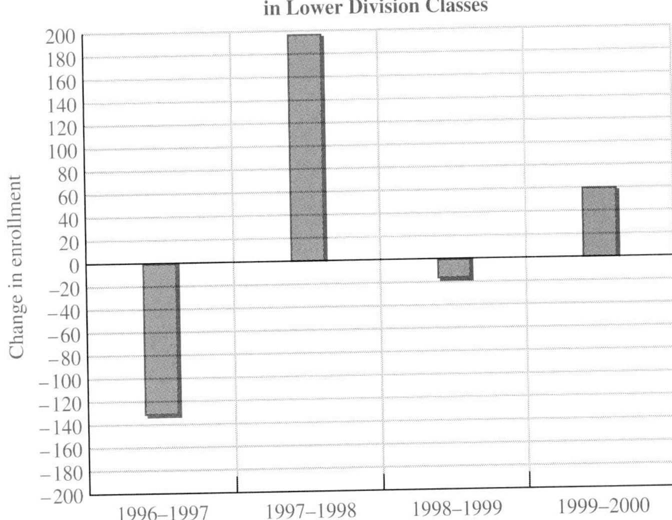

53. *Mistake:* -5^2 is not equal to $(-5)(-5)$. It is equal to $-5 \cdot 5$; **55.** 12 meters.

Section 2.6

1. The volume of this beating block is 588 cubic inches; **3.** The heat of vaporization of alcohol is 204 calories per gram; **5.** The profit per month will be $2,982; **7.** The average speed of the car is 65 miles per hour; **9.** The volume of this worm is about 2,013 cubic centimeters; **11.** The area of this land is 1,550 square miles; **13.** It takes about 60 seconds for a typical person at rest to move the blood in the body through the heart; **15.** The timber contains 117 board feet; **17.** The rough frame opening has a width of 30 inches and a height of 42 inches; **19.** The maximum water use is 720 gallons per hour; **21.** *Mistake:* 3^2 does not equal 6. *Correction:* The safe working load of the rope is 1,350 pounds;

23.

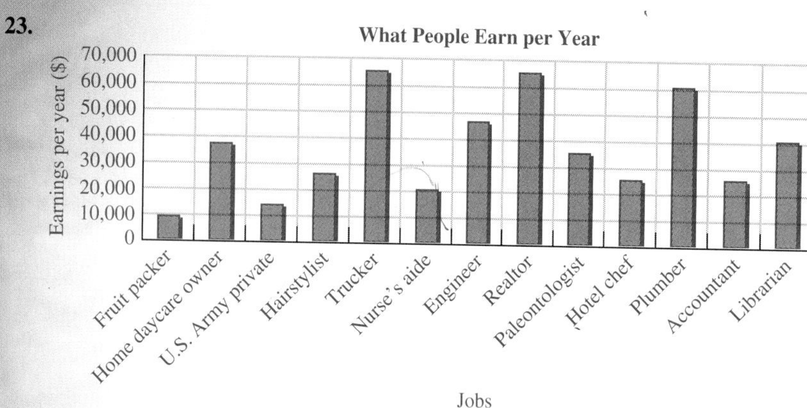

Section 2.7

1. −1; **3.** −2; **5.** 4; **7.** −7;

9a.

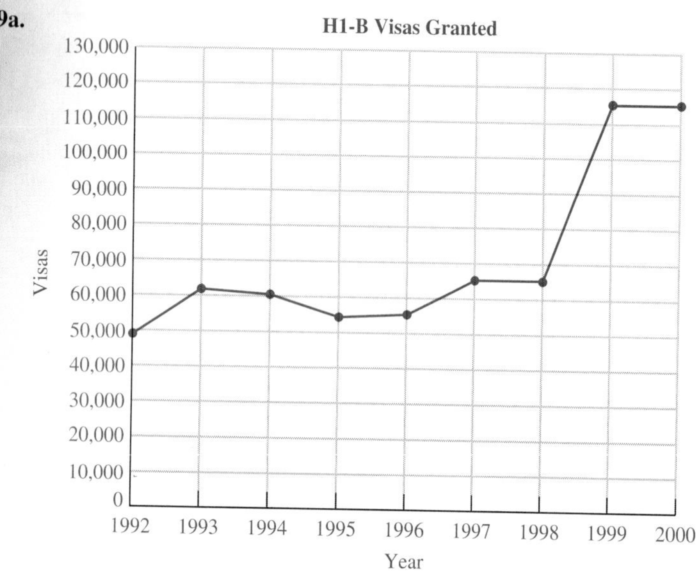

9b. 61,591; **9c.** 71,104 visas; **11a.** 9°F; **11b.** 12°F; **13a.** 2°C; **13b.** 2°C; **13c.** 10°C; **13d.** 4°C; **13e.** −2°C; **15.** *Mistake:* The median was calculated instead of the mean. *Correction:* The mean number of hunting licenses sold was 245,200 licenses; **17.** A product is the result of multiplying two or more numbers. Examples will vary; **19.** {1, 2, 3, 4 . . .}; **21.** A finite set has a limited number of elements. An infinite set has an unlimited number of elements.

Review Exercises

1. −30 points; **3.** +30 points; **5.** $<$; **7.** $>$; **9.** 9; **11.** a depth of 700 feet; **13.** −3; **15.** 6; **17.** 11; **19.** $y = 0$; **21.** $m = 1$ or $m = -1$; **23.** −700; **25.** 0; **27.** −46,000; **29.** −21; **31.** −1; **33.** 1; **35.** −5; **37.** 2; **39.** The population decreased by 1,585 people; **41.** −12; **43.** −2; **45.** undefined; **47.** −35; **49.** 3; **51.** 0; **53.** 36; **55.** 1; **57.** $(-5)^{18}$; **59.** −48; **61.** 36; **63.** −162; **65.** $y = -7$; **67.** $x = 5$; **69.** 28°C; **71.** The density of the block is 6 grams per milliliter; **73a.** 1994; **73b.** 30 percent; **73c.** 35 percent; **73d.** The average percent profit was about 11 percent; **73e.** The median was 15 percent profit.

Chapter Test

1. −10°C;

3. ... −7 −6 −5 −4 −3 −2 −1 0 1 2 3 4 5 6 7 8 9 10 ...

5. $>$; **7.** a gain of 50 points; **9.** −7; **11.** −90; **13.** 3; **15.** $T = -6$°F; **17.** −4; **19.** −11; **21.** $y = -4$; **23.** 0; **25.** $t = -20$; **27.** 1; **29.** 2; **31.** 5^{17}; **33.** 25; **35.** 20; **37.** −2; **39.** 1; **41.** $A = 29$; **43a.** Asia; **43b.** Antarctica; **43c.** Asia; **43d.** 7,500 feet; **43e.** 6,000 feet; **43f.** 6,000 feet; **45.** −48.

CHAPTER 3

Section 3.1

1. Answers will vary but 2, 4, 6, 8, and 10 are five possible multiples; **3.** Answers will vary; 32, 64, 96, 128, 160; **5.** 18; **7.** 120; **9.** 1, 2, 3, 4, 6, 8, 12, 24; **11.** 1, 31; **13.** 1, 7, 11, 77; **15.** 2, 3, 5, 7, 11, 13, 17, 19, 23, 29, 31, 37, 41, 43, 47, 53, 59, 61, 67, 71; **17.** 1, 2, 3, 4, 5, 6, 7, 8, 9, 10, 11, 12, 13, 14, 15, 16, 17, 18, 19, 20; **19.** All prime numbers are greater than 1. Zero is less than 1 so it cannot be prime; **21.** even; divisible by 2; sum of the digits is 3; divisible by 3; **23.** odd; not divisible by 2; sum of the digits is 15; divisible by 3; **25.** even; divisible by 2; sum of the digits is 9; divisible by 3; **27.** not divisible by 2; sum of the digits is 27; divisible by 3; divisible by 9; **29.** not divisible by 2; sum of the digits is 6; divisible by 3; not divisible by 9; **31.** divisible by 2, 3, 6, and 9; **33.** divisible by 2, 3, 4, 5, 6, 10; **35.** divisible by 3 and 5; **37.** divisible by 2, 3, 4, and 6; **39.** Each division could not have had the same number of soldiers because 13,000 is not divisible by 3. We know this because the sum of the digits in 13,000 is 4 and 4 is not divisible by 3; **41.** This distance could not be divided into four equal parts. The last two digits in 1,542 are 42 and 42 is not divisible by 4; **43.** 360 degrees can be divided into three equal parts because the sum of the digits in 360 is 9 and 9 is divisible by 3; **45.** 60; **47.** 40; **49.** *Mistake:* The wrong divisibility rule was used. Divisibility of the sum of the digits by 6 is not the correct rule. *Correction:* 741 is an odd number and is not divisible by 2. The divisibility rule for 6 requires that the number be divisible by 2. So, 741 is not divisible by 6; **51.** *Mistake:* 1 is included in the list but 1 is not a prime number. *Correction:* 2, 3, 5, 7, 11, 13, 17, 19, 23, 29; **53.** The divisor is 3, the dividend is 15, and the quotient is 5.

Section 3.2

1. 

3.

5. The numerator of the improper fraction is the whole number; the denominator of the fraction is 1; **7.** $\frac{3}{4}$; **9.** $\frac{3}{4}$; **11.** $>$; **13.** $>$; **15.** $=$; **17.** $<$; **19.** $\frac{6}{48}$; **21.** $-\frac{36}{48}$; **23.** $-\frac{14}{48}$; **25.** $\frac{9}{72}$; **27.** $-\frac{54}{72}$; **29a.** $\frac{650}{5{,}000}$; **29b.** 650 reporters; **31.** 17; **33.** 2^6; **35.** $2^3 \cdot 3 \cdot 5$; **37.** $-\frac{2}{3}$; **39.** $\frac{1}{3}$; **41.** $\frac{1}{7}$; **43.** 5; **45.** $\frac{3}{5}$; **47.** $\frac{7}{4}$; **49.** $\frac{2}{25}$; **51.** $\frac{121}{219}$; **53.** $\frac{2}{3}$; **55.** $\frac{1}{100{,}000}$; **57.** *Mistake:* The product of 8 and 4 is not 32. *Correction:* $-\frac{15}{24}$; **59.** In words: Changing the grouping in multiplication does not affect the product. In symbols: $a(bc) = (ab)c$; **61.** In words: The order of multiplication does not affect the product. In symbols: $a \cdot b = b \cdot a$.

Section 3.3

1. $\frac{1}{12}$; **3.** $\frac{5}{9}$; **5.** -4; **7.** $\frac{1}{4}$; **9.** $-\frac{1}{21}$; **11.** $\frac{5}{12}$; **13.** $\frac{1}{64}$; **15.** $\frac{1}{16}$; **17.** $\frac{1}{7}$; **19.** $\frac{2}{5}$; **21.** $\frac{4}{3}$; **23.** -3; **25.** $\frac{7}{2}$; **27.** $\frac{8}{27}$; **29.** $-\frac{8}{27}$; **31.** $\frac{5}{9}$; **33.** $\frac{3{,}375}{21{,}952}$; **35.** One-third of this distance is 20 feet 2 inches; **37.** The weight of $\frac{1}{4}$ of a bushel is 15 pounds; **39.** Quincy will be allowed to spend \$800 per month; **41.** The value of these crops is \$288.50 million or \$288,500,000; **43.** There are 400 birds in the Mono Lake Colony; **45.** Four hundred children could recognize their letters; **47.** 588 of the novels were historical romances; **49.** The price of a copy is \$17; **51.** The area of the triangle is $\frac{35}{128}$ square inches; **53.** The area of the window is 12 square feet; **55.** The volume of the pyramid is about 3,900 cubic feet; **57.** *Mistake:* The denominators were not multiplied. *Correction:* $\frac{21}{100}$; **59.** The dividend is 65, the quotient is 13, and the divisor is 5; **61.** In any division problem, the product of the quotient and the divisor must equal the dividend. This is impossible when the divisor is 0 since all the products of 0 are 0.

Section 3.4

1. 3; **3.** $\frac{5}{4}$; **5.** $\frac{1}{8}$; **7.** $-\frac{1}{50}$; **9.** 1; **11.** $\frac{16}{3}$; **13.** $\frac{4}{5}$; **15.** $\frac{25}{9}$; **17.** $-\frac{28}{27}$; **19.** $\frac{15}{16}$; **21.** $\frac{1}{32}$; **23.** $\frac{1}{64}$; **25.** $\frac{7}{36}$; **27.** $\frac{2}{5}$; **29.** $\frac{3}{4}$; **31.** undefined; **33.** $\frac{35}{72}$; **35.** 192 packages can be packed from 3 pounds of herbs; **37.** 210 textbooks will fit on this shelf; **39.** 38 sample boxes can be stacked on a shelf; **41.** $\frac{3}{2}$; **43.** 6; **45.** $\frac{1}{25}$; **47.** $-\frac{196}{225}$; **49.** There were 980,884 inmates held in state or federal prisons; **51.** $\frac{9}{2}$ pounds of the rubber can be recycled rubber; **53.** Connor will need 80 runners; **55.** *Mistake:* The dividend was turned upside down. Instead, the divisor should be turned upside down. *Correction:* $\frac{35}{33}$; **57.** $18 + (-5) = 13$; **59.** 16.

Section 3.5

1. $\frac{1}{2}$; **3.** $-\frac{7}{10}$; **5.** $\frac{7}{10}$; **7.** $-\frac{1}{3}$; **9.** $\frac{1}{3}$; **11.** $\frac{2}{5}$; **13.** About one-half inch of plastic is ground away to make the typical finished lens; **15.** The total amount of jelly is $\frac{5}{4}$ ounces; **17.** $\frac{12}{18}$; **19.** $\frac{45}{72}$; **21.** $\frac{7}{56}$; **23.** $\frac{36}{48}$; **25.** 168; **27.** 360; **29.** 396; **31.** 120; **33.** 323; **35.** $\frac{43}{72}$; **37.** $\frac{1}{3}$; **39.** $\frac{19}{56}$; **41.** $\frac{19}{18}$; **43.** $\frac{13}{18}$; **45.** $\frac{9}{10}$; **47.** $\frac{991}{1,000}$; **49.** The total thickness of the floor is $\frac{7}{8}$ inch; **51.** The minimum length of bolt needed is $\frac{3}{2}$ inch; **53.** The distance between the picture frame and the edge of the photograph is $\frac{23}{16}$ inch; **55.** subtracting; **57.** Answers will vary; **59.** *Mistake:* The denominators were added. The 3's were crossed out. *Correction:* $\frac{19}{15}$;

61.

63. −940.

Section 3.6

1. $\frac{25}{8}$; **3.** $\frac{65}{9}$; **5.** $\frac{39}{16}$;

7.

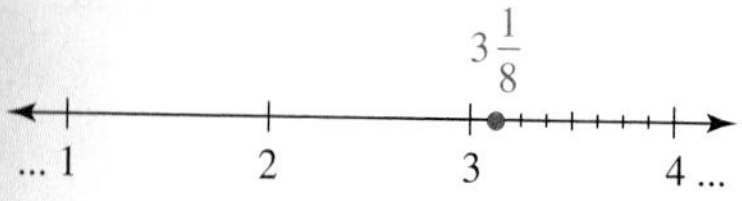

$3\frac{1}{8}$

... 1 2 3 4 ...

9.

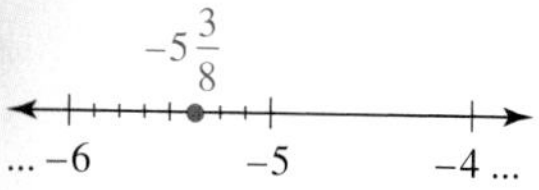

$-5\frac{3}{8}$

... −6 −5 −4 ...

11.

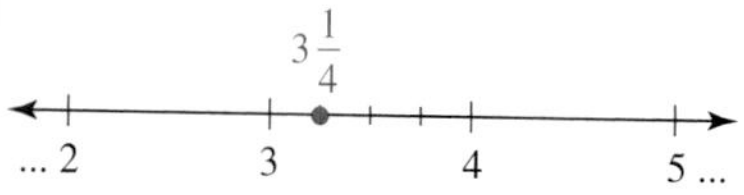

$3\frac{1}{4}$

... 2 3 4 5 ...

13. $2\frac{1}{6}$; **15.** $-3\frac{1}{3}$; **17.** $-83\frac{1}{3}$;

19. 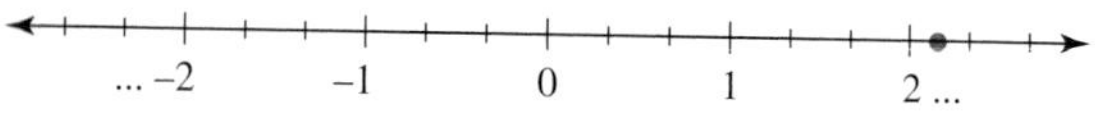 ... −2 −1 0 1 2 ...

21. 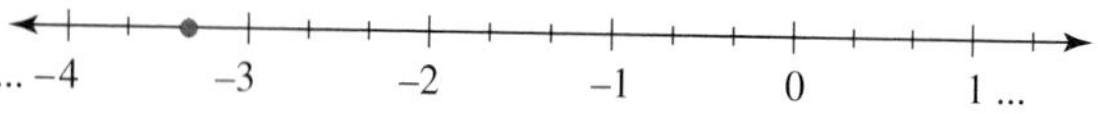 ... −4 −3 −2 −1 0 1 ...

23. 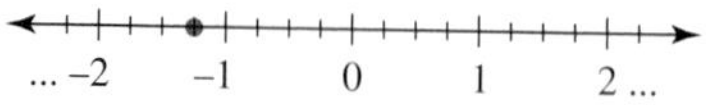 ... −2 −1 0 1 2 ...

25. 3; **27.** 3; **29.** −14; **31.** The ingredients for this chili weigh about 213 ounces; **33a.** $\frac{1}{2} \cdot 3\frac{1}{8}$; **33b.** The product of $\frac{1}{2} \cdot 3\frac{1}{8}$ should be larger because one-half of something is larger than one-fourth of something; **33c.** $\frac{1}{2} \cdot 3\frac{1}{8} = \frac{25}{16}$ and $\frac{1}{4} \cdot 3\frac{1}{8} = \frac{25}{32}$. We can change $\frac{25}{16}$ into an equivalent fraction with a denominator of 32: $\frac{25}{16} = \frac{50}{32}$. Since $\frac{50}{32} > \frac{25}{32}$, then $\frac{25}{16} > \frac{25}{32}$; **35.** $1\frac{9}{16}$; **37.** $4\frac{1}{8}$; **39.** $-14\frac{3}{8}$; **41.** 26; **43.** −13; **45.** $4\frac{73}{82}$; **47.** Fergus will need $4\frac{1}{2}$ cups of processed cheese; **49.** Ben will need $1\frac{1}{3}$ cups of broth; **51.** There are 1,496 squares on a sheet of this graph paper; **53.** The total shipping weight is 75 pounds; **55.** Eight boxes of cereal will fit across the shelf; **57.** The area of the living room is 375 square feet; **59a.** The area of the roof is $710\frac{2}{3}$ square feet; **59b.** There are four triangles of equal size that form the roof. The formula for the area of the roof is 4 · (area of the triangle) or $4 \cdot \frac{1}{2} \cdot$ length · width where the length of the roof is the base and the width is the height; **61.** $\frac{93}{40}$; **63.** $-\frac{91}{40}$; **65.** $-\frac{266}{45}$; **67.** *Mistake:* $-7\frac{1}{5}$ was incorrectly changed into a mixed number. $-7\frac{1}{5}$ equals $-\frac{36}{5}$, not $-\frac{34}{5}$; **69.** Evaluate an expression from left to right, doing operations inside of grouping symbols first. Do exponents first, then multiplication or division, and then addition or subtraction; **71.** When subtracting two numbers, we line up the digits by place value. If the digit in the number being subtracted is greater than the digit in the same place in the other number, borrowing is necessary. Examples will vary.

Section 3.7

1. $11\frac{31}{36}$; **3.** $-1\frac{23}{36}$; **5.** $111\frac{13}{16}$; **7.** $20\frac{889}{1{,}000}$; **9.** $2\frac{17}{45}$; **11.** $-2\frac{25}{28}$; **13.** $-55\frac{1}{2}$; **15.** $-2\frac{1}{21}$;
17. The distance to the top of the stack from the ceiling is $7\frac{5}{8}$ inches; **19.** The measure of the third angle is $67\frac{3}{4}$ degrees;
21. He drank $3\frac{14}{15}$ quarts of liquid; **23.** $11\frac{13}{24}$ bushels of tomatoes were donated that day; **25.** $\frac{25}{12}$ or $2\frac{1}{12}$;
27. 3; **29.** $-9\frac{1}{2}$; **31.** $\frac{271}{18}$ or $15\frac{1}{18}$; **33.** $\frac{8}{3}$ or $2\frac{2}{3}$; **35.** $\frac{2{,}500}{441}$ or $5\frac{295}{441}$; **37a.** The unit run is $10\frac{3}{8}$ inches;

37b.

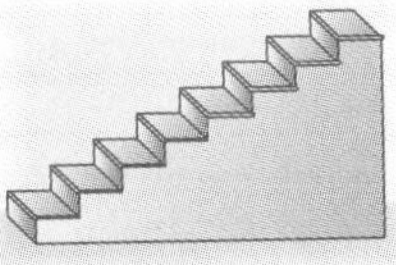

37c. The total run is 83 inches; **37d.** The total rise is $64\frac{1}{8}$ inches; **39.** It will cost about \$1,462.50 to heat the same house in Chicago; **41.** The area of silk in three such scarves is $4\frac{1}{20}$ square feet; **43.** *Mistake:* $\frac{3}{7}$ was incorrectly subtracted from $\frac{1}{7}$. *Correction:* $20\frac{5}{7}$.

Review Exercises

1. 7, 14, 21, 28, 35; **3.** 1, 2, 4, 5, 8, 10, 20, 40; **5.** 6, 12, 15, 20, 21, 22; **7.** prime; **9.** composite;

11.

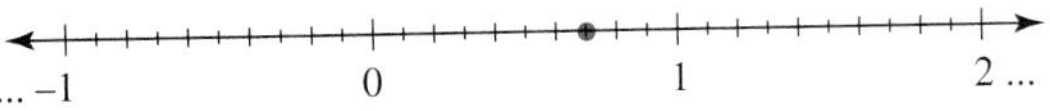

13. $\frac{9}{2}, \frac{8}{8}, \frac{11}{10}, \frac{5}{5}$; **15.** $\frac{6}{7}$; **17.** >; **19.** $-\frac{9}{21}$; **21.** $-\frac{42}{13}$; **23.** $\frac{14}{3}$; **25.** $\frac{7}{12}$; **27.** $5 \cdot 7$; **29.** $2^2 \cdot 5^3$;
31. $\frac{11}{10}$; **33.** 4 hours; **35.** $-\frac{1}{8}$; **37.** $\frac{3}{7}$; **39.** $\frac{21}{25}$; **41.** $-\frac{11}{9}$; **43.** $-\frac{2}{7}$; **45.** $\frac{57}{40}$; **47.** $\frac{5}{18}$; **49.** 504;
51. $2\frac{4}{7}$; **53.** $3\frac{1}{3}$; **55.** $-\frac{51}{8}$; **57.** −4; **59.** −5; **61.** The volume of the sleeping bag is $8{,}761\frac{59}{64}$ cubic inches;
63. $21\frac{24}{35}$; **65.** $-11\frac{9}{50}$; **67.** $-\frac{29}{5}$ or $-5\frac{4}{5}$; **69.** $\frac{7}{40}$; **71.** $\frac{1{,}196}{225}$ or $5\frac{71}{225}$; **73.** $\frac{85}{4}$ square inches or $21\frac{1}{4}$ square inches;
75. The difference in size of the motors is $\frac{7}{24}$ horsepower.

Chapter Test

1. 8, 16, 24, 32, 40; **3.** $\frac{15}{14}$; **5a.** yes; **5b.** no; **5c.** yes; **5d.** no; **5e.** no; **7.** $\frac{4}{5}, \frac{6}{7}, \frac{0}{4}$; **9.** <;
11. 1, 73; **13.** $\frac{1}{4}$; **15.** $\frac{3}{4}$ horsepower; **17.** 9; **19.** \$6 is charged for packaging; **21.** $-4\frac{1}{5}$; **23.** 3^3

25.

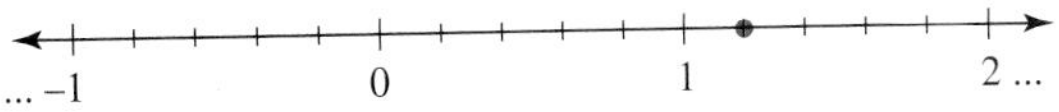

27. $-\frac{4}{7}$; **29.** $\frac{71}{35}$; **31.** 66; **33.** $\frac{53}{9}$; **35.** $15\frac{3}{10}$; **37.** $4\frac{1}{56}$; **39.** $3\frac{14}{25}$; **41.** $\frac{2}{9}$; **43.** $576\frac{1}{3}$ square inches.

CHAPTER 4

Section 4.1

1. hundredths; **3.** ten-thousandths; **5.** tenths; **7.** one thousand, four hundred fifty-nine and nine hundredths dollars;
9. $\frac{3}{4}$; **11.** $\frac{1}{4}$; **13.** $-\frac{3}{5}$; **15.** $\frac{39}{25}$; **17.** $\frac{2{,}007}{20}$; **19.** $-64\frac{1}{500}$; **21.** >; **23.** <; **25.** <; **27.** >;
29. <; **31.** −48.06 degrees; **33.** \$1,067.00; **35.** 71.3 feet; **37.** 0.0 liters; **39.** −\$16.41; **41.** 101 milliliters;
43. 903.55; **45.** 22.71; **47.** −10.2; **49.** 57.0 grams; **51.** 56.99 grams; **53.** 60 grams; **55.** 10; **57.** −20;
59. The total cost of the letter is \$3.80; **61.** On average, Americans eat 3.5 more gallons of ice cream per year than the French;
63. This reading is 0.51 milliequivalent above the normal range; **65.** There was \$4.8443 billion more spent on books for adults than was spent on books for children; **67.** The increase was 0.076 subscribers; **69.** *Mistake:* The decimal points were not aligned. *Correction:* 17.35; **71.** 360.

Section 4.2

1. 1,984; **3.** −1,984; **5.** 0.001984; **7.** −3.995; **9.** 9.9; **11.** −58.764; **13.** 200; **15.** 8; **17.** 0.15; **19.** 50; **21.** 0.00032; **23.** The repairs will cost $24.50; **25.** The property taxes on the house will be $2,540.80; **27.** The U.S. Department of Agriculture sold 15,045,000 pounds of sugar; **29.** The steak will cost about $8.73 at $3.49 a pound, about $11.73 at $4.69 a pound, and $9.95 at $3.98 a pound; **31.** The cost of the ad will be $45; **33.** Malik ate 29 grams of carbohydrate; **35.** The weight of these seizures is 2,177,368.6 pounds; **37.** Stephanie will spend $321.60 less on the car if she pays $4,000 down; **39.** The increase in annual consumption for a family of three was 106.5 pounds of chicken; **41.** Answers will vary; **43.** 432; **45.** 43 R(4).

Section 4.3

1. 2.14; **3.** −214; **5.** −0.04; **7.** 40; **9.** 0; **11.** −3.05; **13.** 9.9; **15.** 555; **17.** 0.29; **19.** 2.5; **21.** 25; **23.** −17.5; **25.** $0.2\overline{6}$; **27.** The amount of operating revenues generated per box was about $1.95; **29.** She has 827.91 euros; **31.** The cost per employee for the event was about $525; **33.** Answers will vary; **35.** 8.87; **37.** 124.3597 centimeters^2; **39.** 356.319; **41.** 46.1785 meter^2; **43.** Carl would save $282; **45.** Daphne's gas bill was $19.91; **47.** Mike's electrical bill in March was $68.99; **49.** The difference in gross wages is $196 per week; **51.** If Manhattan Island was a 13.4 mile by 2.3 mile rectangle, its area would be about 30.8 square miles. This is larger than its actual area of 22.5 square miles; **53.** Answers will vary, depending on the estimated value from the graph. The difference in the amount received by a farmer is about $2,625; **55.** 10^5; **57.** 6,250.

Section 4.4

1. 406.4; **3.** 4.064; **5.** −0.04064; **7.** 5.5992; **9.** −559.92; **11.** The cost of this purchase was $236,300; **13.** The total weight of the mice is 790 ounces; **15.** Each piece of avocado contains about 3.05 calories; **17.** $\$8.29156 \times 10^5$; **19.** $\$1.76 \times 10^{10}$; **21.** 4.5×10^9 years; **23.** 5.9742×10^{24} kilograms; **25.** 7.53×10^1 kilograms; **27.** 2×10^{-3} second; **29.** 4,600; **31.** 50; **33.** 0.000713; **35.** 0.0339; **37.** negative; **39.** 10^{-3}; **41.** 10^{-6}; **43.** 10^{-5}; **45.** 10^{-1}; **47.** 10^{-1}; **49.** *Mistake:* The exponent is incorrect. *Correction:* 7.86×10^{-3}; **51.** 1; **53.** Answers will vary. 1 foot = 12 inches is one example of a conversion factor.

Section 4.5

1. 56.6 liters; **3.** 72.5 feet; **5.** 32 inches; **7.** 10 meters per second; **9.** 0.39 euro per liter; **11.** 19.6 grams per square centimeter; **13.** 39 tonnes; **15.** 176 inches per second; **17.** 1.2 nesting pairs per square kilometer; **19.** $0.58 per day; **21.** 9.20 people per square kilometer; **23.** 9.8 meters per square second; **25.** 300 milliliters; **27.** *Mistake:* 60 seconds does not equal 1 hour. Also, the last fraction is incorrect. It should be $\frac{1.6 \text{ kilometers}}{1 \text{ mile}}$; **29.** the distance around the figure; **31.** the number of cubic units inside the figure.

Section 4.6

1. 37.68 centimeters; **3.** 376.8 centimeters; **5.** 189.97 centimeters; **7.** 2.55 inches; **9.** 2.87 feet; **11.** A Major League baseball must have a diameter between 2.87 inches and 2.95 inches; **13.** The circumference of the orbit is 265,400 kilometers; **15.** Answers will vary; **17.** 176.6 centimeters^2; **19.** 15,386.0 centimeters^2; **21.** 7.1 feet^2; **23.** The cross-sectional area of the large hose is about 0.44 square inch. Each small hose has a cross-sectional area of about 0.20 square inch, a total of 0.40 square inch. This is less than the area of the large hose; **25.** The large pizza is a better buy; **27.** Answers will vary; **29.** 3,589.5 centimeters^3; **31.** 3,589,543.3 centimeters^3; **33.** About 0.16 cubic inch of liquid is contained in this paintball; **35.** The minimum volume of the ball is 333 cubic inches; **37.** 100 feet^3; **39.** 159.0 feet^3; **41.** 804 inches^3; **43.** The volume of this can is 73.5 cubic inches; **45.** The volume of rubber in a hockey puck is 7 cubic inches; **47.** The combined volume of the cone and the ice cream is 102.5 cubic centimeters; **49.** *Mistake:* The radius of the circle is 5 centimeters. The diameter was used instead of the radius. *Correction:* 79 centimeters^2; **51.** $A = s^2$ or $A = lw$; **53.** $4 \cdot 4$.

Section 4.7

1. 5; **3.** 4.12; **5.** no real number; **7.** $\frac{12}{13}$; **9.** 25; **11.** 0.79; **13.** 1.73; **15.** 0; **17.** 4.58; **19.** 4.58; **21.** 12; **23.** 4; **25.** 19; **27.** 3.41; **29.** no real number; **31.** The circumference of the arena is 314 feet; **33.** The drop of water takes about 14 seconds to fall to the bottom of the waterfall; **35.** Each side of the deck should be 7.6 feet; **37.** The diameter of the tank should be 33 feet; **39.** 10.0; **41.** 7.5; **43.** 3.2; **45.** 27.8; **47.** 1.1; **49.** Each side of the ladder is 10.2 feet; **51.** The length of the diagonal is 2.9 meters; **53.** The length of the wire is 242 inches; **55.** It should be attached 5.75 inches from the corner (5.66 inches rounded to the nearest quarter inch); **57.** *Mistake:* The square root of the difference was not found. *Correction:* 64 inches; **59.** $5 \cdot 61$; **61.** $\frac{61}{20}$.

Review Exercises

1. thousandths; **3.** $\frac{59}{100}$; **5.** $\frac{1}{125}$; **7.** >; **9.** $-1{,}041.9$; **11.** 16.24; **13.** \$80.17; **15.** 0.11; **17.** 58.5; **19.** 80; **21.** The wallpaper has an area of 5.33 square yards; **23.** The total expenditure by employers and employees was \$258.6 billion; **25.** -63; **27.** $-2.\overline{6}$; **29.** 20; **31.** Williams' rushing yardage per game was 184.8 yards; **33.** 0.625; **35.** 24.44; **37.** 2.888; **39.** 9,110; **41.** 2.5159×10^6 square kilometers; **43.** 1.675×10^{-24} kilogram; **45.** $70.4 \frac{\text{inches}}{\text{second}}$; **47.** Pi is the circumference of any circle divided by its diameter; **49.** 65.4 inches^3; **51.** 80.1 inches^3; **53.** The volume of Sputnik was 5.572 cubic inches; **55.** The circumference of the compass was 10.4 inches; **57.** The volume of the container is 152 cubic inches; **59.** 5; **61.** no real number; **63.** $\frac{4}{7}$; **65.** 3.9 meters; **67.** 149.57 feet; **69.** Answers will vary.

Chapter Test

1. thousandths; **3.** $\frac{7}{8}$; **5.** 670; **7.** 35.462; **9.** $\$4.65 \times 10^8$; **11.** 5.7 grams; **13.** 0.8; **15.** 4 tablets; **17.** 42.12; **19.** 10.8 m; **21.** 10.292; **23.** \$368.26; **25.** 16.4 milligrams; **27.** \$95,880; **29.** 7.71 feet; **31.** 9 cm; **33.** 13 feet^3; **35.** no real number; **37.** 13.

CHAPTER 5

Section 5.1

1. 13; **3.** 2.83; **5.** 3.25; **7.** 430; **9.** 42; **11.** 3,780; **13.** 520; **15.** 4,300; **17.** 10; **19.** 6; **21.** 36.65 meters^2; **23.** 0 grams; **25.** -252; **27.** $\frac{5}{12}$ or 0.42; **29.** 6.60; **31.** 148; **33.** 1.26; **35.** 1; **37.** 15; **39.** 56.25; **41.** *Mistake:* Addition was done before multiplication. *Correction:* 11; **43.** 6 feet.

Section 5.2

1. 3; **3.** -8; **5.** $41x$; **7.** $-27x + 19$; **9.** $11.5x - 12.8y - 6$; **11.** $46ab$; **13.** $2.6x$; **15.** $9x + 55y - 24$; **17.** $305x - 99y + 20$; **19.** $1{,}110x + 1{,}218y + 1{,}000$; **21.** $2{,}818x$; **23.** $1{,}656x$; **25.** $15x^2 + 11x$; **27.** $2x + 2z$; **29.** $12 + 2x$; **31.** $6x - 12$; **33.** $-2x + 18$; **35.** $4x + 36$; **37.** $-35 + 7y$; **39.** $-12x + 4y$; **41.** $-14x - 126$; **43.** $x - 16$; **45.** $33h + 17$; **47.** $1{,}500a + 1{,}500b + 1{,}500c$; **49.** $194x + 22$; **51.** $10x$; **53.** $10x + 18$; **55.** $13x + 19.5y + 32.5$; **57.** $8x^2 - 7x + 16$; **59.** $17x^2 + 4x + 10$; **61.** *Mistake:* The 2 and 9 were not multiplied. Also, the $7x$ and 6 were combined and they are not like terms. *Correction:* $7x + 15$.

Section 5.3

1. $x = 4$; **3.** $z = 14$; **5.** $y = 7.3$; **7.** $a = 44$; **9.** $c = -\frac{17}{9}$; **11.** $x = 21$; **13.** $a = 4{,}000$; **15.** $x = \frac{43}{6}$; **17.** $y = 79.2$; **19.** $x = 307$; **21.** $x = 30$; **23.** $z = 96$; **25.** $c = 6$; **27.** $x = 275.76$; **29.** $x = 2$; **31.** $x = -8$; **33.** $x = 14$; **35.** $c = 0$; **37.** $m = 72$; **39.** $t = 20$; **41.** *Mistake:* One side of the equation was multiplied by 5. Instead, both sides should have been divided by 5. *Correction:* $x = 3$; **43.** Do all operations inside grouping symbols first. Working from left to right, evaluate exponents and square roots. Do multiplication or division, left to right. Do addition or subtraction, left to right.

Section 5.4

1. $x = 2$; **3.** $w = 2$; **5.** $x = 61$; **7.** $m = 8$; **9.** $y = 2$; **11.** $v = 3$; **13.** $u = 7{,}660$; **15.** $h = 6$; **17.** $x = 12$; **19.** $x = 10$; **21.** $x \approx 11.43$; **23.** $x = -5$; **25.** $x = 0.7$; **27.** $x = 97.9$; **29.** $x = 32.782$; **31.** $x = 6.1$; **33.** $y = -7.34$; **35.** $y = \frac{21}{20}$; **37.** $t = 0$; **39.** $y = -3$; **41.** $x = \frac{35}{4}$; **43.** $x = -\frac{1}{12}$; **45.** $y = \frac{1}{10}$; **47.** $x = -\frac{22}{45}$; **49.** $x = \frac{22}{3}$; **51.** $x = \frac{3}{2}$; **53.** $y = \frac{7}{20}$; **55.** $w = 48$; **57.** $x = -\frac{11}{84}$; **59.** *Mistake:* The distributive property was used incorrectly. 3 should have been multiplied by 6 in the first step. *Correction:* $x = 0$; **61.** Step 1. *I know.* Organize the information in the problem and assign a variable.
Step 2. *Equations.* Write a word equation and a math equation that describe a relationship in the problem.
Step 3. *Solve.* Solve the math equation to find the value of the variable.
Step 4. *Check.* Make sure the proposed answer is reasonable. Look for arithmetic errors.
Step 5. *Final Answer.* Answer the question in the problem using the complete sentence.

Section 5.5

1. To pay for the salary of the inspector, 280 violations will have to be levied; **3.** Mario can stay 7 nights at the motel; **5.** The city had 108,500 rooms before the new rooms were added; **7.** The decrease in the amounts was $1,816,375; **9.** The required area of the vents is 8 square feet; **11.** The loss of biomass is 326 million pounds; **13.** The net gain of the campaign was $689,736; **15.** The shipping and handling charge was $23.65; **17.** He can afford to take 15 credits; **19.** The area of the concrete deck is 650 square meters larger than the area of the pool; **21.** The school district gained $2,737 from this incident; **23.** The full jug contains about 800 milliliters of water; **25.** The trail was about $\frac{3}{8}$ mile long; **27.** The courthouse was $5\frac{7}{10}$ miles from the law office; **29.** Ken actually works 60 hours a week; **31.** $\frac{1}{2}$; **33.** $\frac{60}{100}$.

Review Exercises

1. 124; **3.** 27; **5.** 24; **7.** 8.8; **9.** $36x + 13y + 10$; **11.** $10x + 1$; **13.** $-12x + 8$; **15.** $x - 2$; **17.** $x = -16$; **19.** $x = 9.3$; **21.** $x = -5$; **23.** $y = 648$; **25.** $x = 308$; **27.** $x = 4$; **29.** $x \approx 2.28$; **31.** $x = \frac{64}{15}$; **33.** About 1,500,000 women take parental leave under the act per year; **35.** The cost to rent the car for 10 days is $290; **37.** The person works 1,850 hour per year; **39.** To be fully reimbursed, the employee needs to receive 15 paychecks.

Chapter Test

1. 34; **3.** 2; **5.** $12x + 85$; **7.** 13; **9.** This group of faculty teaches 24 FTE; **11.** $x = 20$; **13.** $x = 54$; **15.** $x = 3{,}555$; **17.** $x = \frac{23}{18}$; **19.** $y = 21$; **21.** The area of this rectangle is 80 square centimeters; **23.** The difference is 73,000 Americans.

CHAPTER 6

Section 6.1

1a. $\frac{3}{5}$; **1b.** $\frac{8}{5}$; **1c.** $\frac{3}{8}$; **3.** $\frac{2{,}584}{1{,}151}$; **5.** $\frac{39}{425}$; **7.** $\frac{161}{176}$; **9.** $\frac{625}{5{,}676}$; **11.** $\frac{3}{10}$; **13.** $\frac{3}{8{,}281{,}250}$; **15.** $\frac{1}{50}$; **17.** $\frac{1}{6}$; **19.** $\frac{1}{9}$; **21.** $\frac{2{,}400 \text{ lipsticks}}{1 \text{ hour}}$; **23.** $\frac{3.2 \text{ minutes}}{1 \text{ exam}}$; **25.** $\frac{\$164.14}{1 \text{ ton}}$; **27.** $\frac{16{,}438 \text{ infections}}{1 \text{ day}}$; **29.** $\frac{20 \text{ drops}}{1 \text{ minute}}$; **31.** $\frac{\$0.85 \text{ returned}}{\$1.00 \text{ deposited}}$; **33.** $\frac{\$1.11 \text{ tax}}{1 \text{ pack}}$; **35.** $\frac{67}{1{,}000}$; **37.** $\frac{179}{209}$; **39.** $\frac{\$1.90}{1 \text{ capsule}}$; **41.** $\frac{424}{191}$; **43.** *Mistake:* The difference of 9,000 and 2,250 was not calculated. *Correction:* $\frac{2{,}250}{6{,}750} = \frac{1}{3}$; **45.** $x = 120$; **47.** $c = \frac{200}{7}$.

Section 6.2

1. 1,410,000 Americans will develop heart valve abnormalities from their use of these drugs; **3.** 8,000 college students were working as well as going to school; **5.** 1,540 of the women are expected to die within a year; **7.** The cost of 3 ounces of truffle will be $107.81; **9.** The *Stars* will win 28 games; **11.** There are approximately 17 slices in a pound; **13.** About 7,875 students eat a la carte at the college each day; **15.** The bracelet contains 10.5 ounces of pure gold; **17.** Woods will have 71 wins in 100 tournaments; **19.** 1,665 students in the survey used tobacco products; **21.** 113 of the horses are receiving Lasix; **23.** 31,188 quarts of milk are needed to make the cheese; **25.** There were 13.45 mishandled bags; **27.** 450 rooms should be nonsmoking; **29.** Tracy will make 15 free throws; **31.** 123 of the offenders will be approved for parole; **33.** 93 children enrolled in the program will have baby bottle tooth decay; **35.** There are 3,185 unemployed people; **37.** 45,087 people were opposed to breaching the dams; **39.** 173 of the crashes are side-impact crashes; **41.** *Mistake:* Both sides were multiplied by $\frac{1}{5 \text{ miles}}$ instead of the correct $\frac{3.5 \text{ inches}}{1}$; **43.** 0.07; **45.** 2.11; **47.** $\frac{7}{100}$.

Section 6.3

1. 0.03, 3%; **3.** $\frac{50}{100}$; 0.5; **5.** $\frac{71}{100}$; 71%; **7.** $\frac{43}{1{,}000}$; 0.043; **9.** $\frac{130}{100}$; 1.30; **11.** $\frac{5}{1{,}000}$; 0.005; **13.** $\frac{100}{100}$; 100%; **15.** 0; 0%; **17.** $\frac{500}{900}$; $0.\overline{5}$; **19.** $0.\overline{428571}$; $42\frac{6}{7}\%$; **21.** $<$; **23.** $<$; **25.** $>$; **27.** Divide by 100. Remove the percent sign; **29a.** 28% increase; **29b.** 14% difference; **29c.** schools that have less than 6% minority students; **29d.** schools that have 50% or more minority students; **31.** $\frac{7}{50}$; **33.** $\frac{59}{250}$; **35.** $\frac{13}{50}$; **37a.** $\frac{3}{20}$; **37b.** 15%; **37c.** 85%; **39a.** $\frac{2}{5}$; **39b.** 40%; **39c.** 60%; **41.** 25% decrease; **43.** 6%; **45.** 24% increase; **47.** 59% decrease; **49.** 14.3% decrease; **51.** 49% decrease; **53.** *Mistake:* 25% is equivalent to $\frac{25}{100}$, not $\frac{1}{25}$. *Correction:* $\frac{1}{4}$; **55.** 0.7.

Section 6.4

1. He should eat 2,240 calories each day; **3.** The sale price should be $51; **5.** This price is 32% below the average price; **7.** 15% of the total price is donated to CARE; **9.** There were 21,632 new infections in the United States; **11.** The percent decrease in lodging rooms was 22%; **13.** 3.6% of the prices were scanned incorrectly; **15.** The new record was a 3.2% increase; **17.** The original average electric bill was $55; **19.** The timber harvest decreased by 69.0%; **21.** The enrollment the previous year was 345,861 students; **23.** The population count should have been 2,094,241 people; **25.** The employee is assuming that only one death occurred per accident; **27.** The markdown was 9%; **29.** The final price of the sweater was $31.49; **31.** The simple interest is $900; **33.** The simple interest is $191.25; **35.** 33% of his total yield was barley; **37.** 37% of individuals served were under age 18; **39.** His state income taxes will be about $2,880; **41.** The percent increase in concrete used was 268%; **43.** The percent increase in power is 47%; **45.** The percent increase in viewers was 128%; **47.** The percent increase in elephants was 57%; **49.** *Mistake:* 16% was incorrectly changed to a decimal. *Correction:* $222.72;

51a. $\frac{1}{4}$; **51b.** 25%; **53a.** $\frac{1}{8}$; **53b.** 12.5%

Section 6.5

1a. 9,408,791 people; **1b.** 38,223,214 people; **1c.** 11,172,939 people; **3a.** 3,292,160 children; **3b.** 3,189,280 children; **3c.** 3,086,400 children;

5.

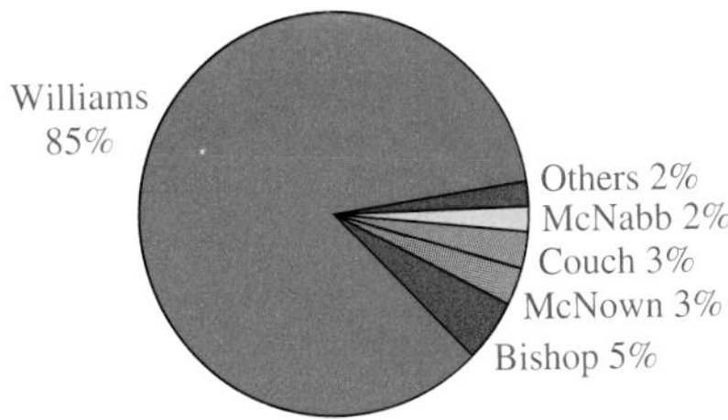

7a. 4% = 14°; 14% = 50°; 3% = 11°; 65% = 234°;

7b.

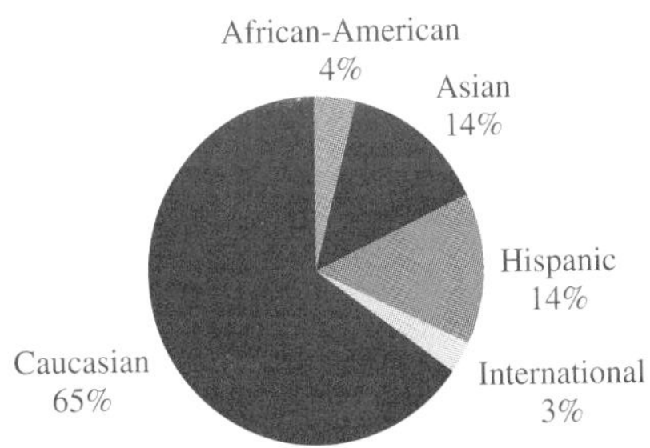

9a. 71%; **9b.** 22%; **9c.** 7%; **9d.** 22% = 79°; 71% = 256°; 7% = 25°;

9e.

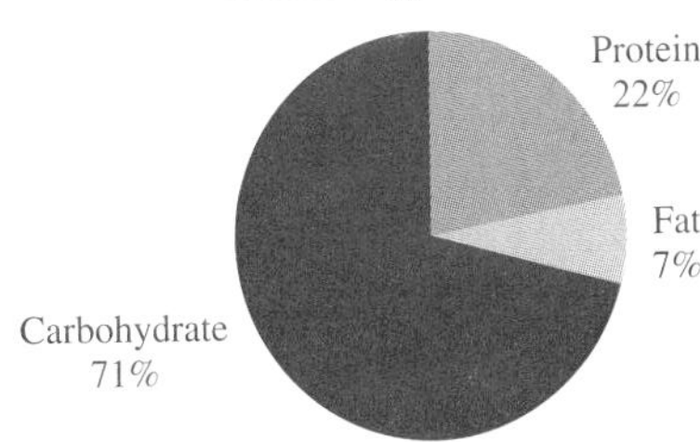

11.

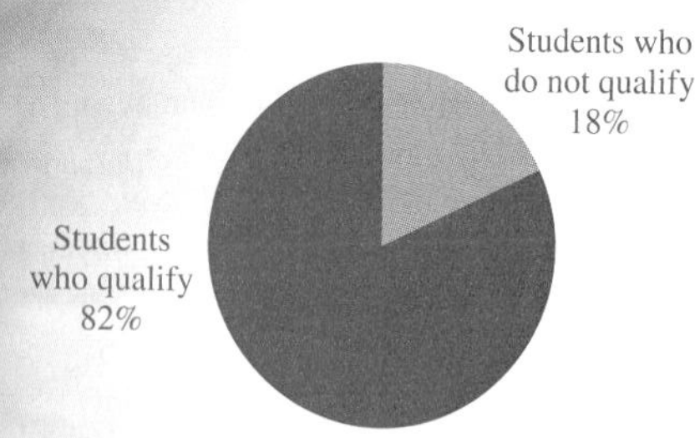

13a. 70%; **13b.** 12%; **13c.** 10%; **13d.** 5%; **13e.** 3%;

13f.

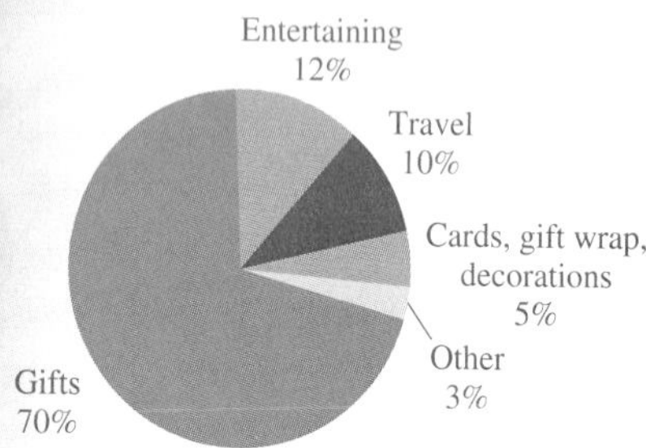

15a.

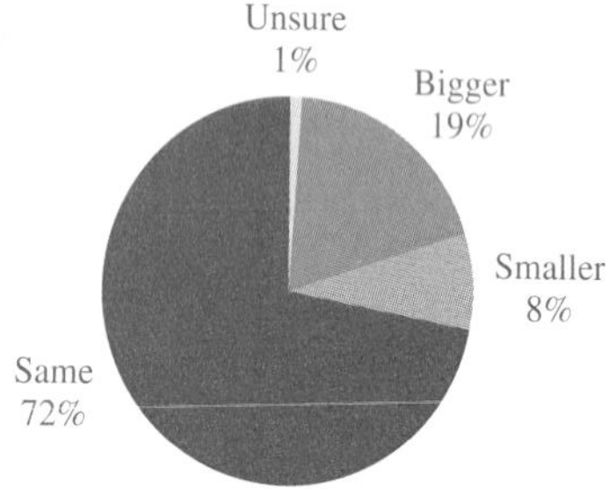

15b. 194 adults; **17a.** $\frac{59}{580}$; **17b.** 10%; **17c.** 17%;

17d.

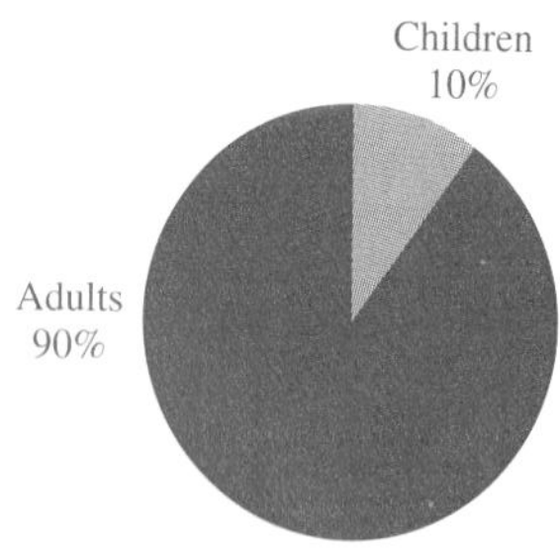

19. *Mistake:* She did not use ratios to find the degrees to mark off in making the circle graph. *Correction:* Beverages: 120°; Tips: 96°; Entertainment: 144°; **21.** $p = 5$; **23.** $y = 75$.

Review Exercises

1. A ratio compares two like units; a rate compares quantities with unlike units; **3.** The rate must have a denominator of 1 unit; **5.** $\frac{1}{10}$; **7.** $\frac{105 \text{ miles}}{1 \text{ hour}}$; **9.** 50,000 winter visitors; **11.** 18 deaths; **13.** 37 food establishments; **15.** Change the percent to a decimal, change the decimal to a fraction, and simplify the fraction to lowest terms; **17.** $\frac{7}{100}$; **19.** 0.003; **21a.** $\frac{11}{30}$; **21b.** 37%; **23.** 33% decrease; **25a.** 10% increase; **25b.** 18% increase; **27.** 2,050; **29.** 5,610 workers; **31.** \$90; **33.** 75% increase; **35.** \$33.80; **37.** \$228.0 billion; **39a.** 35%; **39b.** 74%; **41.** 13%; **43.** 17,861,640 workers.

Chapter Test

1. A ratio compares two like units; a rate compares quantities with unlike units; **3.** $\frac{71}{100}$; **5.** 114,000 people; **7a.** $\frac{79}{213}$; **7b.** 37%; **9.** $\frac{137 \text{ miles}}{3 \text{ hours}}$; **11.** $1,113.75; **13.** 84% loss; **15.** 17%; **17a.** Coke; **17b.** 13 students; **17c.** 2,600 students; **19.** 216 homemakers; **21.** $6,824; **23.** $628.95; **25a.** 61%; **25b.** 21%; **25c.** 18%;

25d.

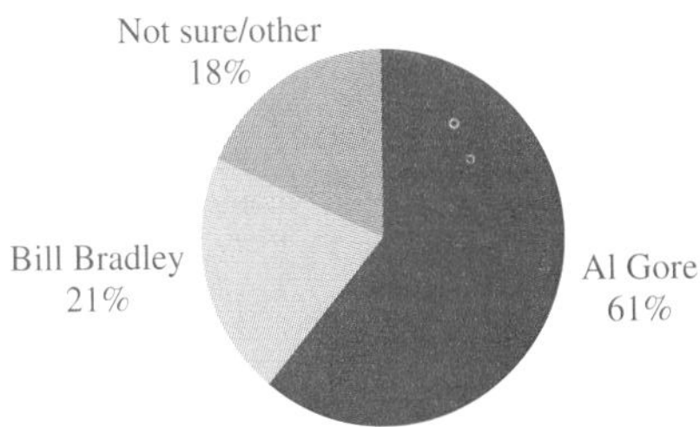

APPENDIX 1

Section A1.1

1. A line segment has endpoints; a line does not have endpoints; **3.** A diameter of a circle must be a chord that includes the center of the circle. All chords do not include the center; **5.** An isosceles triangle has two sides of equal length; an equilateral triangle has three sides of equal length; **7.** 117°; **9.** A trapezoid has exactly two parallel sides and does not contain any right angles; each of the angles in a rectangle is a right angle.

Section A1.2

1. Pi is the ratio of the circumference and diameter of a circle ($\pi = \frac{C}{d}$) and is approximately equal to 3.14; **3.** 12.6 cm; **5.** 60 millimeters; **7.** 9.6 cm; **9.** $5\frac{1}{4}$ inches.

Section A1.3

1. 24 in.^{2}; **3.** 44.0 cm^2; **5.** 78.5 cm^2; **7.** 8.8 meter^2; **9.** 6 cm^2.

Section A1.4

1. 6 in.^{3}; **3.** 523 cm^3; **5.** 2,533.6 in.^{3}; **7.** 480 ft^3; **9.** 804 cm^3.

Section A1.5

1. 32 ft; **3.** 0.9 in.; **5.** 300 ft; **7.** 300 in. or 25 feet; **9.** yes.

Index